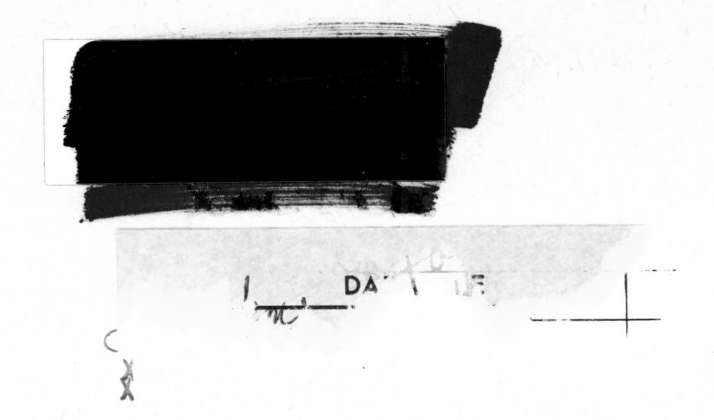

Atlas of the Textural Patterns of Basalts and their Genetic Significance

ATLAS OF THE

TEXTURAL PATTERNS OF BASALTS AND THEIR GENETIC SIGNIFICANCE

S.S. AUGUSTITHIS

Professor of Mineralogy, Petrography, Geology
National Technical University of Athens
Athens, Greece

Elsevier Scientific Publishing Company
Amsterdam - Oxford - New York 1978

Library of Congress Cataloging in Publication Data

Augustithis, S S
Atlas of the textual patterns of basalts and their genetic significance.

Bibliography: p.
Includes index.
1. Basalt--Origin. 2. Petrofabric analysis. I. Title.
QE462.B3A93 552'.2 77-16177
ISBN 0-444-41566-1

ELSEVIER SCIENTIFIC PUBLISHING COMPANY
335 Jan van Galenstraat
P.O. Box 211, Amsterdam, The Netherlands

Distributors for the United States and Canada:

ELSEVIER NORTH-HOLLAND INC.
52, Vanderbilt Avenue
New York, N.Y. 10017

ISBN 0-444-41566-1

With 604 illustrations and 17 tables

Printed in The Netherlands

To my wife
Kammersängerin
Vera Little-Augustithis

Preface

Despite the enormous amount of existing literature on basaltic rocks – the author is acquainted with at least 2000 publications, dealing directly or indirectly with basalts, published in the last 15 years – there is a gap of information regarding the microstructure (textures) of the basaltic rocks. The present atlas aims to present the most important and most common textures of basalts and their genetic significance.

The study of and research on basalts has passed through the evolutionary phases of the microscopic era at the beginning of this century and the physico-chemical approach (1914–1960) and has now entered the stage where the study of the basalts is directly interconnected with the mantle project and with the hypothesis of plate tectonics.

Whereas until now emphasis has been laid on the chemical composition of the basalt and the understanding of the physico-chemistry of the basaltic melts or rather the artificial melts of approximating basaltic composition, the microstructure of the basalt has not been used as a decisive criterion for the understanding of the sequence of crystallization of basaltic mineral components and for the genetic interpretation of the basaltic rocks.

In contrast to the prevailing tendency, nevertheless, and taking into consideration the contribution of physico-chemical studies on the basalt problem, the present textural studies of basalts contradict Bowen's principles of continuous and discontinuous reaction series, fractional crystallization and basaltic magma differentiation.

Instead, emphasis is placed on the approach of "comparative anatomy", i.e. on the comparison of textural patterns and their genetic significance in basalts. It is believed that a lot can be learned from the study of the intimate mineral relations in the basaltic rocks.

In the attempt to study the microstructures of basalts, the author has made use of material collected over a period of more than a quarter of a century and also collections kindly sent by colleagues from the major basaltic occurrences all over the globe. The author especially thanks the following for their kind help:

Prof. Ananda Deb Mukherjee (India); Dr. Aronis, G. (Greece); Prof. Alfredo San Miguel Arribas (Spain); Dr. Backström, J.W. von (South Africa); Dr. Beernevey, R.S. (India); Dr. Einarson, G. (Canada); Mr. Gregory, Bottley and Co. (England); Prof. Grigoriev, D. (U.S.S.R.); Dr. Kochhar Naresh (India); Dipl. Geol. Krantz Renata (Germany); Prof. Manos, C. (U.S.A.); Dr. Malpas, J. (Canada); Prof. Mattoso (Brasil); Dr. McIver, J.R. (South Africa); Dr. Milnes, A.R. (Australia); Dr. Minatidis, D. (Canada); Prof. Mücke Arno (Germany); Prof. Pande, J.C. (India); Mr. Perissoratis, C. (U.S.A.); Prof. Seeliger, E. (Germany); Dr. Shimizu Terno (Japan); Starke and Co. (India); Dr. Strong, D. (Canada); Dr. Venter, T.A. (South Africa); Prof. Vincent Ewart Albert (England); Ward, J. and Associates (U.S.A.).

ACKNOWLEDGMENTS

The author thanks the authors and publishers of the following figures, which have been used in the present Atlas:

Fig. 1a, by A.E. Sviatlovsky in "Regional Volcanology", 1975, published by Nedra, Moscow.
Figs. 1b and 29, by A. Holmes in "Principles of Physical Geology", 1966, published by Nelson, Edinburgh.
Fig. 1c, by Ramos and Fraenkel (1974).
Fig. 1d, by J.H. Illies in "Graben tectonics as related to crust–mantle interaction", 1970, published by E. Schweizerbart'sche Verlagsbuchhandlung, Stuttgart.
Fig. 3, by X. Le Pichon and Morgan, in "Plate Tectonics", 1973, published by Elsevier, Amsterdam.
Figs. 24 and 25, by H. Tazieff in "Tectonics of the northern Afar (or Danakil)", 1970, published by E. Schweizerbart'sche Verlagsbuchhandlung, Stuttgart.
Figs. 31 and 42, by H.M. Geological Survey, Great Britain.
Figs. 32 and 33, by T. Nichols in "Volcanic Landforms and Surface Features", 1971, published by Springer, Heidelberg.
Fig. 200, by G. Schorer in "Neues Jahrb. Mineral. Monatsh., 1970 (7)", published by E. Schweizerbart'sche Verlagsbuchhandlung, Stuttgart.
Fig. 390, by NASA (S-69-47907).
Fig. 590, by Prof. San Miguel Aribbas.
Fig. 595 a and b, by H.S. Yoder Jr. and C.E. Tilley in the "Origin of Basalt Magmas . . .", 1962, published by J. Petrol.,; also in "Basalts", 1968, published by Interscience–Wiley.
Fig. 596, by Hisashi Kuno in "Differentiation of basalt magmas – Basalts, Vol. 2", published by Interscience–Wiley.
Figs. 598 and 600, by T.F.W. Barth in "Theoretical Petrology", published by Wiley and Sons.
Fig. 599, by N.L. Bowen in "The Evolution of the Igneous Rocks", 1928, published by Princeton University Press.
Fig. 602, by J. Makris et al. in "Gravity field and crustal structure of North Ethiopia", 1975, published by E. Schweizerbart'sche Verlagsbuchhandlung, Stuttgart.
Fig. 604 (Simplified after Schmitt), published by Science, 167 (3918), 1970.

Contents

Chapter 1 | The global basalt distribution and geotectonic hypotheses

Great thicknesses of basalts cover large areas in the Columbia and Snake River regions of the northwestern United States, the Deccan of India, Paraná of South America, Mongolia, Siberia, Arabia and Syria, Ethiopia, around Victoria Falls, along Drakensberg, Natal and parts of Australia, see Figs. 1a, b, c, d and e.

The global distribution of basalts is discussed, among others, by Barth (1952), Krishnan (1961), Schneider (1964), Holmes (1965) and Sviatlovsky (1975).

In addition to these land basaltic occurrences, basalts extend over great parts of the oceanic floors and build oceanic ridges.

The global distribution of basalts is directly or indirectly related to global tectonics. Several geotectonic hypotheses have been proposed for the understanding of the basalt occurrences and distribution.

The relation of basalt with geotectonic features is discussed, among others, by Menard (1958, 1960, 1969), Heezen et al. (1959), Eaton and Murata (1960), Joplin (1960), Krishnan (1961), Hess (1962), Bankwitz (1964), Van Bemmelen (1966), Vine (1966), Borcher (1967), Bjornson (1967), Rance (1967), Rittmann (1967, 1969), O'Hara (1968), Isacks et al. (1968), Van Andel (1968), Bullard (1969), Mitchell and Bell (1970) and Sviatlovsky (1975).

Numerous excellent papers and monographs exist dealing either with a specific geotectonic feature or with its worldwide pattern. Several publications exist dealing specifically with the alpine orogenic belt or with parts of it, e.g., Staub (1924), Scheidegger (1953), Benioff (1954), St. Amand (1957) and Ilić (1967).

Similarly, several excellent treatments exist dealing with the rift valleys, their distribution and origin, e.g., Gregory (1896, 1920, 1921), Rüger (1932), Willis (1936), Dixey (1956), Report of the UMC/UNESCO (Seminar on the East African Rift System – Nairobi, 1965), Vine (1966), Wilson (1966, 1967), "Graben Problems" – edited by Illies and Mueller (1970), Girdler (1971), Artemjev and Artyushkov (1971) and "The Afar" (1975).

The realization that the geotectonic features of the earth are to be understood in conjunction with the earth's mantle has resulted in the development of an extensive field of research, i.e. the mantle project on an international basis.

The following are some of the publications which deal with the importance of the mantle in the understanding of geotectonic features and volcanism: Rittman (1958), Uffen (1959), Bose (1967), Vening Meinesz (1964), Wiseman (1966), Gass (1968), Jackson (1968), Ryall and Bennett (1968), Schreyer (1969), Andersen (1970), Bott (1971a,b,c) and "Das Unternehmen Erdmantel", Deutsche Forschungsgemeinschaft (1972).

Geotectonic features related to basalt volcanism

(a) *Mobile belts.* Volcanism, in the sense of synorogenic basaltic effusion, is related to the geosynclinal cycle in the phase of geosynclinal revolution; furthermore, mantle mobilization or mantle diapirism may take place in this phase.

The basement of the continents is built up of remnant roots of successive orogenic belts. The best exposed and best known orogenic belt is the youngest and it comprises two great circles around the earth; one entirely Circum-Pacific and the other at right angles to it, stretching from Indonesia in the east through the Himalayas and the Alps to the Pyrenees and the Atlas mountains on the Atlantic coast. These two zones constitute, at present, the most important unstable areas in the crust, as seismic and volcanic activity indicates. Interrelated with the orogenic belts we have the ophiolitic suite of rocks, consisting mainly of synorogenic volcanism and mantle diapirism. Diabases and dolerites are the most common synorogenic effusives and they can be considered as basaltoid rock types.

(b) *The island arcs.* The island arcs are, according to Krishnan (1961), manifestations of an expanding or sliding continent encroaching upon oceanic areas. The distinguishing features of island arcs are: a chain of islands arranged in an arc; a narrow and deep trench on the convex (ocean) side; a semi-emergent chain of islands may be found in some cases. (This zone shows large negative gravity anomalies, some 100 km behind this, is the zone of active volcanoes, this is underlain by a belt of shallow earthquake foci); a second ridge containing older volcanoes occurs another 100 or 150 km behind the active volcanoes; a fairly broad belt of deep earthquake is still further behind (this zone is about 600–800 km behind the ocean trench). Volcanism is associated with the island areas.

(c) *Mid-oceanic ridges* [1]. As oceanographic research has established, the Mid-Atlantic Ridge is a continuous feature extending over the whole globe (Fig. 2).

The mid-oceanic ridges differ to some extent in character and may be referred to as three types. The first type shows broad mountain ranges with a rugged central ridge and longitudinally faulted "flank provinces". The central ridge shows an almost continuous rift valley at its top, this being about 5–10 km wide and 2–3 km deep (comparable to and of the same origin as the rift valley of East Africa). The Mid-Atlantic Ridge and Mid-Indian Ocean Ridge are examples of this first type of oceanic ridges. It should be noted that the Carlsberg Ridge (an extension of the Mid-Indian Ocean Ridge) merges into the large rift of the Red Sea at its northwestern end. The ridge is composed mainly of basalts but olivine rocks, often serpentinized and brecciated, have also been found.

The second type is the Easter Island Rise in the Southern Pacific, which is a huge feature over 3,000 km broad. It may owe its origin either to horizontal compression of the ocean crust producing a dome-like feature or it may be connected with subcrustal convection currents rising under it. Though it shows seismic activity and occasional volcanic islands on its crest, it does not possess a rift valley.

The third type of ridge system is in the Central Pacific and is comparatively narrow and steep and is dotted by numerous seamounts and flat-topped guyots.

The eastern part of the Central and Northern Pacific is cut by a series of great fault zones which extend from the coasts of North and Central America for a distance of 3,000–5,000 km into the Pacific Basin. The faults expose steep scarps, 1,000–1,500 m high. In some parts these fault zones show "Grabens" in which the floor has sunk. Along these faults, there are extinct volcanoes and occasional guyots. The ridges and the fault zones of the Central and North Pacific have been the loci of volcanic activity mainly in the Cretaceous and perhaps also somewhat later. This volcanic activity is now completely extinct except in the Hawaii Islands.

(d) *The oceanic crust.* Another extensive basaltic occurrence is the oceanic crust. The oceanic crust, in the deep areas of the ocean, has a thickness varying from 5 to 7 km. In general, practically all of this is oceanic basalt, except for 0.5–0.1 km of sediments lying over the basalt.

From seismic data it is clear that, where there are mountains in the ocean basins, the crust is very much thicker than in the usual oceanic crust but all the material is basaltic. The lower section of the basalt (with P-wave velocity of 7.5 km/sec) is found to be frequently transitional to the mantle rock so that in these cases there is no distinct boundary between the crust and the mantle.

(e) *The rift systems* [2] are geotectonic features which are associated with large quantities of basalt effusion. Considering the distribution of basalts, the East African basaltic cover is related to the tectonics of the Great Rift Valley. Similarly, basaltic effusion is associated with the Rhine graben. As a corollary it should be pointed out that the Mid-Atlantic Ridge and the Mid-Indian Ridge have rift systems and that the Carlsberg Ridge merges into the large Red Sea Rift.

Wilson (1967) considers that the major rift systems lie along the mid-oceanic ridges. The visible rift, a few miles wide along the crest are but the latest expression of a spreading, which since the end of the Jurassic Period has produced the Atlantic, Indian and the Southern oceans and the southeastern half of the Pacific Ocean. Supporting Tuzo Wilson's hypothesis is the interpretation that the Red Sea Rift is an initial phase of oceanic spreading accepting, of course, continental drift and migration of the African and Asian cratons.

The realization that the geotectonic features do not represent isolated structures or events and that they continue beyond the boundaries of the land as sub-oceanic geotectonic features, and that they are a part of a global pattern of structures, has opened a new era.

As a result of an increasing knowledge of geology, geodesy, geophysics, tectonophysics and geodynamics, our concepts about the earth's crust and mantle have been rapidly changing. New hypotheses constantly appear with an aim to interpret new facts and to modify old concepts in order to comply with a continuously increasing informational background.

The following are some of the publications which deal with the "continental-drift hypothesis": Wegener (1924), Runcorn (1962), Blackett et al. (1965), Vine (1966), Wilson (1966) and Beloussov (1967, 1969). Accepting the revival of the continental-drift hypotheses and considering a compounding amount of new geo-

[1] There are large mountain ranges in the oceans, forming the mid-oceanic ridges and swells. But there are also vast level surfaces forming the abyssal depths, plains of such dimensions being unknown on land.

[2] Rift valleys, e.g., the Carlsberg Ridge, the South and North Atlantic Mid-Ocean Ridge, the East Pacific Rise, the East African Great Rift (Gregory's Rift); the Bechuanaland, Australian and Antarctic rifts, also the Rhine graben and the Baikal Rift.

physical information, the hypothesis of plate tectonics has been mainly proposed by Le Pichon (1968), Morgan (1971) and others.

Furthermore, a number of publications attempt to explain the geotectonic features and phenomena on the basis of the plate-tectonic hypothesis, among others are the following: Carey (1958), Dietz (1961, 1963), Le Pichon (1968, 1969, 1970), Tazieff (1968, 1970), Francis (1969), Barker (1970), Bird and Dewey (1970), Dewey and Horsfield (1970), Dewey and Bird (1970), Freund (1970), McKenzie (1970), Larson and Chase (1970), Borcaletti et al. (1971), Coleman (1971), Herron (1972), Le Pichon et al. (1973).

Fig. 3 shows a schematic diagram of the tectonic plates according to Le Pichon and Morgan.

On the basis of the plate-tectonic hypothesis, orogenic belts, mid-oceanic ridges and rift systems would represent surface expressions of junctions of adjacent tectonic plates. On this basis a plausible explanation could be found for most of the earth's geotectonic features. As pointed out, the association of basaltic effusions with rift-valley systems, mid-oceanic ridges and orogenic belts coincides with the hypothesis that the basaltic effusions followed the junctions of adjacent tectonic plates. On the basis of this hypothesis an explanation can be provided for the East African basalts associated with the Great East African Rift, for the Rhine-graben basalts, for the synorogenic basaltic flows (ophiolites), etc.

In contrast, a major objection arises by considering the great basaltic covers, e.g., the plateaux basalts of the Columbia and Snake River region of the northwestern United States, the Deccan traps, the basalts of the Paraná Brazil, the Siberian and Mongolian basalts, the basalts of Drakensberg, Natal and the Ethiopian Plateau basalts. All of these occupy regions well within tectonic plates and are not associated with the margins of adjacent plates. In the writer's opinion, the tectonic plate hypothesis (i.e. effusion along junctions of adjacent tectonic plates) does not provide a satisfactory explanation for these huge quantities of basaltic covers.

Chapter 2 | The East African and Ethiopian volcanic series as a model of a major basaltic effusion on the earth's surface

The East African volcanism, which starts from the Red Sea, covers the greatest part of the Ethiopian Plateau, the Somali–Harrar Plateau and furthermore, this volcanism is intimately associated with faults and fissures of the Great Rift Valley. Basaltic rocks occupy a great part of the Great Rift Valley. Illies' map, Fig. 1d, shows the distribution of basic volcanics in East Africa.

In the vast area which is occupied by basic volcanics two distinct second-order relief features are distinguished geomorphologically, the plateaux and the Great Rift Valley.

The Ethiopian Plateau, which is drained by the Blue Nile Basin (a rejuvenated drainage system), is a classical example of a second-order relief feature. The Blue Nile drainage pattern is a system of gorges cut by the river, which provide excellent cross-sections of the sequence and thickness of the basaltic flows. The morphology of the Ethiopian Plateau is also characterized by buttes and mensas, see Figs. 4 and 5. A view of the Ethiopian trap series is shown in Fig. 6, as exposed along the Lalibela Gorge, a branch of the Blue Nile. In contrast, Fig. 7 shows a gorge-terrace; about 600–700 m of basalt is cut by the Mugher Gorge, another branch of the Blue Nile river. In the upper part of the Mugher Gorge the basalt trap-series overlie Cretaceous sandstone, as is shown in Fig. 8. Fig. 9 is a model showing the succession of formations in the Blue Nile Gorge; these consist, from bottom to top, of Permian sandstone, Antallo limestone (the Cretaceous sandstones of the Mugher cross-section (Fig. 8) are eroded prior to the basalt series) and the basaltic trap covers. The basalt erosion terraces are shown in Fig. 10. Repeated tholeiitic basaltic flows of enormous extensions, as exposed along the Blue Nile Gorge – a third-order relief feature, represent a typical picture of the plateau basalts (Fig. 11).

Sedimentary formations such as intercalated sandstones show locally developed lacustrine conditions within a basaltic trap formation (Fig. 12); also locally intercalated within the basaltic flows are chert beds (Fig. 13). An extensive reconnaissance of the Blue Nile Gorge in the years 1959–1960 (Special Fund, Water Resources project) has not shown abundant dykes transversing the limestones and sandstones below the basaltic cover and no feeding conduit of the basaltic flows could be determined with the exception of plugs cutting through the sediments below the basaltic cover, as the sedimentary formations are exposed along the Blue Nile Gorge System (Fig. 9). In addition, basaltic plugs transversing the Precambrian basement and acting as feeding channels of the basaltic flows are known and are shown in Figs. 14 and 15. The plateau volcanism was, through many centres of volcanic activity, often protruding over the general elevation of the plateau, or standing out as well-recognized plugs. In addition, in certain cases, swarm dykes can be considered as feeding channels of the basaltic flows.

Examples of the interrelation of the rift-valley volcanism with fault and fissure lines of the Ethiopian Rift Valley are discussed below.

In contrast to the plateaux, where the tholeiitic series predominate and where the volcanism builds trap-series, mainly consisting of repeated extensive basalt flows (Figs. 6, 8, 11), the volcanic activity of the Ethiopian Rift is associated with the tectonics of the rift valley.

The interrelation between volcanism and geotectonic features such as fault lines and fissures, is characteristically demonstrated by the lines of volcanoes, e.g., the Bishufto volcanic line (Figs. 16 and 17). Also, as the diagram (Fig. 18) shows, the volcanoes illustrated in Figs. 19–23, belong to the Dallol volcanic lines (first reported by Augustithis, 1964 – files of the Ministry of Mines, Ethiopian Government). Some of the volcanic cones indicated in the sketch diagram (Fig. 18) and shown in the photographs, are impregnated with sulphur and melted salt has poured out from one of the volcanic craters, indicating that extensive assimilation of the Dallol evaporites has taken place by the volcanic activity, cutting through the evaporitic series.

The most impressive evidence of the interrelationship between rift faulting and lava outpouring is described by Tazieff (1970) in the region of Borale Ale, Dankalia (Fig. 24). Similarly, Fig. 25 shows the interrelation between volcanism and rift-valley fissuring.

In contrast, the olivine basalt flows of the Debra Sine region at the margin of the Ethiopian Rift are step-faulted (Fig. 26). Similarly, step-faulting of volcanics within the rift is illustrated in Fig. 27.

As pointed out, volcanism is associated with the major fault lines of the Great Rift Valley. Fig. 28a shows a structural map of a part of the Ethiopian Rift which is of particular geotectonic significance; as Fig. 28b shows, the faults mainly follow three directions which correspond to the Red-Sea faulting, to the Carlsberg Ridge and to the Owen's fractures.

Chapter 3 | Basalt structures

Columnar joints (columnar basalts)

The columnar structures in basalts have been considered in the past to be prismatic crystals, crystallized from aqueous solutions. This explanation has dominated for years, Trembley (1656), Henckel (1725) and Welch (1764) until it was disproved.

Holmes has pointed out the similarity between mudcracks and columnar basaltic structures and has proposed contraction under cooling as having been the mechanism responsible for the basaltic columns. Holmes, discussing the mechanism of columnar structure development in basalts, states the following:

> "When a hot homogeneous rock cools uniformly against a plane surface, the contraction is equally developed in all directions throughout the surface. This condition is mechanically the same as if the contraction acted towards each of a series of equally spaced centres. Such centres (e.g., C1, C2, C3, etc., in Fig. 29) form the corners of equilateral triangles, and theoretically this is the only possible arrangement. At the moment of rupture the distance between any given centre C and those nearest to it (e.g. 1–6) is such that the contraction along lines such as C–1 is just sufficient to overcome the tensile strength of the rock. A tension crack then forms half-way between C and 1 and at right angles to the line C–1. As each centre is surrounded by six others (1–6 in Fig. 29), the resultant systems of cracks is hexagonal. Once a crack occurs somewhere in the cooling layer the centres are definitely localized, and a repeated pattern of hexagonal cracks spreads almost simultaneously throughout the layer (Fig. 29). As cooling proceeds into the sheet of rock the cracks grow inwards at right angles to the cooling surface, and so divide the sheet into a system of hexagonal columns".

In addition to the six-sided columnar structures, three-sided and eight-sided columns may develop. Also, cylindrical columns are described by Sirin (1962) in Kamchatka flows. The diameter of the columnar structure may vary from 60 cm to 3 cm in the case of three-sided Greenland's swards (i.e. three-sided long columnar basalt from Greenland), (Fig. 30). In addition to the common columnar structures in basalts, comparable structures are formed in volcanic ashes due to contraction during cooling.

As field studies show, the columnar structures of basalts often develop in the middle of the flow, whereas on the top and bottom, cellular basalt is predominant (see Fig. 31). The most impressive columnar basaltic structures develop in volcanic plugs of basaltic composition (Fig. 32).

Whereas mineralogical and petrographical studies by Symons (1967) have shown no variation in the density, grain size and mineral abundance of the basalt of a single column; systematic variations related to the column's shape, of about 40% were found in the remanence intensities, magnetic hysteresis and thermomagnetic properties, reflecting systematic variations in the O and Ti content of the titanomagnetite.

Symons has proposed continuous contraction as a mechanism of column formation.

Basalts structures due to consolidation (ropy and blocky lavas – pyroclastics – pillow lavas)

The surface structures of effusive basaltic flows depend on the conditions of the basaltic melt consolidation. Two main types of surface consolidation structures are recognized, the Pahoehoe (ropy lavas) – see Fig. 33, and the Aa-type (blocky lava), see Figs. 34 and 35.

According to Washington (1923) the following are the conditions of formation of the two types. The Aa (blocky lava) type is gas-rich, with few big vesicular cavities and little glass; it flows quickly and is poor in FeO. In contradistinction, the Pahoehoe lava is gas-poor, has many small vesicular cavities, a lot of glass, flows slowly and is relatively rich in FeO.

It should be pointed out that in the Aa-lava, t and viscosity are low and it crystallizes quickly. Many gas-cavities are in contact with crystals and, due to high fluidity, unite to form larger gas-cavities. The temperature and the fluidity are diminished first slowly and later quickly as the gas is given out and complete crystallization takes place.

In contrast, the Pahoehoe lavas are of high temperature and are poured out quickly. The gas is given out rapidly with a strong reduction in fluidity and as a consequence, glass is formed. As a result of these conditions, many small gas-cavities are developed.

In contrast to the basic basaltic lavas, acid representatives on consolidation due to crust formation result in the formation of volcanic caves as is illustrated in a case of a pantelleritic flow from the region of Matahara, in the Ethiopian part of the Great Rift Valley (Fig. 36).

Of particular interest are the basaltic pyroclastics which are either interbedded with volcanic ashes (Fig. 37) or basaltic ejected scoria "bombs" which are embedded among pyroclastics of smaller size (Fig. 38). The scoria nature and the abundance of gas-cavities in the pyroclastic "ejecta" are in accordance with their air environment of consolidation.

In contrast to the structures which are primarily dependent on the composition of the effused or ejected basaltic material, most characteristic structures are formed when basalts are effused under subaqueous conditions. The pillow lavas and associated spheroidal forms have been extensively studied by many workers (e.g., Gass, 1958; Alsax, 1959; Bloxam, 1960; Wilson, 1960; Hopgood, 1962; Montgomery, 1962; Gass and Masson-Smith, 1963; Duffield, 1969; Galli, 1963; Vallance, 1965; Moore, 1966; Solomon, 1966; and Rossy, 1969).

A very interesting account, an introduction to the pioneering work on the pillow lavas, is given by Dahm (1967) from which the following is quoted: "Usually pillow lavas develop in basic and intermediate lavas, but pillow lavas acid in composition have also been found."

The pillow lavas are explained as subaquatic effusions, the spheroidal structures of subaquatic effusions are also considered to be the result of the interaction of hot melts and water vapour. The typical dark shells (see Fig. 39), which often surround the pillow structures are due to the chilling of the suddenly cooled lava. The dark shells were originally glass which subsequently has altered into chlorite.

The amygdaloidal cavities, often found in the central part of the pillow structure, or, in cases, in the peripheral part of the structure, are most probably due to gas escape (i.e. gas cavities); in contradistinction, the cracks radiating from the central part of the pillow structure outward are most probably due to shrinkage.

When melts enter into wet, not-yet consolidated sediments, similar spheroidal structures to those mentioned can be developed. The pillow structures and subaqueous formations, and the form and structure of the pillow lavas, are illustrated in Figs. 40–42.

The spaces between the "pillows", as well as the amygdaloids and the cracks within the pillows themselves, can be filled with chlorite, calcite or occasionally, with chalcedony.

Cellular structures of basalts (gas cavities)

Basaltic scoria, and usually the tops and bottoms of basaltic flows (Fig. 43) and ejected basaltic bombs often are characterized by cellular structures (gas cavities) which in the extreme case render a spongy appearance to the basalt. Fig. 44 shows cellular structures developed at the top and bottom of a basaltic flow. Also, the shape of the gas cavities is often dependent on the flow direction of the lava.

The phenomena of the cellular structures would primarily depend on the escape of gasses and on the influence which could be exercised on them by the presence of already crystallized phenocrysts. Figs. 45–47 show the form and shape of the cellular structures dependent mainly on the gas escape. In contradistinction, the shape of gas cavities can be influenced by the presence of already crystallized olivine phenocrysts (Figs. 48 and 49).

Similarly, pyroxene phenocryst can exercise an influence on the shape of the cellular structures (Figs. 50 and 51). In addition to this, early crystallized magnetite can also influence the shape of a cellular cavity (Fig. 52). In addition to these early crystallization components which, as expected, would influence the shape of gas cavities plagioclase groundmass laths can also be partly present, as already-formed crystals which would influence the form of the gas cavities (Fig. 53). Also late-phase plagioclase collocrysts are pre-cellular structures, the presence of which has influenced the gas-cavity forms (Fig. 54).

As expected, the escape of gases has influenced the shape of the gas cavities and their walls. Occasionally, the gas escape has produced an advancing gas front which has entered into adjacent olivine phenocrysts and has caused alterations (Figs. 55 and 56).

The gas-escape effect on adjacent phenocrysts does not assume only the nature of advancing intracrystalline front of alteration but it can cause the disrupture of the already crystallized phenocryst. Fig. 57 shows an augite phenocryst with continuation of the gas cavity also in the augite, extensions of which attain a branching system within the augite. Fig. 58 shows a channeling which has been produced in an augite phenocryst by extension in the pyroxene-intracrystalline of the gas-cavity spaces.

Similarly, Fig. 59 shows small gas-cavity chambers and interconnecting fine channels in an augite-phenocryst; often the fine channels form an interconnecting system with the vesicular gas cavities (Fig. 60).

Figs. 61 and 62 show that the gas cavity, in addition to its extension in the pyroxene as intracrystalline gas cavities, is accompanied by alterations in the surrounding margin of the intracrystalline gas cavities, attaining the character of "scoria" extensions in the intracrystalline of the augite, comparable to the alteration fronts already described (see Figs. 55 and 56).

Cluster structures (or synneusis)

Often pyroxene phenocrysts grown together present cluster structures (Fig. 63). In contradistinction, it should be pointed out that synneusis is the process of drifting-together and mutual attachment of crystals suspended in a melt. According to Vance (1969): "Unions of crystals in synneusis relation normally occur on their broader faces in preferred parallel or twinned orientations, which coincide with positions of low interfacial energy. Synneusis structures have been misinterpreted as epitaxial intergrowths, primary twins, irregular growth forms, or random union of crystals which have grown into contact". As is indicated in Fig. 64, the clustering of pyroxene phenocryst is often associated with early magnetite accumulations which act as a kind of nucleus for cluster pyroxenes. This pyroxene–magnetite interrelation and intergrowth is a characteristic of cluster-pyroxene structures. In cases where there is an intrapenetration of the pyroxenes (see Chapter 8), phenocrysts or hourglass augites are clustered together, as Fig. 65 shows. Fig. 66 shows augite phenocryst with epitactic growths of smaller pyroxenes in a cluster assemblage of augitic phenocrysts.

Comparable to the cluster structure of titano-augites are the cluster structures composed of olivine, pyroxene and prismatic elongated plagioclase phenocrysts, Figs. 67 and 68. In cases, ophitic intergrowth of plagioclase/pyroxene is shown. Fig. 67 shows plagioclase prismatic growth invading pyroxenes in an ophitic intergrowth of phenocrysts in a groundmass background.

It should be noted that whereas the phenocrysts in clusters are early magmatic crystallization and they have crystallized while the groundmass was to a great part liquid; the question "What caused them to cluster together?" arises. Perhaps the same causes that bring about synneusis, i.e. the drifting together and mutual attachment of crystals suspended in a melt, relate to a magmatic turbulence. It should be mentioned that whereas synneusis is the drifting together of already formed crystals, the cluster structures also show mutual intergrowth patterns and crystal penetration.

Chapter 4 | The olivines in basalts

In order to understand the petrogenetic significance of olivines in basalts, it is necessary to consider the wide range of crystallization possibilities of the mineral, on the basis of experimental mineralogy and comparative petrography. Some of the diverse conditions of olivine crystallization are synoptically discussed.

Experimentally, the isomorphous series, forsterite (Mg_2SiO_4) and fyalite (Fe_2SiO_4) shows, according to Bowen and Schairer (1935), that pure magnesium olivine melts at 1890°C and the pure iron olivine at 1205°C, thus suggesting comparable crystallization temperatures.

Physicochemically, forsterite in the presence of water vapour is at an equilibrium stability under considerably lower temperatures, approximately 400°C in the ternary system $MgO-SiO_2-H_2O$ studied by Bowen and Tuttle (1949). In the quaternary system of Yoder (1952) forsterite crystallizes from powdered mixed and glasses of appropriate composition at about 430°C under excess water-vapour pressure.

Forsterite has also crystallized from gels precipitated from solutions of Na_2SO_3 and $HgCl_2$ (Sabatier, 1950). In contrast to the crystallization of olivines from igneous melts, Tilley (1947) has described forsterite porphyroblasts from the dunite mylonites of St. Paul's Rocks, in which the (010) plane of the olivines is parallel to the foliation of the rock. The petrographical investigations of Drescher Kaden (1969) are comparable; olivine porphyroblasts as regeneration blastic growths in a mylonitic rupture, in Prata, South Chiavenna (Italy). Present studies of the olivine porphyroblast from Prata, Chiavenna, show that the blastogenic growth is in a talc zone associated with a rupture of an olivine fels. Figs. 69, 70 and 71 show the relation between the olivine fels and the olivine megablasts. The metasomatic growth of granoblastic olivines in marbles has also been described by Drescher-Kaden (1969) from Malga Travena, Val di Brenguzzo, Adamello. Fig. 72 shows the impregnation of Travena marbles by Mg-rich metasomatic solutions which have resulted in the granoblastic olivine growths, as shown in Fig. 73.

In contrast to Bowen's early crystallization of olivine, Edwards (1938) has described olivine phenocrysts in basalts in which orthosilicate cores are surrounded by fibrous iddingsite which is surrounded by a margin of unaltered olivine. Edward's explanation is that the initial period of olivine crystallization was followed by a period of hydration–oxidation conditions (i.e. the iddingsite zone formation) which was succeeded by a renewal period of olivine crystallization. Comparable observations from basalts of Ankober, Shoa, Ethiopia show the olivine regrowths surrounding the iddingsite formation (see Figs. 74 and 75 and their descriptions).

With regard to crystallization of olivines in basalts and in contrast to Bowen's concept of crystallization series with olivine crystallization as an early phase of basaltic melts, Stanik (1970) recognized different generations of olivine formation mainly on textural evidence and on the Fo content of the olivines.

Olivines of the olivine-bombs and olivine xenocryst, almost pure forsterites and with deformation lamellae, are considered the early generation. Idiomorphic olivines with parallel intergrowths and in intergrowth or surrounding pyroxenes, are considered as belonging to a later generation.

The crystallization of more generations of olivines, particularly the recognition that there is an early olivine crystallization, almost pure forsterite, is in accordance with present microscopic observations.

Fig. 76 shows rounded olivines (forsterite rich, about 92% Fo) in basaltic groundmass. Their early crystallization is supported by magmatic corrosion and by the richness in Fo. Also forsterite rich are the olivine phenocrysts with deformation lamellae (pressure deformation) which suggest mantle-derivation as interspersed mantle-derived xenocrysts (Figs. 77 and 78). The banding exhibited, is due to pressure deformation of olivines. The deformation lamellae, in which the translation plane is parallel to (100) and the translation direction // (001), have been described in dunitic and peridotitic olivines, and are common in olivine bombs and isolated olivine xenocrysts in basalts.

Indeed, comparative petrographical observations show all transitions between olivine bombs (nodules) in basalts and detached olivines (initially belonging to an olivine bomb) as "isolated phenocrysts". In addition to rounding due to magmatic corrosion, reaction margins often surround such olivine xenocrysts (Fig. 79).

In contrast to these mantle-derived forsterite phenocrysts, olivine which is idiomorphic in appearance due to regeneration, and a late crystallization phase are observed in picritic basalts of Cyprus (Figs. 80 and 81). The picritic basalts of Cyprus are genetically interrelated to mantle diapirism. The regeneration of mantle-forsterite to idiomorphic forsterite phenocrysts of the basalt signifies the mantle material in the basalt.

As a corollary to the regeneration of idiomorphic phenocrysts of the picritic basalts of Cyprus are the idiomorphic regeneration phases derived from forsteritic olivines of corroded and resorbed olivine bombs in basalts (Augustithis, 1972). Fig. 82 shows a forsteritic olivine idiomorphic growth as a regeneration growth from mantle olivines (i.e. of forsterites of the olivine bombs and nodules). In support of Stanik's late-phase idiomorphic crystallizations, showing parallel intergrowth of olivines, Fig. 83 shows a twin-intergrowth of idiomorphic olivine individuals, in a Deccan-trap basalt indicating in this case a late-phase olivine crystallization.

In contrast, the majority of idiomorphic olivine crystals in basalts are early crystallizations and they are within the forsteritic range of olivines crystallizing from igneous melts, i.e. Fo_{88}–Fo_{82}. The preservation of olivine phenocrysts is against a reaction resorption of the olivines by the orthopyroxene phase of the discontinuous reaction series (Bowen). Idiomorphic first-phase crystallization olivines without resorption phenomena are shown in Fig. 84. In cases, the early crystallized olivines can show magmatic corrosion and resorption without orthopyroxene formation (Fig. 85).

An indication for distinguishing between an early olivine crystallization from a basic magma and the xenocryst-olivine type or that regenerated from mantle derivation, is the Fo content of the olivine. In this connection it is interesting to note that the composition of the first olivines to crystallize from many basaltic magmas is in the range Fo_{88}–Fo_{82} and pure forsterite is unknown in rocks of igneous origin. In contrast, the composition of dunitic olivines is usually about Fo_{92}. In this connection it should be pointed out that olivines of the olivine bombs (xenoliths) and olivine xenocrysts in basalt of mantle derivation are in the range Fo_{95}–Fo_{90}. The realization that the olivine bombs in basalts represent xenoliths of the peridotitic layer of the earth (Ross et al., 1954; Ernst, 1963; Augustithis, 1972 – see Chapter 5) has formed the basis for accepting a material participation of the mantle in the basalts. From the above consideration it becomes apparent that olivine has a wide range of crystallization possibilities. The crystallization of olivines in dry magmas or melts (e.g., lunar basalts and gabbroes) suggests that crystallization in these cases is not influenced by the presence of water vapour. On the other hand, the crystallization of olivines can be greatly reduced by the presence of the water-vapour conditions that can exist in wet terrestrial basaltic magmas. It should be taken into consideration that the crystallization temperature, as Bowen (1928) pointed out, might be reduced by the presence of other components (e.g., presence of plagioclases).

The crystallization of olivine as a late phase, as is proposed by Evžen Stanik and Edward, is supported by present observations and examples have been mentioned of idiomorphic (due to regeneration) olivines, derived by the resorption of mantle olivines. Furthermore, the textural patterns indicated by Stanik, such as idiomorphism (idiotecoblastesis), intergrowths and inclusion of pyroxenes by the later crystallization of olivine, parallel and twin-intergrowths of olivines, are comparable and commensurable with the textural characteristics of tecoblastic pyroxenes and plagioclase phenocrysts described by Augustithis (1956, 1959/60).

The mantle derivation of olivine bombs, nodules and xenocryst, with olivines rich in Fo and showing textural evidence such as rounding, spinel–olivine symplectic intergrowths, deformation lamellae, strongly supports mantle participation as a contributing source of basaltic magma.

Chapter 5 | Mantle petrography (the petrography of mantle fragments brought up by basalts)

Of particular significance for the understanding of the interior of the earth is a petrographic description of the pieces of mantle brought up to the earth's surface by the basalts.

The samples are brought up from considerable depths and as Ernst (1963) referred to them, they are "letters sent from great depths". A petrographical study of these pieces of mantle, even if they only represent fragmental pieces of the earth's mantle, is of importance for they are the only direct evidence of the mantle available. Their study can yield most valuable information on the mineralogical phases present, the intergrowths and solid reactions between the minerals and the effects of overall pressure and deformations to which the earth's mantle has been subjected.

In spite of this, the pieces of mantle can be regarded as relatively rare occurrences, mainly in alkali-olivine basalts; the information obtained by a petrographic description of as many as possible of such fragments can show the variation in mineral composition and the most representative textural patterns exhibited, so that a sound idea can be formed of what would be the forsterite–pyroxene–spinel layer of the earth's mantle.

Mineralogically, the peridotitic mantle xenoliths in basic volcanics of different occurrences can be distinguished into two broad groups: (a) the olivine–pyroxene–spinel group of the upper mantle, and (b) the olivine–pyroxene–garnet group (with possible transition into eclogite).

The olivine–pyroxene–spinel type is more commonly found in basalts as xenoliths; exceptionally, eclogitic fragments (xenoliths) and garnet peridotites are found in basalts, e.g., Hawaiian basalts. It is possible that peridotic-garnet mantle (with transitions to eclogite) may exist as windows in the more common forsterite–pyroxene–spinel type. It is also possible that the garnet-peridotitic mantle fragments are derivatives of greater depths, e.g., kimberlitic breccia pipes.

The present effort is mainly restricted to a petrographic description of the forsterite–pyroxene–spinel mantle fragments found in basalts. The mineralogical composition is variable, but the most predominant mineralogical phases are olivines (90–95% Fo), pyroxenes, bronzite, enstatites, diopsides, chrome-diopsides and from the spinel group: translucent spinels (brown) chrome-spinels and magnetites. Both the olivines and pyroxenes are well developed with a tendency for equigranular olivine to be a predominating phase (Fig. 86). Pyroxenes are variable in composition. Bronzite is of widespread occurrence and often displays polysynthetic twinning (Fig. 87) and a peculiar circular fracture (Fig. 88). In cases, green-coloured chrome-diopsides are present (Fig. 89). In addition to these predominating phases, symplectic spinel in a graphic-type intergrowth (due to the skeleton shape of the crystals) with forsterite and bronzite is shown in a number of mantle fragments. Comparable spinels and intergrowths are common in ultrabasic peridotites which are believed to be diapiric mantle in orogenic belts (Fig. 90).

The spinels are of petrogenetic significance and are considered to be characteristic of the upper-mantle layer. Detailed microscopic observations reveal that olivine grains can be partly surrounded by spinels (Fig. 91). A characteristic transformation of the margin of the spinels, due to oxidation, into an opaque rim is shown in Figs. 92 and 93. In cases this marginal transformation is due to infiltration of solutions, as is displayed in Fig. 94, and this can be considered as being due to deuteric effects.

In contradistinction to the translucent spinels, granular magnetite is often present; despite its granular tendency, in some cases it attains texturally symplectic forms (Figs. 95 and 96). In addition to spinels and magnetite, chrome-spinels and chromite also attain symplectic intergrowths with the pyroxene (Fig. 97) and in some cases more or less follow the margins between forsterite and pyroxene (Figs. 98 and 99).

In contrast to the equigranular olivine–pyroxene textures, most complex intergrowth patterns are also displayed and they indicate the reactions in solid state between these mineral phases. Fig. 100 shows the symplectic intergrowth of bronzite–forsterite and Fig. 101 illustrates bronzite with an extension protruding into a weak zone of the adjacent deformed olivine. In some cases, the bronzite is in intergrowth with the host forsterite, occupying interleptonic spaces in it (Figs. 102 and 103), indicating a post-olivine mobilization or crystallization of the pyroxene. In cases, the forsterite is surrounded by later-bronzite; however, within the same bronzite there is an olivine lamella orientated parallel to the bronzite's fine polysynthetic twinning (Figs. 104 and 105).

Most interesting intergrowths and intrapenetration textures are displayed between the pyroxenes of mantle xenoliths in basalts. A series of photomicrographs (Figs. 106–108) shows all phases of bronzite infiltrated and replaced by diopside; clearly the diopsidic phase is a later crystallization and has taken advantage of the bronzite's cleavage. Often the diopside attains the form of interleptonic fillings. Indeed, such characteristic infiltration and replacement textures are often exhibited in mantle xenoliths in basalts. Fig. 109 shows two differently orientated bronzites with a diopsidic phase intergranular and forming extensions which follow the interleptonic cleavage spaces in the bronzite. In addition to the described intergrowth of two phases, equigranular forsterites react symplectically at their mutual contact producing "myrmekitoids" (Figs. 110 and 111); in some cases neocrystallizations develop, within these reaction margins (Fig. 112).

Most interesting reaction textures often develop between bronzite and forsterite in mantle fragments (Fig. 113). This reaction margin is in reality a crushed zone between these two mineral phases produced by the overall pressure to which the mantle minerals are subjected. Often recrystallization takes place within these reaction margins. Figs. 114 and 115 show bronzite in contact with forsterite; a reaction zone is developed between the two which extends as a crushed zone within the bronzites as well. Recrystallization and new olivine grains are developed in this zone. Under pressure it is possible that the transformation of forsterite to bronzite is a reversible reaction.

In addition to the high content of Fo which distinguishes dunitic-mantle olivines from basaltic crystallizations, deformation evidence is exhibited by olivines of the bombs. As mentioned, the banding displayed is considered to represent deformation lamellae in which the translation plane is parallel to (100) and the translation direction // (001) of the olivines. There is a variation in the width and shape of the deformation lamellae, as is apparent by comparing Figs. 116 and 117. The deformation lamellae of the olivine shown in Fig. 117 are subparallel and radiating.

In equigranular, differently orientated, olivine grains, the deformation lamellae of the one grain may happen to be orientated perpendicularly to the lamellae of the other olivine individual (Fig. 118); this is consequent to the relation of the deformation lamellae to certain crystallographically defined translation directions within each olivine crystal. In some cases, olivine grains with approximately similar orientations have deformation lamellae, giving the impression that they transgress crystal boundaries (Figs. 119 and 120). Occasionally, the deformation lamellae may exhibit curved boundaries and tend to resemble an undulating extinction effect rather than a well-defined system of deformation lamellae (Fig. 121). In contrast, the most complex deformation lamellae are exhibited in olivine xenocrysts and nodules in basalts (Fig. 122). The deformation has resulted in a more complex pattern than a crystallographically definable orientated pattern of lamellae.

In some cases, the development of deformation lamellae is dependent on the presence of small spinel bodies in the olivines (Fig. 123). As shown, the boundary phase of the deformation lamellae follows the elongated direction of the spinel body in the olivine.

That the mantle fragments have been subjected to deformation is supported by the evidence of the development of the crushed-zones around pressure-resistant minerals *. Therefore, when a spinel mineral is enclosed or partly surrounded by olivines and pyroxenes due to overall pressure exercised at the mantle depth by the overlying crust, fragmentation of the less pressure-resistant olivines and pyroxenes takes place in comparison to the spinels; as a result, crushed zones develop in the olivines and pyroxenes surrounding spinel.

Such crushed zones have been observed around alumino-spinels, magmetites and chromites which are in symplectic intergrowth with the olivines and pyroxenes of the mantle fragments. Impressive "crushed zones" are developed surrounding symplectic spinels, as seen in Figs. 124 and 125. As a corollary, Fig. 126 shows symplectic spinel, intergrown with bronzites and marginal to an intensely crushed zone. It should be noted that parts of the bronzite or olivine enclosed in the symplectic spinel are protected by the pressure-resistant spinel and have survived crushing. Similarly, Fig. 127 shows a symplectic spinel enclosing a bronzite part which is thus protected by the pressure-resistant mineral, and in this case a crushed zone has developed between bronzite and spinel. Fig. 128 shows a symplectic spinel in intergrowth with bronzite and forsterite, a crushed zone is developed mainly in the olivine around the spinel, suggesting a different crushing resistance between olivine and bronzite. Indeed the crushed zones around symplectic spinels mainly attain considerable width and they are observed in almost every case of olivine bombs examined where symplectic spinels are surrounded by olivines or pyroxenes (Figs. 129 and 130). As pointed out, similarly crushed zones are formed around symplectic magnetites in intergrowth with olivines (Figs. 95 and 96).

* The spinel group can be considered as stable under high pressures.

Rarely, the crushed zones are "mobilized" and fine extensions of them extend along thin cracks into the pressure-resistant spinel grains (Figs. 131 and 132). This mobilization is most probably due to plastic "mobility" of the silicates under overall pressure. Often, either in relation with the crushed zones or independently, bronzite and other pyroxenes show a cleavage set caused by deformation.

Fig. 133 shows a symplectic spinel with olivine and bronzite and with a characteristic and well-developed "crushed" zone surrounding the spinel. Adjacent to the crushed zone, the bronzites develop the cleavage set caused by deformation. In mantle fragments and in cases where equigranular olivine–bronzite textures are developed this "deformation cleavage" is often produced (Figs. 134 and 135).

Considering the deformation effects on mantle fragments already described, olivine yields by plastic deformation as is the case with the development of undulating extinction and when a pressure-resistant spinel mineral is surrounded by forsterite; a crushed zone develops in the olivine around the spinel body. Clearly the plastic limit has been surpassed and fracturing has occurred. Pyroxenes and particularly bronzite behave in a commensurable way around spinels and crushing (fracturing) takes place. However, in contrast to the olivines, in bronzites deformation cleavage is often produced.

In contrast to the described deformation textures, in the case of mantle diapirism, mantle bodies occur as dunite intrusions in orogenic zones, where spinels are in symplectic intergrowth with olivines and pyroxenes and no such crushed zones have been observed. However, in cases of tectonic deformation, plastic deformation indicated by undulating extinction, crystal bending and twin lamellar gliding, are developed in the pyroxenes and characteristic plastic deformation, e.g., undulating extinction in the olivines. Micromylonitization takes place along shearing directions and deformation "augen" structures of the more resistant spinels are formed within the micromylonitization zones; in such cases deformation of the spinel can take place.

Despite the intense tectonic deformation caused by directed forces – shearing forces – the deformation effects are different from those described in mantle fragments derived from great depths. Clearly the tectonic deformation effects, characteristic of mantle fragments (as olivine bombs), are caused by the overall pressure exercised by the weight of kilometres of crust above and they differ from the tectonic deformation structures of diapiric mantle bodies in the orogenic belts, where directed tectonic forces play the predominant role.

Magmatic corrosion effects, reaction structures and intergrowths between mantle-derived fragments (xenoliths) and the basaltic melts

The olivine bombs, nodules or mantle fragments often vary in size and form and in their distribution within a basalt. Characteristically at Jato, Lekempti, Ethiopia, mantle fragments in the basalt are very abundant and vary from rounded forms (Fig. 136) to angular pieces, indicating clearly that they are fragments detached from a larger body and brought up with the basalt. In fact, the impression given in the field is that they are fragments of a body bound together by the basalt. Similarly the occurrence of olivine bombs in Lanzarote, Canary Islands again gives the impression of fragments which are in basalt and ejected as volcanic bombs in which a mantle fragment is included (Fig. 137). In contrast, rounded olivine bombs and nodules interspersed and scattered in basalts are common in basalts from Yubdo, Wollaga, Red Sea Islands and elsewhere.

The existence of mantle bodies (olivine bombs, nodules and xenocrysts) in the basalt can be regarded as foreign bodies which are in the process of assimilation by the basalt, i.e. a phenomenology should be expected which shows resorption of the foreign bodies, infiltration of basaltic melt and reaction margins and intergrowths between the foreign mantle bodies and the attacking basaltic melt.

The composition of the olivines (92–95% Fo), the deformation textures and the symplectic high-pressure spinels are all evidence in support of the view that olivine bombs are foreign bodies and not an early crystallization phase of the basaltic melts.

Despite the fact that olivines and pyroxenes could be early crystallization phases of basalts, these fragments, due to their composition and derivation, are treated by the basaltic melt as foreign bodies and are strongly attacked by it.

An attempt is made here to present the phenomenology of magmatic corrosion, synisotropization, resorption, replacement and characteristic textures resultant from this process. Figs. 138 and 139 show olivine xenocryst (Fo_{90}) exhibiting a complex pattern of deformation lamellae, rounded by magmatic corrosion and surrounded by a synistropization margin, i.e. the olivine lattice is broken down and is changed into an isotropic margin: this process possibly has been accompanied by migration of elements. It should be pointed out that, as a result of the olivine reacting with the basaltic melt, no orthopyroxene is formed, as would be expected in accordance with the hypothesis of the discontinuous reaction series of Bowen.

Comparable and commensurable reaction phenomena are developed when diopside is corroded and affected by the basaltic magma. Fig. 140 shows magmatic corrosion effects on diopside by the basaltic melt. Similarly, Fig. 141 and Fig. 142 show diopside invaded and corroded by basaltic melt. Arrow (A, Fig. 141) shows synisotropisation of the pyroxene. A most interesting reaction, phenomenology, is illustrated by Fig. 143 where all transitions are displayed between unaffected diopside and an anisotropic margin caused by basaltic melt. An altering, advancing front is also to be seen in the form of advancing "canals" of alterations into the unaffected pyroxene.

In discussing the reaction phenomenology of pieces of mantle with basaltic melt, one is surprised by the great intracrystalline penetrability of the basaltic melt (see chapter 17). Both pyroxenes and olivines can be infiltrated by basaltic melts, often "melt paths" extending from the phenocryst's external margin are seen to invade the host crystal (Figs. 144 and 145). Fig. 146 shows these "melt paths" in a forsteritic olivine grain of a mantle fragment. It should be noted that the form attained by these "melt-path" bodies resemble "myrmekitoid" intergrowths. In all these cases of intracrystalline melt, penetrability due to the great fluidity of the basaltic melt, should be noted.

In contrast to the "melt paths" which indicate intracrystalline penetrability of the basaltic melt is the infiltration of the basaltic melts along interleptonic spaces, e.g., intergranular spaces, cleavage, fractured zones, cracks, etc. Fig. 147 shows bronzite of the mantle fragment corroded by the basaltic melt (out of which the groundmass crystallized). Infiltration of the basaltic melt has taken advantage of the pyroxene cleavage and by infiltration corrosion and complex replacement processes the groundmass–bronzite intergrowths, shown in Figs. 148 (a detail of Fig. 147) and 149, have been caused.

In cases, the penetrability of the basaltic melt has exploited the intergranular spaces between a pyroxene and an olivine and, taking advantage of the pyroxene's cleavage, has also formed a "myrmekitoid" fringe of bronzite in intergrowth with groundmass infiltration (Fig. 150).

The infiltration of basaltic melts has often resulted in crystallization of magnetite, together with the other groundmass components as interleptonic "infiltrations" within the pyroxenes (Fig. 151).

In special cases, infiltration of the basaltic melts has taken place within a crushed zone of a bronzite and, as a result, neocrystallizations of olivine in these complex reaction margins are formed. Associated with these reaction margins, intracrystalline infiltration paths of the basaltic melts are to be seen – often attaining a myrmekitic symplectic appearance (Fig. 152).

In cases, infiltration of basaltic melts following penetrability directions within olivines of the mantle fragments in basalts, has taken place resulting in the formation of olivine neocrystallization-agglutinations either following olivine cracks (Fig. 153) or weakness in the forsterite (Figs. 154 and 155).

Comparable reaction infiltration and replacement structures result when basaltic melts attack detached or isolated chrome-spinel grains of the mantle fragment that occur as xenocrysts, within the basaltic melt. Fig. 156 shows a chrome-spinel xenocryst with a magnetitic reaction margin–migration of elements has played a role in this reaction process.

In cases, the chrome-spinels show, in addition to the magnetite reaction margin, myrmekitic synantetic intergrowths which are caused by the infiltration of basaltic melts into both the magnetite margin and the unaffected chromite (Figs. 157 and 158).

As Fig. 159 shows, most complex myrmekitoid intergrowths result and, in addition to the symplectic intergrowth, feldspar laths often develop within the magnetite reaction margin.

In cases, a corona structure develops as the result of olivine reaction with the basaltic melt (Fig. 160). Such structures are very complex and consist of fine pyroxene aggregates, often in an orientated intergrowth to the olivine itself (Fig. 161).

As Fig. 162 shows, such corona reactions, consisting of fine aggregates of olivines (perhaps relatively richer in Fo%), are not only present marginally to the olivine but also extend as fine canals from the margins inward.

Another instance of such a corona structure taking advantage of the interleptonic spaces is shown in Fig. 163, where the reaction corona is not only present in the margin of the olivine with the groundmass but also extends into an intergranular space between two olivines. Comparable corona reaction structures as a reaction between olivine and groundmass are described by Augustithis (1965, 1972).

In contrast to the relatively narrow corona structures formed between forsterite and groundmass, a case of a complex bronzite–olivine symplectite is formed as a result of reaction with the groundmass (Fig. 114).

The reaction of groundmass with the forsterite of the mantle fragments, in addition to the reaction margins and coronas mentioned, can also result in the formation of plagioclase laths forming a fringe or, as is the case in Fig. 164, plagioclase laths develop perpendicularly to the walls of the intergranular space between two differently orientated olivines.

Reactions comparable to the coronas formed between

forsterite and groundmass are also shown between bronzite and the groundmass-forming basic melts. Figs. 165 and 166 show a corona growth between spinel and bronzite (partly representing a crushed zone) with symplectic extension of it into the bronzite and also with the olivine (Fig. 167).

As a result of basaltic melts reacting with forsterites and bronzites (mantle fragments) due to melt infiltration along penetrability direction and often extending from the periphery to the centre olivine, neocrystallizations are formed (perhaps a phase richer in Fo) in olivines and particularly in a bronzite host (Figs. 168 and 169).

So far we have considered reactions and infiltrations formed within the orthomagmatic-phase, as the crystallization proceeds to its concluding phase; hydrous solutions play an important role and zeolites are formed by aqueous solutions within the intergranular spaces, in spaces provided by the crushed zones and along deformation weakness-zones.

Figs. 170 and 171 show a forsterite of the mantle fragment with deformation lamellae and with a fracture which starts from the crushed zone at the margin of the olivine and continues into it. Zeolites in vein form infiltrated the olivine following the fracture zone, also zeolites occupy the interspaces in the crushed zone. Other cases of zeolites filling the crushed zone spaces between spinel and bronzite are illustrated in Fig. 172. Comparable is the infiltration of aqueous zeolite forming solutions in a crushed zone marginal to a xenolithic bronzite (Fig. 173).

Chapter 6 | Comparative tables of "xenoliths" in volcanics and their petrogenetic significance

The wide variation of xenoliths in volcanics and particularly in basalts is presented in the series of Tables I–XI.

Also tabulated is information regarding the mineralogical composition of the xenoliths and their possible depths of derivation and origin.

Considering the nature of the xenoliths in basalts (see Table I–XI), it can be seen that they predominantly consist of gabbroic material; xenoliths of the mantle, consisting of olivine–pyroxene–spinel, are also very abundant. Furthermore, as regards the composition of the basalts (i.e. the plagioclases are a predominant mineral constituent) and their differences mineralogically and geochemically (see Chapter 25), it is obvious that the basalt is not a fusion derivative of mantle or more precisely, of the olivine–pyroxene–spinel layer of the mantle (upper mantle).

In contrast, the abundance of gabbroic xenoliths (see comparative Tables I–XI) and the fact that by fusion the gabbro could give basalt (in spite of the fact that a number of other basic rocks could, on fusion, also produce basalt), support the hypothesis that broadly speaking, the basalt is derived by fusion of the gabbroic (protolytic) layer which is also geophysically determined to exist as a transitory layer between the crust and the mantle.

Eclogite xenoliths or garnet periodititic xenoliths are relatively rare in basalts as compared with the gabbroic xenoliths and the olivine–pyroxene–spinel xenoliths found in basalts (see comparative Tables I–XI).

The abundance of mantle xenoliths (olivine–pyroxene–spinel) in basalts can be attributed to mantle fragments which are detached from the mantle layer in the process of the fusion of the protolytic layer which is immediately above the mantle.

The occasional, though rare, presence of eclogite or garnet peridotite xenoliths in basalts could be explained by the fusion of a deeper mantle layer or that the fusion zone has involved eclogitic or garnet peridotic windows within the olivine–pyroxene–spinel layer of the mantle.

TABLE I

VARIOUS NODULES (XENOLITHS) IN BASALTS (INTERPRETED AS DERIVATIVES OF THE UPPER CRUST)

Locality	Author and title of publication	Mineralogical composition	Type of basalts	Mono- or poly-xenolithic	Depth of deviation	Remarks
Volcano Tarumai, Hokkaido, Japan	Ishikawa, Toshio, 1953. Xenoliths included in the lavas from volcano Tarumai, Hokkaido, Japan. *J. Fac. Sci. Hokkaido Univ.*, sec. iv, 8: 225–244	*Accidental xenoliths:* range from non-metaphosed shale to a cordierite-hypersthene plagioclase-glass rock *Cognate xenoliths:* characterized by hypersthene and plagioclase as the principle constituents	Lavas	Poly-xenolithic (a) accidental (b) cognate	Upper crust –crust	It should be noticed that the xenolithic material is characteristic as being of accidental origin (engulfed xenoliths from the upper crust) whereas the hypersthene-plagioclase xenoliths are considered to be cognate, they might actually represent material from the lower gabbroic crust
Nagauruhoe, New Zealand	Steiner, A., 1958. Petrogenetic implications of the 1954 Nagauruhoe (New Zealand) lava and its xenoliths. *New Zealand J. Geol. Geophys.*, 1: 325–363	Contains numerous small quartzose and feldspathic xenoliths and intensely vitrified xenoliths of boulder size	Olivine bearing basaltic andesites	Mono-xenolithic	Upper crust From the granitic-gneissic basement	Structural petrographic and chemical data indicate that the xenoliths are derived from acid gneiss
Puy Beaunit	Brousse, R. and Rudel, A., 1964. Bombs de peridotités, de norites, de charnockites et de granulites dans les scories du Puy Beaunit. *C.R. Acad. Sci. Paris*, 259: 185–188	Poly-xenoliths material of petrographically distinct rock types: peridotite norite gabbro charnockite garnet–silimanite granulite granite	Irregularly bedded scoriae of augite–olivine-basalt	Poly-xenolithic	Upper crust	The granulites and charnockites are unknown in the crystalline rocks of the region
Surtsey (Iceland)	Sigurdsson, H., 1968. Petrology of acid xenoliths from Surtsey. *Geol. Mag.*, 105: 440–453	Acid xenoliths have undergone partial melting extraction of a low melting fraction in places producing residues of plagioclase and tridymite	Icelandic lavas	Mono-xenolithic	Upper crust	Melting and recrystallization of the initial xenoliths has taken place resulting in the formation of plagioclase and tridymite

TABLE II

OLIVINE BOMBS IN BASALTS INTERPRETED AS XENOLITHIC FRAGMENTS OF DEEP SEATED PERIDOTITES

Locality	Author and title of publication	Mineralogical composition	Type of basalt	Mono- or poly-xenolithic	Depth of derivation	Remarks
	Ross, C.S., Foster, M.D. and Mayers, A.T., 1954. Origin of dunites and olivine-rich inclusions in basaltic rocks. *Am. Mineral.*, 30: 693–737	Olivine, enstatite, chrome-diopside and chrome-spinel	Basaltic rocks	Mono-xenolithic	Crust Deep-seated peridotites	Author's interpretation: chemical and spectrochemical analyses of the mineralogical constituents of the xenoliths and of the corresponding basaltic mineralogical components show striking similarities thus suggesting similarity in their sources. However, it is suggested that in some cases the nodules may be the result of differentiation in the basaltic magma
Yubdo, W. Ethiopia	Augustithis, S.S., 1965. On the origin of olivine and pyroxene nodules in basalts. *Chem. Erde*, 1965: 197–210	Olivine (forsterite) diopside, chrome-spinel (chromite)	Plateau-basalt	Poly-xenolithic (a) forsterite nodules (b) diopside nodules (c) chromite xenocrysts (d) rarely sperrylite xenocrysts	Deep-seated dunite–peridotite	The Yubdo basalt with olivine and pyroxene nodules partly covers the dunite–peridotite Yubdo ultrabasic (platiniferous) ring complex
Carlton Hill, Derbyshire	Hamad, S. and Hamad, D., 1963. The chemistry and mineralogy of the olivine nodules of Carlton Hill, Derbyshire. *Mineral. Mag.*, 33: 485–497	Olivine ($Fo_{90.3}$ and $Fo_{91.4}$) Orthopyroxene ($Ca_{1.5}Mg_{89}Fe_{9.5}$ and $Ca_{0.8}Mg_{88.6}Fe_{10.6}$) Clinopyroxene ($Ca_{39.6}Mg_{55.8}Fe_{4.6}$) Chromian spinel (16.85% Cr_2O_3) The clinopyroxene (1.06% Cr_2O_3)	Basalt	Mono-xenolithic	The xenoliths represent fragments from a deep-seated peridotite	Authors interpretation: the high Cr contents of the spinel (16.85% Cr_2O_3) and clinopyroxene (1.06% Cr_2O_3) are taken to suggest that these minerals are not differentiates of the basaltic magma and the nodules are considered to represent fragments from a deep-seated peridotite

TABLE III

VARIOUS NODULES (XENOLITHS) IN BASALTS (INTERPRETED AS DERIVATIVES OF THE GABBROIC–DUNITIC LAYER)

Locality	Author and title of publication	Mineralogical composition	Type of basalt	Mono- or poly-xenolithic	Depth of derivation	Remarks
Nagahama, Japan	Harumoto, Atsuo, 1953. Melilite–nepheline basalt, its olivine nodules and other inclusions from Nagahama. *Mem. Coll. Sci., Univ. Kyoto, Ser. B,* 20: 64–88	Basic xenoliths consisting of olivine, hypersthene, augite and bytownite in variable amounts, also phyllitic xenoliths showing zones of metamorphism	Melilite-nepheline basalts (as small scattered areas in nepheline basalts	Poly-xenolithic	Basic xenoliths from gabbroic layer of lower crust	The phyllitic xenoliths are of the upper crust (metamorphic basement)
Lake Graenavatn Maar, Keykjanes peninsula, Iceland	Tryggvason, T., 1957. The gabbro bombs at Lake Graenavatn. *Bull. Geol. Inst. Uppsala,* 38: 1–5	Olivine gabbro: large crystals of plagioclase enclosed large pyroxene crystals Olivine Fa_{30} Plagioclase 79–85% anorthite content	Icelandic lavas	Mono-xenolithic	Basic xenoliths olivine gabbroic layer of the lower crust	
Lanzarote, Canary Islands	Frisch, T., 1970. The detailed mineralogy and significance of an olivine-two pyroxene gabbro nodule from Lanzarote, Canary Islands. *Contrib. Mineral. Petrol.,* 28: 31–41	Gabbro inclusion containing: 5% olivine (Fo_{76}) 60% plagioclase (An_{81}) 30% clinopyroxene ($Ca_{43}Mg_{46}$ and Fe_{11}) 5% orthopyroxene ($Ca_{1.6}Mg_{76}$ and Fe_{22}) The orhopyroxene occurs also as ex-solved lamellae along (100) of the clinopyroxene	In alkalic basalt	Mono-xenolithic	Basic xenolith from the gabbroic layer of the lower crust	Authors interpretation: this xenolith is thought to have crystallized from tholeiitic melt at a depth of <9 km

TABLE IV

VARIOUS NODULES (XENOLITHS) IN BASALTS (INTERPRETED AS DERIVATIVES OF THE GABBRO PERIDOTITIC LAYER)

Locality	Author and title of publication	Mineralogical composition	Type of basalt	Mono- or poly-xenolithic	Depth of derivation	Remarks
Kerguelen Archipelago	Talbot, J.L., Hobbs, B.E., Wilshire, H.C. and Sweatman, T.R. Xenoliths and xenocrysts from lavas of the Kerguelen Archipelago	Perioditic xenoliths: olivine, green diopsidic clinopyroxene, orthopyroxene plagioclase and spinel and rare phlogopite	Oceanites and limburgites (with xenocryst derived from peridotite)	Mono-xenolithic	Peridotitic layer	Author's remark: The xenoliths in the basalt and related rocks have a metamorphic fabric the evidence indicates they have undergone some deformation and subsequent grain boundary adjustments in the solid state
Grand Canyon, Arizona	Best, M.G., 1970. Kaersutite-peridotite inclusions and Kindred megacrysts in basanite Caras, Grand Canyon, Arizona. *Contrib. Mineral. Petrol.*, 27: 25–44	*Abundant xenoliths:* olivine rich chrome rich chrome rich spinel *Less common xenoliths:* titaniferous black clinopyroxene orthopyroxene olivine amphibole spinel (Fe-rich Cr-poor) (poikilitic amphiboles – kaersutites – are characteristic)	Basanitic lavas	Poly-xenolithic (a) olivine, Cr-diopside, Cr-spinel (b) Ti-clinopyroxenes, orthopyroxenes, olivine amphitiboles, Fe-spinel	As indicated by xenoliths (b) derivation of xenoliths is close to the mantle–crust boundary	Author's interpretation: Composition of pyroxenes and coprecipitation of orthopyroxenes, clinopyroxenes, olivine and Mg-spinel, together with the total absence of feldspar as a cumulate or intercumulate phase are compatible with crystallization near 10 kbar. Pressures of this degree are attained at depths close to the mantle–crust boundary in the western Grand Canyon
Takenotsuji, Iki Island, Japan	Aoki, Ken-Ichito, 1970. Petrology of kaersutite-bearing ultramafic and mafic inclusions in Iki Island, Japan. *Contrib. Mineral. Petrol.*, 25: 270–283	Polyxenolithic material of kaersutite bearing peridotite clinopyroxenite, gabbro and hornblendite. Also xenocrysts of kaersutite, andesine and titano-magnetite	In alkali basaltic Scotia	Poly-xenolithic		Author's interpretation: From the petrography and chemistry of the inclusions and recent experimental data from high pressure high-temperature investigations it is suggested that kaersutite-bearing inclusions and megacrysts have been produced from alkali basalt magmas under hydrous conditions at 25–30 km in the lowest part of the crust

TABLE V

OLIVINE BOMBS IN BASALTS INTERPRETED AS MANTLE DERIVATIVES

Locality	Author and title of publication	Mineralogical composition	Type of basalt	Mono- or poly-xenolithic	Depth of derivation	Remarks
Three German occurrences, Germany	Ernst, T., 1965. Do peridotitic inclusions in basalts represent mantle materials? I.U.G.S. *Upper Mantle Symp. New Delhi, 1964,* pp. 180–185 (publ. in Copenhagen)	Peridotitic inclusions in basaltic rocks		Mono-xenolithic	Their primary origin as part of the mantle is considered to be plausible	Author's interpretation: The nodules often have angular shape and the olivine nodules show signs of mechanical stress
	Ernst, Th., 1962. Folgerungen für die Entstehung der Basalte aus dem speziellen Auftreten der Pyroxene in diesen Gesteinen. *Geol. Rundsch.,* 51: 364–374	Olivine, bronzite; "the olivine nodules occurring in alkali basalts contain bronzite although orthopyroxene is normally absent from alkali basalts (exception very rare enstatite crystals)"	Alkali basalts	Mono-xenolithic	Probably derived from the outer parts of the Earth's mantle	The olivine nodules, due to the presence of bronzite, are interpreted as foreign components which also show metamorphic associations and which can probably be derived from outer parts of the Earth's mantle
Gough Island, South Atlantic	Le Maitre, R.W., 1965. The significance of gabbroic xenoliths from Gough Island, South Atlantic, *Min. Mag.,* 51: 364–374	(Gabbroic xenoliths) Olivine (Fa_{23}–Fa_{28}) Augite Orthopyroxenes (Fs_{25}–Fs_{30}) Plagioclase The pyroxenes contain exsolved yellow-green spinel, with a 8.10 Å and the augite may also contain exsolved oriented magnetite (a 8.40 Å) or ilmenite	In basalts and trachybasalts	Mono-xenolithic	The xenoliths are believed to be derived from the mantle above the magma source area, the mantle in this region being considered to have an olivine-tholeiitic composition	Oceanic basalts and trachy basalts with gabbroic in composition xenoliths

TABLE VI

OLIVINE BOMBS IN BASALTS INTERPRETED AS AN EARLY CRYSTALLIZATION PHASE

Locality	Author and title of publication	Mineralogical composition	Type of basalt	Mono- or poly-xenolithic	Depth of derivation	Remarks
New Zealand	Brothers, R.N., 1960. Olivine nodules from New Zealand. *Rep. Int. Geol. Congr., 21st Sess., Norden,* Pt. 13: 68–81	Olivine (Fo_{92}–Fo_{82}) Orthopyroxenes (En_{92}–En_{85}) Clinopyroxenes ($Ca_{46}Mg_{45}Fe_9$ $Ca_{11}Mg_{41}Fe_{11}$) Dark brown spinels Laths of zeolites	Basalt	Mono-xenolithic	Upper mantle (spinel bearing peridotites)	Petrographic analysis gives diagrams similar to those for olivines with preferred orientation in intrusive peridotites
Hualalai Volcano	Richter, D.H. and Murata, K.J., 1961. Xenolithic nodules in the 1800–1801 Kaupulehu flow of Hualalai Volcano. *U.S. Geol. Surv., Prof. Pap.,* 424-B: 215–217	Olivine (Fa_{12-18}) Clinopyroxenes Plagioclase (An_{60-70})	Tholeiitic Basalt	Mono-xenolithic	Lower crust (gabbro–peridotite)	Author's interpretation: They represent consolidations of depth of crystals that effected the fundamental change in magma type
Totatoka Northland	Black, P.M., Brothers, R.N., 1965. Olivine nodules in olivine nephelinite from Tokatoka, Northland. *New Zealand J. Geol. Geophys.,* 8: 62–80	Olivine (the olivine in olivine-rich nodules occurs as both first generation – average composition Fo_{97} and second generation – average Fo_{76} crystals; enstatite (2V 76°–88°) diopside pyroxene (B 1.683–1.684, 2V 54° and 58°–60°). Accessory spinel (pale to dark brown picotite) Andesine and fibrous zeolites. Salite (B 1.690, 2V 58°) also occurs as a second generation mineral	Olivine nephelinite	Mono-xenolithic		Author's interpretation: A cognate origin for these nodules is considered to be most probable

Faial, Azores	Baker, M.J., 1966. Blocks of plutonic aspect in a basaltic lava from Faial, Azores. *Geol. Mag.*, 103: 51–56	Varying proportions of forsteritic olivine, chrome-diopside, calcic plagioclase and chromite	Ankaramitic Basalt	Mono-xenolithic	Author's interpretation: The textures and gradations in composition suggest that these blocks were formed by crystal accumulation
Kilaue	Murata, K.J.R. and Richter, D.H., 1966. The settling of olivine in Kilauean magma as shown by the lavas of the 1959 eruption. *Am. J. Sci.*, 264: 194–203	Olivine ($Fa_{12.5}$) separates at this stage, leading to a series of lava types ranging from picrite-basalt to olivine poor basalt	Fractional crystallization of tholeiitic magma		Author's interpretation: The temperatures of the erupting lavas varied with the olivine content, suggesting that the magma body is zoned into a cooler, olivine-depleted upper part and a hotter, olivine-enriched lower part. The part that the most picritic lavas were erupted during periods of a high rate of lava discharge, together with a consideration of the hydraulics of olivine settling, lead to the conclusion that unusually strong magma currents may occur at such times and pick up sedimented olivine in the lower part of the chamber

TABLE VII

ECLOGITES, LHERZOLITES AND WEHRLITE-XENOLITHIC NODULES IN BASALTS

Locality	Author and title of publication	Mineralogical composition	Type of basalt	Depth of derivation	Remarks
	O'Hara, J.M., 1969. The origin of eclogite and ariegite nodules in basalts. *Geol. Mag.*, 106: 322–330	Eclogitic xenoliths in basalts			The origin of eclogite nodules in basalts is reviewed in the light of experimental studies. Many such nodules are considered to have been precipitated from basic magma as garnet–pyroxene accumulates at high pressure, but some may represent exsolved clinopyroxenitic accumulates
Auvergne, France	Collee, A.L.G., 1963. A fabric study of lherzolites with special reference to ultrabasic nodular inclusions in the lavas of Auvergne (France). *Leidse Geol. Meded.*, 28: 3–102	Lherzolites	In lavas of Auvergne	Since the Mg/Fe distribution in the pyroxenes of the nodules suggests that these minerals crystallized at temperatures well above those of magmatic assemblages it was concluded that the studied specimens are likely to be fragments of the earth's peridotitic shell	Interpretation of textures by author: Recent theories and experiments lead to the conclusion that the orientation of crystals during growth is governed by their lattice; the crystal structure and not the grain shape governs the fabric of many ultrabasic rocks

	Den Tex, E., 1963. Gefügekundliche und geothermometrische Hinweise auf die tiefe, exogene Herkunft iherzolithischer Knollen aus Basaltlaven	Lherzolites	Basalts	
Hawaii	Kuno, Hisashi. Mafic and ultramafic nodules in basaltic rocks of Hawaii	Poly-xenolithic nodules (a) a lherzolite series (b) a wehrlite series comprising dunite, wehrlite and pyroxenite (d) an eclogite series (d) a group of gabbro	In basaltic rocks	The wehrlite series possibly represents the lower portions of solidified magma reservoirs of the Hawaiian volcanoes. The eclogites are supposed to have constituted pockets within the mass of the lherzolites series and the gabbro nodules are fragments of the crust
Nye County, Nevada	Trask, N.J., 1969. Ultramafic xenoliths in basalt, Nye County, Nevada. *U.S. Geol. Surv., Prof. Pap.*, 650-D: 43–48	Poly-xenolithic material (a) lherzolites from ejecta and flows around two vents (moderately to strongly deformed) (b) olivine-rich wehrlites and dunite (intensely deformed) (c) clinopyroxene-rich wehrlites, clinopyroxenites and gabbroes are undeformed or slightly deformed	In basalts	It is suggested that the lherzolites are from the upper mantle and the clinopyroxene rocks from the lower crust

TABLE VIII

GARNET–PERIDOTITE XENOLITHIC NODULES IN BASALTS

Locality	Author and title of publication	Mineralogical composition	Type of basalt	Mono- or poly-xenolithic	Depth of derivation	Remarks
Salt Lake Crater, Oahu Honolulu volcanic series	Beeson, M.H. and Jackson, E.D., 1970. Origin of garnet pyroxenite xenoliths at Salt Lake Crater, Oahu. *Mineral. Soc. Am., Spec. Pap.*, 3: 95–112	All xenoliths are now garnet pyroxenite, but textural reconstruction permits them to be divided into four original rock types: (1) clinopyroxenite (2) websterite (3) garnet websterite (4) garnet pyroxenite	Honolulu volcanic series	Poly-xenolithic	Mantle-derived garnet	Author's interpretation: Comparison of rock composition and mineralogy of the xenolith with those expected for fractionation in the experimental system $CaSiO_3$–$MgSiO_3$–Al_2O_3 at 30 kbar suggests that fractional fusion, rather than fractional crystallization was the dominant process in the origin of the xenoliths
Hawaiian Islands	Jackson, E.D., 1968. The character of the lower crust and upper mantle of the Hawaiian Islands, *Rep. Int. Geol. Congr., 23rd, Prague*, 1: 135–150	Poly-xenolithic material (a) coarse-grained xenoliths of layered feldspathic cumulates (b) xenoliths of deformed or recrystallized dunite and garnet peridotite	In tholeiitic lava basalts ankaramite, trachyte, hawaiite and nephelinite	poly-xenolithic	The textures and mineralogy suggest that they are fragments of sub-crustal rocks	Author's interpretation: Seismic data place the origin of tholeiitic magma as deep as 60 km; the presence of sub-crustal fragments in alkali basalt and nephelinite would require a mantle origin for these magmas as well
Honolulu volcanic series, Hawaii	Jackson, E.D. and Wright, T.L., 1970. Xenoliths in the Honolulu Volcanic series, Hawaii. *J. Petrol.*, 11: 405–430	Poly-xenolithic material (a) dunite (b) lherzolite (c) garnet pyroxenite and peridotite	(a) melilite–nepheline basalts (b) nepheline basalts (c) alkalic olivine basalts	Poly-xenolithic	There is some evidence that the dunite xenoliths are mantle residua, produced during the generation of the tholeiite	Author's interpretation: The rocks of the series are compositionally zoned with respect to the shield, those nearest the Koolan caldera being melilite–nepheline basalts grading outward to nepheline basalts and then at the apron of the shield to alkali olivine basalts. The xenoliths in those rocks are similarly zoned: those in the caldera area are mostly dunite, followed outwards by lherzolite and then by garnet pyroxenite and peridotite

TABLE IX

OLIVINE NODULES IN VARIOUS ROCKS

Locality	Author and title of publication	Mineralogical composition	Type of basalt	Mono- or poly-xenolithic	Depth of derivation	Remarks
Jos Plateau, Nigeria	Wright, J.B., 1969. Olivine nodules in trachyte from the Jos Plateau, Nigeria. *Nature*, 223: 285–286	Olivine nodules ranging from small fragments to ovoids 15 cm long	Trachyte plug in the Cenozoic volcanics of the Jos Plateau			Probably first reported occurrence in trachyte
East Otago, New Zealand	Wright, J.B., 1966. Olivine nodules in a phonolite of the East Otago Alkaline province, New Zealand. *Nature*, 210: 519	Olivine (B 1.672) Clinopyroxene (B 1.685) Orthopyroxene (B 1.672) Spinel (n 1.80 and 1.83)	Phonolite	Mono-xenolithic	Sub-crustal levels	Author's interpretation: The occurrence suggests that phonolite may be generated at sub-crustal levels, and is consistent with strontium isotope ratio in carbonatites and other rocks from alkali igneous complexes
Upper Winterton River, Awatere Valley, South Island, New Zealand	Challis, C.A., 1969. Layered xenoliths in a dyke, Awatere Valley, New Zealand. *Nature*, 109: 11–16	Cognate xenoliths of layered pyroxenite and anorthosite – the pyroxenite contains a deep pink titano-augite. It is associated with a brown Kaersutitic amphibole, ilmenite and minor labradorite	Camptonitic dyke	Poly-xenolithic		
Beu Cleat area, Skye, Gt. Brit.	Gibb, F.G.F., 1969. Cognate xenoliths in the Tertiary basic dykes of South West Skye. *Min. Mag.*, 37: 504–514	Ultrabasic xenoliths composed of essentially the same minerals as the dykes	Ultra-basic dykes in the Beu Cleat area of Skye		These xenoliths are attributed not to primary upper-mantle material but to the disintegration of layered ultrabasic rocks genetically related to the dykes	

TABLE X

XENOLITHS IN KIMBERLITES

Locality	Author and title of publication	Mineralogical composition	Type of kimberlite	Mono- or poly-xenolithic	Depth of derivation	Remarks
Obnazhennaya Olenck basin	Milashev, V.A., 1960. Cognate inclusions in the kimberlitic pipe "Obnazhennaya" (Olenck basin). *Mem. All-Union Soc.*, 89: 204–299	Contains 30–40% xenoliths. These are mostly of country rock but include cognate inclusions of earlier formed kimberlites and nodules of ultrabasic rocks, dunites, peridotites, harzburgites, lherzolites, pyroxenite enstatites and eclogites. On the mineralogy of the xenoliths: Analysis is given of olivine dunite and garnet lherzolite; pyrope from dunite, lherzolite and eclogite, omphacite-diopside from eclogite; Cr-omphacite from garnet–websterite and picotite from lherzolite	Altered kimberlitic breccia	Poly-xenolithic (a) country rock (b) peridotite–harzburgite pyroxenite–enstatite dunites (c) lherzolite websterites (garnet) (d) eclogites	Tentatively suggested: The xenoliths are derived from material of the crust (a) upper-mantle (b–c) and mantle (d)	The derivation of xenolith from different depths as shown from the poly-xenolithic nature of the xenoliths is in accordance with concept of "breccia"
Siberian kimberlites	Kuryleva, 1958. On the petrography of Siberian Kimberlites. *All-Union Min. Soc.*, Ser. 2, 87: 233–237	Limestone and other rocks (country rock derivatives) micaceous picrite porphyry, griquaite, eclogite	Kimberlites	Poly-xenolithic (a) Country rocks limestones (b) micaceous picrite etc. (c) eclogite	Polyxenolithic material derived from various dykes and zones of the crust and mantle	
Roberts-Victor mine South Africa	Kushiro, Ikuo and Aoki, Ken-Ichiro, 1968. Origin of some eclogite inclusions in kimberlites. *Am. Mineral.*, 53: 1347–1367	Eclogitic xenoliths (garnets, clinopyroxenes and phlogopite)	Kimberlites	Poly-xenolithic phlogopite bearing peridotite and eclogites	It is suggested that Kimberlites rich in K_2O and H_2O may have formed from "phlogopite-bearing peridotite in the upper mantle"	
Obnazhennaya kimberlite pipe, Olenck basin	Sobolev, N.V., 1965. Xenolith of eclogite with ruby. *Dokl. Acad. Sci. U.S.S.R., Earth-Sci. Sect.*, 157: 112–114 (*Dokl. Akad. Nauk S.S.S.R.*, 157: 1382–1384)	Eclogitic xenoliths: pale green monoclinic pyroxene, pale orange xenomorphic garnet (The garnet has n 1.737 and a 11.600 Kx and is composed of 47% pyrope, 15% almandine and 38% grossular). Pinkish violet idiomorphic ruby (the corundum proves pressures of formation less than 30 kbar)	Kimberlite	The discovery of corundum eclogite and the presence of corundum in grospydite proves that there is a break in the pyropegrossular series for those rocks and that they must have formed at pressures of less than 30 kbar since experimental work shows that the continuous pyrope-grossular series forms at 30 kbar		Author's interpretation: Diamond-bearing eclogite must have formed at pressures greater than 40 kbar and should be considered a deeper facies than the ruby-bearing eclogite xenoliths

TABLE XI

XENOLITHS IN KIMBERLITES

Locality	Author and title of publication	Mineralogical composition	Type of kimberlite	Mono- or poly-xenolithic	Remarks
Northern Yakutia	Sobolev, V.S. and Sobolev, N.V., 1965. Xenoliths in kimberlite of northern Yakutia and the structure of the mantle. *Dokl. Acad. Sci. U.S.S.R., Earth-Sci. Sect.*, 158: 22–26 (Trans. from *Dokl. Akad. Nauk S.S.S.R.*, 158: 108–111)	Xenoliths of banded corundum eclogite	Kimberlite		Derived from the mantle and formed by metamorphism of pre-existing rock at increase pressure
South African kimberlitic pipes	Garswell, D.A. and Dawson, J.B., 1970. Garnet peridotite xenoliths in South African kimberlite pipes and their petrogenesis. *Contrib. Mineral. Petrol.*, 25: 163–184	Polyxenolithic: (a) xenoliths of garnet peridotite (b) garnet pyroxenite (c) garnet-free peridotite (d) highly altered garnet peridotite from the surroundings of Kimberly at Matsuko, Lesotho, South Africa Coexisting olivine compositions are Fa_{7-9} for garnet peridotites and Fa_{14-18} for garnet pyroxenites	Kimberlite	Poly-xenolithic	The evidence at present available suggests that in the upper mantle region garnet peridotite, specifically garnet lherzolite is an important and probably predominant rock

Chapter 7 | The gabbroic lower crust (protolyte?)

THE SIGNIFICANCE OF A HOLO-PHENOCRYSTALLINE "GABBROIC ROCK" OCCURRING IN A VOLCANIC LINE OCCUPYING A FISSURE OF THE GREAT RIFT VALLEY

A holo-phenocrystalline "volcanic rock" occurring within a volcanic line of the Akaki–Duncan region, south of Addis Ababa, and following a fissure of the Great Rift Valley has been described by Augustithis (1964). Of particular petrogenic significance is that despite its volcanic mode of occurrence it shows gabbroic textures, such as mutual plagioclase (Ca-rich) pyroxene (augitic) contacts (Fig. 174), and intimate intergrowth patterns (Fig. 175) of zoned plagioclases in contact with gigantic pyroxenes. In some cases, elongated plagioclases are in gabbroic ophitic intergrowth with twinned pyroxenes (Fig. 176). Occasionally, pyroxenes and olivine in contact with plagioclases show undulating extinction due to deformation effects (Fig. 177a, b). In contrast to these gabbroic features in certain cases, most impressive corrosion-resorption and reaction phenomena are observed which support a partial and "restricted" melting of this rock (topo-tectically) into a glassy phase out of which a microcrystalline groundmass has subsequently crystallized.

Fig. 178 shows plagioclase and pyroxene "phenocrysts", which are corroded and rounded and partly surrounded by the "topotectically" formed groundmass. Of particular interest are the synisotropization margins exhibited by plagioclase phenocrysts. Similarly, Fig. 179 shows plagioclases, pyroxenes and olivines, rounded and surrounded by this groundmass which locally has a glassy character (Fig. 180). In contradistinction to the fine groundmass, the glassy phase which formed topo-tectically is very mobile and extends along cracks into the pyroxene phenocrysts (Figs. 181a, b). This topotectically formed phase gave this rock the ability to behave as a volcanic material that could move along the volcanic fissures and still maintain its gabbroic characteristics. The evidence of tectonic influences (undulating extinction of the pyroxenes) and the holo-phenocrystalline nature of the rock, despite its volcanic occurrence, supports a deep crustal origin.

The volcanic fissures of the Great Rift are tectonic features which extend at great depths; furthermore, the tremendous quantities of basalts that have poured out of them support the idea that the basalt could be easily mobilized from its depth of origin, along these deep features to the surface.

The question remains whether this gabbroic rock could be the gabbroic protolyte of the lower crust out of which basaltic melts could be generated by melting. In addition, and considering the following deep-rock features exhibited by this rock: (a) gabbroic size of the mineral components; (b) gabbroic textural intergrowth patterns; (c) tectonic deformation effects, the questions arise as to whether this rock is an incompletely melted part, a relic of the gabbroic "protolyte" brought up to the surface, and how by its melting, the generation of basalt could take place?

Chapter 8 | The pyroxene (augite) phenocrysts in basalts

(ZONING, OSCILLATORY ZONING, TWINNING AND CLEAVAGE, "HOURGLASS" STRUCTURE IN AUGITES)

The significance of the pyroxene in basalts and particularly the composition of the pyroxene in relation to the basaltic melts has been discussed by Poldervaart and Hess (1951), Kuno (1954), Wilkinson (1956a), Sarbadhikari (1958), Oji (1961), Yashima (1961), Tilley (1961), Kushiro (1962), Aoki (1962), Le Bas (1962), Huckenholz (1964/65), Muir and Tilley (1964) and Binns (1969), amongst others.

On the basis of these investigations, the significance of the pigeonites as an important pyroxene phase in basalts is unquestionable; however, in terms of textural patterns the augites (another important pyroxene phase in basalts) exhibit a most interesting phenomenology which is emphasized in the present Atlas.

The zonal growth of pyroxenes shows great variation from simple to most complex cases.

A case of simple zoning is shown in Fig. 182, the pyroxene section is cut perpendicularly to the c-axis and in addition to the central part, which exhibits the characteristic augitic set of cleavage (110)($\bar{1}$10), an external overgrowth zone is also indicated which essentially follows the central part.

Another instance of simple augitic zoning is illustrated in Fig. 183. In the central part forsterites used as a nucleus of crystallization (often idiomorphic olivines without corrosion appearances) are included in the augites (Fig. 184). The outer zone of the augite is characterized by Ti-pigmentation and its development resulted in idiomorphic megaphenocrysts.

In contrast, the crystallization of titano-augites in basalts essentially free of olivine and in basalts belonging to the Ethiopian plateau series (such as the Mt. Selale basalts) is of petrogenetic significance and shows the great complexity in the formation of the phenocrysts. The megaphenocrysts often show a nucleus relatively poor in Ti (about 1%) and an outer part consisting of fine rhythmical (oscillatory) zoning and having an average titanium content of about 3%.

Different explanations can be advanced to account for the difference in the Ti content between the nucleus (I generation) and the outer Ti-rich augite. It is possible that the nucleus, which represents the first generation of augite crystallization, might be a high-pressure phase and this will account for the relatively small content of TiO_2 in comparison to the second generation which is a later phase and which is formed under different pressure conditions from the nucleus. It has been proposed that the crystallization of plagioclases in the basaltic melt greatly influences the Al and Ti content of the melt in the sense that if the feldspar crystallization starts in the melt, the Al content (also Ti) will decrease. In contrast, in the studied cases, the Ti most probably follows the Fe. The increase of the Ti from 1% in the nucleus, to 3.5% in the outer augite is independent of plagioclase crystallization in the melt and most probably is related to a tendency for Fe increase in the outer augite. A series of photomicrographs (Figs. 185 and 186) show the first generation of augitic nucleus free of Ti pigmentation and often exhibiting oscillatory zoning, and the second generation of augite, consisting of fine rhythmical (oscillatory) zoning and characterized by strong Ti-pigmentation.

Similarly, Fig. 187 shows an augite nucleus with oscillatory zoning and with a corroded outline included by the augite overgrowth composed of fine rhythmical zoning.

A further example of an augitic nucleus surrounded by a later generation of augite is illustrated in Fig. 188. It should be pointed out that the augitic nucleus shows a pressure-caused undulating extinction. Considering the differences in composition and the phenomenology, it can be concluded that the first generation was formed under diverse conditions but in both cases the oscillatory zoning formation was the mechanism for the augite growth.

Commonly, the first generation of augitic nuclei shows a complex textural pattern and they are composed of patches having different orientation "shadow patterns", and often irregularly shaped magnetite grains are associated with the augitic nucleus (Fig. 189). In this case, the augitic nucleus does not indicate rhythmical zoning. In contrast, the nucleus of the gigantic augitic phenocryst, shown in Fig. 190, shows a corroded outline and is composed of oscillatory zoning and contains magnetitic grains. A contemporaneous magnetite crystallization is tentatively proposed for the association of magnetite–augite in the nucleus.

In contrast, observations show augite phenocrysts composed of broad zones which are regular and devoid of sectoral reversing of the zones of growth, i.e. normal zoning is exhibited (Figs. 191 and 192). In such cases it should be noted that the growth of the outer augitic zones continued with or took place after the crystallization of the groundmass, since components of it, particularly iron minerals, are enclosed or partly engulfed by the outer augitic zone of crystallization (Figs. 191 and 192).

In contrast, the central part of the normally zoned augite shows irregular patches, which perhaps indicate compositional differences which could be attributed to groundmass assimilation by the early phase of the phenocryst development, suggesting a possible contemporaneous crystallization of groundmass and the central part of the augite.

Oscillatory augite-zoning

Gigantic oscillatory-zoned augitic phenocryst exists in the plateaux basalt series of Ethiopia, in the region of Mt. Selale (Fig. 190). Often the phenocryst consists of a nucleus or central part and of an outer part, both showing fine oscillatory zoning.

Fig. 190 shows an augitic central part (A) with zonal growth; the traverse marked (a) has been analysed by microprobe point-analysis by J. Ottemann; the results are shown in Fig. 193a, b, c in terms of % content of the elements Mg, Fe, Ti. Corresponding to the fine-zonal growth there is an oscillation in the percentage content of each of these elements. The oscillations of Mg and Fe are reversed and that of the Ti in general line follows that of Mg.

The outer part of the gigantic augite which surrounds the central part is again fine zoned (marked B). The traverse marked (b) has been similarly analysed by microprobe point-analysis. The results, as shown in Fig. 194, show an oscillation in the Mg and Fe contents, corresponding to the fine-zones of the augite.

Here again, there is a reversal in the oscillations of Mg and Fe. In contrast to the crystal central part (A), where the Ti content shows oscillatory fluctuations, in the case of the outer part of the gigantic augite, the Ti content is noticeably increased but it remains more or less constant without pronounced oscillatory fluctuations, corresponding to the fine zoning. Also whereas the Ti content of the central part does not result in a colouration of the pyroxene, the outer part (B) is pinkish coloured due to evenly distributed fine Ti pigmentation (see Fig. 194). Perhaps the outer crystal part is saturated in respect to Ti. Whereas the Mg content in the central and in the outer crystal parts does not show great differences, the Ti has increased from 1% in the central part to more than 3% in the outer. Despite the possibility that the Ti and Mg can be interchangeable in the pyroxene lattice, the increase of Ti in the outer pyroxene parts does not show a sufficient corresponding reduction in the Mg content to account for the Ti increase.

Perhaps Al and its interchangeability with Ti in the lattice of pyroxenes has played a role in the zonal distribution of Ti in the pyroxene. The distribution of Al in these pyroxenes is a subject of further investigation.

Comparing the results of the microprobe point-analysis along the transverse (a) of the augite central part (A) to the similar results of the transverse (b) of the outer crystal growth, we see that there is no noticeable decrease in Mg along the transverse (b) which is actually the outer part of the crystal. Also Fe is about the same, only Ti is present in a noticeably higher percentage than in the central part.

These results are clearly in contradiction to the statement of Bowen that the Mg content in pyroxenes due to the concept of the continuous series and magmatic differentiation, decreases in the outer zones of pyroxene. As a comparison of Figs. 193 and 194 shows, the Mg content of the outer pyroxene is not noticeably less than that of the central part.

An explanation of this oscillatory zoning is indeed a very difficult matter since changes in pressure, composition fluctuation, temperature differences and other possible factors could effect oscillatory zoning.

An attempt is merely made here to provide an explanation of this phenomenology. As in the case of comparable plagioclase oscillatory zoning (Augustithis, 1963), it is suggested that these oscillatory zoned phenocrysts have been formed while in movement in the basaltic melt. Starting with a "nucleus" of crystal growth out of an added supply of material, a relatively Mg-rich zone will crystallize followed by a zone relatively poor in Mg but relatively Fe-rich. Thus, out of the same supply we will have a pair of oscillations. As the phenocryst is in constant movement within the basaltic melt, more material will be added which is subject to the same process.

Thus, a non-perfect rhythmic repetition of relatively magnesium-rich, iron-poor and Fe-rich, magnesium-poor zoning will be formed. Despite the crystallization of a Mg-rich zone followed by a relatively Mg-poorer zone for each added supply of growth, the insignificant decrease of Mg in the outer crystal part (see Figs. 193 and 194) shows, as far as the development of these gigantic augitic phenocrysts is concerned, that we do not have an instance of a continuous series of crystallization and of a magmatic differentiation.

A magmatic differentiation would have produced a significant Mg-poorer outer crystal part and the outer zones of these gigantic phenocrysts would have shown an appreciable decrease in Mg.

Regarding the Mg distribution in zonal pyroxenes, it is interesting to note the reverse zoning in lunar pigeonites determined by Johnston and Gibb (1973) whereby a decrease in Fe/Mg ratio is determined from the crystal centre outwards.

Irregularities in the growth of the augitic pyroxenes

In some cases an "unconformity" is noticeable between the zoning of the central part of an augite and the zoning in its exterior part. There is an interruption in the growth followed by magmatic corrosion and regrowth; a marked discontinuity in crystal growth has taken place (see Fig. 190). A comparable case is illustrated in Figs. 195 and 196 where the arrow "*a*" marks the zoning of the augite crystal part A and arrow "*b*" shows the direction of zoning in the crystal part B. Similarly, the zoning in the crystal part C is indicated by an arrow "*c*" and in part D the direction of zoning is marked "*d*". The zoning in parts A and C is parallel and also in parts B and D, as the arrows indicating the zoning show. What is interesting is that B and D are interrelated with a peripheral part F (which is in continuity with B and D and shows zoning with the same orientation). This phenomenology is interpreted as indicating that the zonal growth A was interrupted and B followed, which in turn was similarly followed by C and D which also formed an overgrowth (F) to the parts A, B, and C.

Regarding the growth of the zones and their irregularities, often a much wider growth parallel to certain crystallographic faces is noticeable as compared with the same zone parallel to other crystal faces. In Fig. 197 arrow "a" shows the width of the zone parallel to the crystal face (001) and arrow "b" shows the zone parallel to crystal face (111).

Another interesting instance is the change of the parallelism of the zones of growth within the same augite phenocryst. As Fig. 198 shows, the zones marked by arrow "a" are parallel grown whereas the zones marked "b" are not parallel due to interzonal "unconformity" marked by the interrupted line.

Very often rutile needles are parallel to the zonal growth of augitic pyroxenes (Fig. 199). In some cases the rutile needles are resorbed and assimilated by the pyroxene growth.

The oscillatory zoning and "hourglass structure" of basaltic augites

In discussing the formation of "hourglass structure", Schorer (1970) points out that the different sectors have crystallized simultaneously. According to him there is a strong compositional difference (as determined by microprobe point-analysis) and to account for this, he has proposed that different growing faces are thought to have incorporated material of different composition at the same time; for example, the (111) sectors are much richer in Si and poorer in Al and Ti than the (100) sectors.

Schorer also explains their good visibility in microscopy by the strong dependence of the optical properties on composition.

Also Downes (1974) supports a sectorial variation of composition in the case of titano-augites showing an hourglass structure. According to him the principle intersectoral chemical variation is shown to be that of Ti and Al by as much as 39% (cation %) change in each and the least variation to be that of Ca (showing a 0.77 to 3.20% change). Also, for each of three isochronous surfaces the Ti/Al ratio is shown to be individually constant no matter what the sector and this is considered to indicate near-equilibrium conditions of the crystal–liquid interface. Furthermore, Downes explains the oscillatory zoning and the sectorial differences of the titano-augites as being dependent on relative diffusion rates of cationic spaces in the melt and due to growth rates of the crystal faces.

In contrast to the above "sectorial" explanations for the hourglass structure in titano-augites, Augustithis (I.M.A. meeting, Berlin, 1974) presented the suggestion that the hourglass structure is due to oscillatory and crypto-oscillatory zoning and that the sectorial differences in the optical behaviour are due to the reversing of each segment of the fine or crypto-zone which corresponds to a sector. Also it is pointed out that the sectorial differences indicated by the microprobe point-analysis are perhaps due to the fact that different zones have been analyzed in each respective sector (see Schorer, 1970, Fig. 200). It is rather the differences in composition of the oscillatory zoning and the differences in orientation of each segment of the zoning corresponding to a sector than the vectorial supply of melts with different compositions (as suggested by Downes) that explains the hourglass structures.

Fig. 201 shows that the width of the zones of growth determine the shape of the "hourglass structure", i.e. the width and shape of the sector. Furthermore, as can be seen in Fig. 201, certain zones show reversed orientation in the parts corresponding to adjacent crystal faces, thus the "hourglass structure" is produced (sectors of different orientation are formed). In certain cases the orientation of the zonal growth is not reversed in the parts which correspond to adjacent crystal faces and thus a continuous "belt" is formed transversing the "hourglass" sectors (Figs. 202 and 203). This zone has an independent orientation, is wider and does not conform to the general pattern of the oscillatory zoning which is responsible for the formation of the "hourglass structure".

Comparable to the "hourglass structures" are the patches of different orientation which can exist within an augite, as shown in Fig. 201. In this case reversing of a part of the same zone, independently of crystal face orientation, has taken place.

A series of photomicrographs shows the variation in size and other peculiarities of augites exhibiting "hourglass structure". In contrast to the megaphenocrysts shown in Figs. 201 and 202, Fig. 204 shows an augite, approximately the same size as the augites of the groundmass, exhibiting "hourglass structure".

Despite size and the non-apparent character of the zoning building up the augite, in this case, the "hourglass structure" is caused by oscillatory zoning which is, in fact, crypto-oscillatory in nature.

In cases, in contrast to the fine oscillatory zoning associated with and responsible for the "hourglass structures" shown in augites, the zoning of a particular sector might be composed of broad zones (which in turn might be composed of cryptozoning, Fig. 205).

In support of the hypothesis that the "hourglass structure" is formed by the reverse orientation of each segment of a fine oscillatory zone that corresponds to a sector, is the case of the multi-sectoral augite shown in Fig. 206.

The twinning of the augitic megaphenocrysts in basalts

Twinning is a tendency of nature in the micro and macro cosmos and it is expressed by many "laws" of duplication, intergrowth and co-existence.

Augites often show simple or polysynthetic twinning. Twins with (110) as twin plane are common. Polysynthetic twins with (001) as twin plane are found occasionally. Combining (100) with (001) polysynthetic twins gives what is known as a herringbone (fishbone) structure. Figs. 207 and 208 show augites exhibiting idiomorphism and with polysynthetic twinning running parallel to the (100) of the crystal face of the augite.

In contrast to the simple twinning, fine lamellar polysynthetic twinning is exhibited in some cases (Fig. 209). The polysynthetic twinning can show variations in the width of the lamellae from very fine to broad, as shown in Fig. 210 where the twin lamella is parallel to the (100) and only one of the cleavage set is developed as has been determined by U-stage measurements by E. Boskos.

Occasionally, in simple twinning the twin plane can be diagonal and intersecting the cleavage set as is shown in a section parallel to (010) and in this case the twinning plane is // (100).

In contrast to the described planes of twin intergrowth, cases exist indicating a zig-zag intergrowth of two augites, as is shown in Fig. 211.

Irregularities of polysynthetic twinning

The polysynthetic twinning of basaltic titano-augites often shows irregularities both in the continuation of lamellae and in their form.

Fig. 212 shows a cross-section of a titano-augite illustrating an eucrystalline form with traces of the cleavage pattern (110 and $\bar{1}10$). Two thin lamellae are to be seen parallel to the (100) of the augite, whereas the one is continuous the other is only half-developed.

Similarly, in Fig. 213 there is an interruption in the continuation of a particular twin lamella; in contradistinction to the continuation of other adjacently lying twin lamellae.

In other instances the development of a twin lamella is only restricted in its development within a simple twinned augite, in which case the plane of simple intergrowth coincides with and is parallel to the partly formed twin lamella.

The above observations clearly show that twinning is a secondary regeneration process within the augitic growths and it develops after the crystallization of the pyroxene; this is also indicated by the relation of twinning to crystal zoning. The partial development of a twin lamella and its sudden disruption suggests that these are instances showing incompletely developed lamellae in the process of twin-lamellae formation within an augite. Perhaps twinning could be regarded as an internal ex-solution phase development within the augitic host-crystal.

In addition, irregularities are shown in the form of the twin-lamellae. Fig. 214 shows a malformation in the shape of an augitic lamella which attains a "club" shape. These anomalies should be seen rather as a loss of control of the "laws" governing the plane of intergrowth within the host augite. In certain cases, there is a sudden change in the width of the augitic lamellae as shown in Fig. 215 where the lamella's width is suddenly changed (see arrow "a" in Fig. 215).

Again seen as an irregularity in the shape of twin lamellae is the "knick" which appears as a sudden bending of the twin lamellae (Figs. 216 and 217). The case illustrated in Fig. 216 can particularly be seen as being due to post-crystallization strain effects.

In contradistinction to the irregularities discussed, a disturbance of a twin lamella can be caused due to forcible intrusion of another augitic pyroxene perpendicular to the one with the lamellae. As a result of this intrusion and the force of crystallization exercised by its growth, bending and fracturing of the twin lamellae takes place (Fig. 218). In other comparable cases a disruption (interruption) is produced in an augitic twin-lamella (see Fig. 219) due to the later growth of an "intrusive" augitic crystal grain.

Zoning and twinning

The augitic twinning often traverses the fine oscillatory augite zoning as illustrated in Fig. 190. In such cases it is obvious that the crystal twinning, and often an augitic polysynthetic twinning, is developed after the formation of zoning. A comparable case (example) is the relation of plagioclase twinning and oscillatory zoning (see Augustithis, 1963).

More evidence of the post-zonal development of polysynthetic augite twinning is provided by instances where the twinning as a post-zonal development has influenced the pyroxene zones. Often a "knicking" is produced (i.e. an interruption in the smooth running of the zones) due to the influence of the polysynthetic twinning (Fig. 220).

In some cases, the post-zonal development of the twinning is shown by the rotation of the twin individuals perpendicular to the twinning plane (Fig. 221). In such cases of post-zonal twinning a "swallow tailed" pyroxene is produced.

Another interesting case of post-zonal twinning is shown in Fig. 222. Twin lamella (a) traverses the augitic zoning, whereas twin lamella (b) is developed at the outer zone (C) of the zoned pyroxene, the succession of the zonal growth of augite being A, B and C.

The relation of zonal growth to twinning is not always post-zonal twin development. In certain instances the augitic twinning can be post-zonal in respect to some zones and older in respect to others; Fig. 223a shows a zoned augite with zonal growths A, B, C and D. The twin lamella "a", because it traverses zones A and B, is post-zonal in respect to those zones and in turn, zones C and D are formed later than the twinning.

It is possible that the outline of zone B, as shown in Fig. 223b indicates a pronounced interruption in crystal growth.

The relation of crystal cleavage to twinning

Crystal cleavage can be seen as a disruption in the continuity of a crystal-lattice; in some cases, this disruption is not accompanied by a displacement of the crystal lattice, i.e. the cleavage is a fracture or a system of fractures without block displacement. Fig. 224 shows an augite showing twinning and exhibiting a set of cleavages without displacement of the crystal along the cleavages.

In contrast, when twin lamellae of the augite are intersected by cleavage, a displacement of the crystal parts is associated with the cleavage microfractures, i.e. the cleavage has acted as a series of microfaults with repeated small-scale displacement of the crystal along the cleavage fracture (Fig. 225).

Similarly, Fig. 226 shows that the set of cleavages has acted as micro-faults and has caused crystal micro-displacement along the cleavage fractures.

Chapter 9 | Pyroxene tecoblasts and late phase pyroxene growths

Comparable to the late-phase olivine growths in basalts (Eržen Stanik and Edward), late-phase pyroxene tecoblasts in basalts have been described by Augustithis (1956/60). Considering the crystallization sequence of basalts from Kindane, Meheret, Addis-Ababa and the Ethiopian Rift Valley, late-phase pyroxene and plagioclase tecoblasts crystallize after the groundmass, components of which can be enclosed, partly assimilated (see Chapter 10) or autocathertically pushed out of the blastic pyroxene growth. Particularly the pyroxene blasts seem, due to autocathersis, to have cleaned themselves from inclusions, though occasionally, groundmass plagioclases and magnetite may be engulfed by the blastic growths. In the sequence of the basalt's crystallization, the pyroxene blastesis follows the groundmass crystallization which consists of magnetites, small pyroxenes and plagioclase laths (anorthite content 52–75%). A post-pyroxene plagioclase tecoblastesis gave rise to plagioclase phenocrysts (anorthite content 35–50%) which assimilated and invaded the pre-existing pyroxene tecoblastesis.

In contradistinction to the pyroxene tecoblasts, zoned pyroxenes often enclose groundmass components which are rounded and corroded or reduced to pigment size. Comparable to the late-phase olivine growths in basalts are the tecoblastic pyroxene growths described by Augustithis (1956/1960).

Fig. 227 shows a zoned augitic pyroxene with olivine inclusion, rounded by the late pyroxene and with smaller pyroxenes and groundmass, reduced to pigment size, interzonally incorporated by the late pyroxene growth. Similarly, Fig. 228 shows feldspar laths interzonally incorporated in the augite. As a corollary to these, observations (Chapter 8) show magnetites of different size and forms incorporated by the later pyroxene growths. In some cases plagioclase laths can be orientated in the host pyroxenes parallel to its crystal faces (Fig. 229).

Chapter 10 | Plagioclase tecoblasts

According to orthodox views, the feldspar phenocrysts are considered to represent an earlier crystallization phase and they are believed to be richer in anorthite content than the groundmass feldspars. The basaltic plagioclase phenocrysts have been studied, among others, by Vance (1962), Sarabadhikari (1965) and Stewart et al. (1966), who have pointed out the differences between phenocryst plagioclases and groundmass feldspars.

In contrast, observations by the author have shown a phenomenology of the feldspar phenocrysts that is not in harmony with these generally accepted views.

Detailed microscopic observations have revealed plagioclase phenocrysts assimilating and digesting components belonging to a pre-existing crystallization phase. Examples have been described showing a tecoblastic nature for these feldspars, that is, the plagioclase has corroded and assimilated pre-existing pyroxene phenocrysts and groundmass components. The assimilation of the groundmass can be followed in all its transitions from unaffected minerals (grains) to pigment rests.

Considering the importance of the tecoblastic phenomenology, a new series of observations illustrating the assimilation and digestion of pyroxene phenocrysts and of the groundmass is introduced.

A type of tecoblastic (post-pyroxene) plagioclase is illustrated in Figs. 230 and 231. As a result of the force of the crystallization of its growth, it has penetrated the pre-existing pyroxene, probably along a primary crack. The plagioclase shows evidence of pyroxene assimilation which influences the zonal composition of the tecoblast.

The An-content varies in different parts of the plagioclase (Figs. 230 and 231) which are recognized as zones I–IV. The An-content of these different zones is as follows:

Zone IV	34%	An.
Zone III	31–32%	An.
Zone II	44%	An.
Zone I	38–39%	An.

Zones IV and III are independent of the presence of the pyroxene. In contrast, zone II is as wide as the pyroxene; this zone shows the greatest An-content, which is due to the assimilation of the pyroxene by the tecoblastic feldspar; pigment-rests of the pyroxene are present. Zone I has a lower An-content than zone II, but higher than zones III and IV. It is probable that the plagioclase has grown in the direction from zone IV to zone I.

During the assimilation of pyroxene by the tecoblast, isotropisation of the pyroxene has occurred. Final products of "assimilation" are isotropic pigment-rests which represent elements which could not be assimilated by the plagioclase lattice.

Another interesting case of pyroxene being partly assimilated by a tecoblast is shown in Fig. 232. Here again, the pyroxene shows all the phenomenology of assimilation; rests of pyroxene and pigment-like bodies are present in the plagioclase part which is a product of the reaction of the two phases.

A most characteristic intergrowth of later tecoblastic plagioclase enclosing groundmass and replacing pyroxene is shown in Fig. 233. As the arrows indicate, the plagioclase has engulfed pyroxene parts and is in the process of "digesting" them. These resemble an ameboidal "digestion" of the pyroxene by the later plagioclase tecoblast, which has also parts of the groundmass.

In volcanic rocks the groundmass is often seen penetrating or "infiltrating" into the pre-existing phenocrysts. As a result of the later association of the groundmass components with the phenocrysts, it frequently appears as though groundmass has been included by the phenocrysts.

Such inclusions of groundmass components have often been explained as liquid drops of the magma enclosed by the crystallizing phenocrysts; the liquid drops, in turn, having crystallized and produced components identical to the groundmass. The above explanations are in accordance with the idea that the phenocrysts have crystallized first, and it is frequently the case that they show all the appearances of magmatic corrosion.

However, observations of plagioclase phenocrysts of lower An-content than the groundmass and indicating all phenomena of having assimilated pre-existing mineral phases, have been described by the author (1956/1960).

Figs. 234 and 235 show most interesting examples of plagioclase tecoblasts enclosing groundmass components. Small pyroxene and groundmass feldspars are enclosed in a later-formed plagioclase tecoblast. The groundmass mineral components are equally distributed in a frame of plagioclase.

The amount and extent to which the groundmass can be enclosed, assimilated or "autocathartically" (by self-cleaning processes) pushed along lines or margins of the host plagioclase, can vary.

Often and as a result of compositional differences, the groundmass components in the host feldspar are surrounded by reaction margins. Such reactions and a type of zoning surrounding the inclusions are seen as being due to the degree of the assimilation of the groundmass by the building-up of the later feldspar tecoblast.

Fig. 235 shows a great amount of groundmass mineral components enclosed by a feldspar tecoblast. Differences in the plagioclase (interference colours) represent reactions of the building-up of the feldspar with the pre-existing groundmass components. Similarly, Fig. 236 shows groundmass components surrounded by a tecoblast feldspar with a marked zoning due to compositional differences.

The cases so far described have shown groundmass components distributed in the plagioclase tecoblast. However, additional observations show that pockets of groundmass can also exist in the tecoblast. Fig. 237 shows an example of such a pocket with nearby and also enclosed in the tecoblast, corroded and partly assimilated groundmass components. Due to the assimilation of the groundmass, pigment-rests are present in the feldspar.

The corrosion and assimilation of the groundmass components can vary from grain to grain enclosed in the tecoblast: often idiomorphic groundmass components, with little or no corrosion appearance, are enclosed, as shown in Fig. 238.

More often, however, the mineral components of the groundmass enclosed in the tecoblast can show all the appearances of corrosion and assimilation. Fig. 239 shows groundmass components in a plagioclase tecoblast where although the groundmass is to a great extent unaffected it also contains mineral components which are clearly corroded and reduced to the form and appearance of pigments. It is of interest to mention that the pigments, as seen in the detailed photomicrograph of Fig. 240, follow the cleavage direction of the plagioclase tecoblast. It is assumed in this case that the groundmass components have been corroded and assimilated by the tecoblast: within the complex processes of pigment production, pigment-rests representing elements that could not be assimilated or be incorporated in the lattice of the plagioclase partly show evidence of "assimilation" and partly are pushed or confined along the cleavage of the tecoblast.

Within the same tecoblast, as seen in Figs. 238 and 239, mineral components of the groundmass can be unaffected with an obvious crystalline outline, while adjacent to them groundmass components can be affected and reduced to a pigment form.

As evidence that these are components of the groundmass which were already crystallized before the tecoblastic plagioclase is the olivine crystal grain which is obviously a pre-plagioclase crystallization. Indeed, the phenomenology of tecoblasts can show most convincing examples of groundmass mineral components assimilated and greatly corroded by the later-formed feldspar. Fig. 241 shows a large tecoblast full of inclusions of groundmass, and the following series of illustrations (Figs. 242–244), gives more detailed information of the form and corrosion of these components. In Fig. 242 rounded and corroded groundmass components are shown enclosed in this tecoblast and also, in contrast, the mineral components of the groundmass exterior to the tecoblast are seen to be larger and more crystalline in form. The groundmass enclosed in the tecoblast is corroded and reduced to pigment size, as is shown in a detailed view in Fig. 244.

Figs. 245 and 246 show pyroxene and groundmass enclosed in a tecoblast. As a result of corrosion and assimilation, the pyroxene shows a rounded outline due to corrosion and parts of the groundmass are reduced to rest-pigments. Similarly, Fig. 247 shows a pyroxene enclosed and corroded by a tecoblast. Another plagioclase tecoblast is shown in Fig. 248 with the groundmass assimilated and reduced to pigment-rest size. Due to "autocatharsis" the rest pigments are restricted within the central part of the plagioclase tecoblast.

The corrosion of the groundmass can, in turn, result in differences in the anorthite content of the tecoblast part immediately surrounding the groundmass inclusion, or the rest-relics of it, respectively. In addition to Fig. 235, which clearly shows differences in the plagioclase parts surrounding the enclosed groundmass, further observations (Fig. 249) show that variations in the composition of the tecoblast feldspar exist depending on the reaction of the groundmass with the feldspar. Fig. 249 shows initial groundmass mineral components enclosed in the tecoblast and reduced to pigment-rests, surrounded by a reaction zone of the feldspars.

From additional observations, an interrelation between pigment distribution and composition of the tecoblast is shown in Fig. 250. Here, pigment-rests of assimilated groundmass are distributed in a zonal manner within the tecoblast, and a zone of the tecoblast free of pigment-rests is also distinguishable. (Such a process of self-cleaning from inclusions is known as "autocatharsis".)

Despite the fact that the series of observations introduced shows to some extent the post-groundmass age of the tecoblast, in some cases the relationship of the tecoblast with the groundmass is still difficult to explain and uncertain. For example, Fig. 251 shows groundmass extending into a tecoblast. (This observation can be regarded as dubious, since it could be argued that it is a part of the groundmass partly enclosed by the later tecoblast but still retaining its connection with the groundmass outside the tecoblast.)

In contrast to the usual statement that the plagioclase phenocrysts are richer in anorthite and older than the groundmass components, the tecoblasts described are younger than the groundmass and their composition shows an anorthite content lower than the groundmass (pre-pyroxene) feldspars (Fig. 252).

The tecoblasts often show reverse zoning, and they resemble the blastic plagioclases in metamorphogenic gneisses. Frequently the tecoblasts show local variation in the anorthite content of the patches or zones building up these crystals. Figs. 253a and b show a tecoblast with zones of variable An-content.

Zone I	25% An.
Zone II	15% An.
Zone III	24% An.
Zone IV	25–27% An.

It is interesting to note that the central zone I is richer in An-content than zone II. Pigment-rests of assimilation are present in zone I. Also noteworthy is the fact that zone III is richer in An-content than zone II, thus showing a type of reversed zoning. This example shows too that the An-content also depends on the extent of assimilation, as shown by the presence of assimilation rests in zone I.

The photomicrograph of Fig. 249 shows local differences actually conditioned by differences in the An-content which, in turn, depends on the reaction of enclosed groundmass components with the surrounding tecoblast.

Analogous to the tecoblastic phenomenology, is the plagioclase pseudomorphism after augite, introduced by Pieruccini (1961); ("Su di un particolare processo pseudomorfico: la transformazione dell'augite in termini plagioclasici.").

Further, the pseudomorphic transformations take place by hydration processes in tuffs, whereas the tecoblastic growths have been observed in effusives (lavas, often with cellular structures), and have taken place within the consolidation period of the melt.

Professor R. Pieruccini shows, in a series of original illustrations, the transformation of augite into plagioclase, whereby the crystal form, or outline, and the typical cleavage of the augite is taken over by the "neogenic" plagioclase. Also, additional observations showing relics of augite, as well as details of the gradual transformation of plagioclase after augite (through a phase of synisotropisation), have been given extensive treatment by Pieruccini. This transformation is seen as an example of the process of "grado-separation".

In a comparison of the tecoblastic growths with the pseudomorphic transformations, it should be emphasized that whereas the former are blastoid growths the latter are replacements. It must be noted that despite the differences between these processes of tecoblastic assimilation and pseudomorphism, isotropization of the affected pyroxene is observed as a phase. As shown in Figs. 230 and 231 and in their captions, in zone II of the plagioclase tecoblast enclosed pigments indicate the phenomenology of isotropization.

Chapter 11 | The oscillatory zoning of plagioclase phenocrysts in basalts

The growth of plagioclase phenocrysts in volcanic rocks has been interpreted as being due to crystallization at depth during the earlier stages of the liquid-melt cooling. As a consequence, the phenocrysts are richer in anorthite content than the plagioclase building up the groundmass of the rock.

In contrast to this general concept, the author believes that special conditions of crystallization can also exist, namely those of tecoblastesis, as a result of which growths of phenocrysts of lower anorthite content can take place later than the groundmass plagioclases. ("Über Blastese in Gesteinen unterschiedlicher Genese Migmatit, Granit, Metamorphit (Smirgel), Basalt." Augustithis, 1956/1960).

An additional case of special crystallization conditions are the oscillatory zoned gigantic phenocrysts of Debra-Sina–Debra-Braha, Ethiopia (Augustithis, 1963).

The plagioclase phenocrysts in the olivine basalt of Debra-Sina attain sizes of 4–5 cm (Fig. 254), are plate-like in form, and show a definite "flow-orientation". But in one case these crystals were only developed locally within the consolidated lava mass.

The polysynthetic twinning varies from broad to extremely fine lamellae (Fig. 255). Other types of twinning (probably resorption lamellae?) are illustrated in Fig. 256. It is interesting that these feldspars often show twin lamellae without sharp boundaries. This phenomenon is believed to be caused by thermal influences during or after the plagioclase crystallization.

In addition to the above-mentioned twinning, zonal structures of a most complex type are characteristic for the Debra-Sina plagioclase phenocrysts. Often a combination of twinning and extremely fine zoning results in a parquet-like structure (Fig. 257, arrow).

The composition of these plagioclases can only be understood by a study of their zonal structures. These large phenocrysts are built-up of fine zones which in some cases number more than 100. A study of these zones with the universal stage often reveals that they, in turn, are built up of more minute and extremely fine zones which, as a group, yield an average-optical value (An-content) for that zone.

The compositional variations of the feldspars were studied on the basis of universal-stage determination of An-content. In one case the An-content of the feldspar zones varied from 40 to 50%. Determinations of another plagioclase showed variations round 56–70% for anorthite. Figs. 258 and 259 show the oscillatory zoning of two measured plagioclases.

The zoning in these plagioclases attains most unusual forms for feldspars. Figs. 260 and 261 show that this zoning often attains a form resembling colloform structures. It should be pointed out, however, that in this case we are dealing with a crystal plate and not with a colloform body, though such instances do exist and are described under the heading of "colloform tecoblastoids". Often groundmass components are included in the plagioclase feldspars and they are surrounded by the fine zoning typical of the Debra-Sina plagioclase phenocrysts (Fig. 261). But such cases are dubious in character. It is difficult to say whether the included groundmass was formed from liquid drops that later crystallized or whether there were crystallized components of the groundmass round which the zoning of the phenocrysts took place. Another case which illustrates the effect that the groundmass had on the zonal formation, is illustrated by Figs. 262 and 263. One could explain this case as being the result of an interaction between the plagioclase phenocryst and the groundmass which must have been, to a great extent, crystallized before the crystallization of the outer zones of the phenocryst illustrated in Figs. 262 and 263.

On the basis of these observations it is possible to consider that the groundmass, enclosed and surrounded by the oscillatory feldspar zoning, must also have crystallized or had partly crystallized before the phenocryst. In contradiction to these observations is the "flow orientation" of the phenocrysts, which supports the theory that a certain fluidity must have existed to allow the "flow orientation" of the phenocrysts. In some cases, these oscillatory zoned plagioclases are only clusters within a fine crystalline basaltic mass.

Discussion of causes of the oscillatory zoning of the plagioclase phenocrysts

Theoretically the chemical composition of the plagioclase phenocrysts in basaltic rocks is dependent on the chemical composition of the melt and on the temperature–pressure conditions. Changes in the above-mentioned factors influence the phenocryst composition and can result in growth zones of different An-contents.

As mentioned, changes in the pressure can influence the composition of plagioclases. Carr (1954), in discussing the oscillatory zoning of the plagioclases in the gabbroic rocks of the Skaergaard intrusion, states that changes in pressure, caused by the circulation of magma under convection-current conditions, are responsible for the oscillatory zoning in the plagioclases studied. "Theoretically the effect of pressure changes on plagioclase composition is governed by change in shape of the liquidus and solidus curves with pressure, regarding which no experimental evidence is available. Existing data suffice only to afford an estimate of the increase in melting-point of the pure end members with pressure: the value dT/dP for albite is thereby indicated to be approximately four and a half times that of anorthite. Using this relationship, Fig. 264 shows diagrammatically the manner in which the required change towards a more albitic solid phase might occur as pressure increases". Thus, on the sinking and rising of the feldspars due to convection-current circulation, an oscillatory zoning is believed to have taken place.

This hypothesis is not, however, enlightening in the case of the Debra-Sina oscillatory zoning of the phenocrysts. In contrast to the limited number of zones described by Carr, the number of zones in the Debra-Sina phenocrysts may exceed 100. Furthermore, no convection-current circulation can be proved for the volcanism in this area, as is supposed in the case of the Skaergaard gabbroic intrusion.

Variations in temperature could also be considered as a possible cause that might bring about such oscillatory zoning. Such temperature variations can be caused by the gas explosions of volcanic eruptions which are connected with high temperature release.

Unfortunately, our knowledge of the volcanism in the Great Rift Valley (Ethiopian part) is limited. One important observation, however, can be made, i.e. that in general the volcanism and faulting of the Great Rift Valley seem to be related. It is perhaps so that friction, resulting from the deep faulting (the block-faulting of the Great Rift), has played a role in the genesis of the melt which came up as lava. This is in accordance with the hypothesis of the melting of the lower-crust gabbroic and upper mantle peridotic layers of the earth (see Chapter 32).

However, the origin of the melt due to friction (as a consequence of major block-faulting) and the alternative magmatitic explanation cannot provide temperature variations that will provide a satisfactory explanation for the oscillatory zoning of the plagioclase phenocrysts. Therefore, factors other than pressure and temperature variations must be considered as responsible for the oscillatory zoning of these feldspars. The fact that these large feldspars show a flow orientation strongly suggests that their crystallization began, and was to a great extent completed, while at depth and in a melt environment. The flow orientation of the phenocryst plagioclase emphasizes the fact that these feldspars have crystallized at depth and in a melt environment (i.e. before the crystallization of the groundmass). Furthermore, the crystallization of the feldspars was not a single event but, as is evident from the zonal structures, the material building up these gigantic basaltic plagioclases was supplied at intervals.

Considering the conditions of crystallization, the plagioclase formation occurred in a dynamic and mobile system rather than in a static one; consequently the feldspars must have been in motion within the melt and could have come in contact with portions of it which had different compositions.

On the basis of the above explanation, the compositional variations of the plagioclase phenocrysts, as shown in Figs. 258 and 259, become understandable, as does the fact that the first zone can be anorthite poor. The oscillatory nature of the zoning should be understood as being due to the fluctuation in the composition of the supplied material building up the zones. The building up of the "unit of oscillation" begins with an An-rich crystal grading into an An-poor one. This cycle is repeated every time new material is supplied by the movement of the crystal through the melt. The growth of these plagioclase phenocrysts is therefore not a single event but rather a series of crystallization stages combining to give the cumulative oscillatory zoning effect.

Chapter 12 | Deformation effects on augite and plagioclase phenocrysts in basalts

Microscopic observations show unmistakeable evidence of deformation effects in augite and plagioclase phenocrysts in basalts. Fig. 265 shows a gigantic titano-augite consisting of radially arranged segments, which have been formed by the fracturing of the augite. These radially arranged segments show differences in their orientation, e.g., Fig. 266 shows that one augite segment shows the cleavage set $(110,\bar{1}10)$, whereas the adjacent segment exhibits the cleavage (010). Similarly, another augitic phenocryst (Fig. 267) shows fracturing which resulted in radially arranged, differently orientated segments. In contrast to the radially arranged segments there are cases where fracturing and rotation of a "block" within a phenocryst can take place. Fig. 268 shows a zoned titano-augite with a differently "orientated block" which is delineated by a twinning plane and an irregularly running crack.

As Figs. 269 and 270 show, polysynthetically twinned augites may include "block fragments" which are differently orientated within the pyroxene phenocryst. Furthermore, it should be pointed out that within this disorientated "block" there is a dislocation of twinning due to the fact that the cleavage-set acts as a microfaulting system (Fig. 270). Comparable disorientation "blocks" occur within plagioclase phenocrysts. Fig. 271 shows disorientation plagioclase "blocks" within a plagioclase phenocryst. Whereas the cleavage direction is maintained in the disorientation "block" and in the surrounding plagioclase, there is a distinct disorientation of the crack-pattern of the "blocks" and of the surrounding feldspar.

Crystal fracturing and microdisplacement of twin lamellae are rarely evident in augite phenocrysts. Fig. 272 shows a titano-augite phenocryst with a twin lamella displaced along an irregularly running "fracture" which has now healed. It should be pointed out that the cleavage pattern has developed after the microdisplacement, as the cleavage is transversing the entire structure undisturbed.

Both augitic and plagioclase crystal outlines can distinctly show displacement along microfractured cracks which have developed within the phenocrysts (Figs. 273 and 274). Clearly there is a microdisplacement along the microfaulting: actually the crack system mainly developed at the peripheral areas of the phenocrysts. Whereas the deformation lamellae in forsterite phenocrysts in basalts can be explained by presuming a xenocrystalline origin for the olivine (mantle derivation), the unmistakable phenomenology of deformation effects exhibited by augites and plagioclase phenocrysts in basalt (which have not been subjected to deformation after their consolidation) is difficult to explain and remains problematic.

The radial segmentation of an augite and simultaneous rotation of the segments in respect to one another, the "block" disorientation and the microdislocation along microfaults are all unmistakeable evidence of crystal deformation. The fracturing and the crystal deformations mentioned have taken place while the phenocrysts have crystallized (were solids), i.e. at an early phase of the basalts's consolidation. The following suggestions are tentatively put forward as possible explanations for the fracturing and crystal deformations:

(1) Gas-explosion within volcanism.

(2) Sudden temperature changes could produce fracturing and dislocations.

(3) Magmatic "attrition" (collision) of the phenocrysts in lava stream motion.

(4) The early phenocryst could form a crystal cummulate which could transmit deformation.

(5) Temporary formation of "crust" and its subsequent fracturing due to renewed lava streams; the crust fracturing could result in crystal fracturing.

Chapter 13 | Plagioclase collocrysts in basalts and basaltic rock types

The significance of colloids and the transformation of colloform structures into crystals has been discussed by many workers, particularly in the case of hydrothermal deposits, e.g., the relation of colloform pyrite to pyrite crystallization by Ramdohr (1960) and Suzuki (1962); gel-pitchblende to crystalline uraninite, by Augustithis (1964); gel silica to quartz, by Levicki (1955), Augustithis (1967, 1973). The significance of colloids and metacolloids in endogenic deposits has also been extensively discussed by Lebedev (1967).

Elliston (1963), Nashar (1963) and McNiel (1963) present elaborate, pioneering work concerning the importance of colloids in petrogenesis in their descriptions of quartz and feldspar collocryst in sediments.

Parallel to these publications, Augustithis (1963) pointed out (though without proposing a colloidal origin) the resemblance of colloform structures to a certain complex oscillatory zoning indicated by plagioclase phenocrysts in basalts. Drescher-Kaden (1969) has proposed a colloidal origin of zoned plagioclase in granites and Augustithis (1973) has supported this explanation for certain zoned crystalloblasts in granites. So far, no claim has been made concerning the existence of colloform feldspars in basalts and, indeed, the predominant orthomagmatic interpretations seemed to have discouraged such a possible interpretation. In contrast, present microscopic observations show colloform plagioclase growths in basalts and basaltoid rocks.

Plagioclase collocrysts in basalts

Colloform plagioclase spheroids, indicating all the characteristics of colloform structures and collocrysts, are illustrated in a series of photomicrographs. Fig. 275 shows a plagioclase collocryst which shows relics of typical colloform structures, such as colloform zonal growth, colloform zones turbid with rest pigments (Fig. 276) and a typical colloform spheroidal general appearance.

As is illustrated in Fig. 275, the collocryst shows a typical undulating extinction with the beginning of development of polysynthetic twinning which can be attributed to the tendency of individual crystal development from a collocryst.

Another collocryst (colloform blastoid) exhibiting colloform zonal growth marked by incorporated groundmass pigment-rests and having a spheroidal appearance is shown in Fig. 277. The plagioclase collocryst shows undulating polysynthetic twinning. Another typical characteristic of the collocrysts is the non-definite margins, with groundmass marginally incorporated by the collocryst. In some cases, most intricate colloform structures develop which differ in appearance from the spheroids. Fig. 278 and Fig. 279 show a plagioclase, with a polycolloform structure, in the sense that an intricate colloform zonal pattern develops around the rest of the groundmass (e.g., pyroxenes) or in cases around zeolitic enclosures with pyroxenes of the groundmass as particularly indicated in Fig. 279.

In contradistinction to the above observation, Le Bas (1955) has described the development of a glomeroporphyroblastic mosaic, and aggregates of small homogenous Ca-plagioclase crystals as a result of the transformation of zeolite-filled amygdaloids, due to thermal metamorphism. These observations differ from the plagioclase collocrysts which, as shown in Fig. 54, have been formed before the vesicular cavities of the basalt.

Another characteristic of the colloidal derivation of the collocrysts is the undulating extinction and groundmass rest pigments incorporated in collocrysts. In addition to the intricate pattern, the general outline is rather that of a collocryst in development.

In some cases, the collocrysts may exhibit an intricate colloform zoning which may be preserved as relic after the transformation from gel→ collocryst (Fig. 280). However, it should be pointed out that the basic colloidal patterns often tend to disappear after crystallization due to the fact that the growing crystal tends to assume a shape dictated by the "lattice-strength" of the crystal. The initial colloform zoning which is a contemporary growth pattern of the colloid, if preserved, represents colloform relics.

Occasionally, most impressive collocrysts develop in basalts, with zones turbid with groundmass inclusions consisting of the ore-minerals (magnetites) and groundmass components reduced to pigment size (Figs. 281 and 282).

In other cases large twinned plagioclase colloform growths are present in basalts, exhibiting intricate colloform zonal patterns, with a characteristic topozonal growth development around groundmass relic components engulfed by the collocrysts. In addition, embayments of the groundmass in the collocryst are also indicated (Fig. 283).

The collocrysts do not represent a post-consolidation amygdaloidal filling of the vesicular basalt (gas-cavity fillings) but rather a solution phase at and after the concluding phase of the orthomagmatic phase of crystallization and indeed a plagioclase developmental phase inherent to the wet-magma crystallization. Fig. 54 shows collocrysts of plagioclase as "phenocrysts" in fine crystalline groundmass with vesicular cavities (gas-cavities) and indeed, the arrangement and shape of the vesicular cavities can be influenced by the pre-existing plagioclase collocryst.

Colloform zoning transgressing to crystal-zoning

As pointed out, basic colloidal patterns, such as the colloform zoning tend to disappear and the growing crystal tends to assume a shape dictated by the lattice strength. The change from a colloid or colloform body through a collocryst to a crystal is often accompanied by transformations in the basic colloform textures such as zonal growth and shape. Transition phases of these transformations should be expected.

Another possible explanation for collocrysts exhibiting amphoteric characteristics (i.e. typical colloform features and simultaneously typical crystalline characteristics, such as zoning parallel to developed crystal faces), could be attributed to the co-existence and simultaneous development of the colloform and crystallization phases. Fig. 284 shows a plagioclase with a zone exhibiting typical colloform characteristics which is nevertheless surrounded by an external broad zone consisting of fine zoning and being parallel to well-developed crystal faces.

In contradistinction, a collocryst can show colloform characteristics in its central part and crystal faces externally. As is indicated in Fig. 285, the central part of the collocryst is turbid, with groundmass inclusions, which are often reduced to pigment size. A colloform zoning is followed and at the margin the collocryst attains a crystalline outline, with groundmass embayment. A rather more advanced case of collocryst, where both the zoning and the shape conform more to a crystal, is indicated in Fig. 286. However, in the external zones groundmass relics of irregular shapes hint at a collocryst derivation.

A rather interesting case, with amphoteric characteristics, is shown in Fig. 287. An intricate colloform central zoning with crystal-developed faces. Similarly, Fig. 288 shows a plagioclase with a central part with colloform characteristics followed by a fine crystal zoning and well-developed crystal faces.

The development of polysynthetic twinning in collocrysts

As is often the case, colloform spheroids (due to syneresis) tend to separate into parts, which because of the subsequent transformation of colloform to collocryst, give rise to separate individual crystals. The development of separate quartz crystals from a silica colloform spheroid is described by Augustithis (1967). Comparable and commensurable is the formation of a separate collocryst or twins from a plagioclase spheroid as a result of its transformation to crystalline (collocryst). The colloform phase is regarded as metastable, out of which the more stable crystalline phase develops.

Augustithis (1973), in discussing the development of colloform blastoids in granite rocks, states the following: "In contrast to the gel-silica and the colloform relic structures in crystalline quartz, colloform structures in feldspars are more uncertain and they can only be inferred on the basis of the textural patterns of their zonal growth".

The growth of the colloform-blast has taken place under static conditions. It should be pointed out that the zonal growth is transversed by later-formed twinning which is developed subsequently to the zonal growth (the zonal growth represents the stage-by-stage development of the colloform phase). The twinning is developed during the transformation of the colloform to the crystalline phase.

Comparable to and commensurable with the plagioclase colloform blastoids in granites are the "collocrysts" in basalts. A series of microscopic observations indicates that in the case of the plagioclase collocryst, the development of twinning is related to the tendency of the colloform structure to separate, during the transformation, due to syneresis, and to produce distinct individuals. This, in the case of plagioclase collocrysts is achieved by the development of polysynthetic twinning which transverses the colloform zonal relics.

Figs. 289 and 290 show the transformation of a zoned colloform structure to a plagioclase collocryst. As relics of the colloform structure there is the colloform fine zoning, as is indicated in Fig. 290. As a result of the transformation from a colloform to crystalline state, the undulating extinction is substituted by the development of polysynthetic twinning which is developed as a consequence of the transition of colloform to crystalloid. Furthermore, the collocryst has attained a crystalline outline. Figs. 291 and 292, two different orientations, depict a plagioclase collocryst in basalt, showing polysynthetic twinning transversing the relic colloform zoning. The development of the polysynthetic twinning is not definite and distinct and this is typical of twin lamellae developed in a collocryst as a part of the process of a colloform structure being transformed to collocryst. As a corollary to these observations, Figs. 293 and 294 show basic colloform structures, such as zoning transversed by later-developed polysynthetic twinning. Characteristically, Fig. 294 shows that, as a result of the colloform structures' transformation to collocryst, in addition to polysynthetic twinning separate plagioclase individuals also developed, though their boundary transverses the relic colloform zoning.

An additional case indicating both colloform characteristics and crystalline features is shown in Fig. 295. The collocryst shows a typical colloform zoning and its outline is transitory between a colloform structure and a crystal. Furthermore, it can be seen that, whereas half of the collocryst still maintains a colloform general appearance and has complex colloform fine zoning and shows undulating extinction, the other half of the collocryst attains a crystalline outline and shows polysynthetic twinning.

Colloform blastoids (collocrysts) enclosing a pre-existing plagioclase phase

In some cases the growth of a colloform blastoid can commence around a single pre-existing plagioclase or by engulfing an aggregate of pre-existing plagioclases. Fig. 296 shows a collomorph blastoid surrounding, corroding and partly assimilating a preexisting plagioclase. The overgrowth of plagioclase indicates a pattern of colloform zonal growth; however, it has attained a crystalline outline and secondarily developed twinning transverses both the blastoid colloform overgrowth and the pre-existing nucleus.

Comparable cases are shown in Figs. 297 and 298, where pre-existing plagioclases, with polysynthetic twinning exhibited, are rounded and enclosed by a later-formed colloform blastoid, the zonal growth of which is influenced by the presence of the enclosed (generation I) plagioclase. It should be pointed out that in these cases the plagioclase blastoids, apart from their colloform zonal growth, show characteristic features of a crystalline phase, rather than the typical collocrysts shown in Figs. 275 and 277.

In contrast, Fig. 299 shows a plagioclase exhibiting both colloform and crystalline features and enclosing a corroded and partly assimilated pre-existing plagioclase (generation I) phase. The blastoid shows a partly colloform outline and is turbid, with inclusions also indicating a zoning which, in parts, is typically colloform; simultaneously crystalline characteristics, such as crystal faces and growth zones parallel to them, are shown. As already mentioned, Fig. 284 shows a plagioclase collocryst (plagioclase generation II) which surrounds plagioclase generation I.

Considering the above observation, the transformation from a colloform to a crystalline body undoubtedly takes place, however, the detail and mechanism of this complex process of colloform to crystalline is difficult to explain. Augustithis (1964) discusses the transformation of gel-pitchblende to well-crystalline uraninites and states the following:

The colloid conditions from which gel-pitchblende is formed are believed to originate rather from condensation than from dispersion; moreover, such systems have the property of reversibility, consequently the alteration or partial alteration of gel into a sol will result in the formation of true solutions, in which rearrangement of the atoms can take place and a crystalline structure can be produced.

It must be emphasized, however, that the reversibility of gel to sol is often incomplete and gradual, and the formation of true solutions can be local and restricted, so that the change of gel-pitchblende into crystalline can take place and yet the initial gel features can be maintained. It is clear that on the basis of the above suggestion, the transformation from colloform to crystalline is through a sol phase, that is through a solution phase and a solution front.

A rather different mechanism is proposed by Elliston (1963) in the formation of quartz and feldspar collocrysts: "In the very slow process of crystallizing from a gel, the material loses water. First, the solution shell is spontaneously reduced by syneresis and finally the amorphous material or metacolloid becomes crystalline in character through a molecular re-arrangement. Heat would be liberated very slowly and because of the hydrated semi-solid nature of the original colloid material, it would not be expected to be comparable in quantity to the latent heat when crystallization is achieved from the liquid state."

A rather more extreme position regarding the transformation of colloform bodies to crystalline is taken by Boydell (1924, in Elliston, 1963) who has suggested that crystallization from solutions can only proceed via the gel phase.

Despite the difficulties in understanding the mechanism of this transformation, the phenomenology as described by independent workers, Lebedev (1967), Elliston (1963), Augustithis (1964, 1967, 1973) and

Dresder-Kaden (1969) shows similarities which could perhaps, be summarized as follows:

(1) There is a transformation of the body from colloform to crystalline. The growing crystal tends to assume a shape dictated by the lattice strength and short, incomplete or partly developed crystal faces are very common.

(2) The colloform shape, that of a spheroid, may be maintained depending on the development of crystal-faces by the transformation of a colloform to a crystalline body. Many of the collocrysts have gradational crystal boundaries, and the deep embayments and inclusions of the "matrix" or "groundmass" within the collocryst are a further indication of its presence during their formation.

(3) Basic colloidal patterns, such as colloidal zoning, inclusions distribution and pore-distribution, may be preserved after the transformation and can be used as evidence of a colloidal origin.

(4) Due to syneresis, there is a tendency for cracks to develop which may lead to separate individual crystals. Interrelated with this tendency and occurring after the transformation from colloform to crystalline, is the tendency for polysynthetic twinning, particularly in plagioclase collocrysts.

(5) In quartz and feldspar collocryst, undulating extinction may be preserved and, in the crystalline phase, there is a graduation from undulating to diffuse twinning which is particularly noticeable in feldspar collocrysts.

Plagioclase phenocrysts in a glassy colloform groundmass

Plagioclase phenocrysts in a complex glassy groundmass consisting of colloform structures, show "amphoteric" phenomenology, i.e. characteristics which support a phenocrystalline interpretation and other textural appearances which could support a more or less contemporaneous growth of the plagioclase and the hyalo-colloform groundmass.

In support of the phenocrystalline interpretation are: idiomorphic shapes (Fig. 300); the formation of colloform margins surrounding both large and smaller plagioclases present in the groundmass (Fig. 301), and the infiltration, corrosion and replacement of the large feldspars along their polysynthetic twinning by the glassy colloform groundmass (Figs. 302 and 303).

In contrast, glassy substance appears to be enclosed by plagioclases interzonally, since the shape of the feldspar zones depends on the engulfed glassy mass (Figs. 304 and 305). Also, colloform and irregularly shaped glassy inclusions of the plagioclase effect the composition of the host that enclosed them (Fig. 306). In cases, a phase of "mesostasis" between glass and feldspar may be present in plagioclases (Fig. 307).

The above amphoteric characteristics suggest that most of the plagioclases are early-formed phenocryst, though in some cases, there has been a prolonged crystal development which has possibly involved the glassy phase as well.

Chapter 14 | Leucite tecoblasts

In basaltoid leucite-bearing tephrite rocks from Travolata, Rome, well-developed leucite icositetetrahedra (112) often are present which are paramorphosed to rhombic β-leucite, which consists of a complex network of lamellae with straight extinction and with their axial plane parallel to (100).

As a result of the inversion of the high-temperature cubic to the rhombic leucite, the crystal separates into smaller "blocks" with lamellar twinning. The inversion temperature of the cubic to the rhombic leucite is about 606°C.

Cundari and Le Maitre (1970) have proposed that the leucite-bearing rock of the Roman and Birunga volcanic regions belong to a trend which can be interpreted in terms of low-pressure crystal differentiation and that the leucites are true phenocrysts.

Microscopic observations show that the leucite phenocrysts are in reality gigantic tecoblasts, in the sense that they have been crystallized as high-temperature modification forms, enclosing and corroding pre-existing pyroxene phenocrysts (Figs. 308 and 309) and the groundmass components, a part of which must have crystallized before the leucite tecoblast.

The abundance, in certain growth zones, of minute apatite needles, which appear as dots in cross-section (Fig. 310), is of significance in the understanding of the development of the initial high-temperature cubic modifications.

The apatite-abundant zones often do not follow definite crystallographic faces of the leucite (Fig. 311) and, in some cases, they appear to follow in their arrangement enclosed and corroded pyroxenes and groundmass feldspar inclusions (Fig. 312).

In contrast, in certain cases the leucite zoning, as is indicated by the abundance of apatite needles, may follow crystallographically orientated directions which coincide with definite crystal faces of the leucitohedron (Fig. 313).

It is obvious that the crystal developed as an icositetrahedron, since growth zones follow the initial icositetrahedral crystal development, which at low temperatures inverted to the rhombohedric (pseudo-tetragonal) modification, consisting of individual "blocks" built up of minute rhombohedral lamellae.

Additional evidence of the tecoblastic origin of the leucitoheders is the abundance of groundmass olivines, pyroxenes and plagioclase laths which are included in the later-formed tecoblasts (Fig. 314). The groundmass feldspar laths and the other components are not groundmass embayments in the leucite. As Figs. 315 and 316 show, the groundmass was engulfed by the growing tecoblast, and in addition to corrosion appearances, plagioclase laths have disturbed the zonal development of the leucite.

As it can be seen in Fig. 316, particularly a plagioclase lath has influenced the development of an apatite-needle-rich zone of the growing leucite tecoblast, clearly indicating that it must have existed as a foreign body and its presence has disturbed the smooth running of the leucite zone. It is therefore clear that the groundmass must have been crystallized to some extent as the leucite tecoblast developed.

Despite the tecoblastic genesis of the leucite megacryst, no idioblastic crystal forms (outlines) have been maintained and this is due to the continuous reaction between the already formed tecoblast and the groundmass which crystallized after the tecoblast. The existence of groundmass components as engulfed and corroded components in the leucite tecoblast is not in contradiction to the corrosion outline which is shown in Fig. 313 by the leucites resulting from magmatic corrosion of the leucite tecoblast and the groundmass that crystallized and consolidated well after the tecoblast. This is possibly due to the growth of the tecoblast at a certain depth and its transportation to the surface in a nonconsolidated melt, or not-completely consolidated one.

Chapter 15 | Hornblende, biotite and magnetite phenocrysts in basalts

Hornblende and biotite phenocrysts in basalts

Comparing lunar with terrestrial basalts, the absence of minerals containing (OH) in their lattice in the moon rocks has been explained by the lack of atmosphere around the moon. Consequently, the presence of (OH) in the lattice of minerals in terrestrial basalts could be attributed to an atmospheric origin due to water "assimilation", despite their idiomorphic form in basaltic rocks.

The composition and structure of basaltic hornblende from Cernosin, Czechoslovakia, was studied and described by Heritsch and Riechert (1960), and is explained as a crystallization phase of the basalt.

In contradiction to the above hypothesis, Oxburgh (1964) has suggested that the presence of amphiboles in basalts could be considered as petrological evidence for the presence of amphiboles in the upper mantle.

In contradistinction to the formation of hornblende due to uralitization (Fig. 317), hornblende can be formed surrounding the pyroxene and in association with magnetite crystal grains (Fig. 318). This pyroxene–hornblende association and also the reverse case, where hornblende is surrounded by pyroxene, rarely occurs (Fig. 319 and 320).

In contrast to these pyroxene–hornblende associations is the idioblastic basaltic hornblende, which indicates blastic growths with autocathartic properties (Fig. 321). Figs. 322 and 323 show idioblastic hornblende with interzonally arranged apatites and groundmass components. Furthermore poikiloblastic hornblendes have been observed. A poikiloblastic hornblende with iron oxides is shown in Fig. 324. Another type of hornblende occurrence in basalts is an aggregate consisting of hornblende crystals radiating in their arrangement and associated with early olivine crystallization and included basaltic groundmass (Figs. 325 and 326).

In contrast to the idiomorphic hornblende in basalts, the biotite is mainly poikiloblastic enclosing pyroxenes, feldspars and magnetite crystal grains (Figs. 327 and 328). In other instances the biotite forms a network occupying intergranular spaces between pyroxene crystal grains (Fig. 329). In this case, the biotite is an anchi-poikiloblastic phase. In contradistinction to these intergranular and poikiloblastic biotites, idiomorphic laths with idioblastic characteristics can also be found (Figs. 330 and 331).

In addition to the textures described, biotite as symplectic intergrowth with hornblende and pyroxene exists in basalts, as shown in Figs. 332 and 333 respectively.

In certain cases, mica can grow blastically in chloritic aggregates of basalts, either in chlorite of amygdaloidal fillings or in intergranular chloritic aggregates (Figs. 334 and 335).

On the basis of these textural patterns the crystallization of biotite appears to be more dependent on a "solution" phase than on a direct crystallization from a wet-melt phase. However, certain textural associations of mica are difficult to explain, such as, for example, the symplectic intergrowth of phlogopite with magnetite skeleton crystals (Fig. 336).

Magnetite phenocrysts in basalts

Magnetite is a common opaque mineral in basalts, in addition to the groundmass-abundant magnetite, phenocrysts are also common. Ore-microscopic and X-ray data indicate that the magnetite is a high-temperature titanomagnetite in the sense that the ilmenite crystal molecule is in solid solution within the lattice of the magnetite. A pinkish colour is often characteristic of these basaltic titanomagnetites (Fig. 337).

Ade-Hall and Lawley (1970) have pointed out the petrological differences between Tertiary basaltic dykes and lavas. There are differences in the opaque minerals of basaltic dykes and lavas, both in the relative abundances of phases crystallizing from the magma (separate ilmenite is low in dykes and abundant in lavas) and in the relative development of deuteric oxidation (somewhat higher in lavas), as shown by the type of titanomagnetite present. The frequency of occurrence and average amounts of primary iron sulphides are much greater in basaltic dykes than in basaltic lavas.

Magnetite phenocrysts are often enclosed or partly enclosed by pyroxene phenocrysts (Fig. 338) and they are definitely an early crystallization phase. The relation of the magnetite phenocrysts with the basaltic groundmass can be variable. Magnetite phenocrysts can be later than the groundmass feldspars, enclosing plagioclase laths or partly enclosing them (Fig. 339).

In contrast, a symplectic intergrowth of magnetite phenocrysts with groundmass exists, suggesting a simultaneous magnetite-phenocryst-groundmass crystallization (Fig. 340). Most interesting symplectic magnetite–phlogopite and feldspar intergrowths are shown in Fig. 336, the magnetite phenocrysts attain well-developed skeleton-crystals or are idiomorphic.

However, in most cases, the magnetite phenocrysts crystallized before the groundmass, which has invaded the magnetite in some cases (Fig. 341).

Occasionally, special cases of magnetite crystallization may exist with magnetite agglutinations in a chloritic mass, filling a basaltic cavity (Fig. 342). The above textural patterns indicate the wide spectrum of textural patterns exhibited by magnetite phenocrysts and that their crystallization is not restricted to the early phases of basaltic crystallization, but can also take place within the "hydro" phases during a concluding phase of the basalt's crystallization or after its consolidation.

Chapter 16 | The basaltic groundmass

Flow textures

Whereas the basaltic phenocrysts exhibit most impressive textural patterns and are of significance for understanding the derivation and crystallization sequence, the groundmass comprises by far the greatest part of the earthly basalts. While discussing the significance of groundmass, it should be pointed out that the lunar basaltic types do not show development of phenocrysts and this has given rise to the hypothesis that they are impactites, i.e. gabbroic rocks melted due to meteoritic impacts. In contrast, most of the earthly basalts are composed of phenocrysts and groundmass. In the present effort we have emphasized the textural patterns of the phenocrystalline phase of the basalts, though it should be pointed out that the groundmass is the general background in which the phenocrysts play their individual role. Whereas previously the groundmass was considered to be the concluding phase of the basalts' crystallization and the phenocrysts were believed to be an earlier crystallization phase, the tecoblastic, collocrystalline and multiple phases of crystallization of the phenocrysts are evidence that the groundmass could be an even earlier crystallization phase than certain phenocryst phases (see Chapters 9 and 10). It should also be emphasized that the groundmass components do not all crystallize simultaneously, e.g., magnetite, rutiles and apatites can be interzonally orientated within phenocrysts, clearly indicating that the crystallization of groundmass components had commenced before the crystallization of certain phenocrysts.

On the other hand, groundmass embayment into the phenocryst, corrosion, infiltration reactions, symplectic intergrowths and replacements of phenocrysts by the groundmass, are very common and are in fact always present. The phenocrysts and the groundmass are of the same or of different derivation, reacting and competing within the same system.

The derivation of the phenocrysts and the crystallization can be variable (e.g., xenocrysts, early crystallizations, tecoblastic and collocryst phases) and so can also be the crystallization of the basaltic groundmass, feldspar, pyroxenes, olivines and particularly the accessories.

Despite all of this, it is true that the crystallization of the groundmass coincides with the consolidation of the basalt. Its textures are of primary significance for the understanding of the crystallization of the basalt.

The fluidity of the basalts depends on their composition. The concept that the more basic a rock is, the greater its fluidity, should be qualified in basalts. Plateau basalts and tholeiitic types show a tremendous fluidity (e.g. the trap structures). Both the phenocryst and groundmass components of a basaltic rock can exhibit flow textural patterns. Fig. 254 shows flow orientation of the gigantic, oscillatory-zoned plagioclase plates of the Debra-Sina basalts. Fig. 343 shows a basaltic groundmass consisting of tabular, medium-grained plagioclases and plagioclase laths both showing flow orientation. In addition to the plagioclases of the groundmass components showing a flow orientation, prismatic pyroxenes can also follow the general flow-orientation of the plagioclase groundmass feldspars. Fig. 344 shows a polysynthetically twinned, augitic-prismatic, coarse-groundmass-size component showing the same flow orientation as the plagioclase laths of the groundmass.

Fig. 345 shows groundmass laths having an arrangement parallel to a pyroxene phenocryst's outline, due to flow orientation. The groundmass has also followed directions of weakness of the phenocryst. A most impressive case of groundmass flow orientation is indicated around the periphery of a magnetite xenocryst in basalt (Fig. 346). The magnetite is a component of an olivine bomb (xenolith) in the basalt and the flow orientation and shape of the feldspar following the xenocrysts outline is impressive.

Again showing a flow orientation, is a plagioclase (considerably larger than the average plagioclase lath of the groundmass) which is parallel intergrown and with a reaction margin with a basaltic augitic phenocryst (Fig. 347).

In some cases, the fluid basaltic groundmass gives rise on consolidation to minute groundmass components (see Fig. 51). There is a flow arrangement of the trichites comprising this glassy groundmass.

Intrapenetration (ophitic) and engulfment textural intergrowths

In contrast to the flow or fluidal textural patterns, under more static conditions of basalt consolidation intersertal or ophitic textures develop. There is a graduation and transition of intersertal to ophitic textures. On the other hand, whereas micro-ophitic textures are exhibited by earth "basalts" and particularly in dolerites and micro-gabbroic types, e.g., Karroo dolerites, Palisade Sill, etc., the gabbroic ophitic textures are common and dominant in both terrestrial and lunar gabbroes.

In addition to these transitions and variations in size, there is a wide field of textural intergrowths which show the characteristics of intrapenetration or engulfment of one phase by another.

The interpretation of whether an intergrowth is due to intrapenetration (i.e. caused by infiltration "intrusion" into the host along weak directions by the crystallization force of the penetrating phase, or penetration due to replacement) or due to engulfment of phase "A" by a younger phase "B" is often dubious and only if the phenomenology supports a particular interpretation, can such an explanation be tentatively suggested.

The following intrapenetrations and engulfment textural intergrowths are recognized in the basaltic groundmass. However, some of these textural patterns are not restricted to groundmass size but are equally recognized as intergrowth patterns of the phenocrysts as well; furthermore, comparable and commensurable textures are common and dominant in the microgabbroic and gabbroic rocks.

(1) Augite engulfing olivines.

(2) Augite penetrating into augite (augite to augite intrapenetration textures).

(3) Intersertal plagioclase to plagioclase.

(4) Plagioclase in ophitic intergrowth with hornblende.

(5) Plagioclase in ophitic intergrowth with magnetite.

(6) Plagioclase in ophitic intergrowth with olivine.

(7) Plagioclase in ophitic intergrowth with pyroxene (interstitial to micro-ophitic and gabbroic textures).

Fig. 348 shows augite phenocrysts exhibiting hourglass textures and enclosing or partly engulfing corroded and idiomorphic olivine grains. Similarly, Fig. 349 shows corroded olivine grains included and partly enclosed by augitic phenocrysts. In both cases the later augitic crystals have corroded and enclosed the olivines.

In contrast, augites exhibiting hourglass structure show intrapenetration intergrowth (Figs. 350 and 351) which could be interpreted as being due to simultaneous crystallization of the augites. In contradistinction, the large-zoned augite has most probably engulfed the smaller-zoned augite and other pyroxenes, as is shown in Fig. 352.

In contradistinction, Fig. 353 shows a later augite, intruding perpendicularly to the elongated prismatic direction of a preexisting augite; most probably the intrusion followed a crack in the host.

Similarly, Fig. 218 shows a forcible intrusion of an augite into its augitic host, most probably due to the crystallization strength of the former.

Fig. 354 shows groundmass plagioclases with polysynthetic twinning and with an intersertal textural pattern exhibited; a most interesting case of plagioclase penetration by a plagioclase lath is exhibited in Fig. 355 in which case the force of crystallization of the plagioclase lath (lamellae) was sufficiently strong to allow the growing plagioclase to penetrate through the augite and its enclosed plagioclase.

As microscopic observations show, the crystallization strength of the plagioclase lamellae can cause plagioclase embayments and penetration of pre-existing mineral phases by the later plagioclase growths. Fig. 356 shows plagioclase laths penetrating into a basaltic hornblende. Comparable to this is the penetration and embayment of magnetites by plagioclase laths, exhibiting crystallization strength (Fig. 357).

A rather rarer case of ophitic intergrowth of olivine phenocryst and basaltic groundmass plagioclases is illustrated in Figs. 358a and b. The plagioclase groundmass has penetrated into the olivine along canals formed by magmatic corrosion of the phenocryst.

Intersertal to micro-ophitic and gabbroic-ophitic textures

Intersertal mutual relationships between pyroxenes and plagioclase laths are observed in consolidated tholeiitic flows (Deccan, Oregon, etc.) and in the upper part of hypabyssal sills (e.g. Palisade Sill). When static conditions of cooling prevail, such intersertal textural patterns are common. Figs. 359 and 360 show pyroxene–plagioclase mutual intersertal textures. Comparable and commensurable are the intersertal textures from the upper part of the Palisade Sill (Fig. 361). Under more hypabyssal conditions of cooling, micro-ophitic and gabbroic ophitic intergrowth patterns develop. An analysis of these textural patterns based on the phenomenology reveals a wide spectrum of mutual intergrowth possibilities between pyroxene and plagioclase in basaltic rocks. An attempt is made here to present the most common ophitic patterns and to interpret their genetic relationships.

Figs. 362 and 363 show pyroxene phenocryst invaded

by plagioclase laths which extend along either crack or the cleavage directions of the host. As is particularly shown in Fig. 363, the plagioclase laths invade the peripheral part of the pyroxene, extending from the "environment" of the groundmass inwards into the pyroxene host.

Similarly, Fig. 364 shows a pyroxene which is invaded by plagioclase laths. The feldspar invades the host, extending from a groundmass "environment" into the peripheral part of the pyroxene. In some cases, the plagioclase laths extend from the groundmass into the pyroxene without pronounced changes in the lath form (Fig. 365). Also, ramifications of the plagioclase laths, as they extend from the margin of the host into the pyroxene, are common (Fig. 366). In contradistinction to the cases where the shape of the plagioclase lath is maintained outside and inside the host, frequently there is a distinct change of the shape and width of the plagioclase laths within the pyroxene (Figs. 367 and 368). On the other hand, the plagioclase laths in intergrowth with the pyroxene frequently do not show "conspicuous" reaction margins, particularly with the feldspars invading the pyroxene cleavage or cracks (Fig. 369).

In contradistinction, plagioclase laths invading the pyroxene along crystal penetrability directions other than the cleavage pattern, often show reaction margins (Fig. 370) and occasionally undefined and straight contacts, again supporting a reaction between the host pyroxene and the "invading" plagioclase (Fig. 371). Indeed, the shape (the general form) and the outline of the plagioclases in ophitic intergrowth with the pyroxene deviate from the accepted pattern of feldspar laths in intergrowth with the pyroxenes.

In addition to the reaction contacts and branching of the plagioclase within the pyroxene (Fig. 372), variations in the general form of the feldspars are occasionally observed. Fig. 373 shows a basaltic pyroxene with undulation due to "deformation", in intergrowth with a "curved" plagioclase lath which is thinner in its central part in comparison to its margins. Similarly, a plagioclase in ophitic intergrowth with the pyroxene shows a "curvature" (Fig. 374). As Fig. 375 shows, in some cases the plagioclase lath in ophitic intergrowth with the pyroxene shows a margin indicating pyroxene corrosion and simultaneously prolongation of feldspar indentations into the pyroxene. Another case supporting pyroxene corrosion by a plagioclase lath intergrown perpendicular to the elongation direction of the pyroxene is shown in Fig. 376.

In contrast to the rather exceptional cases of curved plagioclase laths, corrosion and indentation intergrowth margins, etc., there is clearly a very common and indeed predominant textural pattern of plagioclase laths invading the pyroxene along the cleavage directions of the host pyroxene.

Fig. 377 shows a plagioclase lath clearly following pyroxene cleavage; similarly, Fig. 378 illustrates a pyroxene phenocryst with plagioclase laths in the peripheral part of the pyroxene and invading the host along the two cleavage directions. In addition, Fig. 379 shows the unmistakeable orientation of certain of the plagioclase laths parallel to and following the host's cleavage. A comparable example is the textural pattern shown in Fig. 380, in which the plagioclase laths follow two prevailing directions of orientation within the host, corresponding to the penetrability direction of the pyroxene.

In contrast, the plagioclase laths, despite the existence of a pronounced cleavage set and of cracks within the pyroxene host, often transverse these planes or directions of penetrability without taking advantage of and following them (Fig. 381). It seems that independently of the penetrability directions of the host, the growth of the plagioclase laths equally depends on the crystallization strength ("Kristallization Kraft") of the growing plagioclases which can penetrate the host's lattice and can show variable interpenetration patterns with other feldspars laths in the pyroxene host.

Within the ophitic intergrowth, the intrapenetration plagioclase textures, i.e. plagioclase laths, invading the pyroxene and in turn invaded by plagioclase laths belonging to the same sequence of crystallization, show the significance of the crystallization strength of the plagioclases (a tremendous vectorial strength of crystallization of the plagioclase laths along their elongation direction).

Fig. 382 shows a plagioclase transversing a pyroxene in an ophitic intergrowth; similarly plagioclase laths invade the pyroxene and partly extend well into the first plagioclase lath with which they are intergrown perpendicular to its elongation direction.

In contradistinction, plagioclase laths show interpenetration intergrowths within the pyroxene host. The plagioclase laths exerting a crystallization force have simultaneously invaded the pyroxene host and have intrapenetrated one another (Fig. 383). Fig. 355 shows plagioclase laths orientated parallel to cleavage directions, and other feldspar invading the pyroxene host from its margin inwards. A plagioclase lath, due to its vectorial strength of crystallization, penetrates through another feldspar lath.

As additional evidence of the post-pyroxene nature of the plagioclase laths, in ophitic intergrowth with the pyroxene crystal, grains are included in the plagioclase which invades the pyroxene host (Fig. 384). Such textural patterns support the assertion that the feldspar laths are at least younger than the pyroxenes enclosed by them. All transitions from intersertal to ophitic intergrowth are also exhibited in this case (Fig. 385).

Some cases of symplectic plagioclase–pyroxene intergrowths, ophitic in character (plagioclase invading a pyroxene), are exhibited in Figs. 386 and 387. Occasionally, the symplectic ophitic intergrowths are difficult to interpret, as is shown in Fig. 388. However, in this case a later feldspar * has invaded and replaced a pyroxene.

Lunar research has revealed the great significance of ophitic gabbroic intergrowths of pyroxene and plagioclases. Professor San Miguel (1972) has pointed out the similarity between and comparable origin of the terrestrial and the lunar ophitic intergrowth. ** Furthermore, he explained the classical ophitic gabbroic and subophitic textural intergrowth of the lunar samples (Fig. 390) as being due to pyroxene replacement along a crack pattern by the later plagioclase crystallization phase. Undoubtedly, Professor San Miguel's explanation satisfactorily accounts for the orientation of feldspars which clearly follow a crack system in the pyroxene host. Furthermore, terrestrial research (Augustithis, 1956/1960) shows that plagioclase tecoblast can invade pyroxenes along cracks and replace the host (see Figs. 230 and 231).

As a corollary to the tecoblastic plagioclases in intergrowth with the pyroxenes, Fig. 391 shows a plagioclase lath invading a pyroxene along a crack in the host.

Topo-concentrations of augites, olivines and magnetites

Within the basaltic groundmass, "agglutinations", segregations of specific minerals, can often take place and, as a result, topo-concentrations (local concentrations) can take place. Figs. 392 and 393 show topo-concentration of prismatic pyroxenes within the groundmass. The idiomorphic nature of the prismatic microcrysts and their arrangement as occupying a micro-geode indicate that their crystallization is within the crystallization process of the groundmass, i.e. the topo-concentrations are not due to xenolithic derivation. Similarly, Fig. 394 shows a cumulation of zones augites with the groundmass occupying the intergranular spaces.

Comparable to olivine, cumulates (or topo-concentrations, local segregations) of early eucrystalline to granular olivines are less frequently observed in basaltic rocks (Fig. 395). The eucrystalline to granular nature of the cummulate crystals and the absence of spinel intergrowths, as well as the absence of the characteristic reactionary phenomenology that is often exhibited between xenoliths and groundmass, support the cumulate origin of these topo-concentrations of early olivines within the groundmass.

As has been observed, the local topo-concentration of magnetite within the basaltic groundmass can lead to the local formation of an iron ore with idiomorphic apatites frequently present (Fig. 396). In some cases, the topo-concentration of magnetite can take place adjacent to pyroxene phenocrysts or within a pyroxene, as is seen in Figs. 397 and 398.

Often topo-concentrations of magnetite occur around hornblende in basalts. The cumulates of early crystallized phases within the basalt show a type of crystal segregation rather than differentiation due to fractional crystallization and gravitative separation; these local segregations exist interspersed within the basaltic groundmass and do not form magmatic layers within the basalt.

Radiating and globular textures of basaltic groundmass

Very often, acicular plagioclase laths with a radiating arrangement change on microscopic scale to radiating elongated plagioclase laths. Fig. 399 shows that the crystallization of fine acicular plagioclase and the abrupt change from fine acicular to elongated laths are textural patterns of groundmass consolidated under oceanic conditions of crystallization. These basaltic textures are comparable to the spinifex textures found in the Archaean greenstones and they are believed to be due to superheated basaltic melt and its subsequent rapid cooling under oceanic floor conditions.

Hyalo-colloform trichites, glassy amorph and undifferentiated basaltic groundmass

Often the glassy basaltic groundmass shows hyalo-colloform structures (Fig. 400) which surround the plagioclase phenocrysts. Fig. 401 shows a glassy basaltic groundmass with prismatic plagioclases and with a spheroidal hyalo-colloform structure.

In contrast to the glassy and hyalo-colloform basaltic groundmass consisting of micro-trichites, fine minute acicular crystals are shown in Fig. 51.

Frequently, the basaltic groundmass is "undifferentiated". Fig. 402 shows "undifferentiated" basaltic groundmass with prismatic pyroxenes, microcrysts and magnetite crystallization (nonidiomorphic in shape).

* In some cases a selective alteration of the plagioclases in ophitic intergrowth with the pyroxenes can take place, as shown in Fig. 389.

** According to Mason and Melson (1970) the plagioclase in the pyroxene–plagioclase intergrowth is much more calcic (bytownite to anorthite) than in common terrestrial basalts (labradorites).

Chapter 17 | The relation of groundmass to plagioclase augite and olivine phenocrysts

Groundmass invading the phenocrysts along the twinning, zoning or crystal directions with greater penetrability

The basaltic groundmass, due to its great fluidity being a rather basic liquid, infiltrates along cracks and twinning planes of the phenocrysts. Fig. 403 shows a zoned augitic phenocryst infiltrated along fine cracks by the basaltic groundmass. Irregular bodies of basaltic groundmass, which have infiltrated the augite, in some cases follow the twinning planes of the polysynthetic augitic twinning. However, it should be pointed out that the infiltrated groundmass often attains irregular shapes within its host, frequently transversing the twin planes. In contrast, Fig. 404 shows twinned augites with the basaltic groundmass invading the pyroxene along the twin-plane and being exactly delimited by the twinning planes.

Similarly, plagioclase phenocrysts can be infiltrated, invaded and replaced by groundmass following the polysynthetic plagioclase twinning (Figs. 405 and 406). In certain cases, the groundmass extends as embayment with the host plagioclase following the twinning and producing fine reaction alteration margins within the plagioclase. Fig. 407 shows a groundmass embayment as an enclosure within the plagioclase and producing reaction margins.

In contrast, groundmass infiltration within the phenocryst may exploit the inter-spaces between the zonal growth (Fig. 408).

In addition to twinning, cleavage may be a penetrability direction which is followed by the invading basaltic melts. Fig. 409 shows an augite which is invaded by basaltic groundmass following the pyroxenes cleavage and replacing the host along these planes.

In contradistinction, embayment of the groundmass in the augitic phenocryst may not be associated with any obvious channels of crystal penetrability (Figs. 410 and 411). In contrast, the groundmass embayments in the olivine phenocryst, shown in Fig. 412, follow clearly defined cracks within the olivine. Similarly, a plagioclase phenocryst is invaded by basaltic melts along a crack system of the feldspar (Fig. 413).

Intracrystalline penetration of melts (penetration paths through a phenocryst)

In cases such as intracrystalline extensions of basaltic melts, which have corroded and marginally altered augitic phenocrysts, infiltration paths can be traced from the augitic margins extending into the pyroxene without taking advantage of an obvious penetrability direction (Fig. 414). Often these melt paths within the host olivine assume myrmekitoid appearances (Fig. 415). They are due to intracrystalline penetration of basaltic melts and the myrmekitoid shapes are due to the reactions involved between the olivine lattice and the intracrystalline penetration of melts. Fig. 416 shows that these myrmekitoid bodies roughly follow preferential directions within the host.

Comparable basaltic-groundmass intracrystalline infiltrations are also present in plagioclase phenocryst, in some cases extending from the feldspar's margin and following roughly the twinning of the host (Fig. 417).

In some cases, though these bodies are following random directions within the plagioclase and, as a result of groundmass melt and plagioclase reaction, distinct mineral phases can be produced, prismatic forms commonly have their elongation perpendicular to the penetrability direction followed by the melt path (Fig. 418). Plagioclase magnetite and pyroxenes may be formed in the intracrystalline penetration paths and resorption and reaction shapes often characterize the mineral phases formed in these "paths" of melt penetration.

The reaction of phenocrysts with the groundmass

One should see the reaction phenomenology of phenocrysts–groundmass as an amphoteric reaction, i.e. the groundmass can corrode, infiltrate and replace the already crystallized phenocrysts. On the other hand, tecoblastic phenocryst, collocryst and late crystallizing phenocrystalline phases can include, corrode and assimilate groundmass components. Their relation should thus be as that of a component reacting in a system.

As already described, early formed forsterite, particularly components of mantle xenocrysts, can be corroded and infiltrated and reaction-margins are formed when the olivines are attacked by the basaltic melt, i.e. olivine–groundmass reactions (see Chapter 17). A corresponding and, to a certain extent, comparable phenomenology is produced when olivine phenocrysts are corroded, infiltrated and replaced by the basaltic groundmass. Fig. 419 shows rounded olivine phenocryst and indentations due to magmatic corrosion. Rather more pronounced are the effects of magmatic corrosion on the olivine phenocrysts, shown in Figs. 420 and 421. As is shown in Fig. 420, a former idiomorphic olivine is corroded, invaded and replaced by the basaltic melt to produce the groundmass.

In this case, the olivine–groundmass reaction did not produce the orthopyroxene of the reaction series; we simply have a replacement of the olivine by the groundmass. As Fig. 422 particularly indicates, idiomorphic olivine phenocrysts, that is, early crystallization phases of the same basaltic melt, are corroded and replaced by the basaltic groundmass at a later phase. The corresponding phenomenology of olivine (mantle fragment, i.e. foreign to its basaltic melt) xenocrysts and groundmass is indicated in Figs. 138 and 139.

Occasionally embayments of groundmass can take place in an eucrystalline olivine invading well-formed crystal faces. Often, within this groundmass embayment, which clearly produces corrosion boundaries in the olivine, prismatic plagioclases are formed in the microcrystalline in a nature groundmass embayment (Fig. 423).

Comparable are the corrosion and reaction phenomenologies of augitic phenocrysts in basaltic groundmass. Fig. 424 shows an augitic phenocryst with corroded outlines and extensions of groundmass into the pyroxene. A reaction margin has been formed on the augite, clearly depending on the influence exercised by the groundmass on the phenocryst. A comparable phenomenology is shown in Figs. 425 and 426. Fig. 425 shows an augitic phenocryst, magmatically affected. A reaction zone is produced in the pyroxene and has extended into the unaffected augitic central part, which indicates the advancing alteration front. Comparable reaction margins are illustrated in Fig. 426. However, in this case the alteration margin of the augite is invaded by magnetite pigments of the groundmass. These alteration margins should be seen as being primarily caused by the migration of the elements.

Plagioclase phenocrysts crystallized at an early phase of the crystallization history of a basalt are subsequently corroded, and marginally altered by the groundmass crystallization.

Fig. 427 shows a plagioclase phenocryst, with corroded outline and with reaction margin, approximately following the phenocryst's corroded outline. Inversion of the plagioclase twinning has taken place in the reaction margin. A comparison of the An percentage content of the unaffected plagioclase and of the alteration margin shows an average of An percentage between 55–60% for the margin and 75–80% for the unaffected part.

Fig. 428 shows a plagioclase phenocryst with a reaction margin caused by magmatic corrosion. Furthermore, an embayment of basalt groundmass has taken place in the feldspars, producing a corresponding reaction margin within the host. Plagioclase laths as components of the basaltic embayment, are crystallized within the reaction margin of the plagioclase, indicating intracrystalline penetration and replacement.

Occasionally typical corrosion margins can be caused on plagioclase phenocrysts in contact with the basaltic melt. In addition to the indentation and groundmass embayment in the feldspar, infiltration of the basaltic melt, "paths" of infiltration are indicated (Fig. 429).

Due to magmatic corrosion of the plagioclase phenocrysts, alteration margins can be produced, internally delined by the zonal growth of the plagioclase phenocryst and still showing corrosion outlines with the basaltic groundmass, which partly extends into these alteration margins (Figs. 430 and 431).

A case is observed (Fig. 432) where the reaction margin, caused by the magmatic influence on the plagioclase phenocryst, coincides with a perfect marginal zoning of the plagioclase which is parallel to the crystal face of the feldspar; within this reaction margin, inversion of the polysynthetic plagioclase twinning is noticeable.

Groundmass components incorporated in the phenocryst or engulfed by external growth zones

In contradistinction to the corrosion, infiltration and, in general, embayment and invasion of the phenocryst by the basaltic groundmass, cases are observed where groundmass components are included, assimilated and often reduced to pigment size by the later phenocryst crystallization (see tecoblast and collocrysts). Fig. 433 shows a plagioclase with an external zone which is engulfing and enclosing groundmass components (despite similarity, it should be distinguished from Fig. 430 which shows a corrosion outline of the feldspar).

Often, pyroxene phenocrysts have crystallized after the magnetite of the groundmass; Figs. 434 and 435 show magnetite enclosed and interzonally incorporated by augitic phenocrysts.

Occasionally, the growth of the external zones of augites has taken place after a great part of the groundmass has crystallized. Fig. 436 shows a zoned augite phenocryst with the external zone enclosing groundmass components. As we have seen (Chapter 8), apatite, rutiles and magnetites, which are early crystallization components of the groundmass, can be enclosed by augite phenocrysts. Fig. 437 shows fine apatite needles incorporated by zoned augite phenocryst; similarly, rutile of the groundmass is zonally incorporated, again by augitic phenocryst (Figs. 199 and 438).

As a corollary to the plagioclase tecoblast (Chapter 10), subphenocrystalline plagioclase may include groundmass plagioclase (Fig. 439). In other cases, the growth of the plagioclase phenocrysts was not a single process. Figs. 440 and 441 show plagioclase phenocrysts assuming a ghost crystal character, in the sense that the early growth of plagioclase was interrupted by a phase of magmatic corrosion (rounded outline of the crystal's central part) and infiltration, which was followed by an overgrowth of plagioclase which attained an eucrystalline shape and form.

In addition to these observations, Fig. 442 shows a two-stage crystallization of the plagioclase phenocryst. The first phase is delineated by a rounded and corroded outline and groundmass, marginally incorporated or invading the first phase of plagioclase crystallization. An overgrowth phase of the phenocryst is shown in the same photomicrograph with marked polysynthetic lamellae. Most probably an inversion of the twinning has taken place in the overgrowth zone. Also, Fig. 443 shows a two-phase growth of the plagioclase phenocrysts. The first phase is turbid with groundmass enclosed. The overgrowth is marked by a pronounced corrosion outline and is twinned and relatively free of groundmass inclusions.

On the basis of the above observations and supported by the idea of plagioclase tecoblastesis, it can be seen that plagioclase phenocrysts can crystallize contemporaneously with or later than the basaltic groundmass.

Chapter 18 | The apatite in basalts and its petrogenetic significance

Apatite occurs as an accessory in basalts, and most probably fluorapatite $Ca_2F(PO_4)_3$ is the most common representative of the apatite group. Considering the P content of igneous rocks, Goldschmidt (1954) stated "basalt is believed to be the principal carrier of phosphorus on the earth's lithosphere". P occurs in iron meteorites in the mineral schreibersite $(Fe,Co,Ni)_3P_1$, in the silicate phase of meteorites and in the lithosphere "the oxidation potential is high enough to bring phosphorus to its quinquevalent oxidation state, and consequently the only stable occurrence in the earth's lithosphere is a phosphate ion $(PO_4)^{-3}$". So far the most significant P mineral in basalts is apatite and its different topo-paragenetic intimate association with certain minerals (locally within the basalt and in certain intergrowths) associations are of interest in understanding its derivation and its position in the rock's sequence of crystallization. The topoparagenesis of apatite is not governed by a single process or a geochemical–mineralogical control; many factors influence its presence. Apatite occurs as inclusions either as random or zonal intergrowths with most of the mafic minerals of basalts.

Figs. 444 and 445 show apatite as random inclusion within a basaltic olivine; in contradistinction, Fig. 446 shows an apatite accessory of large size and smaller prismatic bodies also in an "idioblastic" later-formed olivine generation (see Chapter 4).

The smaller apatites are zonally incorporated in this basaltic olivine. This topo-paragenetic association with the olivine is rather rare, but it signifies the possible relation and incorporation of P in the lattice of the orthosilicate group. In contrast, fine apatite needles and small prismatic bodies are abundant in pyroxenes and especially in zoned augites which can occur at random or interzonally incorporated within the zoned titano-augites (Fig. 447). Here again, this topo-paragenetic association is more understandable on the basis of the association of P with the augite lattice (in the sense that P can be incorporated in the augitic lattice).

Another "mafic" mineral with which apatite is often associated or in intimate intergrowth with, is hornblende. The topo-paragenetic association of apatite and hornblende is not only found in basaltic rocks but also in granites (Augustithis, 1973). Since hornblende in basalts is almost always a blastoid, the apatite associated with it could represent assimilation produced by the basaltic melt, which on its way up, transgresses other rocks.

Fig. 448 shows prismatic apatites at random within the hornblende host; also Fig. 322 and its detail Fig. 323, show blastoid hornblende in basalt, including apatite, which is included in a zone within the host that happens to be turbid with other groundmass-assimilated components. Often in basaltic hornblendes (as well as in granites) the apatite prismatic bodies, included in the amphibole, show corroded and assimilated outlines (Fig. 449).

The topo-paragenetic association of apatite with mafic minerals, as discussed so far, indicates its affinity with this group of minerals in the basalt and it also explains the greater P abundance in basic rather than in acid rocks.

In addition to the interzonal apatite in the mafic basaltic components, apatite occurs as an interzonal prismatic inclusion in collocryst – with recognizable colloform zoning and crystalline faces developed (Figs. 450 and 451). Also fine needle-form apatites occur as interzonal inclusions in leucite tecoblasts (see Figs. 310 and 313). However, the abundance of apatites with plagioclases and feldspathoids is subordinate to that of the mafic components.

In contrast, apatite occurs as needles (in cases with groundmass feldspars) in the basaltic groundmass and is rather a common constituent (Fig. 452).

Wyllie et al. (1962) have pointed out that apatite crystals, co-existing with the liquid or vapour in the system $CaO–CaF_2–P_2O_5–H_2O–CO_2$, are equant whereas those formed by quenching from a liquid are acicular and/or skeletal. Apart from these fine crystalline forms, idiomorphic apatites occur in the Selale Mt. augite-rich basalts. Fig. 453 shows sub-phenocrystalline idiomorphic apatite with augites in basalt. (It should be noted that the apatite abundance in the Selale Mt. basalts could be partly attributed to limestone assimilation, since the feeding pipes have transversed some hundreds of metres of limestone).

In contrast to the idiomorphic sub-phenocryst described, apatite occurs in basalts which are sub-phenocryst to phenocryst in size, with distinct magmatic corrosion appearances. Fig. 454 shows an apatite sub-phenocryst, with rounded and corroded outline; the striations in the crystal are due to orientated inclusions (Fig. 455). Another comparable case is illustrated in Fig. 456. The corroded apatite sub-phenocryst has a crystal cavity in which pyroxene and magnetite are included (probably a subsequently crystallized liquid drop enclosed in the apatite). Often apatite sub-phenocrysts develop in the groundmass (Fig. 457) and are abundant topo-paragenetically with hornblende in basalts; Fig. 458 shows idioblastoid hornblende and an idiomorphic apatite associated with it.

The association of apatite and magnetite is well known from ore-deposit studies, e.g., in the ores of the Kiruna district, North Sweden and elsewhere. Comparable paragenetic associations can be produced topo-paragenetically in basalts. Often within a basalt there can be a "concentration" of magnetites and with it there is sub-phenocrystalline, often idiomorphic occurrence of apatite. Figs. 455, 459 and 460 show magnetite "concentration" locally within a basalt and with it there are idiomorphic and magmatically corroded (Fig. 455) apatite sub-phenocrysts.

Particularly, Fig. 455 shows both idiomorphic apatite (with inclusions appearing as dots parallel to the prismatic faces of the apatite), and a larger apatite with a developed crystal face and with a strongly magmatic corrosion appearance. Apart from these topo-paragenetic associations, in some cases, there is a more intimate intergrowth of apatite and magnetite. Figs. 461 and 462 show apatites enclosed by magnetite phenocrysts in basalts. In particular, Fig. 463 shows a symplectic-synantectic intergrowth of apatite which is marginally invaded, corroded and replaced by the later magnetite. Goldschmidt (1954), quoting Oelsen and Maetz (1940), points to the area of unmiscibility between the oxide phase and the phosphate phase, reacting as a maximum when the ratio $C_9O/P_2O_5 = 3/1$. However, the present observations and particularly Fig. 464 show that these synantectic-symplectic intergrowths are not due to the unmiscibility, as the pattern of melt-infiltrations within the apatite shows, and also in this case, we are dealing with the intracrystalline penetrability of melts.

In considering the geochemical distribution of phosphorus in igneous rocks and particularly in basalts (the arithmetical mean of P in basalts according to Goldschmidt (1954) is 2,440 ppm) – which are considered to be the principal "carrier" of phosphorus – and the topometasomatic and intimate intergrowths with other basaltic minerals, one is tempted to suggest that the phosphorus in basalts is partly primary, i.e. derived from the "protolyte" which gave rise to the basalt by melting, and that a considerable part of it could be attributed to assimilation by the basaltic magma rising through the crust.

Blastogenic apatite in basalts

Almost all the cases described so far deal with an early apatite crystallization, yet, in contrast, a series of observations shows later blastogenic apatite in the basalt from Totenkappel b. Meiches, Vogelsberg, Germany.

Fig. 465 shows elongated apatites in leucite and extending into adjacent pyroxenes. Similarly, blastogenic apatite transverses the boundaries of two pyroxene phenocrysts and clearly indicates that it is a post-pyroxene blastogenic phase with an immense crystalloblastic force. As is indicated in Fig. 466, the elongated apatite crystalloblasts are orientated in the leucite phenocrysts, showing directions of orientation intersecting one another at an angle. This textural pattern of the apatites and the fact that interpenetrations are exhibited, has, as a prerequisite, the previous existence of leucite and the blastogenic growth of the apatite.

As a corollary to the late blastogenic growth of the apatite, Fig. 467 shows elongated apatite growths transversing granophyric quartz which is in intergrowth with the leucite and penetrating into the pyroxene as well. In contradistinction to the apatite growths described so far, apatite can crystallize as a later hydrous phase in amygdaloidal calcitic cavities of basalts (Fig. 468).

The above observations show a wide range of apatite textural patterns and crystallization possibilities.

Chapter 19 | Granophyric quartz in basalts

(MICROPEGMATITIC QUARTZ–FELDSPAR INTERGROWTHS IN BASALTS)

Rosenbusch (1898), explaining the presence of quartz in basalts, states the following: "the most probable explanation of this abnormality is through the supposition that we are dealing, in this case, with a mixture of dacitic and basaltic magmas which are extruded simultaneously . . ." ". . . the olivine belongs to the basaltic magma and the quartz to the dacitic . . ."

In contrast, Bowen (1928) explains the presence of quartz (particularly the micropegmatic quartz–feldspar intergrowth found in basalts) as being due to an enrichment of silica in the last portion of the basaltic melt to crystallize, which, in turn, is due to the early separation of the olivine by fractional crystallization.

On the other hand, Shand in his attempt to explain the presence of free silica in basalts, has introduced the concept of "oversaturated basic rocks". In contrast, Daly attributed the presence of free quartz to the assimilation of silica-rich sediments by the ascending basaltic melt.

The micropegmatic quartz of Bowen (1928) is most often quartz in granophyric intergrowth with plagioclase and occupying spaces between the plagioclase phenocrysts (Figs. 469 and 470). Occasionally, the granophyric quartz in intergrowth with feldspar surrounds and encloses idiomorphic plagioclases (Figs. 470 and 471).

Fig. 472 shows an idiomorphic plagioclase surrounded by a granophyric quartz/feldspar intergrowth occupying the intercrystalline spaces surrounding the plagioclase. As is shown by arrow "b" in Fig. 472, the granophyric quartz has taken advantage of the intergranular space marginal to the plagioclases.

As Figs. 473 and 474 show, the quartz attains most typical hieroglyphic forms and is comparable and commensurable with graphic quartz intergrowths with K-feldspar and plagioclases in granitic rocks (Drescher-Kaden, 1948, 1969; Augustithis, 1960, 1962, 1964; 1967, 1973). The granophyric quartz intergrowths in basalts are metasomatic infiltrations and replacements and their phenomenology presents comparable characteristics with the metasomatic graphic quartz in granites, granitized conglomerates and gneisses.

Granophyric quartz extends into a plagioclase of the basalt, metasomatically invading the feldspar and replacing it along twin lamellae (Fig. 475). Such textural intergrowths clearly indicate a later quartz which has replaced the plagioclase and shows micrographic textural patterns. Similarly, interstitial quartz extends marginally into a plagioclase of the basalt and the quartz prolongations attain granophyric character (Fig. 476).

The granophyric quartz in basalts is not restricted to plagioclases but occasionally infiltration and replacement of biotites by granophyric quartz is shown. (Comparable graphic quartz intergrowths in biotite in pegmatite are of metasomatic origin, see Fig. 477.) As illustrated in Fig. 478, the granophyric quartz is in intergrowth with feldspar and extends along into an adjacent hornblende. In addition to this granophyric quartz in intergrowth with hornblende, granophyric quartz in intergrowth with plagioclase similarly has prolongations which invade the basaltic biotite along the directions of crystal penetrability (Fig. 479). Occasionally, interstitial quartz has extensions which invade the adjacent basaltic biotite. The quartz extensions clearly attain granophyric character in the mica (Fig. 480).

The granophyric quartz–feldspar intergrowth often surrounds idiomorphic pyroxenes, and in rare cases quartz extensions from a granophyric quartz–feldspar intergrowth invade adjacent pyroxenes along cracks (Fig. 481). From the above textural pattern, it is obvious that the granophyric quartz–plagioclase symplectic intergrowth is not an eutectic system since, theoretically, only quartz K-feldspar would crystallize simultaneously under eutectic conditions of crystallization.

Moreover, the infiltration of granophyric quartz in adjacent biotites and along cracks of pyroxenes here again supports a metasomatic replacement origin for the granophyric quartz. In some cases free quartz, in addition to the interstitial quartz which attains granophyric character, can show transition to a granophyric quartz–feldspar symplectite. Fig. 482 (a and b) shows free quartz in intergrowth with an eucrystalline pyroxene, showing transitions and gradations to granophyric quartz–plagioclase symplectite. Another photomicrograph shows the association of free quartz and the quartz–feldspar symplectite (Fig. 483).

The granophyric symplectite in basalts shows a variation of forms and types and in cases, the micrographic quartz attains quartz "skeleton shapes", i.e. skeletons of incompletely developed quartz crystals (Fig. 484).

This micrographic quartz develops in interspaces between phenocrysts. In contrast, an intimate symplectic intergrowth of quartz–feldspar, as shown in Fig. 485, is common in basaltic rocks.

Very occasionally a paquet-type of quartz–feldspar intergrowth with two prevailing directions of quartz orientation (perpendicular to one another) can be observed within the same feldspar host. Here again, the quartz shapes are hieroglyphic in form (Fig. 486).

In contradistinction to these granophyric metasomatic intergrowths (which are most probably due to "hydrothermal" solutions operating in a concluding phase of the basalts consolidation), are the lunar granophyric intergrowths (Fig. 487) which are due to feldspar–quartz crystallization in dry melts (melts which are basaltic in composition and without water).

Despite their resemblance, the granophyric quartz intergrowth of the terrestrial basalts are as their detailed phenomenology supports, formed by "hydrothermal solution" out of which quartz crystallized and which operated at the concluding phase of the consolidation of the basalt or basaltic type of rock. In this connection it should be emphasized that these granophyric textures are most abundant in basaltic sills such as the Karroo and the Palisade Sill. Silica migration by "hydrothermal" solutions, from the country rocks into which the basaltic melts invaded, is most probable. This would also explain the co-existence of the incompatible phase of olivine in basalt and the adjacent existing granophyric quartz–feldspar intergrowths (Figs. 488 and 489).

Granophyric quartz in intergrowth with leucite and with pyroxene phenocrysts in basalts

A most interesting case, genetically, is the granophyric (micrographic) quartz in intergrowth with leucite in the case of holophenocrystalline basalt from Totenkappel b. Meiches, Vogelsberg, Germany. Considering the phase rule of crystallization, leucite and quartz should be seen as incompatible crystallization phases in the sense that silica should have reacted with leucite to produce feldspar.

The co-existence of leucite and granophyric quartz indicates that these incompatible mineral phases belong to two independent mineral phases which, in turn, belong to two independent systems; i.e., the leucite is a concluding phase of the orthocrystalline basalt's crystallization and the quartz has infiltrated the pre-existing leucite and pyroxene crystal phases of the basalt, resulting in the formation of granophyric quartz in intergrowth with leucite (Fig. 490), and similarly, granophyric quartz in intergrowth with the pyroxene phenocrysts (Fig. 491).

Chapter 20 | Alteration and weathering of olivines

Olivine is a mineral susceptible to alteration processes and particularly to weathering. Microscopic observations indicate that olivine is one of the first minerals to be altered, e.g., when a basaltic rock is subjected to weathering. Experimental work by De Norske Saltwerker Av Bergen and Dittmann (1960) have demonstrated alteration possibilities of olivines which simulate natural conditions of olivine alteration.

Despite all our knowledge of olivine alteration, detail microscopic observations reveal that the alteration of olivines in basalts involves element-migration processes far more complex than was believed. Also, there is a wide spectrum of alteration possibilities depending on the composition of the olivine, on the nature of solutions that bring about an alteration and on the topo-environmental conditions prevailing at the time of the alteration complex processes.

Whereas in the dunites and ultrabasics in general antigoritization of olivines (Mg-rich) is a very common and widespread process, in basalts, it is of restricted importance and relatively rare in comparison to other alterations.

Fig. 492 shows an idiomorphic olivine changed to antigorite in margins and along cracks. Fig. 493 shows that the serpentinization of the basaltic olivine mainly extends along cracks in it. It is believed that antigoritization is rather restricted to forsterite-rich basaltic olivines and that it is taking place under reduction in volume. Schematically the process can be expressed:

Olivine $(MgFe)_2(SiO_4)$

$\rightarrow$ antigorite $[(OH)_4Mg_3(Si_2O_5)]$.

Considering the composition of the end product and the fact that microscopically no iron minerals are produced associated with the antigoritization, it is probable that the olivine involved is rich in forsterite.

In contradistinction to antigoritization, a rather common alteration of the basaltic olivines is chloritization. According to Brown and Stephen (1959) if alteration without volume change takes place it would require nine cells of olivine to give four cells of a layer-lattice silicate. If the alteration is chlorite the volume equivalent is found to be:

$9\,(Mg_8Si_4O_{16}) \rightarrow 4\,[Mg_{12}Si_8O_{20}(OH)_{16}]$
olivine chlorite

The above schematic reaction does show the volume problem involved in the alteration of olivines to chlorite; however, it is an oversimplification of the process. Nevertheless, Brown and Stephen, interpreting this schematic reaction, point out that large amounts of Mg are expelled from the olivine regions which alter into a layer-lattice silicate. Leaching out of Mg is experimentally proved to be a primary cause of olivine alteration (Dittmann, 1959/1960). In his gradoseparation hypothesis Pieruccini also believes that in femisialic compounds (e.g. olivines) the femic compounds are solved out and the sialics are residual. According to Deer et al. (1962) the main characteristic of the chemical composition of the olivine alteration products is the high Fe/Mg ratio (the average Fe/Mg ratio is 8 : 1 compared with the highest ratio 1 : 2 reported in the association unaltered olivine).

In contrast to the schematic olivine-chlorite alteration of Brown and Stephen, Smith (1959) has found that olivine in basalt changes into haematite, chlorite (pennin Sp_3-At_2–Sp_1-At_1) and a small percentage of quartz. Of particular interest is the alteration of olivine into chlorites that contain At (= Amesitsilicate, $H_4Mg_2Al_2SiO_9$). This means that, in addition to elements contained in the lattice of the unaltered olivine on its alteration to pennin (which contains Al), migration of elements takes place in the alteration of olivine to chlorite (pennin). In addition to the leaching out of Mg, which is expelled from the olivine region that changes to chlorite, we have a migration into this region of Al. Considering the general formula of chlorites to be $\frac{2}{\alpha}H_4(Mg,Al,Fe)_3(Si,Al)_2O_9$, the alteration of olivines into chlorite inevitably involves Al migration into the region of olivine alteration.

However, considering that serpentine (Sp = H_4Mg_3-Si_2O_9 = serpentine-silicate) is a chlorite free of Al, antigoritization of olivine takes place without Al migration in the region of olivine alteration. In contradistinction, chloritization proper of the olivines also involves Al migration.

Considering the possibilities of Al in solution, it

should be pointed out that the aluminium remains dissolved both in acid solutions with pH less than 4 and in basic solutions with $pH > 9$. It is possible that solutions in basalts would provide a satisfactory pH milieu so that aluminium would remain in solution.

Microscopic observations (Fig. 494) show green chlorite formed as an alteration of olivine following cracks in the host. In contradistinction, green chlorite can be an alteration product of olivine following cracks in it and consisting of chloritic colloform banding with the minute fibrous phase developing perpendicular to the crack walls of the olivine (Fig. 495).

The green chlorite can also develop as a marginal alteration of the olivine (Fig. 496) or could entirely replace the olivine, with the chlorite aggregates showing a zonal pattern due to the advance of chlorification fronts within the olivine (Fig. 497).

In contradistinction to the green chlorite, a common alteration of the basaltic olivines to brown, strongly pleochroic chlorite often takes place. Figs. 498 and 499 show brown chlorite formation, both marginal and associated with a crack pattern of the basaltic olivine. It should be noted that the increase in the pleochroism of the chlorites is often interpreted as replacement of Mg by Fe and possibly of Al by Fe.

A microscopic study of the alteration front of olivine to chlorite is of particular petrogenetic significance. Figs. 500 and 501 show the alteration front of the olivine and particularly the migration of Al in the region of olivine alteration. Theoretically, Mg is expelled from this region of olivine alteration to chlorite (brown pleochroic). As the strong pleochroism indicates, Fe has also migrated in addition to Al in this alteration front region (Figs. 500 and 501).

Often brown chlorite invades, i.e. forms along a microfissure system of the olivine host (Fig. 502). Fig. 503 shows that in addition to the brown chlorite, limonite is also occupying the central part of the microfissures.

In contrast to the green or brown chlorite alterations, occasionally olivine alteration shows a brown-chlorite alteration front followed by green chlorite formation (Fig. 504). Other cases of olivine alteration into green chlorite and hornblende formation have been but rarely exhibited (Fig. 505).

As already mentioned, antigoritization is rather restricted in basaltic olivines; however, special cases of alteration of olivine mega-phenocrysts in picritic basalts show that both antigoritization and chloritization of olivines can take place concurrently within the same olivine crystal grain. Fig. 506 shows olivine megaphenocrysts transversed by antigorite veinlets as well as a circular cross-section of one. Similarly, Fig. 507 shows a cross-section of a composite veinlet consisting of green chlorite in the centre (radiating) and with radiating peripheral antigorite. In this particular case, this "veinlet" is associated within the intergranular olivine space, but veinlets of comparable composite character can also transverse the olivine phenocrysts, as Fig. 508 shows. Figs. 509 and 510 show composite chlorite–antigorite veinlets (a longitudinal section of them) extending from the margin into the olivine host. In rare cases neocrystallization of biotite can take place within the chlorite veinlets (Fig. 511).

Iddingsitization

In contrast to the chlorite formation as an alteration product of olivines, in some cases where Mg is expelled from the region of olivine alteration and when opposite migration of iron and, to a lesser extent, of Al (also traces of Ca) takes place towards the region of olivine alteration, iddingsitization occurs.

According to Sun (1957) iddingsite consists essentially of goethite, amorphous silica and magnesia. In contrast, Wilshire (1958) showed that iddingsite consists predominantly of smectite (monmorillonite group of minerals) and chlorite (usually together with goethite, also commonly with quartz and calcite and more rarely with talc and mica). Furthermore, according to Brown and Stephen (1959), despite its polycrystalline nature, iddingsite apparently shows optical homogeneity due to the orientation of the crystals of goethite throughout a single olivine grain and partly of those of the layer-lattice silicate, which inherit the oxygen framework of the original olivine.

On the other hand, Smith (1961) points out that there are three alteration products in the pseudomorphs after olivine in the Dunsapie Basalt, Edinburgh, which may be orientated so that saponite (100) // goethite $(0\bar{6}0)$ or $(0\overline{16})$ or haematite (1010); and saponite (001) ⊥ goethite (100) // haematite (0001).

In contrast to these explanations, which support an orientation of the sub-microscopic components of the polycrystalline iddingsite, Gay and Lemaitre (1961) consider that the formation of iddingsite is due to a continuous transformation from the solid state to a disordered irregular arrangement, not a single submicroscopic mineral intergrowth. Furthermore, the iddingsite is found to consist of "relict" olivine structure, goethite, haematite, magnetite and a poorly crystallized layer silicate.

In addition, microprobe studies of iddingsite transformation from olivine by Lemaitre et al. (1966) showed that the iron contained in the iddingsite has been introduced from the outside and that iddingsitization is a hydrothermal metasomatism of a lava transversed by volcanic gas and is not due to isochemical evolution.

As a corollary to the different explanations proposed, which fundamentally agree with Mg-leaching out and with a migration of essentially Fe and to a less extent of Al (and traces of Ca) towards the region of olivine alteration, are the present microscopic observations which present the phenomenology of iddingsitization in basalts.

Fig. 512 shows an idiomorphic olivine phenocryst with an iddingsite alteration margin. Similarly, Fig. 513 shows an olivine phenocryst with resorption "groundmass" embayments with iddingsitization margins, which are also formed around the groundmass "enclave" in the phenocryst.

In contrast to the iddingsitization margins, the alteration of olivine often results in complete iddingsite pseudomorphs after olivine and titano-augitic phenocrysts are replaced by iddingsite (Fig. 514). Often unaltered olivine relics are left within the iddingsite and, in some cases, limonite fills the original cracks of the olivine that are now occurring within the iddingsite pseudomorph (Figs. 515 and 516).

As is often mentioned, there seems to be an orientation between iddingsite and olivine host. Iddingsitization follows preferential solution-penetrability directions within the olivine. Particularly, Fig. 517 shows an iddingsite margin of an olivine with extensions orientated parallel within the olivine host.

As mentioned, Edwards (1938) has described olivine phenocrysts in basalts in which the olivine cores are surrounded by a fibrous iddingsite which, in turn, is surrounded by an unaltered overgrowth of olivine which is post-iddingsite in the crystallization sequence, thus supporting a late phase of olivine crystallization. Similarly, Sheppard (1962) has described olivine phenocrysts with iddingsite margins and olivine overgrowths which have the same composition as the groundmass olivines (which are more ferran than the olivine cores of the phenocrysts). As a corollary to these observations, which show that iddingsitization may be followed by a late olivine crystallization, are the following observations: Fig. 74 shows an olivine phenocryst, the core of which has been completely iddingsitized where an overgrowth of olivine has taken place (free of iddingsitization) and which attains idiomorphic shape; Fig. 75 also shows an olivine overgrowth which has enclosed a groundmass magnetite (no magnetites are present in the olivine cores). Here again, a post-iddingsite olivine overgrowth has occurred.

Occasionally, under extreme conditions of oxidation and due to iron migration which occurred with the complete depletion of the original olivine of Mg (Mg almost completely leached out), magnetite pseudomorphs can be formed after olivine (Fig. 518).

In some cases, a marginal chloritization of the olivines can take place, accompanied by an intracrystalline clouding of the phenocryst which could be attributed to the development of minute chlorite spots or dots within the olivine (Fig. 519). In contradistinction to the limited distribution of this cloudening effect, there are other cases where the entire olivine is changed by this process (Fig. 520).

In contrast to the alterations of basaltic olivines, as described so far, alteration of olivines in ultrabasic rocks show two important alterations which have not yet been observed on a comparable scale in basaltic olivines.

Often, olivine in ultrabasics is, in addition to antigorite, also changed to magnesite which can be seen as alteration margins of olivine which is surrounded by antigoritization (Fig. 521). The magnesite, which forms a network within the olivine grains, can give rise to magnesite "veinlets".

Schematically, the alteration of forsterite to magnesite involves the following processes (Augustithis, 1965): Weathering of olivines due to hydration – with the participation of carbonic acid (CO_2) of the air, results in the formation of magnesite pseudomorphs after olivine.

As already mentioned, among the polycrystalline mineral constituents which comprise the iddingsite, quartz has often been formed, which is to be understood as residual silica after the leaching out and depletion of Mg from the olivine.

This differential leaching process (Pieruccini's gradoseparation, 1961; Augustithis' birbiritization, 1965; Augustithis' differential leaching, 1967), it is of particular significance in ultrabasic rocks since through its operation dunites can change into birbirite (essentially consisting of chalcedony, micro-crystalline quartz, chromitic relics and limonite). The residual silica crystallizes, through a colloform phase, to chalcedony – and microcrystalline quartz. Fig. 522 shows a birbirite with the typical cellular texture representing the initial granular olivine texture which, by differential leaching and complete depletion of Mg, has been altered into a microcrystalline quartz aggregate and limonite.

Chapter 21 | Alteration of augites, plagioclases and basaltic groundmass

The augite phenocrysts in basalts often show alteration to green chlorite (Fig. 523). According to Machatscki (1953) "... die Grünerde oder Saladonit ist zu einer weichen feinstschuppig erdigen grünen Masse von chloritischer oder glaukonitischer Zusammensetzung umgewandelter Basaltischer Augit ..." – the green earth or saladonite is a soft fine-grained earthy green mass, chloritic or glauconitic in composition, of altered basaltic augites.

Of particular interest are the micro-geods within the augite phenocrysts, i.e. intracrystalline cavities filled with minute radiating aggregates of chlorite (Fig. 524). In some cases, cracks in the pyroxenes form feeding channels for the chloritic solutions which invaded the augites (Fig. 525).

In contrast to the phenomenology described, complete chloritization of the augites can take place, as is shown in Fig. 526.

Furthermore, and in contrast to chloritization, iddingsitization of the pyroxenes can take place. Fig. 514 shows complete iddingsitization of two adjacent pyroxene and olivine phenocrysts.

In addition to the pyroxenes, the plagioclases in basalts tend to be altered. In contrast to the sericitization and kaolinization of the granitic feldspars in the basaltic rocks, chloritization of the feldspars is the most predominant alteration process. Fig. 527 shows chloritization of basaltic tecoblast, in which case, the chlorite alteration is rather restricted in the central plagioclase zone. Also Fig. 528 shows chloritization of plagioclase phenocrysts in a completely chloritized basaltic groundmass. It should be pointed out that whereas the chloritization of the groundmass is complete, the plagioclase phenocrysts are maintained, despite intense alteration.

Comparable to the intracrystalline microcavities (microgeods) of the pyroxene, similar structures exist within plagioclase phenocrysts, with intracrystalline microcavities which are lined by fine chlorites; Figs. 529 and 530 show a chloritic microcavity lined with chlorite; also visible is the feeding channel of the microcavity. Similarly, Fig. 531 shows a plagioclase microcavity filled with radiating chlorite aggregate.

The plagioclase chloritization can show preferential directions of plagioclase alteration; Fig. 532 shows chloritization following polysynthetic twinned lamella of the plagioclase. A relic unaltered part of the plagioclase free of chloritization is also indicated.

Most impressive chloritization–glauconitization phenomena of plagioclase are shown in phenocrysts of sub-oceanic basaltic effusions. Fig. 533 shows a plagioclase phenocryst of sub-oceanic basalt with chloritization–glauconitization replacing the phenocryst; a reaction margin between the glauconite and the unaltered feldspar is often noticeable. Fig. 534 shows chloritization–glauconitization extending across and following the plagioclase twinning.

One of the most significant processes of the alteration of basaltic rocks is the alteration of the groundmass. Particularly, the mafic groundmass components are most susceptible to alteration processes and especially to chloritization, which is the ultimate phase of the alteration of basalts.

Fig. 535 shows a preferential alteration of mafic components of the basaltic groundmass into colloform chloritic aggregates which surround the phenocrysts and the basaltic groundmass feldspars which are themselves more or less unaltered at this stage of alteration. Fig. 536 shows orientated chloritic cavities within a picritic basalt. Radiating chloritic aggregates, initially colloform, are present in the groundmass of picritic basalt. Colloform chlorite aggregates are also common as alteration products of the basaltic groundmass; Fig. 537 shows colloform chlorite aggregates in basaltic groundmass. Occasionally magnetite crystal grains act as nuclei of the chloritic colloform structures (Fig. 538).

In some cases, alternating brown and green chlorites occur as space filling in a "cavity" of the basaltic groundmass. Occasionally, colloform chlorite structures are exhibited as groundmass space-fillings with feldspar laths perpendicular to the banded chlorites (Fig. 539).

Chapter 22 | Weathering and weathering forms of basalts (lateritic covers of basalts)

Due to iron mobilization under chemical weathering of basalts, diffusion rings of different types may develop. Fig. 540 shows diffusion rings of microscopic scale due to limonite concentration. In contrast, Singer and Navrot (1970) have described a basalt boulder, buried at depth of 20 cm (in a fossil latosol of the Negev) and which has altered, forming differently coloured concentric diffusion rings. Due to leaching out of elements and due to beidelitic smectite which is more abundant in the central part of the weathered boulder, a white core is produced with brown diffusion rings surrounding it which contain haematite. The following order of depletion of the major elements is proposed: Fe, Mg, Na, Ca, Si, K; moreover the minor elements were depleted in the order Mn, Zu, Co, Cu, Cr, Ni.

In contrast, the reddish-brown rings which are rich in iron, are attributed to a rise in the *Eh* due to fissuring of the rock. Comparable are the results of the alteration of the basaltic cover in Galillee (Singer and Navrot, 1970). Due to the alteration of basalts, large losses of Mg, Ca, Na, K and smaller losses of Si are reported, together with the disappearance of mafic minerals and the formation of smectite, helloysite and kaolinite. Comparable diffusion rings are described in microgranites by Augustithis and Ottemann (1966). However, it should be emphasized that in contrast to the reported leaching in basalts, enrichment of Al, Si, K, Zr, Y and Rb has been found in the leached microgranite (white rings) and enrichment of Ca and Fe in the brown rings of the diffusion-ring structure of the altered microgranite.

However, the mineralogical and geochemical changes accompanying the lateritization of basalts are very complex and they are not included in the scope of the present Atlas. Nevertheless, a series of illustrations is introduced showing the stage-by-stage alteration of basalts to laterite. Fig. 541 shows spheroidal weathering in basalts. The alteration is due to a front of solutions which advances from the joints (columnar and its perpendicular set) towards the centre; as a result of this alteration front spheroidal weathering occurs, with unaltered basalt as relics in the central part of the spheroids. Fig. 542 shows a structure of spheroidal weathering preserved in a more or less altered basalt.

Fig. 543 shows a joint system in basalt and the formation of weathering spheroids with central unaltered boulders of basalts. The chemical alteration of basalts is mainly due to alteration of basaltic mineral components and due to migration of elements, lateritic basalts, i.e. covers may be produced and thicknesses of 20–30 m are not rare. Fig. 544 illustrates an advanced phase of laterite formation with spheroidal relic structures and with a rounded basalt relic. Fig. 545 shows a typical red-soil lateritic cover of basalts which, in the particular case, is more than 20 m thick.

"Sonnenbrenner effects" – a physical alteration of basalts

"Sonnenbrenner" (gray spots and development of microfissures in basalt)

In contrast to the chemical weathering and disintegration of basalts is the "Sonnenbrenner" effect which is most probably caused by physical factors (Ernst and Drescher-Kaden, 1941). "Sonnenbrenner" develops mainly in olivine-basalts, nephelinites and nepheline basalts and is believed to be caused by the presence and distribution of analcime in basalts. If analcime occurs, scattered or in isolated crystal grains, the "Sonnenbrenner" develops, but if nests and zonal concentrations occur, there is a tendency for "Sonnenbrenner" development. Furthermore, it is believed that the analcime aggregates are weak spots in the rock because they are in an unbalanced state of tension. If this state of equilibrium is influenced by external factors, disintegration of the rock takes place. By quarrying and fracturing of the rock or by rock exposure, liberation of free energy may take place which influences the state of tension of the analcime which can in turn influence its coherence.

Fig. 546 shows "Sonnenbrenner" microfissures in basalt; a general view showing their distribution in basalt.

In contrast to this physical explanation of "Sonnenbrenner" effect, which is mainly attributed to the distribution of analcime and its state of tension, microscopic observations of "Sonnenbrenner" fissures, transversing the groundmass and phenocrysts of the basalt, show chemical alteration effects as the microfissures transverse the phenocrysts. Fig. 547 shows a "Sonnenbrenner" microfissure transversing an olivine phenocryst.

Alteration of the olivine marginal to the microfissure is noticeable (see arrow "a", Fig. 547). Comparable effects are noticeable in feldspar phenocrysts transversed by "Sonnenbrenner" microfissures. Arrow "a" in Fig. 548 shows an alteration effect (a shadow) in the feldspars marginal to the microfissure. Similarly, a limonitization effect is produced in the feldspar marginal to the "Sonnenbrenner" microfissure (Fig. 549).

In contrast, and in rare cases, the "Sonnenbrenner" microfissures may be partly occupied by iron oxides (Fig. 550). In such cases, it is dubious whether the chemical alteration margins adjacent to microfissures or the occupation of the microfissures by iron oxides or other material is of any genetic significance. In certain exceptional cases microfissures in basalts which transverse plagioclase sub-phenocrysts, may be occupied by chloritic material (Fig. 551). However, the typical "Sonnenbrenner" structure is an open-fissure, in cases exhibiting a "conhoidal pattern".

Chapter 23 | Cavity fillings (amygdaloidal fillings)

As already described, intracrystalline cavities of plagioclases and pyroxenes are often lined or occupied by chloritic minerals and often the channels of the infiltration are noticeable; in other instances cracks and the cleavage are solution channels (see Chapter 21). As a corollary to intracrystalline cavities are groundmass extensions, i.e. groundmass "embayments" within olivine phenocrysts, with the feeding channels (cracks of the olivines) clearly to be seen (Fig. 552). Occasionally, in these groundmass enclaves within the olivine phenocrysts, zeolites have also infiltrated, in addition to the groundmass components, possibly after the crystallization of the groundmass – which often is mainly lining the enclave walls (Figs. 553 and 554). Also, intracrystalline cavities in pyroxene phenocrysts are occasionally filled with glass (Fig. 181a) and in some cases, cracks are present as feeding channels (Fig. 181b). It should be noted that the intracrystalline microcavities are either due to solution of the host or they are spaces in the crystal growth. Often though, as the groundmass "embayments" show, they are due to magmatic resorption and infiltration. A rather impressive case of intracrystalline cavity filling is illustrated in Fig. 555 which shows zeolites occupying the greater part of a pyroxene crystal. A reaction margin is noticeable between the pyroxene and zeolites.

In contrast to the intracrystalline microcavities, there are the vesicular and gas cavities which are due to gas escape from the consolidating lava (see Chapter 3). In some cases, the cavities have walls consisting of prismatic pyroxenes perpendicular to the walls of the cavities or as separations of adjacent cavities (Fig. 556). In contradistinction to the cavities with prismatic pyroxene wall lining, there are the cavities which are occupied and filled by prismatic pyroxenes and feldspars (Fig. 557). In other instances, the gas cavities have prismatic pyroxenes perpendicular to the wall with quartz in the central part (Fig. 558). However, more common are the vesicular cavities with prismatic pyroxenes perpendicular to the walls of the cavities and filled with zeolites (Fig. 559). Occasionally, gas cavities with a lining of elongated prismatic pyroxenes perpendicular to the walls of the gas cavity are occupied by a semi-translucent substance which has small prismatic pyroxenes interspersed (Figs. 560 and 561); also within this substance there is a zeolitic mass as an enclave. In some cases which are in contradistinction to the prismatic pyroxenes lining the walls of the gas cavity, there are elongated groundmass components, mainly plagioclase laths, adjacent to the vesicular structure and partly forming its walls (see Fig. 562).

Gas cavities or irregular spaces in the basaltic groundmass are often occupied by monomineralic substances. Fig. 563 shows zeolites as irregular space fillings exhibiting an interlocking mosaic intergrowth structure; in other instances, the zeolites show a radiating textural pattern (Fig. 564 a and b).

Also belonging to the monomineralic gas-cavity fillings are calcitic amygdaloids (Figs. 565 and 566). Green chlorite surrounding individual groundmass components and showing colloform structural patterns, often occupies gas cavities in basalts (Figs. 567 and 568).

The most impressive basaltic groundmass gas cavities are filled with colloform or microcrystalline quartz which presents fascinating patterns, often illustrating all the transitions from colloform to crystalline quartz (Figs. 569–572).

In contrast to the monomineralic gas-cavity fillings are the polymineralic vesicular cavity fillings, consisting of rhythmical or interreacting mineral phases, often of contemporaneous or repeated supply. Fig. 573 shows a basaltic gas cavity with the wall feldspar laths surrounded by chlorite. The centre part of the gas cavity is occupied by calcite. In contrast, rhythmical chlorite–calcite structures are often exhibited in amygdaloids with alternating chlorite–calcite phases. The solutions which gave supply to the amygdaloidal structures were colloidal. Most impressive colloform structures and their subsequent transformation to crystalline is almost the rule within the amygdaloidal structures.

Fig. 574 shows colloform chlorite–calcite alternating structures with calcite occupying the central part of the cavity. Similarly, Figs. 575 and 576 show colloform calcite, followed by colloform chlorite and with a calcite central part with colloform chloritic spheroids. As these textural patterns show, there is an alternation of colloform calcite and colloform chlorite.

A common association of amygdaloids is quartz and calcite (Fig. 577). There is the tendency for calcite to crystallize without maintaining colloform relic structures. Figs. 578 and 579 show rhythmical colloform bands of quartz and calcite (i.e. initially colloform bands of silica structures now transformed into crystalline). A most impressive rhythmical calcitic-chalcedonic banding is illustrated in Fig. 572. A more common pattern is the intergrowth and reaction of quartz and calcite (Fig. 580).

In some cases, radiating prismatic quartz, which may attain idiomorphic shapes, occurs in association with calcite in basalt amygdaloids (Fig. 581a, b, c). Reaction between the quartz and the calcite has taken place and is shown in this case. Among the most common multimineralic amygdaloids are the chlorite–quartz associations. Fig. 582 shows chlorite marginal with a diffused transition to quartz which occupies the central part of the gas cavity. Often the quartz in the amygdaloids (despite its crystalline appearance even if it is exhibited by a granular structure) shows colloform relic patterns in the form of interspersed granules (Fig. 583).

Amygdaloids in basalts often have an external cover of a fine-grained (sub-microscopic) green mineral aggregate with the central part of the amygdaloidal structure consisting of crystalline zeolites (Fig. 584). In some cases, the margins of the amygdaloids consist of quartz followed by a diffuse green cover of the sub-microscopic substance and with zeolites occupying the central part of the amygdaloidal structure (Fig. 585). In addition to the minerals mentioned, it is occasionally found that plagioclase, in association with quartz, occurs as a vesicular mineral filling (Fig. 586). Also rarely, secondarily developen mica occurs associated with the zeolites. Other minerals such as epidote may occasionally occur in basaltic amygdaloids (Fig. 587). Lindgreen (1933) reports that in amygdaloids in basalts from Lake Superior, native copper occupies the central part of the amygdaloids (Fig. 588). Also known as amygdaloidal fillings are calcite, epidote and pumpellyite. In addition, Lindgreen mentions the economic significance of basaltic flows in the Houghton, Michigan region, which contain copper–silver-rich amygdaloids. The amygdaloids contain calcite, epidote and zeolites. Copper occurs in these but also replaces the rock itself. Some native silver, bordering copper with sharp contacts, also occurs in the amygdaloids.

Walker (1960) has reported zonal variation in the mineralogical composition of basaltic gas cavities (vesicules) in Tertiary basalt flows from eastern Iceland. In the Tertiary olivine basalts mesolite and scolecite are the predominant gas-cavity fillings in a lower zone more than 2500 ft thick. Analcime predominates in the following 500 ft and chabazite and thomsonite in the upper zone. In contrast to the above-mentioned vesicular mineral parageneses, are the Tertiary tholeiitic basalts which also show a zonal variation in the mineralogical composition of their gas-cavity fillings. A lower zone more than 1000 ft thick has vesicular fillings consisting of stibite, lenlandite, scolecite and epistibite accompanied by quartz and chalcedony. The gas-cavities of the next 1000 ft are filled with modernite, quartz, chalcedony, chlorophaeite and caledonite. Above this the vesicules are empty.

The variation in the mineralogical composition of the gas-cavity fillings of the olivine basalt and of the tholeiitic basalt is far from being accidental. It is possible that the gas-cavity fillings might bear some relation to the basalt in which they occur, though often the solutions are hydration-solutions of metasomatic derivation.

Chapter 24 | Metamorphism and diaphthoresis of basalts

Basalts can be dynamically metamorphosed and dynamically deformed; Fig. 14 shows a basaltic plug transversing basement granites of the east African basement, at Harrar-Ethiopia, which is tectonically fractured (Fig. 15). Similarly, Fig. 589 shows the dynamic metamorphism of the Borrodale basalts (Lake District, England) indicating that the basalt has attained the character and appearance of slate due to dynamic metamorphism.

In contrast to the dynamic effects, basalts can be subjected to metamorphic–metasomatic processes and due to blastogenic growths, new mineral phases may result. Fig. 590 shows an initial basalt changed by metamorphism into a rock predominately consisting of hornblende (from La Gomera, Canary Islands). As a result of metamorphism and diaphthoresis, an entirely new rock type has been formed with very few of the initial basaltic mineral components and textures preserved. The hornblende blastesis has played a predominant role; hornblende megablasts develop in a groundmass consisting mainly of chloritic micaceous mineral-aggregate, a diaphthoretic alteration of the initial basaltic groundmass (Fig. 591).

In some cases, aggregate relics of the initial basaltic plagioclase may be preserved within the chloritic micaceous groundmass, suggesting rather an incomplete diaphthoresis than an initial phase of metamorphism (Fig. 592). Apatite is present either within the diaphthoretically formed groundmass or within relics of the initial basaltic groundmass (Fig. 593). In contrast, most of the apatite shows a preferential textural association with the hornblende megablasts.

Apatites with rounded corroded outlines which are perhaps remnants of the initial basalt, may be found within the megablastic hornblende (Fig. 594). In some cases the hornblende megablast may have suffered diaphthoresis. This metamorphic alteration of the basalt is due to a diaphthoresis in which H_2O solutions have played a predominant role. In particular, the hornblende blastesis, parallel with the diaphthoretic processes (which have rendered the chlorite-micaceous aggregate), has played the most predominant role.

A rather interesting case of metamorphism is described by Harker (1950) under the chapter heading "Repeated Metamorphism – Mechanical Generation of Heat"; according to him "Basic dykes within the same belt of country are not only crushed and sheared but show a development of new minerals of metamorphism, such as epidote, actinolite and sphene". Also, an example at Scourie in Sutherland was described by Teal who traced the gradual transformation of the dolerite into a hornblende schist (plagioclase amphibolite). Teal pointed out that there may be a total reconstruction of the rock without the setting up of any parallel structure.

Chapter 25 | The petrogenetic significance of trace elements in basalts

The trace-element differences in basaltic rocks, are due to the small differences in the chemistry of the rocks of the basalt clan and to the fact that we have (in the case of the terrestrial basalts) a case of interpolation; in certain cases only, variations may occur which can be related to differences in derivation. On the whole, variations in trace-element content and changes in their ratios cannot be taken as distinct and exclusive criteria of the origin and depth of derivation of the basaltic material. In contrast, trace-element geochemical differences are most significant in comparing lunar and terrestrial basalts (a case of extrapolation), e.g., the Ti and lanthanides (see Chapter 34), which show distinct differences in abundance in terrestrial and lunar basalts.

Nevertheless, in certain selected cases, the trace-element geochemistry in terrestrial basalts has provided data of petrogenetic significance.

As a corollary to the evidence already introduced supporting assimilation as a process which greatly influences the composition of basaltic magma, the presence of certain trace elements in basalts also indicates assimilation. The presence and geochemical relations of Zr, which varies in alkali basalts (Borodin and Gladkikh, 1967) with the K + Na content and the (Si + Al)/(K + Na) ratio of the rock, is here interpreted as indicating crustal assimilation.

Of petrogenic significance in the basaltic rocks is the K/Rb ratio which shows great variations; from 3000 ppm in alkali-poor abyssal tholeiitic basalts to 100 ppm in K-rich leucitites. The average K/Rb ratio in alkali-rich basalts is about 300 ± 100 (Agiorgitis et al., 1970).

Of significance for the K/Rb ratio in basaltic rocks is its dependance above all on the minerals hornblende and biotite (Shaw, 1968). Hornblende and biotite can be blastogenic phases in basalts indicating assimilation of crust (see Chapters 15 and 32); consequently the K/Rb ratio in basalts will depend on the blastogenic hornblende–biotite phases and on the possible derivation of hornblende from the hornblende rocks of the lower crust or upper mantle.

Characteristic of the complexity of trace-element geochemical data is the concluding statement of Agiorgitis et al. (1970) in discussing the K/Rb, Ca/Sr and K/Sr relations in basaltic rocks of the East Alpine region. "Some basalts of the orogen border seem to be less alkalic. A higher contamination with sialic material may be possible in the region of the orogen. Basalts from the region of the orogen and from the Hungarian Plain show a strong tendency towards the formation of alkali-rich rock types. A relation to the thickness of the sial seems to be probable".

Basalts which indicate an increase in alkali, U, Th also support an assimilation of sial. However, in this connection, it should be pointed out that the formation of alkali basalts by assimilation of carbonate rocks, which would bring about the desilicification, is not supported by the geochemical studies of Wedepohl (1963, 1967) and Savelli and Wedepohl (1967), since the high contents of Sr, Ba, Lan, Zr, Cu, Se, V and Rb in alkali basalts could not be explained by the assimilation of carbonate rocks.

In contradistinction, by comparing the abundance of trace elements in tholeiitic abyssal basalts and in alkaline, Gast (1968) indicates that the high abundance of elements with large ionic radii in alkaline basalts cannot consistently be explained by fractional crystallization processes. The abundance of these trace elements can be explained by considering alkali basalts as a partial (3–7%) melting of accepted "upper mantle" mineral assemblages. The high K/Rb and low Ba/Sr and La/Yb ratios in abyssal basalts indicate partial melting of the upper mantle.

A comparable explanation regarding the depth of origin of basaltic magmas is based on the presence of Pb as traces. Masuda (1964) found that pigeonitic, hypersthenic and alkaline lavas, mainly from Izu Hakone region, have a characteristic field on the 20-Pt/^{208}Pb vs ^{207}Pb/^{208}Pb diagrams, which is interpreted as indicating differences in the original source of these three types. Furthermore, considering the variation of radiogenic Pb in the crust and in the mantle and the possibility that alkali olivine basalts, the high-alumina and the tholeiitic basalts, represent different zones of melting, then their comparative depths of derivation are given as follows: alkali olivine basalt (greatest depth), high-alumina basalt and the tholeiitic which is considered to represent the smallest depth (see Chapters 30, 31 and 32).

Table XII, by Taylor et al. (1969), shows a compari-

TABLE XII

COMPARISON OF MAJOR AND TRACE ELEMENTS BETWEEN ALKALI AND THOLEIITIC BASALTS, AND HIGH-Al BASALTS

	Alkali and tholeiitic basalt (%)	High-Al basalt (%)
SiO_2	49.2	51.7
Al_2O_3	15.7	16.9
FeO	11.2	10.4
MgO	8.7	6.5
CaO	10.8	11.0
Na_2O	2.3	3.1
K_2O	1.0	0.4
TiO_2	1.8	–
	(ppm)	(ppm)
Ni	120	25
Co	50	40
Cr	120	40
V	200	250
Se	40	40
Ni/Co	2.4	0.63
V/Ni	1.7	10.0

son of trace and major elements between alkali and tholeiitic basalts on the one hand and high-Al basalts on the other.

The trace-element comparisons of alkali and tholeiitic basalts with high-Al basalts indicate that Ni, Cr, Co are relatively more abundant in the first group and that the relatively smaller content of the high-Al basalts is probably due to crust assimilation, as a consequence of which there is a reduction in the abundance of Ni, Cr, Co (see Table XII).

Considering the Ni and Co atomic radii and that of Mg, the incorporation of Ni and Co in the forsterite substituting Mg is well to be understood. Furthermore, the fact that Ni, Cr, Co are abundant as traces in fragments (i.e. olivine bombs) of mantle derivation (see Tables XIII and XIV) supports the concept that the Ni, Co and Cr trace elements of basalts indicate mantle derivation (also see Chapter 33).

In contrast to the above interpretation, which supports a crust or mantle derivation for certain trace elements in basalts and which is in agreement with the theoretical interpretation of the derivation of basalts from the lower crust and fragments of it from the mantle (see Chapters 6 and 32), geochemical studies by other authors indicate considerably greater depths of basalt derivation (Masuda, 1966). The presence of La, Ca, Nd, Sm, Eu, Cd, Dy, Er and Yb in Japanese basalts was quantitatively determined by isotopic dilution methods. It is also suggested that high-alumina olivine basalt was a primary liquid-type material, the primary tholeiitic basalt was a primary solid-type material and the alkali-olivine basalt, which has been supposed primary, was a secondary liquid-type material. The original depths of these magmas were estimated from the relative enrichment of lanthanides as 200, 150 and 220 km respectively.

Undoubtedly, Ti is a trace element with petrogenetic significance, since Ti contents (high values) are a characteristic distinguishing lunar from terrestrial basalts (see Chapter 34). In addition, Ti contents are a characteristic distinguishing between circum-oceanic basalts (with less than 1.5% TiO_2) and intra-oceanic basalts (with relatively high Ti contents). On the basis of the sequence of basaltic flows, Noe Nygaard (1967) has concluded that the "intra-oceanic" basalts are from shallow levels below the oceanic crust and the "circum-oceanic" ones are from deeper levels.

Furthermore, in discussing the distribution of titanium between silicates and oxides in igneous rocks, Verhoogen (1962) points out that in igneous rocks more Ti should generally occur in oxides than in silicates if the partial pressure of oxygen is low. In known rocks, in

TABLE XIII
TRACE-ELEMENT CONTENT (ppm) IN THE FORSTERITE OF MANTLE XENOLITHS IN BASALTS (OLIVINE BOMBS IN BASALTS)

Trace-element determinations by A. Vgenopoulos

Locality	Ni	Cr	Co	Cu	Zn	Mn	Ti	Zr	Sr	Ba
Lekempti, W. Ethiopia	2410	145	54	52	202	2020	215	45	–	100
Lanzarote, Canary Islands	2820	95	42	55	220	960	180	25	40	85
La Gomera, Canary Islands	3020	1000	29	41	197	3420	205	45	197	145
Gerona, Spain	2710	168	94	42	250	2090	435	75	–	145
Red Sea Islands (offshore Assab)	2420	2770	106	58	250	2080	470	85	–	145
Eifel, W. Germany	2440	140	28	53	142	1990	620	62	–	80
Olivine phenocryst picrite basalt, Cyprus	1150	–	35	28	232	1220	915	50	610	–

TABLE XIV

TRACE-ELEMENT CONTENT (ppm) IN FORSTERITES, PYROXENES AND SPINELS OF MANTLE FRAGMENTS IN BASALTS
Trace element determination by A. Vgenopoulos

Locality	Mineral	Ni	Cr	Co	Cu	Zn	Mn	Ti	Zr	Sr	Ba
Lekempti, W. Ethiopia	forsterite	2410	145	54	52	202	2020	215	45	–	100
	pyroxene (bronzite)	265	2610	7	17	115	1100	780	9	–	850
	Spinel	80	27935	–	32	180	915	620	180	–	–
Canary Islands, Lanzarote	forsterite	2820	95	42	55	220	960	180	25	40	85
	pyroxene	1250	3400	85	56	225	450	345	–	–	175
	spinel (chromite)	2200	9880	99	58	220	140	465	250	565	312

which the opposite is true, higher temperatures of crystallization and/or low silica and iron oxides explain the difference. The distribution of Ti in basaltic titanoaugites is discussed in Chapter 34.

Chapter 26 | The geological spiral and the significance of basalts (The history of the earth starting from a "basaltic initial crust")

In extrapolation with the moon's crust the initial star crust of the earth may have been peridotitic–gabbroic in composition and comparable to the gabbroic rocks of the moon, i.e. the initial crust of the earth was proto-basaltic (anchi-basaltic in nature).

The formation of atmosphere–hydrosphere around the earth's lithosphere can be seen as the commencement of the geological history of the earth, the beginning of geomorphological and geological cycles and the concurrent exogenetic and endogenetic operation of processes.

However, the formation of the earth's atmosphere is intimately related to the formation of the earth (Mason, 1952): "Our ideas as to the composition of the primeval atmosphere are conditioned largely by the mode of origin that we ascribe to the earth. On the planetesimal hypothesis, the particles aggregating to form the earth had no atmosphere associated or combined with them; the primeval atmosphere originated from the gases occluded or combined within the planetisimals and released by heat and chemical reactions accompanying and following aggregation. There seems little question now that the constituents of the atmosphere have been largely if not entirely exhaled from the interior of the earth".

Considering the gradoseparation hypothesis of Pieruccini (1961) whereby the femic components in femisialic compounds (e.g. olivines, pyroxenes) are solved out and the sialic are residual, and certain aspects of the differential leaching of elements in which element-leaching "hydration" is considered to be a secondary process often involving mineral and rock disintegration, the release of elements from a crystal lattice gives rise to crystal changes which may be noticeable as changes in the physical properties and composition. With progressive element-leaching, more pronounced alteration takes place and, as a result, new minerals and textures may develop. Element-leaching is essentially a chemical process, but only a postulation of the reactions is possible, based on mineralogical and chemical comparison of the mineral phases involved. In addition, it is a chemical differentiation, with preferential mobilization of some elements over others from a mineral or rock. No "rules" or "laws" seem, so far, to explain overall differential leachings of elements; however, the "kinetic" potential of the element from a crystal lattice and valency seem to play a role (Pieruccini, 1961; Augustithis, 1964, 1965, 1967; and Augustithis et al., 1974).

Indeed, an impressive case of differential leaching of elements is the transformation of dunite into birbirite, a compact rock essentially consisting of colloform chalcedony, limonite accumulations and residual chromites. The rock is an example of Mg leaching out due to differential leaching and as a result we have the transmutation of an ultrabasic rock into an acid rock. Considering the aforementioned processes, it is most likely that due to chemical weathering, grado-separation of elements, differential leaching and geochemical mobilization of elements, the initial peridotitic–gabbroic crust of the earth produced acid and intermediate residuals. The total sum of these processes can result in the formation of acid to intermediate anchi-sediments (proto-sediments).

In accordance with the principle of "uniformitarianism", the anchi-sediments would be subjected to the processes of the geological cycle: denudation, aggradation, metamorphism, ultrametamorphism (granitization, gabbroization, ultrabasic-formation). Parallel with these processes, mantle fusion and diapirism took place in geotectonically mobile zones.

The evolution of the earth's upper crust, under the concurrent exogenetic and endogenetic influences, deviates considerably from the picture which would be given if the principle of "uniformitarianism" were applied, since we are forced to accept the existence of an initial gabbroic–peridotitic crust, something very different from the present configuration of the earth's surface; we have to go back to a picture as monotonous and as different as the moon's surface.

The celebrated principle of Huttonian geology, the principle of "uniformitarianism", on which geology has been based for the last two centuries, in my opinion, no longer holds good in the sense that there is no cyclical repetition of processes, but the unfolding of a spiral.

It should be added that the repetition of processes, as defined in the concept of the geological cycle, should be seen as part of an unfolding spiral, in the sense of successive diverse phases.

In the beginning there was the peridotitic–gabbroic initial crust, which gave rise to the anchi-sediments, which, by the operation of simple, multiple and multifold geological cycles, including the processes of metamorphism, ultrametamorphism, granitization and anatexis through successive orogeneses, built up the basements of the continents.

With the further unfolding of the spiral and due to the "indefinite" repetition of processes in time, a phase of evolution of the earth's upper crust might be produced, which might be unpredictably different from the present one; as different as, say, the surface of the moon is from that of the earth. Apart from the initial crust of the earth which may have been gabbroic–peridotitic in composition, basalts might have been formed by impact or fusion on the initial star crust of the earth. Both the initial peridotitic–gabbroic crust – no vestiges of which are left – and the proto-basaltic flows which might have existed are actually postulated in extrapolation with the moon's crust.

In contrast to these postulated basalts, basaltic volcanism interconnected with geodynamic events in the geological history of the earth, is often associated with epeirogenic and orogenic movements. Of particular significance are the rift systems transversing cratogens and oceanic floors. Out of these fissure systems huge quantities of basalts have been poured out.

The basaltic flows are of variable geological age *. The palaeo-basalts (Precambrian to Mesozoic) are mainly represented by melaphyres, basanites, diabases and metabasalts. Both the relatively quick chemical weathering and alteration susceptibility of the basaltic rocks in general, as well as the fact that they have often been subjected to metamorphism, have as a result the relatively restricted abundance of palaeobasalts, particularly in comparison to the younger basalt volcanism. It is also possible that the younger and recent basalt volcanism is the result of a more active period of the history of the earth's crust.

The continental drift, which started with the Mesozoic, and the fact that some of the biggest basaltic outflows date from the Jurassic (e.g. Parana basalts, South America) or commenced in Jurassic time and continued into the Tertiary (e.g. Mongolian–Siberian, East African, Arabian, Drakensberg Natal basalts) or up to present, support a relation between crust–mantle activity and the younger, huge basaltic outflows.

Parallel to the unfolding of the geological spiral, there have been major geochemical mobilizations of elements. The main rock-forming minerals characteristic of the initial star crust have been the source of the formation of new assemblages, by the unfolding of the geological spiral. These geochemical changes involved the major elements of the most abundant rock-forming minerals as well as the redistribution of the trace elements.

Considering the trace elements in basalts, we have seen that their derivation depends partly on the earth's mantle and partly on the earth's crust (see Chapter 32). It is therefore clear that the basalts cannot be taken as a model for understanding the abundance and distribution of the trace elements in the initial earth's crust since the trace elements of basalts are partly derivations of the mantle and partly derivations of the crust which is itself a derivation of the initial earth's crust through the unfolding of the geological spiral.

* In contradistinction to the basaltic flows, the Archean green rocks represent Precambrian ultramafic flows.

Chapter 27 | On the classification and definition of basalts

(PLATEAU BASALTS, OLIVINE BASALTS, ALKALINE-OLIVINE BASALTS, THEOLEIITES, ALUMINOUS BASALTS, ETC.)

A classification of the basaltic rocks should take into consideration the narrow limits of variation of the composition from one type to another. An attempt is made here to present and discuss the most important basalt classifications which have been proposed.

The term basalt is used by Pliny and is considered to be a derivative of the Ethiopian word "basal" used to describe the black volcanic rocks of Ethiopia. Agricola (1530) also used the term basalt to describe a fels-type of rock from Saxony. However, a petrographic-microscopic study of the basaltic rocks is first presented in Rosenbusch's "Elements der Gesteinslehre" (1898). In addition to olivine augite and opaque ores, which are almost always present, some basalts contain plagioclase, others nepheline, leucite or melilith, so we distinguish plagioclase or feldspar basalt, nepheline-basalt, leucite-basalt and melilith-basalt. If we have the co-existence of plagioclase with nepheline or with leucite the rock type is distinguished as "basanite". The terms "Melaphyr" and "Diabas" have been introduced by Alex Brongniert and correspond to labradorite–porphyrite and to dolerite, respectively.

In addition to the mineralogical classification of basalts, Rosenbusch describes geological time of formation as a criterion of basalt classification. "Basalts are called the Tertiary and Recent rocks of this family, melaphyre the rocks older then Tertiary usually the Carboniferous or permian and diabase the old palaeozoics." The age differentiation between metaphyre and diabase is tending to disappear and there is a tendency to speak about a meso-diabase, etc.

As another criterion Rosenbusch uses "textures", e.g., porphyric, holocrystalline, hypocrystalline, vitrophyric and hyalo-basalts. These textures are characteristic of the conditions of crystallization and consolidation of the basalts. Furthermore, in his "Elemente der Gesteinslehre", Rosenbusch uses the presence of characteristic minerals to distinguish certain basalt types, e.g. hornblende, graphite, iron (magnetite), quartz and olivine (picrite or oceanite) basalts.

It should be noted that special reference is made to the term "tholeiitiobasalt". "The intersertal diabasic basalts (dolerites and anamesites) correspond to the tholeiitic and olivine–tholeiite type of the melaphyre, which occurs as intrusive in the upper Cuseler, Lebacher and tholeier beds of the Saar–Nahe region and Pfalz".

Another definition of a basaltic type which can be considered to be of importance in the development of our knowledge is that of a plateau basalt by Washington (1922). According to him lavas forming the basaltic plateaux, basalt trap-series, are the "plateau basalts" with a small total range in variation, and they are characterised by absence of phenocrysts, general absence of olivine, the presence of a pyroxene of pigeonite-type, and a low content of MgO. Several of the characters reflect the conditions of uprise and outpouring of the magma, rather than distinctiveness of magma type. The plateau basalts, because of their geographical distribution and composition, are believed to represent magma from the universal basaltic substratum. Clearly Washington introduces "magma derivation" as a new criterion for his definition of "plateau basalt".

Kennedy (1933) can be considered responsible for a basic theoretical step forward in the definition and understanding of basaltic rocks. He distinguished between the olivine basalt and the tholeiitic magma types and has related their derivation from a basaltic substratum dual in character. The recognition of these two main basalt types by Kennedy has formed the basis for further developments in the study of basalts.

Parallel to this development in thinking, Bowen (1914) introduced physico-chemistry and with that, experimental mineralogy as an approach to the investigation of silicate melts and their solidus phase. This undoubtedly marked a new era in petrological thinking and has greatly influenced subsequent thought. In his "Evolution of Igneous Rocks", Bowen (1928) introduced the hypothesis of magmatic differentiation which he considered as the mechanism of the genesis of rock series, starting from a parental basaltic magma (see Chapters 30 and

31). Most of the modern classifications are based on the study of silicate melts and on the differentiation hypothesis of Bowen.

Based on the Bowen hypothesis of differentiation and on silicate melt-studies, Yoder and Tilley (1962) classified basalts with Q in the C.I.P.W. norm as tholeiites, those with Ol and Hy are called olivine tholeiite and those with Ne are called alkali basalts (Fig. 595a, b).

Again based on the differentiation hypothesis and silicate melt studies, Poldervaart (1964) divided alkali basalts into (a) Ol and Hy normative group and (b) Ne normative group (both having no reaction relation between olivine and pyroxene). Despite the fact that there is a basic difference between Yoder and Tilley's classification and that of Poldervaart, both conform to the broad concept of "olivine and tholeiitic basalt" of Kennedy.

A more elaborate classification of basalts was introduced by Kuno (1968). He follows Kennedy in recognizing two main basalt types, the alkali olivine and the tholeiite, which correspond to the olivine and to tholeiite of Kennedy, respectively. Kuno recognizes a transitional between the two, the high-alumina basalt with an Al_2O_3 higher than 16.5% in aphyric rocks and the $Na_2O + K_2O$ contents lying between those of the other two basalt types for a given SiO_2 content.

Schematically, the three-basalt types of Kuno can be distinguished by plotting their $Na_2O + K_2O$ contents against SiO_2 (see Fig. 596).

Also using the content of soda plus potash in relation to the silica content, Middlemost (1972) recognizes the alkali-olivine basalts and the tholeiites; furthermore, by taking into consideration the Al_2O_3 content, he recognizes the following types of basalts: (1) tholeiitic basalt; (2) high-alumina basalt; (3) alkali basalt; (4) trachybasalt; (5) hawaiite and (6) leucitite.

Apart from these mineralogical and chemical classifications, Barth (1952) introduced a broad classification of basaltic rocks based on their geotectonic setting and petrographical province concept. * The idea of a petrographical province was first introduced by Judd (1866), in which the members making up a province are believed to be co-magmatic or consanguinious, implying that they are all derived from a hypothetical common magma.

Barth recognized the following main groups of basalts:

(a) The oceanic basalts which are subdivided into the Central Pacific province, the Mid-Atlantic Ridge and the Indian Ocean province.

(b) The basalts of the continents and their shelves, which are further divided into the non-orogenic (epeirogenic) and the orogenic groups. The non-orogenic basalts are, in turn, subdivided into the plateau basalts and the shield and multiple-vent basalts. In contrast, the orogenic basaltic rocks comprise the olivine group of ophiolites.

In addition to the concept of petrographical province, and the fact that a petrographic province can contain a differentiated series of related rocks, Barth introduced in his classification the relation of the basalt types to the main geotectonic features of the earth, i.e. cratogenic blocks (epeirogenic tectonics) rift systems, oceanic ridges and orogenic belts.

The classification of basalts is interrelated with the evolutionary trends of petrological thinking and, in fact, developments in basalt have greatly influenced the formation of broad geological concepts.

At the beginning of the century, an attempt was made to classify basalts on the basis of microscopic investigation – Rosenbusch's microscopic era in petrographical science, as a consequence of which we have the mineralogical and textural concept of the classification of basalts. Interconnected with the mineralogy and textures, geological age has also been used as a criterion for distinguishing the basalts.

In the twenties and thirties, Washington and Kennedy introduced and elaborated the idea of plateau basalts and the derivation of basalts from a basaltic substratum. This concept was developed into the dual character of sima. Kennedy introduced the fundamental concept of dividing basalts into the tholeiite and olivine basalt groups and explained their derivation from an olivine-poor and olivine-rich layering of the sima. Thus, Kennedy emphasized the importance of the subcrust sima and opened the way for the derivations of basalts from the earth's lower crust–mantle. The concept that basaltic magma exists as isolated pockets within the crust or below it eventually died out.

Also starting in the early twenties, "experimental mineralogy", based on progress in the physico-chemistry of the silicate melts, provided the theoretical background for a chemical classification of the basaltic rocks with an attempt to bridge together pure experimental mineralogical models with natural basalt occurrences. We thus entered the "chemical analysis" era of the basalt problem.

Progress in geophysics and geotectonics and the expansive knowledge due to world-wide geological exploration both on land and under the oceans, and indeed particularly the advancement in the exploration of the oceanic floors, revealed the wide distribution of basalts and their interrelation with the main geotectonic

* Recently Miyashiro (1975) has emphasized the significance of volcanic-rock series and tectonic setting.

structures of the earth. The mantle project and the plate theories, which reflect the advancements in geophysics, geodesy and geotectonics, have enlarged the basalt concept and have shown its importance. We thus have entered the "lower crust–mantle era" of the basalt problem. The derivation of the basaltic melts by fusion of the lower gabbroic crust is the mechanism and the source of the basaltic magma derivation.

Considering the evolutionary history of our knowledge and classification of basalts, it is clear that whereas we are in the "lower crust–mantle era" of our basalt problem, we maintain as a classification system the analytical approach and the relics of mantle (the olivine bombs in basalts) are acknowledged as the only evidence of the participation of mantle as a parental source of basalt derivation.

Olivine bombs and xenocrysts do not represent early crystallizations of the basaltic magma consolidation. Besides mantle participation, assimilation is another equally important factor which has been disregarded or greatly underestimated by chemical classifications. Assimilation has played a far more important role in the formation of the basaltic series and in the basalts themselves than is widely accepted.

A study of the basalts and of the basaltic trends should be primarily based on the idea that basalts represent a fused and extruded portion of the lower crust and mantle which, in ascending through the upper crust, has been not merely contaminated but seriously changed, so that the final basaltic products consist in part of the mantle and of the crust. Therefore, it might be useful that, in addition to the analytical approach, an understanding of the basalts should be based on the textural patterns, with an attempt to understand the presence and the degree of preservation of mantle relics and assimilation components (see Chapter 6), on the development of the crystal phases as often exhibited by their zonal growth, the sequence of crystallization, the reactions and intergrowths of the mineral phases, the late-crystallizations and on all the peculiarities and evidence that can be helpful in deciphering the crystallization and origin of the basaltic melts. The present atlas on the textural patterns of basalts and their genetic significance is a preliminary attempt to that goal.

Chapter 28 | Discussion of the Bowen-Niggli physico-chemical principles of basaltic magma crystallization

Perhaps one of the books that has influenced petrographical thinking the most is "The Evolution of the Igneous Rocks" by Bowen (1928). Despite the fact that half a century has elapsed since its publication, its principles and fundamental thoughts still govern today's petrological thinking. The principles: continuous reaction series; discontinuous reaction series; fractional crystallization and magmatic differentiation, which were assumed to be true beyond doubt and unquestionable, are used not only as a basis for further thinking, but also deductions based on them are pushed beyond ultimate consequences to the extent that there is no correspondence of observation with deduction.

Taking into consideration the historical value of Bowen's work and the fact, however, that today its principles are extensively used without any inhibitions, one feels that it is necessary to reconsider and re-evaluate Bowen's principles as stated in Chapters V (the reaction principle) and VI (the fractional crystallization of basaltic magma) of his book.

Furthermore, this reconsideration is necessary because, since the publication of "The Evolution of the Igneous Rocks", a great deal more has been learned about the subject and also new research methods and techniques (e.g., microprobe analysis, high-voltage transmission electron microscopy) have been developed and these can throw light on many of the problems connected with the subject.

Furthermore, the conquest of the moon (1968) has provided another ground for research and for testing the correctness of some of Bowen's principles, e.g., particularly that of magmatic differentiation.

A discussion, therefore, of some of the fundamental principles of the Bowen hypothesis, in the form of a point-by-point critical reconsideration, based on his original text, would seem to be useful.

". . . The change of composition of the crystals is a perfectly continuous one, taking place by infinitesimal increments. Such a solid solution series may therefore be termed a continuous reaction series . . ."

". . . The continuous reaction series of the plagioclases is the best-understood series of rock minerals . . ."

Here again, one feels that this statement by Bowen was indeed written half a century ago. There is serious doubt whether an intermediate composition plagioclase is, in reality, a single crystal intermediate in the series albite–anorthite in the sense of Bowen. In this connection it is interesting to quote H. Eckermann (1947): "Although the system anorthite–albite is generally considered a perfect example of an isomorphous series, the X-ray data do not verify a corresponding gradual change of the size of the unit cell, but indicate a sudden jump in size at an intermediate point, suggesting two converging solid solution series, anorthite respectively albite being the solvent".

Increasingly more light is thrown on the plagioclase problem by the application of high-voltage transmission-electron microscopy, which permits the study of the plagioclases in Unit Angstrom size. A series of available publications; Laves et al. (1965), Olsen (1974), Nissen (1974), McLaren (1974) and Hashimoto et al. (1961, 1974, 1975) indicates a sub-structure consisting of microlamellae, either of ex-solved phases or micro-twinning.

Perhaps more work should be done in the field of fine zoning in plagioclases. Particularly the fine oscillatory zoning of basaltic plagioclases and indeed the fact that such a fine zone in turn consists of sub-microscopic fine zones, which give a middle-optical value of anorthite content under the universal stage, suggests that what appears to be a plagioclase "zone", "intermediate" in composition; in reality shows a sub-structure of still finer zonings (Augustithis, 1963).

The pyroxenes provide another group of minerals which has been extensively studied which often shows micro-substructure of ex-solutions, twinning and zoning (Chapters 8 and 34). With the increasing application of modern techniques and research, our basic concept of what is believed to be a single-crystal phase or an intermediate-crystal phase of an isomorphic series, has fundamentally changed.

As a derivative of the continuous-series hypothesis, the following statement by Bowen can be considered:

"In the crystallization of the plagioclase feldspars, when accomplished by simple cooling, a plagioclase always separates before any other plagioclase that is less calcic".

Here again there are many exceptions and in many

cases there is no correspondence between Bowen's deductions and more recent observations. Considering the zones of plagioclase feldspars as distinct individuals, it is clear that plagioclases with oscillatory zonings (rhythmical repetition of anorthite-rich and anorthite-poor zones) are in contradiction with the Bowen's above statement, since less-calcic plagioclases will crystallize before zones richer in anorthite content (see Figs. 258 and 259).

Another instance is the tecoblastic growths * of plagioclase phenocrysts, which show either reversed zoning or zones with irregular anorthite contents. In some cases, zones with higher anorthite content crystallize later than less calcic zones (see Figs. 253a and b).

In certain exceptional cases, the anorthite content of the basaltic feldspars can be strongly influenced by the assimilation of pyroxenes. As Figs. 230 and 231 show, due to pyroxene assimilation, the zone I (An 38–39%) is richer in Ca than zone III (An 31%) and similarly zone IV (An 34%) is richer in Ca than zone III due to reverse zoning. In contradistinction, zone II has the highest An% content, due to pyroxene assimilation.

A further basic principle of Bowen's hypothesis is that of the *discontinuous reaction series*: "Forsterite and clino-enstatite may therefore be styled as a reaction pair. By this is meant that crystals of the first compound react with the liquid to produce the second during the normal course of crystallization. A reaction of this latter type may exist between three or more compounds and the compounds, arranged in proper order, may then be said to constitute a discontinuous reaction series . . ." "The detection of the discontinuous reaction series is not always so easy, and the element of judgement enters to some extent. It has been stated above that pyroxene, amphibole and mica constitute a discontinuous reaction series, the conclusion being based on the fact that in certain rocks they show the corona relation".

Considering the discontinuous reaction series pyroxene → amphibole → mica, as their crystallo-chemical formulae show, are of commensurable composition:

$\frac{1}{\alpha} R^{[8]}R^{[6]}(SiO_3)_2$
(pyroxene)

$\rightarrow \frac{1}{\alpha} (OH,F)_2 R_2^{[8]} R_5^{[6]}(SiO_8O_{22})$
(amphibole)

$\rightarrow \frac{2}{\alpha}, K, Na, Ca(OH,F)_2\ R_2^{[6]}(Si_4O_{10})$
(mica)

The reaction can be seen as oxidation–hydration (with K mobilization in the case of mica formation), with mobilization of the R cations. ** That this reaction series is a progressive oxidation–hydration process is seen by its absence in lunar material. Oxidation–hydration is therefore responsible for the formation of the series, the reverse reaction: mica → amphibole → pyroxene is possible under metamorphism. The alteration of pyroxene (mainly augite to hornblende) and the amphibole subsequently to biotite is to be understood, on the basis of their crystallochemical formulae, as a chain of processes due to increasing oxidation–hydration.

The role of forsterite as a member of the discontinuous series and its petrogenetic significance as an early crystallization phase is emphasized in the work of Bowen: ". . . forsterite, however small an amount, always separates first. Not only does forsterite begin to crystallize early but it also ceases to crystallize early, an impossible condition in any eutectic system".

Bowen emphasized the reaction step olivine → pyroxene by pointing out that early crystallized olivine is often surrounded by later-formed pyroxenes, corona reaction-structures have commonly been explained as evidence of the discontinuous reaction series.

In contradistinction to Bowen's explanation regarding the formation of corona structures surrounding olivines, Deer et al. (1962) present the following elaborate account on the olivine corona structures, mainly taking into account the views prevailing in the nineteen-fifties:

"Corona structure. In some metamorphosed basic igneous rocks the olivine is mantled by pyroxene and amphibole. These rims have been variously described as coronas, kelyphitic borders, reaction rims and corrosion mantles. The coronas usually consist of orthopyroxene, diopsidic augite, amphibole, spinel or garnet, these minerals generally are not all developed in an individual corona and the two common sequences (Shand, 1945) are: olivine–orthopyroxene–(amphibole + spinel)–plagioclase and olivine–orthopyroxene–garnet–plagioclase."

Other variants have been described, but the rim adjacent to the olivine always consists of a pyroxene or amphibole, and where an outer rim is also present it consists of amphibole, or amphibole and a green spinel, usually as a symplectic intergrowth. In coronas in which garnet is present the garnet, or garnet–amphibole intergrowth, is adjacent to the plagioclase. Murthy (1958) has described coronas developed around olivines ranging in composition from Fo_{96} to Fo_{26}.

The formation of coronas has been ascribed to both thermal (Huang and Merrit, 1954; Friedman, 1955) and regional (Gjelsvik, 1952) metamorphism, as well as to

* Also the very existence of tecoblasts, i.e. An-poor phenocrysts in basalts with anorthite-rich groundmass feldspars, is in contradiction to Bowen's hypothesis.
** R = divalent or trivalent cations: Ca, Mg, Fe^{II}, Fe^{III}, Al^{III}.

late magmatic reactions (Herz, 1951). The amphibole of some coronas has been identified as cummingtonite, which together with garnet is a typical metamorphic association, and it is evident from the field relationships of some rocks in which the olivine is mantled by coronas that they have been metamorphosed. Shand (1945) considered that the formation of coronas is due to the instability of olivine and its conversion to orthopyroxene, a reaction which is accompanied by the expulsion of Mg and Fe^{2+} ions in the pressure–temperature environment of thermal metamorphism. The inner corona of orthopyroxene developed by the peripheral conversion of the olivine is related to the high degree of disorder and chemical potential along the original interface of olivine and plagioclase, and to the action of water (Murthy, 1958). During the conversion two oxygens are released from the SiO_4 tetrahedra of the olivine and are changed to hydroxyls:

$$(Mg,Fe)_2SiO_4 + H_2O$$

$$\rightarrow (Mg,Fe)SiO_3 + (Mg^{2+},Fe^{2+}) + 2\,(OH)$$

The formation of the outer rim of amphibole (+ spinel) and/or garnet takes place by the replacement of the plagioclase adjacent to the pyroxene rim, the Mg^{2+} and Fe^{2+} ions essential to their composition being derived from the olivine during the formation of the orthopyroxene corona. Although these reactions take place in an open system, the following equations illustrate some of the possible relationships of corona formation:

$$\underset{\text{olivine}}{(Mg,Fe)_2SiO_4} + \underset{\text{anorthite}}{CaAl_2Si_2O_8} \rightarrow \underset{\text{garnet}}{Ca(Mg,Fe)_2Al_2Si_3O_{12}}$$

$$2\,\underset{\text{olivine}}{(Mg,Fe)_2SiO_4} + \underset{\text{anorthite}}{CaAl_2Si_2O_8}$$

$$\rightarrow \underset{\text{diopside}}{Ca(Mg,Fe)Si_2O_6} + 2\,\underset{\text{orthopyroxene}}{(Mg,Fe)SiO_3} + \underset{\text{spinel}}{(Mg,Fe)Al_2O_4}$$

The typical corona succession, olivine, orthopyroxene, amphibole (cummingtonite), plagioclase, has been related by Ellis (1946) to the different response of these minerals to directed stress, and the olivine, orthopyroxene, amphibole and plagioclase structures are considered to offer increasing resistance to the effects of shearing stress.

Osborn (1949) has described in detail the products of reaction between magnesium-rich olivine and plagioclase An_{85} developed during the conversion of the New Glasgow, Quebec, troctolite to coronite. In the troctolite the olivine is surrounded by an inner rim of enstatite showing a conspicuous preferred orientation and an outer rim of fine-grained diopside–spinel symplectite. In the originally olivine-poor variety of the troctolite, the plagioclase is surrounded by a zone of diopside which grades outwards to a diopside–spinel symplectite. The enstatite rims around olivine are commonly continuous and uniform in width and are considered to have formed largely from the orthosilicate. Spinel is rare in the orthopyroxene rims, and the diopside–spinel symplectites are believed to have replaced the plagioclase. In the synthetic diopside–forsterite–anorthite system, the association of clinopyroxene and spinel is incompatible above solidus temperatures. Osborn and Tait (1952), however, have suggested that in the dry system at temperatures below approximately 900°C the crystallization fields of spinel and diopside probably adjoin, and that the fields of anorthite and forsterite cease to be contiguous. Thus at temperatures below 900°C it is possible that the high-temperature assemblage, calcium-rich plagioclase and magnesium-rich olivine, is not in equilibrium and may be transformed to one or other of the stable assemblages pyroxene, spinel and plagioclase, or pyroxene, spinel and olivine.

Kelyphitic coronas round fresh olivine, the inner rim of which consists of an optically positive chlorite, probably penninite, and an outer rim of tremolite adjacent to plagioclase, have been described from the St. Stephen, New Brunswick, norite by Dunham (1950).

In contrast to the assumption that the olivine present in basalts represent early crystallization, rounded olivine phenocrysts in basalts are considered as "refractory", resistant mantle relics in the basaltic melt. Their rounded form, undulating extinction and deformation-twinning suggests mantle xenocrysts (see Chapter 4). Resorption of olivine by later-formed pyroxene is possible, however, due to their refractory nature, the early-formed olivines are often preserved in the magmatic environment. Of course, a great part of the idiomorphic olivine in basalts is undoubtedly an early phase of crystallization, due to the preservation of the eucrystalline forms, and this indicates that no olivine resorption has taken place, as the olivine–pyroxene reaction would lead us to assume.

Bowen's statement that olivine crystallized first and ceased to crystallize early is in contradiction to the studies of Evžen Stanik (1970) who, on chemical and textural basis, has identified four generations of olivines in some basalts from Bobrna; pyroxene and groundmass are considered to represent inclusions in some of the subidiomorphic olivine phenocrysts of the third generation. These observations are clearly in contrast to Bowen's deductions that olivine crystallizes first and ceases to crystallize early.

An additional evidence of the late growth of olivine is the post-iddingsitic overgrowth of olivines described by

Edwards (1938), see also Chapter 4. Considering further the reaction of early crystallized olivine and basaltic magma, Bowen's statement: "Again, in the crystallization of any system exhibiting a compound with an incongruent melting point, reaction between liquid and crystal is a constant factor" needs discussion. Also in this connection Bowen has stated "Forsterite and clinoenstatite may therefore be styled as a reaction pair. By this is meant that crystals of the first compound react with the liquid to produce the second during the normal course of crystallization".

As pointed out (see Chapter 4), the olivine phenocryst in basalts may be of multiple origin. Rounded forsterites or forsterites showing undulating extinction or deformation twinning are considered to be mantle-derived xenocrysts, particularly when olivine bombs are included in the basalt. Often in such cases, the olivines represent interspersed olivine-bomb components in the basalt.

In basalts, these olivines are frequently forsterites and they do not show an increase in the fayalite content from the crystal centre to the margins. This lack of zoning of the olivine phenocryst supports that no continuous reaction has taken place between the olivine crystal nucleus and the liquid to produce zonal olivines, in accordance with the continuous reaction series.

In support of the multiple origin of the olivine-phenocryst, one has to mention the eucrystalline olivines which often exhibit no corrosion phenomena whatsoever, they represent a stable phase and perfect shapes of crystals are preserved. In this case these olivines contradict the statement of Bowen that "reaction between liquid and crystal is a constant factor".

As pointed out above, Evžen Stanik has described generations of olivines in basalts and, in particular, he has described a third sub-idiomorphic olivine generation with pyroxene inclusions; indeed, this again is in contrast with Bowen's discontinuous series of crystallization. Again in contrast to the discontinuous reaction series is the case that when already crystallized olivine reacts with the basaltic liquid the product is not what is expected in accordance with the discontinuous series (clino-enstatite), but, as shown in Figs. 138 and 139, a "translucent" synisotropic margin is produced. In those cases when already crystallized olivine reacts with the basaltic melt, a complex corona reaction structure is produced (Figs. 162, 163 and 167). So it is obvious that corona structures are also formed when an already crystallized phenocryst reacts with the basaltic melt.

When reacting with the basaltic liquid, olivines often show a corroded outline and the groundmass commonly replaces the olivine (Figs. 419–423).

Commensurable are the results of pyroxenes in contact with the basaltic liquid; that is, no reaction products are formed in accordance with the discontinuous reaction series (i.e. no amphiboles are formed), but when pyroxenes react with the basaltic liquids synisotropization of the diopside takes place (Figs. 141 and 142) and an extensive phenomenology of pyroxene corrosion, replacement and infiltration by the groundmass is exhibited (Figs. 140, 144, 148 and 149). In cases where bronzite reacts with the basaltic liquid, complex corona reaction intergrowths are formed (Figs. 150 and 151). In this complex reaction zone of bronzite and basaltic liquid, and as a result of the reaction bronzite–basaltic melt, olivine grains can also be produced (Fig. 152).

As is discussed in Chapters 5 and 17, the reaction product of the basaltic liquid with the already crystallized phenocrysts differs entirely from the concept of Bowen's discontinuous reaction series. Another statement of Bowen provoking criticism is the following:

"A criterion for the reaction series, common to both the continuous and discontinuous type, and serving to show their fundamental likeness, is simply the tendency of one mineral to grow around another as nucleus. In the case of the continuous series, this is commonly known as zoning of mix-crystals and in the discontinuous series, as the formation of reaction rims, coronas, etc. Thus we have plain evidence of this kind, from a wide range of rocks, that the plagioclase constitute a continuous reaction series and that pyroxene, amphibole and mica form a discontinuous series". In particular, the point "the tendency of one mineral to grow around another as nucleus" which is raised requires careful consideration. This type of mineral association is not by far a proof of the existence of the discontinuous series of crystallization. The genetic association of the series olivine → pyroxene → amphibole → mica, and particularly the more precisely defined association series forsterite → augite → hornblende → biotite, is rather to be understood as the result of increasing oxidation and hydration as already pointed out.

That depending on their commensurable crystallochemical formulae, a reaction and a relation between these minerals takes place and exists is to be understood. However, unless increasing oxidation or oxidation–hydration takes place, there is no indication that transformation from one phase to another takes place, i.e. "Uralitisierung", the well-known process where the transformation of augite to hornblende takes place by hydration of the pyroxene and the comparable replacement of hornblende by biotite.

Present microscopic observations show olivine grains and cases of eucrystalline olivines included in titanoaugites (Figs. 183, 184 and 227). In such cases the crystallization of the augite round a nucleus of olivine (eucrystalline in shape and devoid of corrosion appearances) is rather to be understood as it has been defined by Breithaupt (1849): "Unter der Paragenesis der Mineralien ist die mehr oder weniger ausgesprochene Weise des Zusammenvorkommens, der Assoziation der selben zu verstehen. Man hat dabei auf das relative Alter der Körper, da wo eine Sukzession derselben zu erkannen ist, einen besonderen Wert zu legen, weil in diesen Verhalten die meiste Belehrung liegt."

Indeed, a paragenetic association would rather explain the association of olivine surrounded by augite, particularly when reaction rims, coronas and corrosion appearances are missing. ". . . The disappearing of minerals in the order in which they appear, which is of the very essence of the reaction series . . ." is considered by Bowen as a proof for the order of crystallization according to the discontinuous crystallization series.

In this connection, the existence of abundant idiomorphic olivine phenocrysts together with pyroxenes suggests that the formation of the pyroxene was not at the expense of the olivine crystallization.

On the other hand, the order of crystallization of the basaltic rocks is not to be strictly defined as Bowen's hypothesis assumes, i.e. the many crystallization generations of olivine, the tecoblastic growths, the xenocrystalline derivation of mineral components, etc.

The alteration-resorption of one mineral phase depends on the changing of the chemistry and physical conditions of the magmatic liquid during the crystallization of a melt. The preservation of olivine bombs, nodules and xenocrysts (mantle derivatives) indicate that the resorption of an early-formed mineral phase by the basaltic liquid melt is not the rule of the crystallization procedure of a basaltic melt.

On the other hand, as is to be understood, the relation between the mineral phases crystallized and the rest of the liquid is complex. Chapters 4 and 17 show the phenomenology and the relationship between the first phase of crystallization and basaltic melt as it is to be deciphered by mineral to mineral relation and by the relation of the phenocrysts to the groundmass.

The concept of the discontinuous crystallization series, forms the basic concept on which Bowen tries to understand the sequence of the minerals and which in turn is used for arriving at a conclusion as to the existence of the reaction series in rocks. Bowen also claims that he has considered the textural relations of minerals on the basis of observations (the present Atlas shows that Bowen did not have an extensive knowledge of basaltic textures).

As we can see from the following quotation from

TABLE XV

Olivines	Calcic plagioclases
↘	↙
Mg-pyroxenes	Calcic–alkalic plagioclases
↘	↙
Mg–Ca pyroxenes	Alkali–calcic plagioclases
↘	↙
Amphiboles	Alkalic plagioclase
↘	↙
Biotites	
↘	
Potash feldspar Muscovite Quartz	

Bowen, the schematic (Table XV) presentation of the series is, even for him, an oversimplification of the complex nature of the problems involved. "By piecing together the information to be obtained from the examination of such sequences and from observations of the structural relations of the minerals, a conclusion as to the reaction series in rocks is to be arrived at. Without going into further details as to the evidence, an attempt is made below to arrange the minerals of the ordinary sub-alkaline rocks as reaction series. The matter is really too complex to be presented in such simple form. Nevertheless the simplicity while somewhat misleading may prove to serve in presenting the subject in correct form".

Another sentence from Bowen's text that might make a reader sceptical is "It is the common procedure of science, given indication that a certain general relation is true, to assume that it is true, to push deductions to their ultimate consequences in all directions and to make the degree of correspondence of observations with deduction the measure of the probable truth of the original assumption."

One has to regard the approach of Bowen with even greater doubt, particularly if one considers the observations presented in the present Atlas regarding the crystallization of the basaltic rocks, and furthermore, if one takes into consideration that the great bulk of the present observations has no correspondence with Bowen's assumptions.

One of Bowen's most fundamental concepts and the one that has exercised the greatest influence on petrology is undoubtedly the concept of fractional crystallization. ". . . Of all the hypotheses of differentiation of magmas, none except the hypothesis of fractional crystallization can be checked against observation in any detail. It is, therefore, the only one that can be regarded as having any sound scientific basis".

Fractional crystallization is considered as a fundamental principle for magmatic differentiation, particularly in the ultrabasics (i.e. Skaergaard complex, Bushveld complex, banded chromites in dunites, etc.). The significance should be discussed of fractional crystallization in basalts, and whether the process has played a significant role in the crystallization and differentiation of the basaltic rocks, in general and in specific cases, in influencing the formation of the basaltic textures.

In general, most of the basaltic flows and the melts out of which basalts crystallized should not be considered to be "static" but either to be melts under flow motion or basic melts with great fluidity. Consequently the fractional crystallization or the gravitative separation by sinking of an early crystallized phase will be of subsidiary significance; due to the "kinetic" moving conditions of the melt no fractional crystallization will take place.

The scattered distribution of "olivine bombs" and nodules in basalts suggests that these relatively heavy blocks of rock did not separate out of the basaltic melt by which they were carried in accordance to "fractional crystallization" or better stated in accordance with gravitative separation.

A corollary to this is the interspersed distribution of olivine in crystallized basaltic flows, particularly rounded olivine phenocrysts or idiomorphic olivine phenocrysts interspersed in their distribution in basaltic flows. Further evidence of the non-significance of the role played by the mechanism of fractional crystallization in melts under motion, out of which basalt crystallized, is the flow textures of early crystallized mafic components or particularly the cluster structures of the gigantic augitic phenocrysts often observed in basalts, e.g., at Mt. Selale, Ethiopia and in Siberia (see Figs. 63, 66 and 350). Clearly a sinking during crystal fractionation of the agglutinations of the pyroxene should be expected. Also the non-separation of early-formed magnetites and ilmenites further supports that segregation of bands or differentiation of materials was insignificant in such cases. Furthermore, the flow structure of early-formed gigantic plagioclase phenocrysts which occur together with gigantic pyroxene phenocrysts, shows that no fractional crystallization and gravitative sinking or floating of the early-formed mineral phases of a basaltic melt (which was in motion, and out of which there crystallized a basalt with feldspar plates of exceptional size, and pyroxene phenocrysts) has taken place (see Debra-Sina, Ethiopia).

Regarding the origin of basaltic rock types and particularly concerning the hypothesis of basaltic trends – namely the olivine basaltic trend (olivine basalt → trachybasalt → trachyte) and the trend non-olivine basalt → andesite → rhyolite – fractional crystallization and magmatic differentiation has taken place on a much larger scale and extent within the magmatic chamber at depth. Indeed, this hypothesis of basaltic trends is interconnected with the hypothesis of magmatic chambers, the existence of which is not supported by geophysical evidence.

Another significant line of evidence against the operation of the mechanism of fractional crystallization is the proportion of the differentiation members of the trends. The apparent scarcity of rock members intermediate between basalt and trachyte has been pointed out by geologists (see Barth, 1961–1962); Chayes (1964, 1965) also points to the composition "gap" of the oceanic basalt-trachyte suites from the Pacific, Indian and South Atlantic oceans. In addition, Arne Noe-Nygaard has pointed out the composition "gap" in the basalt–rhyolite association in the Brito-Artic basalt province. Noe-Nygaard, in pointing out the composition "gap", gives the following figures:

Normal basaltic	85%
Almost normal basaltic	6%
Basalto-andesitic	1%
Dacitic	4%
Rhyolitic	3%

The above figures show a very marked preponderance of rocks with basaltic composition (91%), the acid members amount to 7% whereas the intermediate groups, basalto-andesitic and andesitic are as low as 2%.

Furthermore these data point out that the hypothesis of magmatic differentiation does not explain the composition "gap" since in accordance to the crystal-fractionation and magmatic differentiation, the intermediate fractions should have been far more abundant. Comparable is the lack of lunar intermediate members (see Chapter 34).

Regarding the origin of rhyolites and independent to the hypothesis of magmatic differentiation trends, there is the hypothesis of Keller (1969). Keller, in discussing the origin of rhyolites and rhyolitic pumice-tuffs from the Greek Islands of Kos, Kalymnos and Tilos, makes the following interesting remarks: "The question, whether acid magmas erupting in great quantities on the earth's surface originate by differentiation of a basaltic parent magma or by anatectic crustal melting, is a basic problem of petrological volcanology. The close relation in space and in time from rhyolites to intermediate and basaltic rocks led many authors to the conclusion that rhyolites are derived from basalts". Furthermore, regarding the origin of his studied rhyolitic material Keller concludes: "The rhyolitic pumice-tuffs from the islands of Kos, Kalymnos and Tilos are of anatectic origin and all belong to a Quaternary eruption with centre Kos. The continuous steps of anatectic melting are shown by granitic xenoliths in all states of fusion."

Reading Bowen's book one has to realize that some

of the petrogenetic problems discussed represent the views and problems which were prevailing half a century ago. The excess of olivine in basalts is treated as an integral problem of the basaltic magma. Regarding the crystallization of olivine and its reaction with the magmatic liquid he states:

"In certain investigated systems it has been found that in the olivine, forsterite has a reaction relation to the liquid. It separates from the liquid in amounts greater than its actual stoichiometric proportion, later to react with the liquid to be partially or wholly converted to pyroxene . . .".

"The question now to be considered is whether the olivine of rocks and especially basaltic rocks bears similar reaction relation to pyroxene and to magmatic liquids . . .".

"Among the criteria of the reaction relation is the formation of reaction rims or coronas – the formation of coronas of pyroxene around olivine in rocks is too well recognized to require discussion here. Another indication of the reaction relation of a mineral is the fact that it crystallizes out at an early stage, then ceases to crystallize and is wholly unrepresented among the later crystallization products of the magma. Olivine meets this requirement eminently."

As pointed out, the excess of olivine in basalts may be due to mantle xenocrysts or an idiomorphic early olivine crystallization as it is present in picritic basalts from Cyprus (see Fig. 80). In both instances the olivine can survive the reaction with the basalt liquids, and notably no pyroxene formation surrounding olivine "nuclei" is developed. There exists evidence, despite Bowen's assumption, that early-crystallized olivine does not enter into the discontinuous reaction series. Rounding, corrosion and reaction margins can be formed in early-crystallized olivines reacting with the basaltic melt, but in most cases the phenomena do not indicate pyroxene formation (see as pointed out in Chapter 4). Furthermore, the crystallization of idiomorphic olivine as a nucleus, round which later pyroxene crystallized, supports, as already pointed out, the non effectiveness of the discontinuous reaction series. In addition the formation as discussed, of more generations of olivines and, in particular, of sub-idiomorphic, late olivine generation, contradicts Bowen's assumptions.

In accordance with the views prevailing at his time, Bowen considers excess of olivine in basalts as an inherent cause of basaltic magma crystallization. "Fortunately, there is a means of verifying the excess separation of olivine in a large number of olivine basalts. The normative proportion of olivine is perhaps not an exact value but is certainly a very close approximation to what may be called the stoichiometric proportion of olivine. It has been repeatedly noted in olivine basalts, especially where these have a very fine-grained or partially hyaline base, that olivine has separated in amounts far in excess of the normative olivine. The fact has been commented upon by Cross and by Washington in connection with their investigation of Hawaiian and other basalts, by Lacroix, especially in his studies of basalts of Madagascar and Reunion, and by Tyrrell."

It should be pointed out, however, that olivine excess in the Hawaiian basalts should be seen rather as an abundance of interspersed olivine xenocrysts of mantle derivation (see Chapter 6). Here again, the assumption of Bowen and his contemporaries is in contradiction with the present trend of thinking. Also the excess of idiomorphic olivine in the Cyprus picritic basalts should be attributed to mantle derivation even if they cannot be considered as mantle xenocrysts.

It is on the early crystallization of olivine and its subsequent gravitative separation that Bowen actually bases his fundamental concept of fractional crystallization which in turn is the key to his magmatic differentiation.

"The reaction relation of olivine to liquid carries with it an inevitable consequence. The early separation of olivine, in excess of its actual stoichiometric proportions, necessitates the late formation of free silica, if for any reason, such a relative motion of crystals and liquid, the olivine fails of complete reaction . . .".

"The residius formed by fractional crystallization should have a tendency to attack earlier minerals when these minerals are members of reaction series, and it is the failure of completion of this reaction that constitutes fractional crystallization".

Combining his principles of discontinuous series and fractional crystallization, Bowen quotes the Palisade sill as an example of magmatic differentiation:

"The example just given of a case of differentiation (Palisade Sill) in which the early separation of olivine has augmented the late development of quartz has been shown to present a certain parallelism with investigated liquids which have a slight excess of silica and yet may precipitate olivine in early stages".

Since the Palisade Sill intrusion is often quoted as an example of fractional crystallization and magmatic differentiation, it is interesting to consider its case in more detail.

Assuming that a melt which is basaltic in composition starts to crystallize in an environment of hypabyssal conditions; the olivine will crystallize first and by fractional crystallization and gravitative separation will sink and by crystal accumulation will form the olivine-rich band of the sill.

As a result of the olivine separation and fractionation, the remainder of the melt will be enriched in silica and free quartz will crystallize at a concluding phase of the basaltic melt's solidification, which will be at the top of the sill.

In contrast to this hypothesis, microscopic and textural studies reveal that crystallization of the top and bottom portions of the sill was fast. due to rapid cooling, both top and bottom portions show a microcrystalline plagioclase–pyroxene intersertal crystallization (Fig. 361).

The central portions of the sill show micro-gabbroic textural patterns comparable to and commensurable with the microgabbroic rocks of the Karroo dolerites. Of particular significance is the relation of plagioclase laths to pyroxene phenocrysts. Here again, the plagioclase laths crystallize after the pyroxene, often including small olivines and pyroxenes (see Fig. 384) and often texturally extending from the margin of the pyroxene following weakness and cleavage planes of the augites (see Fig. 385). These intersertal intergrowth patterns of plagioclases in relation to the early-formed pyroxenes is extensively discussed in Chapter 16.

The crystallization of quartz and its textural pattern is of particular interest in the case of the Palisade Sill. As in the case of the Karroo, the granophyric quartz is in graphic intergrowth with plagioclases and pyroxene and in rare cases it is in proximity to olivine grains in the top of the olivine-rich band of the Palisade Sill (Figs. 488 and 489). Texturally the quartz is in intergrowth with the plagioclases. Fig. 475 shows quartz replacing the feldspar and extending interlamellarly parallel to the plagioclase polysynthetic twinning. Such metasomatic textural patterns are comparable to granophyric intergrowth patterns in granitic rocks (see Augustithis, 1973).

Additional evidence in support of a metasomatic origin of the granophyric quartz is shown in Fig. 481, where the granophyric quartz extends from its association with the plagioclase into an adjacent pyroxene, following a crack in the latter.

Considering the abundant occurrence of interstitial quartz in the top and bottom of the Palisade Sill and also the abundant granophyric quartz (graphic quartz in intergrowth with plagioclase) in the central micro-gabbroic portions, it is clear that the quartz is not the crystallization product of a residual silica-rich small fraction of the residual basaltic melt of the Palisade Sill (as would be expected on the basis of the fractional crystallization hypothesis), but is a "hydrothermal" post-orthomagmatic phase with metasomatic silica-crystallization.

The fact that granophyric quartz–plagioclase intergrowths are present in the part of the olivine-rich band of the sill is evidence that the quartz-olivine incompatible phases co-exist, because the quartz-forming solutions belong to a post-magmatic (orthomagmatic) metasomatic phase. This is further supported by the interstitial nature of the quartz in the top and bottom zones of the Sill (Fig. 597).

One of the textural patterns often exhibited in basaltic rocks, particularly in micro-gabbroic types, is the ophitic intergrowths of plagioclase laths and pyroxenes. The following quotation is from the treatment of the subject by Bowen:

"Fenner made an investigation in 1910. Concerning this view he states: 'It is only necessary to devote a little study to the question to determine that this is emphatically not the case.' He then describes his observations proving mutual interference of the crystals of the two minerals throughout the period of their growth."

"... It would appear then that, while a cursory examination gives the impression that ophitic texture in basalts is the result of late crystallization of pyroxene, a more detailed examination indicates simultaneous crystallization with plagioclase or even continuance of the outer rims of plagioclase after the pyroxene. On this point we may quote the authors of the Mull memoir, who say: 'It is noteworthy that the ophitic augites of the Mull Plateau Type often completed their growth well within the crystallization-period of the associated feldspar' ...".

"To be sure, when the crystallization had advanced practically to completion, an ophitic texture results and Lacroix concludes from this texture that pyroxene crystallized later than plagioclase, a conclusion which does not appear to be supported by his own observations of the stages of development of crystallization in the partly glassy types."

"... This objection is not, however, likely to favour the conception that the plagioclase separates out completely at an early stage in the more slowly cooled masses, for the more slowly a basaltic liquid is cooled the more its texture approaches the gabbroid, which indicates essentially contemporaneous crystallization of pyroxene and plagioclase ...".

Such textural patterns can be variably interpreted, depending on the detail consideration of the textures exhibited and on the generations of mineral formations. Augustithis (1956/1960) has described an ophitic plagioclase–pyroxene intergrowth where a pre-pyroxene plagioclase generation (laths of plagioclase), with an anorthite content about 60%, is enclosed and surrounded by a tecoblastic pyroxene. Also a post-pyroxene plagioclase generation, markedly more Na-rich (saudic) and represented by developed tecoblastic phenocrysts, is also exhibited (see Chapter 10).

In contradistinction, in Chapter 16 micro-ophitic and ophitic textural patterns are described, where the plagioclase laths clearly follow weak directions and cleavage boundaries of the pyroxenes (see Figs. 362, 377 and 378). Such textural patterns are clearly in contrast with the concept of simultaneous crystallization for the plagioclase and pyroxene as stated by Bowen.

Indisputable cases of post-pyroxene plagioclase formation are shown in Figs. 230, 231 and 233, where the tecoblastic plagioclases have corroded and assimilated the pyroxene.

In addition to the ophitic intergrowths of plagioclase and pyroxene, most interesting ophitic intergrowths are shown in Fig. 358, where plagioclases of the groundmass are in ophitic intergrowth with an olivine phenocryst. Also in this case, the feldspar crystallization is most probably post-olivine and indeed the plagioclases extend into and partly occupy cavities in the olivine produced by magmatic corrosion.

Admiring the versatility of Bowen, one notices that he did not completely disregard assimilation of sediments as a cause of magma type variation, or the views which were presented by Daly – but which did not gain sufficient importance and acceptance by Bowen and his followers. "Possibly the change from the one magma-type to the other was due in part to assimilation, as Professor Daly has argued in comparable cases. There is again no direct evidence bearing upon this point; all that can be said is that if assimilation has been of importance in modifying the Mull magma, it must have been accomplished at a high temperature under conditions admitting of complete admixture of melted sediments and original magma. There is no inherent impossibility in this conception."

It is a pity that Bowen did not see the importance of Daly's explanation regarding the interstices (the granophyric intergrowths of quartz–plagioclase often existing as fillings of interstices in microgabbroic rock types, see Chapter 19).

"Another notion of the origin of micropegmatitic interstices is disposed of definitely as far as the Asklund rock is concerned. On account of the strong contrast between such interstices and the basic rock in which they are found, some writers and notably Daly, have concluded that the basic magma has been contaminated with wholly extraneous salic material; in other words that it has assimulated or dissolved a sialic rock."

Bowen, insisting on his mechanism of early crystallization of olivine, olivine fractional crystallization, enrichment of residual liquid in silica-free quartz, did not recognize the petrogenetic significance of the granophyric quartz–plagioclase intergrowths and the possibility that they represent a hydrogenetic symplectic intergrowth at a concluding phase of the microgabbroic consolidation with a possibility of a metasomatic origin (due to assimilation) of the quartz-forming solutions (again, see Figs. 475, 478 and 481 and their descriptions).

It should be noted though, that granophyric quartz–feldspar intergrowths have been described from moon samples which preclude assimilation of sediments. However, the textural patterns exhibited are distinct and are described in Fig. 487.

In contrast to Bowen's interpretation that the hornblendes and micas in basaltic rocks represent products of the discontinuous reaction series, the textural behaviour of these minerals in basalts indicates frequent blastogenetic origins (see Figs. 322, 327, 328 and 330).

The association of the hornblende with the pyroxene and the fact that the amphiboles often surround the pyroxene, is to be understood on the basis of their crystallochemical relationship and due to oxidation–hydration processes. Furthermore, the presence of these (OH)-containing minerals in terrestrial basalts and their notable absence in corresponding lunar rocks is undoubtedly interconnected with the assimilation of atmospheric (OH) in the case of the terrestrial basalts.

Concluding the discussion of Bowen's book, one has to admit that he has provided a hypothesis which has influenced past and present petrological thinking. The objections and criticisms pointed out in the present discussion do not represent uncommon cases or exceptions to the rule, but instances from the complex of problems which are inherent to the crystallization of basalts.

Based on Bowen's principles, the magmatists in the past have suggested that intermediate and acid rocks are differentiation products of basic magmas, i.e. that granites are differentiation products of basic magma. However, the modern and prevailing trend is that granites are products of sediments (ultrametamorphism) either by granitization or by anatexis (Sederholm, 1910; Drescher-Kaden, 1948, 1969; Read, 1957; Mehnert, 1968 and Augustithis, 1973).

Chapter 29 | Barth's hypothesis for the crystallization of basaltic lava

Barth, both in his original paper (1936) and in his textbook ("Theoretical Petrology", 1952), has proposed a generalized working hypothesis explaining the crystallization of basalts. His hypothesis is based on the two fundamental concepts of Bowen; the crystallization of the continuous series, anorthite–albite and the discontinuous series, to be more specific on the crystallization of diopside–hypersthene. Barth uses the end-members of the two series as a quaternary system, which is at best, presented as a tetrahedron, with the end-members being the co-ordinates, see Fig. 598.

Bowen's triangular equilibrium diagram of the system diopside–anorthite–albite (Fig. 599) is taken as a basal face of the tetrahedron. The relation of the series diopside–hypersthene is best shown in a triangular diagram based on Barth (1936), Wahlstrom (1950) and Hess (1941), which also shows the paths of crystallization of pyroxenes from basalts collected from several localities (Fig. 600).

As illustrated in Fig. 598, Barth uses the series di–hy as an edge of his tetrahedral representation of his model of basalt crystallization.

In the ternary diagram, Fig. 599, there is shown the boundary curve which separates the field of feldspar crystallization from that of the pyroxene; along the curve, simultaneous crystallization of pyroxenes and feldspars takes place.

In the four-component system (diagram, Fig. 598), which is represented by a tetrahedron, the boundary curve of the basal face of the tetrahedron becomes a boundary surface which for simplification purposes can be thought of as a "plane" along which pyroxenes and feldspars crystallize simultaneously.

Barth believed that the original composition of the basaltic lava determines whether pyroxene or plagioclase shall crystallize first. As Wahlstrom (1950) mentions "Barth found several basalts in which pyroxene and plagioclase appeared to have crystallized simultaneously. By plotting the proportional normative amounts of hypersthene (Hy), diopside (di), albite (ab) and anorthite (An) of these rocks on a quaternary diagram (tetrahedral presentation) he found that the selected rocks gave points lying in or close to a plane with the following equation expressed in terms of co-ordination:

$$Ab^1 + 2di^1 + 2.3hy^1 = 123$$

ab^1, di^1 and hy^1 are the symbols resulting from the calculation of the normative ab, an, di and hy to 100%, An^1 is omitted."

Barth shows in his diagram (Fig. 598) that the Deccan trap rock, the Oregon trap and the Karroo dolerites indicate composition marked on the boundary plane that separates pyroxene from feldspar crystallization, in other words, these rocks were chosen as showing textural patterns of simultaneous crystallization of pyroxenes and feldspars.

This is indeed the critical point of Barth's hypothesis. Are the ophitic and micro-ophitic textural intergrowths exhibited by these basalts a product of simultaneous crystallization of pyroxenes and feldspars?

As mentioned in the discussion of Bowens hypothesis, reference has been made to the ophitic textures. Barth, a follower of Bowen's school, considers the ophitic textures exhibited in basalts of the Deccan and Oregon traps and in the Karroo dolerites as evidence of simultaneous crystallization. In contrast, present microscopic observations show (Chapter 16) that the ophitic textures are not due to simultaneous crystallization of pyroxenes and feldspars but due to feldspar-forming melts invading the pyroxenes along cracks, cleavages and other penetrability directions ("Wegsamkeit") of the host's lattice.

A series of illustrations (Figs. 362, 377 and 378) are presented from the Deccan and Oregon trap and from the Karroo dolerite, illustrating a post-pyroxene textural behaviour of the plagioclase in ophitic intergrowth with the pyroxene.

Chapter 30 | Modern hypotheses of basaltic magma crystallization

In addition to the pioneering work of Bowen (starting 1914 and continuing in the fifties) and of Barth (1936, 1952), the crystallization of basalts has been a subject of physico-chemical investigations by many workers. Kenzo Yagi (1967), summarizing these investigations in his treatment "Silicate Systems related to Basalts", states the following:

"Most of the important silicate systems relating to the basaltic rocks were studied in the second stage, chiefly by the scientists at the Geophysical Laboratory, among whom Bowen and, later, Schairer, have made outstanding contributions. Some of the important systems such as diopside–albite–anorthite; anorthite–forsterite–silica, or diopside–forsterite–silica were studied from 1910 to 1920, and a series of papers on the ferrous iron-bearing systems in the 1930's marked the highlight of the investigations in this period.

The study of anhydrous silicate systems at atmospheric pressure has been most successfully applied to the problems of basaltic rocks. Among many reasons for this, the following are important: (1) the principal rock-forming minerals of basalts can be easily synthesized, with few exceptions, in the laboratory; (2) most of the basalts have crystallized at or near the earth's surface, where the effect of pressures on crystallization can be neglected without losing much theoretical precision; (3) most melts of basaltic composition attain equilibrium promptly owing to their low viscosity.

The main nonvolatile chemical constituents of basalts are expressed by the following eight oxides: SiO_2, Al_2O_3, Fe_2O_3, FeO, MgO, CaO, Na_2O, K_2O and sometimes TiO_2. Therefore, if we can understand the phase-equilibrium relations in such an eight-component system as K_2O–Na_2O–CaO–MgO–FeO–Fe_2O_3–Al_2O_3–SiO_2, it will be of the greatest help in interpreting basalts. Since we are far from this goal, we should make a more steady approach to the problem by accumulating data on many ternary or quaternary systems, and this is feasible in the laboratory at present."

The main ternary and quaternary systems related to the crystallization of basalts and the main investigators of each system are:

1. Albite–anorthite–diopside (Bowen, 1915, 1945; Osborn, 1942; Hytönen and Schairer, 1961).
2. Diopside–forsterite–silica (Bowen, 1914; Kushiro and Schairer, 1963).
3. MgO–FeO–SiO_2 (Bowen and Schairer, 1935; Muan and Osborn, 1956).
4. Anorthite–diopside–forsterite (Osborn and Tait, 1952).
5. Anorthite–enstatite–diopside (Hytönen and Schairer, 1961).
6. Anorthite–forsterite–silica (Andersen, 1915).
7. Diopside–nepheline–silica (Bowen, 1922; Schairer et al., 1962).
8. Leucite–diopside–silica (Schairer and Bowen, 1938b).
9. Fayalite–leucite–silica (Roedder, 1951b, 1956).
10. Kalsilite–nepheline–silica (Bowen, 1937; Schairer, 1957).
11. TiO_2–SiO_2 (De Vries et al., 1954).
12. Diopside–hedenbergite–enstatite–forsterite (Bowen and Schairer, 1935; Yagi, 1953; Turnock, 1962; Yoder et al., 1963, 1964).

By utilizing various ternary diagrams, it is now possible to discuss the equilibrium relations in some quaternary systems that have petrogenetic applications to the problem of the basaltic rocks e.g.:

a. Anorthite–forsterite–diopside–silica (Osborn and Tait, 1952; Coombs, 1963).
b. Diopside–forsterite–albite–anorthite (Yoder and Tilley, 1962).
c. Diopside–forsterite–nepheline–silica (Yoder and Tilley, 1962).
d. CaO–MgO–SiO and the more generalized system CaO–MgO–Al_2O_3–SiO_2 (Ferguson and Merwin, 1919; Ricker and Osborn, 1954; Rankin and Wright, 1915).
e. Na_2O–Al_2O_3–Fe_2O_3–SiO_2 (Schairer and Thwaite, 1950, 1952; Yoder, 1950; Schairer and Bailey, 1962).

As has been seen from the discussion of Barth's quaternary system, anorthite–albite–diopside–hyperthene, one sees that the criterion of simultaneous plagioclase–pyroxene crystallization is the ophitic textures of basalts having compositions falling on the boundary plane of his model. Also it should be pointed out that the ophitic textures and, particularly the ophitic textural patterns of basalts from the Deccan and Oregon traps and from the Karroo dolerite, despite their simulation, on the whole contradict a simultaneous pyroxene–plagioclase crystallization (see Chapter 16). The above example casts a shadow of doubt as to whether the crystallization of the basaltic melts actually follows the picture which is obtained by the ternary and quaternary systems.

Regarding the "applicability" of the ternary and quaternary systems in the understanding of basalts, complications and difficulties arise, since the presence of water in magma can have an effect on the physical properties of the magma and can influence its trend of evolution. The following are some of the main influences of the presence of water in the magma: (a) decrease the viscosity; (b) lower the temperature of crystallization; (c) bring about the crystallization of hydrous minerals; (d) control the rate of fall of Po_2 with temperature.

In discussing the problem of the presence of water in basaltic melts, Hamilton and Andersen (1967) state: "Many systems having water as a component have been studied, ranging in complexity from uni- to multicomponent rock systems. A survey of the literature makes it apparent that there is a distinct lack of data for hydrous synthetic systems which are applicable to basalts". Unfortunately there is no ternary or even quaternary system which is a good approximation to basaltic composition, and the importance of controlling the oxidation state of iron is another complicating factor.

Despite the attempt to classify water into magmatic, connate and metamorphic waters by their isotopic contents by Craig et al. (1956) and White (1957), there are difficulties in assigning a water content to the parent magma. In addition to these difficulties, if one compares the lunar basic rocks with the terrestrial, no hydrous minerals have been found in the former and it is most probable that all that is understood to be water of the parental magma could, in reality, be assimilation of meteoric water.

One must not think of the basaltic magma that transverses the earth's crust before it is extruded as lava in terms of a laboratory pure melt. Daly has suggested that assimilation can take place to such an extent that it can modify the composition of the melt and give rise to diverse rocks.

Water is, to a great extent, an assimilation component that, as mentioned, plays an important role in the crystallization of the hydrous minerals crystallized or in minerals that can crystallize out of hydrous solutions. It also influences the textural intergrowths such as the interstitial "micropegmatitic" (i.e. granophyric and micrographic, see Chapter 19), colloform tecoblastoids (see Chapter 13) and, at a concluding phase, the cavity fillings (Chapter 23).

Daly's argument that in certain basaltic rocks an excess of silica is determined could be attributed to silica assimilation by ascending magma. In contrast to the physico-chemical ternary and quaternary systems, which attempt to explain the crystallization of free quartz in conjunction with olivine crystal fractionation or without it, the textural patterns show either interstitial later quartz (Fig. 601) or granophyric quartz, again representing symplectic quartz–plagioclase intergrowths which indicate evidence of hydrogenetic, often metasomatic quartz formation.

Chapter 31 | Volcanic trends and provinces - (basaltic differentiation series)

Fractional crystallization is believed, by Bowen, to be the fundamental mechanism for the basaltic differentiation series. He also recognized a single parental basaltic magma which, mainly by fractional crystallization, differentiates to produce more acid derivatives. In discussing the differentiation of the basic rocks, Bowen puts forward the concept of a single polycomponent system, which is dominated by reaction series.

Discussing the diversity of igneous rocks by magmatic differentiation, Barth (1930, 1952) accepts a basaltic parental magma as a starting point of rock differentiation. Furthermore, his recognition of oceanic and continental basalts and the differentiation of the latter into basalts associated with epeirogenic and orogenic movements introduces the relationship that exists between basaltic flows and geotectonics. The basalts of these geotectonic provinces are often subjected to differentiation processes.

In contrast to the single parental basaltic magma concept of Bowen and Barth, Kennedy and Anderson (1938) introduced the dual character of the parental basaltic layer, having the composition of olivine basalt below and the tholeiitic (pyroxene basalt) above, and they explained igneous rocks as products of differentiation of these two types of parental basalts. In his elaborate treatment of basaltic magma differentiation, Hisashi Kuno (1967) recognizes three parental basaltic types which give rise to three distinct basaltic differentiation series; the tholeiitic, the high-alumina and the alkali rock series. A fourth series is recognized, the calc-alkali, which is formed by water assimilation of any of the three previous series.

Kuno is a supporter of Bowen's fractional crystallization hypothesis and in his treatment he disregards assimilation as a major cause of igneous-rock formation. "In this paper, assimilation is considered as a process whose effect is subordinate to that of differentiation by fractional crystallization." Synoptically the differentiation series of Kuno are as follows:

(a) Tholeiitic series:
Basalt → andesite → dacite → rhyolite.

(b) High-alumina basalt series:
Basalt → andesite → rhyolite.

(c) Alkali-rock series:
Basalt → hawaiite → mugearite → trachyte → rhyolite trachyandesite → trachyte → alkali.

(d) Calc-alkali rock series:
Basalt → andesite → dacite → rhyolite.

As pointed out, the calc-alkali series can be a derivative of the other series and, as Kuno mentions, the high-alumina basaltic series is an intermediate between the tholeiitic and the alkali-rock series.

Furthermore, considering the tholeiitic series to be essentially olivine poor, Kuno's scheme resembles Kennedy and Anderson's dual character of parental basaltic layer. Wilkinson (1967) also recognized distinct parental basaltic types which gave rise to distinct differentiation series:

(a) alkali-olivine basalt → hawaiite
mugearite → trachyte → phonolite
↘ pantellerite

(b) tholeiite → tholeiitic andesite → dacite → rhyolite

(c) Central basalt (calc. alkali basalt) Calc. alkali andesite dacite → rhyolite

Wilkinson's and Kuno's schemes of basaltic magma di-differentiation are very close to one another and essentially are based on multi-parental basaltic magma types which, by fractional crystallization, would give rise to different rock series.

In addition to the criticisms raised regarding the basic concept of Bowen, petrographic and field studies of the Ethiopian Plateau and certain areas of the Ethiopian rift valley raise further doubt as to whether some of the previously considered "differentiation series members" are products of magmatic differentiation or due to large-scale assimilation processes.

Tholeiitic basalts predominate on the Ethiopian Plateau. The trap series as exposed in the Blue-Nile Canyon system consist of repeated tholeiitic flows (see Figs. 6, 8 and 11). In contrast, the Selale Volcanics on the Ethiopian plateau are characterized by large megaphenocrysts of augites (see Figs. 190, 198 and 206) but nevertheless are within the tholeiitic range.

Certain andesites and some rhyolites are known in the marginal region of the Addis Ababa rift. The tholeiitic basalts of the plateau are mainly free of olivine but, in certain occurrences of the Ethiopian Plateau olivine bombs and nodules are found in the basalts and they are clearly attributed to mantle fragments, e.g., in the Lekempti region (Augustithis, 1972) and to the dunitic material in derivation xenoliths in the region of Yubdo-Wollaga (Augustithis, 1965).

In contrast, there is a predominance of alkali-basaltic series within the rift valley (Ethiopian part), characteristically the olivine basalts of the Khidane Meheret St Michael, Doddo/Addis Ababa (Augustithis, 1959/1960), Ducam, Bishofto and Awash lines of volcanoes (Augustithis, 1963, 1964). There appears to be a certain relationship between the type of the basaltic series and the geotectonic environment with which it is associated. The tholeiitic basalt most probably represents melting of a gabbroic lower crust with the exception of the olivine bombs and nodule-bearing basalts, which probably indicate a peridotitic mantle involvement.

The olivine-basaltic lines of Duncan, Bishufto, Adama and Awash represent basalts in which the mantle is greatly involved. The olivine-basalt volcanism is associated with deep tectonic faulting (see Chapter 2). Concerning the rift trachytes, e.g. Mt. Zukala and Fontale, and the abundant pantellerites, Rosenbusch, 1898 refers to a pantelleritic zone in the rift from Obock to Djibouti – more than 500 km long – often having proportions far beyond those which could be explained by the concept of the differentiation hypothesis, thus indicating assimilation.

Additional evidence of large-scale basement assimilation is the andesitic volcanism of the Afar and the peculiar anomalous mantle described by Makris et al. (1974), both of which could perhaps be attributed to crust assimilation sinking in the mantle due to the rift-depression formation.

The Ethiopian-rift region can be seen as a region where a large scale of basement assimilation has taken place; petrographically this is supported by the abundance of the Afar andesites, the enormous proliferation of pantellerites between Addis Ababa and Djibouti and the extensive trachyte volcanism, e.g. the volcanoes Zukala and Fontale. Of particular petrogenetic significance is the presence of aegerine rhyolites (a Na-rich phase) in the region of Kuni/Asba Tafari in the Chercher margins of the rift (Augustithis, 1964).

The down-sinking of the basement in the rift region and the intense tectonic shattering of the region are, as the faulting and fissuring indicate, processes which would enhance assimilation.

An additional line of support for the assimilation hypothesis is, as already mentioned, the origin by anatexis of the Aegean Sea islands rhyolites and the composition "gap" previously discussed (Chapter 28).

Chapter 32 | Source material and depth of basaltic magma generation

Many views have been expressed regarding the source and depth of the origin of the basaltic magma (Washington, 1906; Kennedy, 1933; Kuno, 1959; Clark, 1961b; Urry, 1949; and others). The subject has been treated to some extent by Yoder and Tilley (1962) and, mainly on the basis of experimental mineralogy and silicate melts, they have drawn certain conclusions regarding the material (source) and the depth of the generation of the basaltic magma. Since their paper has influenced petrographic thinking and because the nature of their conclusion seems to be mainly physico-chemical, it is considered necessary to discuss some of their conclusions.

Yoder and Tilley consider that basalts can be formed by the melting of a wide range of rock types besides basalt itself and gabbroes. Pyroxenite, amphibolite, pyroxene hornblendite and eclogite are considered to be potential sources of the basaltic magma.

They believe that since every major type of eclogite has a corresponding type of basalt, it is the parental rock from which the basaltic magma originates.

Furthermore, Yoder and Tilley chose eclogite as the parental rocks for the generation of the basaltic melts considering the depth of generation of the basaltic magma. Taking into account that the basaltic melt is generated at a depth of about 40–50 km and that according to a possible zoning of the upper mantle in the order gabbro, amphibolite, hornblendite, peridotite–eclogite, the depth of basalt generation corresponds to the depth of eclogite, even if it is considered a product of a garnet–peridotite as a more primitive source. It is therefore clear that the source of material out of which the basaltic magma is generated is dependent on the depth of basalt-magma generation. Yoder and Tilley have dealt with this point at some length and have reached certain conclusions which are discussed below:

On the basis of the seismic record, the depth of basaltic-magma generation in Kilauea (Hawaii) is considered to be between 45–60 km and that of Kinchevsky volcano, Kamchatka, about 50–60 km.

The depth of basaltic-magma generation is a function of the geothermal curve and liquefaction of the basalt. Since magma arrives at the surface at essentially an all-molten liquidus temperature, taking into account its pressure dependency, until it intersects the geothermal curve to obtain the maximum depth of liquefaction.

Furthermore, Yoder and Tilley believe that there is reason to believe, taking into consideration the arguments of Verhoogen (1960), Urry (1949) and Clark (1961b), that magma generation may take place at depths as shallow as 50 km in specific areas and that the process is accompanied by seismic disturbances.

Although no systematic attempt has been made with seismic methods to outline liquid masses in the earth, it is generally held that initially the region of magma generation is all crystalline.

It should be clarified that Yoder and Tilley base their conclusion that basalt magma originates from eclogites, mainly on the basis of the above considerations. *

In contrast to the aforementioned conclusion, the present studies are not in harmony with the above hypothesis. As Chapter 6 shows, the mantle xenoliths which are common in basalts are mainly olivine–pyroxene–spinel fragments and gabbroic bombs and nodules, of mantle and upper-mantle derivation, respectively. Xenoliths of crustal derivation are also reported. No eclogite or garnet peridotites are common in basalts as mantle-derived xenoliths. The absence or non-common abundance of eclogitic xenoliths in basalts seems to indicate that despite the normative possibilities as far as the chemical derivation of basalts from eclogites is concerned, the eclogite most probably should not be considered as the actual parental rock of basaltic magma derivation. It should be brutally stated that if eclogite were the parent rock of basaltic magma, eclogite xenoliths should have been far more abundant in basalts.

---

* In support to his previous views, Yoder (1976) discusses the different sources of basaltic magma and refers to the garnet peridotite as the key position for its generation.

According to him the garnet peridotite is the preferred source of basaltic liquids because:

(1) The occurrences of garnet peridotite are appropriate to deep-seated environments.

(2) The close compositional relationship to meteorites supports its derivation from material accumulated in the primordial earth.

(3) Partial melts of natural samples at high pressure have basaltic compositions.

(4) The mineral assemblage was found experimentally to be stable at high pressures and temperatures.

(5) It has appropriate densities and seismic velocities.

The presence of gabbroic xenoliths and the great abundance of olivine–pyroxene–spinel bombs and nodules in basalts have a genetic cause and are the clue to the generation of the basaltic magma at the junction of a gabbroic lower-crust layer with a spinel–peridotic-mantle substratum of worldwide distribution (see Chapters 5 and 7).

Yoder and Tilley's consideration of the depth of basaltic magma generation should also be discussed, as basalt must be a product of the melting of a crystalline rock (as pointed out, no liquid masses have been determined below the crust by seismic methods). The two alternative ways of melting a crystalline rock at depth are increase of temperature or pressure release, which is alleged to give rise to the formation of liquid from a crystalline source.

With the increase of the understanding of geotectonics, i.e. the tectonics of the cratogens and orogenic belts, and in particular with the increase of our knowledge of the rift and sub-oceanic rift systems and ridges, the interrelation between geotectonic zones of weakness of the crust and basaltic effusions or intrusions is becoming apparent. It should be considered that these zones correspond at depth to zones of pressure release, as a consequence of which melting of the crystalline substratum occurs.

As has been pointed out in Chapters 1 and 2, the largest basaltic covers which occur on cratogenic regions of the continents cannot satisfactorily be explained by the plate-tectonics hypothesis. However, as the case of the Deccan traps shows and in the case of these basaltic covers (on the cratogens), effusion of basalts along fracture zones is possible.

Large quantities of basaltic flows are poured out from open fissure systems or fault-lines which are often marked by lines of volcanoes and radiating and swarm dykes. However, it should be remembered that these fault and fissure systems do not continue, as such, in the zone of basaltic-magma generation.

It is pointed out that these zones of weakness of the crust correspond, at depth, to the junction planes between adjacent tectonic plates. The depth of basaltic-magma generation is given by Yoder and Tilley, and by most other workers, on the basis of theoretical consideration of the melting of the solids at depth or by seismic studies below volcanoes.

In contrast to the above estimations, which suggest a depth of 40–60 km for basalt generation, other estimations are arrived at by considering the depth of the Moho below some basalt occurrences which contain olivine bombs (mineralogically consisting of forsterite, bronzite, spinel). It is believed that the Moho discontinuity represents the contrast boundary between the crust and the mantle substratum. The depth of the Moho discontinuity will give the limit of the theoretically possible zone of basalt generation. The transition from the granitic to the peridotitic mantle is through a zone of gabbroic composition which belongs to the lower crust.

The depth of this peridotitic mantle substratum most probably represents the limit of the crystalline gabbroic

TABLE XVI

TRACE ELEMENT COMPARISON BETWEEN MINERAL COMPONENTS OF THE MANTLE FRAGMENTS IN BASALT AND THE BASALT IN WHICH THEY OCCUR (By A. Vgenopoulos)

Mineral or rock and locality	Ni	Cr	Co	Cu	Zn	Mn	Ti	Zr	Sr	Ba	
Forsterite Lekempti (W. Ethiopia)	2410	145	54	52	202	2020	215	45	–	100	Main mineral component of mantle fragments in basalt
Bronzite Lekempti (W. Ethiopia)	265	2610	7	17	115	1100	780	9	–	850	
Spinel Lekempti (W. Ethiopia)	80	27935	–	32	180	915	620	180	–	–	
Basalt Lekempti (W. Ethiopia)	6	15	7	16	–	957	11050*	205	445	345	Basalt (enclosing mantle fragments)
Forsterite Canary Islands	2820	95	42	55	220	960	780	25	40	85	Main mineral components of mantle fragments in basalts
Pyroxene Canary Islands	1250	3400	85	56	225	450	345	–	–	175	
Spinel Canary Islands	2200	9880	99	58	220	140	465	250	565	312	
Basalt Canary Islands	350	–	7	7	–	948	12200*	235	598	350	Basalt enclosing mantle fragments

* The high Ti values in basalts indicate a marked difference with the peridotitic mantle fragments enclosed in the basalts.

lower crust from which the basaltic magma is generated by fusion. This peridotitic mantle substratum will not yield basalt by direct fusion (see Table XVI); however, portions of it are mobilized in the fusion zone and render the olivine bombs which actually are mantle fragments which, despite magmatic corrosion on the whole, represent refractory resistant relics in the form of "xenolithic" olivine bombs, nodules or xenocrysts in the basalt.

Regarding the thickness and distribution of this gabbroic lower-crust layer, which is considered to be transitory between a sialic upper crust and the mantle, its existence is postulated by the presence of abundant gabbroic xenoliths in basalts (see Chapters 6 and 7) and it has also been found under the Afar by Makris et al. (1975) using geophysical methods.

This zone is theoretically a parental layer of basalt and its depth is given by the Moho-discontinuity which marks the peridotitic mantle substratum, which has contrasting physical properties. As pointed out, the mantle peridotite participates particularly in the generation of olivine basalt. In contrast to Yoder and Tilley, according to whom basalt derives at a depth of about 40–60 km and by fusion of eclogite, the basalt-generation zone is here proposed to be at approximately the depth of the Moho and it involves the gabbroic lower crust and the upper mantle. The mineralogical composition of the upper mantle is a function of the pressure exercized by the thickness of the overlying crust.

The plateau traps and the tholeiitic series would mainly be the fusion derivatives of the gabbroic lower crust which, by ascending and upper-crust assimilation, would give rise to the tholeiitic assimilation series; the alkali-olivine basalt would be generated at the depth of the lower gabbroic crust with an involvement of the peridotitic mantle substratum, as indicated by the abundance of olivine bombs and olivine xenocrysts in the alkali-olivine basalts *. Due to the assimilation of the upper crust, the alkali-olivine basalt assimilation trend will be produced, e.g. the volcanics of the Ethiopian rift. The present explanation is essentially a return to Kennedy's and Kuno's dual character of the basaltic substratum, which was rejected by Yoder and Tilley. They explained that both trends, the tholeiitic and the alkali-olivine, are produced from the eclogitic layer, due to differences in pressure. It should be pointed out though that, whereas Kennedy emphasized the dual character of his basaltic substratum, in the present suggestion we have the fusion of the lower gabbroic crust and the possible involvement of the upper mantle along friction zones **.

The peridotitic layer (consisting of forsterite, bronzite, spinel), evidence of which is found in the form of "bombs" in most of the major olivine-alkaline basaltic effusions of the earth, strongly suggests the existence of a forsterite–bronzite (pyroxene) spinel mantle-substratum of worldwide distribution.

As an example of basalt with olivine bombs and below which the Moho discontinuity is at a depth of about 34 km, there is the occurrence of the Jota basalt at Lekempti (or Nekempti). These olivine bombs are considered to represent fragments of the forsterite–pyroxene–spinel mantle (Augustithis, 1972).

As the Moho map of Ethiopia by Makris et al. (1975, Fig. 602) shows, the Moho-discontinuity most probably marking the presence of the mantle, is about 34 km in depth at Nekempti, thus indicating that the mantle fragments in basalt are at least of this depth.

Since direct fusion of the mantle (forsterite–bronzite–spinel) would not produce basalt, the gabbroic layer above the mantle could be regarded as a parental rock for the basalt; however, as the "olivine bombs" show, the mantle was involved in the fusion zone.

Also using the Makris Moho-depth map, another olivine-bomb basalt occurrence is located on a volcanic island offshore of Assab in the Red Sea. The basalt belongs to the alkaline-olivine series and the depth of the Moho discontinuity is estimated to be 14 km. This basalt occurrence belongs to the rift eruption series which are following rift-valley fissures and faults.

Both the Nekempti (Lekempti) and the Red Sea island basalt contain fragments of the peridotitic mantle substratum, but their depth of generation is noticeably different. As indicated by the different depths of the Moho discontinuity, the depth of the basalt generation zone is ±34 km in the plateau, whereas in the rift, this depth is only ±14 km. These depths are in contrast with those obtained on seismic evidence below the Hawaiian volcano (Kilauea) and the volcano of Kiuchevsky at Kamchatka. It seems that the depth of basalt generation is a function of the nearness of the peridotitic-substratum to the surface. The picritic basalt of Cyprus (see Figs. 80 and 81) and the olivine-bomb-containing basalts of Lanzarote in the Canaries should be considered as additional examples of basalt generation at small depths. Also the oceanic zone of basalt generation is a function of the depth of the Moho-discontinuity which is negligible.

* Exceptionally, the peridotitic mantle substratum is mobilized in basalts of the tholeiitic series, e.g. mantle fragments in the Lekempti basalt of the Ethiopian plateaux, Augustithis, 1972.

** As a corollary to the above hypothesis and particularly in support of the generation of basaltoid rocks from gabbroes, is the surface of the moon, with a gabbroic–peridotitic crust. Melting takes place due to meteoritic impacts. The terrestrial equivalent is the fusion of the lower crust due to pressure release and basalt generation.

Chapter 33 | The significance of assimilation in basalts and associated rocks

One of the most fervent supporters of the hypothesis of basaltic-magma differentiation by fractional crystallization, Barth (1936, 1952) considered the total amount of weathering during all geological time to correspond to a depth of erosion of 30–60 km and concludes that the rocks of the continents are not "juvenile", they all are derived from sediments. Barth, presenting his new petrographical thinking in 1961/1962, states: "In this way the great diversification of the igneous rocks is better explained than through the conventional assumption of crystallization differentiation of a homogeneous magma."

Considering the great contribution by Barth in supporting the Bowen school of magmatic differentiation, one is surprised by these views presented in 1961/1962. The question arises, what has transformed Barth from an extreme magmatist to an extreme transformist? The answer is perhaps to be found in the following statement which is included in his paper:

"Die erstaunlich grosse Variation der chemischen Zusammensetzung wurde durch magmatische Differentiation vornähmlich nach dem Prinzip der fraktionierten Kristallisation erklärt. Man glaubte dass das Ausgangsmaterial ein homogenes Magma basaltischer Zusammensetzung war. Unter Verwendung einer grossen Anzahl experimentell ausgearbeiteter Gleichgewichts-Diagramme von Silikatschmelzen mit zwei, drei oder sogar vier Komponenten wurde angenommen dass Gesteine der verschiedensten Zusammensetzung aus primärem basaltischem Magma stammen. Zumeist bleib es aber unbeachtet, dass die geologischen Verhältnisse oftmals entscheidend gegen einen solchen Entwicklungsgang sprechen. Die Mannigfaltigkeit der Eruptivgesteine zu erklären, ist somit ein Problem geblieben.

Die Mannigfaltigkeit der Sedimente aber ist einfach zu erklären, denn sedimentäre Differentiation ist ausserordentlich wirksam; Quarzite, Kalksteine, Eisenlagerstätten und andere Gesteine spezieller Zusammensetzung werden durch sedimentäre Prozesse erzeugt. Hierdurch eröffnet sich uns ein Weg zum besseren Verständnis der Eruptivgesteine. Wir können mit jeder beliebigen Art sedimentären Materialen anfangen: Sandstein, Kalkstein, Dolomit, Ton oder Eisenstein. Sie sind mögliche Ausgangsmaterialen; denn alle Eruptivgesteine stammen aus irgendeinem Sediment. Sie können erhitzt werden, der Metamorphose unterworfen sein oder teilweise geschmolzen werden unter Bildung von Porenlösungen, Ichoren oder Magmen. Dadurch wird die Mannigfaltigkeit der resultierenden Gesteine sehr gross. Das Ausgangsmaterial gibt potentielle Möglichkeiten für alle rätselhaften Gesteinstypen".

As pointed out, the extensive andesite occurrences in the Afar region of the Eritrean part of the rift valley indicate assimilation of the basement (upper crust) by sinking, due to the formation of the rift formation. As a corollary to the hypothesis of basement assimilation is the geophysical determination by Makris (1974) of a highly attenuated sialic crust and an anomalous mantle beneath. However, as is suggested in Chapter 6, upper-crust xenoliths are often found in basalts, indicating that assimilation can also take place in the region (depth) of the crust and upper crust. Most impressive evidence of this are the sulphur occurrences in the volcanic line of Dallol (Augustithis, 1964) – unpublished report of a helicopter reconnaissance in the Afar Dallol – Ministry of Mines, Addis Ababa). The sulphur is due to the gypsum and anhydrite layers of the Dallol evaporite series, through which has cut the line of volcanoes shown in Fig. 18. Another more impressive piece of evidence of upper-crust assimilation is the out-pouring from a volcanic crater of Dallol molten salt (Fig. 603), this again is due to melting of halite beds of the evaporite series transversed by the line of volcanoes. Both the dis-association of anhydrite and gypsum and the melting of the halite beds are within the temperature range of the ascending lava.

As already pointed out, the alkali-olivine trend of the volcanics of the Ethiopian part of the rift conform rather to the assimilation hypothesis than to a magmatic differentiation one. Also, as mentioned in Chapters 19 and 28, the excess of silica, in the form of free quartz in basalts and the micropegmatitic and interstitial quartz in basalts, and particularly in the Karroo dolerite, and in the Palisade Sill, is attributed on chemical and particularly on textural evidence to metasomatism of SiO_2.

Daly (1924) was first to explain the excess of silica in the form of free quartz due to assimilation of acid rocks by the ascending basaltic magma, rather than by silica enrichment by fractional crystallization of early fractionated olivines, as was proposed by Bowen.

So far, only isolated examples of assimilation by the basaltic magma have been discussed; nevertheless, it can be seen that a variety of rocks can originate within a petrographic province by assimilation of sedimentogenic and metamorphic rocks.

Differentiation through sedimentation (Barth, 1961/1962) and subsequent assimilation by the basaltic ascending melt can result in a differentiation by assimilation; an example of this is the olivine basalt → trachybasalt → trachyte → pantellerite series of the volcanic trend within the Ethiopian rift, e.g. trachytes of Zukala, trachy-basalts at Mojo and the extensive pantellerite ranges from Addis Ababa to Djibouti and the most recent volcanism of the olivine-basalt volcanic lines of the Bishufto, Duncan, Awash, etc. It should be noticed that the time sequence of the assimilation series is the reverse of the theoretical magmatic differentiation succession since the olivine-basalt volcanic lines represent the latest volcanic activity. The sequence of the assimilation trend is to be understood by assuming that the first members to be extruded will represent the more intense assimilation phase, i.e. the order of succession is, trachytes → pantellerites → olivine basalts; this actually happens to be the succession of volcanism in the Ethiopian rift.

Assimilation as a mechanism of producing igneous rock variety has been proposed by Daly. Both Bowen (1928) and Kuno (1968) disregarded assimilation as a significant mechanism of rock differentiation and considered it subordinate to fractional crystallization differentiation.

In contrast, Barth (1961/1962) has proposed a sedimentogenic origin for all igneous rock in the sense that the 30–60 km of the upper crust represents a sedimentogenic thickness. "Most of the faults and fissures out of which igneous activity can effuse do not surpass the depth of the sedimentogenic crust (exceptions might be provided by the central volcanism of Hawaii). Consequently, the source of igneous activity has to be derived from within the sedimentogenic crust, melting by anatexis will give rise to the 'magma' ".

Barth jumped from the extreme position of a basaltic parental magma (1952) to the other extreme position of the transformation of sediments to igneous rocks. Now we know that, whereas the fundamental mechanism of Barth's transformation hypothesis, i.e. melting of the crust by fusion along friction zones, is basically sound, the tendency is to involve the lower crust and the peridotitic mantle in the genesis of basalt magma-generation. It is the involvement of the lower crust and mantle that is missing from Barth's mechanism of igneous-rock genesis.

Examples of igneous-rock bodies, involving the partial anatexis or preservation in the form of volcanic agglomerate fragments derived from the greatest part of the rock-profile transversed by a conduite, are the Kimberlitic breccia pipes often comprising garnet fragments, eclogite xenoliths (in cases diamondiferous) marble-fragments, peridotite–serpentine and gneissic picked-up fragments. All this agglomerate material is often found in a mobilized calcitic "groundmass". As the variety of the agglomerate ingredients shows, the deep mantle and upper crust contributed to the formation of the Kimberlitic pipe-breccia. In contrast, a plateau basalt poured out as effusive eruptions and covering extensive dimensions is a homogeneous cover, with a great extent of lateral distribution and vertical thickness. These two examples represent diverse extreme conditions: (a) the plateau basalt is the product of the melting of the lower gabbroic crust and if assimilation has taken place, homogenization of the melt has been attained, the mantle involvement is restricted (only exceptional olivine-bomb xenoliths and xenocrysts have been found); (b) in contrast the Kimberlitic pipes represent a breccia without homogenization but regrowths of biotite and calcite are exemplified.

It is, therefore, between these two extreme diverse examples that we have to understand the complex variety of volcanic rocks; fusion of the lower-crust–mantle substratum with homogenization and complete assimilation of the crust material and the opposite extreme case of a Kimberlitic breccia pipe which is greatly inhomogeneous.

Chapter 34 | Comparison between lunar and terrestrial basalts

THE DIFFERENTIATION TREND HYPOTHESIS AND THE LUNAR ROCKS

Lunar research has revealed the predominance of basaltic, in fact gabbroic rocks with a complete lack of intermediate differentiates, such as andesites. The lunar rhyolites reported by Kushiro et al. (1970) * are in insufficient quantities and extent on the lunar surface and could hardly be referred to as a rhyolitic occurrence as Kushiro et al. state; the residual liquids found as mesostasis are rhyolitic which suggests that fractional crystallization of some magmas can generate rhyolitic magmas.

This lack of extensive andesite or rhyolite occurrences on the moon could hardly be in accordance with Bowen's differentiation hypothesis, according to which more andesites and rhyolites would be expected. As is the case with earthly basaltic complexes and also in the case of the moon, no differentiation of parental basaltic magma has produced andesitic or rhyolitic differentiation products in the quantities that would be expected by the differentiation hypothesis.

Whereas gabbroes are abundant on the moon, the "equivalent" acid plutonic member, granite, is not reported on the lunar surface; the "rhyolitic liquid" reported by Kushiro et al. is certainly far from being a granite. The abundance of gabbroes and the lack of granites on the moon shows that Bowen's magmatic differentiation hypothesis is unsatisfactory, both in explaining the lack of granites on the moon and for the understanding of the granites on earth. Indeed, granite is a terrestrial rock, i.e. the product of ultrametamorphism of sedimentary rocks (Augustithis, 1973).

* Kushiro et al's paper "Crystallization of some lunar mafic magmas and generation of rhyolitic liquid" was originally presented at the Apollo 11 Lunar Science Conference at Houston, 1970 under the title "Fractional crystallization of some lunar mafic magmas and generation of granitic liquid".

The continuous crystallization concept in lunar and terrestrial rocks

The application of modern techniques in the study of lunar pyroxenes and feldspars has revealed that the intermediate member of a continuous series in reality often consists of lamellar intergrowths of two phases, thus allowing the interpretation of a host and an ex-solved phase or of micro-lamellar twinning (micro-lamellar parallel orientations). The application of "high-voltage transmission electron" microscopy has enabled the study of crypto twinning or of crypto ex-solution phase in the magnitude of Å units, thus revealing a new micro-textural field which was beyond the resolving capacity of the ordinary polarizing microscope. Whereas ordinary polarizing microscopy accompanied by X-ray diffraction studies yielded pictures and X-ray patterns of single crystals, the application of the high-voltage transmission electron microscope shows that what was apparently a single crystal, in reality consists of different phases of micro-individuals.

Studies of single crystals by Radcliffe et al. (1970) using high-voltage transmission electron microscopy showed an extensive and complex pattern of parallel striations on submicron scale, probably representing a submicron scale ex-solution in a single pyroxene lunar crystal. In contradistinction, similar studies of plagioclases showed extensive micro-twinning appearing in groups or bundles which may correspond to the "lamellar twinning" optically observed.

Similar studies by high-voltage transmission electron microscopy by Ross (1970) of a small crystal fragment from moon rock (10047 Sea of Tranquility) revealed the presence of thin bands 60–100 Å thick, which are interpreted as pigeonite unmixed on (001) from augite host. And, in this case, detail studies revealed a submicron textural intergrowth of two crystal phases which, under normal microscopy, appear to be a single crystal phase.

More evidence in support of the true composite nature of single pyroxene crystals from lunar samples is provided by Bailey et al. (1970). X-ray diffraction studies of some pyroxene single crystals showed them to be unmixed into monoclinic Ca-rich and Ca-poor components; the unmixing was further confirmed by electron microscopy which showed (001) lamellae approximately 500 Å apart.

Using electron microprobe analysis (beam size 2 μm) Malcolm Brown et al. (1970) showed that a pale-brown clinopyroxene was strongly zoned from augite to subcalcic ferro-augite and was accompanied by a decrease in Cr, Al and Ti. Pale-yellow rim and interstitial patches were found to be chemically, but not necessarily crystallographically, subcalcic ferro-augite. In addition to the crypto ex-solution phases and crypto-twinning, both terrestrial and lunar pyroxenes and plagioclases may consist of crypto-zoning which, in cases, can be detected by ordinary polarizing microscopy as for example, the fine oscillatory zoning of Ti augites from Selale Mt., Ethiopia, and the comparable fine oscillatory zoning of the gigantic plagioclase phenocrysts from Debra-Sina, Ethiopia, respectively described in Chapters 8 and 11.

In contrast to the fine oscillatory zoning of the Ti augites from the Selale Mt. basalt, Ethiopia, studies by Hargraves et al. (1970) of zoned lunar pyroxenes showed a pyroxene differentiation trend, unknown in terrestrial pyroxenes, towards extreme enrichment in ferrosilite molecules. It should be pointed out that the zoning analysed by Hargraves et al. was of a coarse nature; due to the lack of high-voltage electron microscopic studies, no detailed information was given about the possible relation between crypto-zoning and enrichment in ferrosilite.

In contradistinction to the studies by Hargraves et al., Johnston and Gibb (1973) showed that the Fe/Mg ratio decreased in reversed zonal lunar pigeonite, from the crystal centre outwards. Here again, the composition changes of zoned pyroxenes are not in accordance with the continuous reaction series hypothesis.

Laves et al. (1965), in their attempt to explain the "Schiller" effect in intermediate plagioclases (i.e. labradorite with An 53%), proposed "unmixing" of submicron lamellae with different Na/Ca and Si/Al ratios as a possible explanation. Their studies revealed that labradorite, with "Schiller" and exhibiting a "single crystal" pattern, showed the submicron lamellae under transmission electron microscopy. These results and the discovery of comparable lamellar patterns in lunar plagioclases clearly support that intermediate plagioclases are not members of continuous series of crystallization, but in the cases studied consist of ex-solution and a host or are due to crypto-twinning.

Most of the above observations tend to support the hypothesis that both the lunar and terrestrial basaltic pyroxenes and plagioclases are frequently not representing intermediate members of a continuous series of crystallization but crypto-ex-solutions, crypto-twinning or submicron crypto-zoning of two mineral phases.

These submicron structures should not be seen as "freaks" of nature but as being inherent to crystal growth and development.

The titanium content of lunar and terrestrial basalts

As a consequence of the lack of a comparable atmosphere to that of the earth, the moon basalts are anhydrous, formed under extreme conditions of reduction. The lack of OH-containing minerals and the limitation of Fe oxides are in accordance with this hypothesis. It is not, therefore, surprising that the lunar basalts are characterized by the absence of hornblende, micas and chlorites; these minerals indicate atmospheric assimilation by the terrestrial basalts.

Another distinguishing feature of lunar basalt is the higher Ti content. Table XVII shows analyses of lunar and terrestrial basalts in respect to their Ti-content. It can be seen that, on the whole, the lunar basalts are distinguished by higher contents of titanium.

However, analyses of terrestrial basalts, prior to the landing on the moon, show that certain terrestrial basalts are also characterized by abnormal contents of titanium. Analysis of a titaniferous basalt from the island of Pantelleria by Washington (1914) showed 6.43% TiO_2, the same basalt re-analysed by Zies (1962) showed 3.94% TiO_2. Also high titanium contents are reported by Noe Nygaard for a non-porphyritic basalt from Vaag Fjord, Suduroy and for a light plagioclase porphyritic basalt from Nessitindar, Kallosy which respectively contain 4.83% and 4.42% TiO_2.

Regarding the petrogenic significance of titanium in basaltic rocks, Chayes (1964, 1965) found that most of the "interoceanic" basalts contain more than 2.0% TiO_2 in contrast to the "circum-oceanic" which contain less than 1.5%.

Present microprobe point analyses (see Chapter 8) along traverses perpendicular to fine oscillatory zoning of titaniferous augite phenocrysts, show that the pyroxene generation I which forms the nucleus has about 1% Ti. In contrast, the outer zones of the augite phenocryst show as much as 3.5% Ti. The basalt in turn showed 2.2% of titanium and in addition to the titaniferous augite, ilmenite was also abundant.

Also, the high Ti content of the lunar basalt is attributed to the presence of Ti-clinopyroxenes and to the presence of ilmenite. However, in addition to ilmenite,

TABLE XVII

TITANIUM CONTENT OF TERRESTRIAL AND LUNAR SAMPLES

Terrestrial basalts	TiO_2 (%)	Analyst
Basalt of Pantelleria Island	6.43	Washington (1914)
Basalt of Pantelleria Island	3.94	Zies (1962)
Basalt from Vag Ford Suduroy	4.83	Noe Nygaard
Basalt, Nessitinder, Kalloy	4.42	Noe Nygaard
Alkali-olivine basalt from north central Oregon	3.6	Robinson (1969)
Augitic basalt, Mt. Selale, Ethiopia	2.2	Vgenopoulos (1976)
Mean analysis of 1966 basalts	1.9	Manson (1967)
Lunar basalts (Apollo 11 Lunar samples numbers)	**TiO_2 (%)**	**Analyst**
10017	11.71	Maxwell et al. (1970)
10020	10.72	Maxwell et al. (1970)
10072	12.28	Maxwell et al. (1970)
10084	7.54	Maxwell et al. (1970)
10017-29	11.16	Maxwell et al. (1970)
10020-30	10.28	Maxwell et al. (1970)
10084-132	7.19	Maxwell et al. (1970)
10084-14	7.75	Agrell et al. (1970)
10045-24	11.10	Agrell et al. (1970)
10044-39	9.18	Agrell et al. (1970)
10060-25	9.02	Agrell et al. (1970)
10084-65	7.42	Hiroshi Haramura et al. (1970)
10085-13-1	11.93	Hiroshi Haramura et al. (1970)
Average analysis of lunar crystalline rocks	**TiO_2 (%)**	**Analyst**
Group I	11.94	Compton et al. (1970)
Group II	11.88	Compton et al. (1970)
Group I	11.72	Essine et al. (1970)
Group II	10.42	Essine et al. (1970)
Type I (vesicular)	11.00	Taylor et al. (1970)
Type II (crystalline)	9.00	Taylor et al. (1970)
Basalt and gabbro	9.55–13.2	Rose et al. (1970)

pseudo-brookite is reported and a new mineral related to the pseudo-brookite ($Fe_{0.5}Mg_{0.5}Ti_2O_5$) was found by Haggerty et al. (1970). Another Ti mineral in lunar samples is a chrome–titanium spinel.

Taking into consideration that some lunar basalts can contain as much as 7% Ti, the high Ti content of lunar basalts can be taken as a distinguishing characteristic and an explanation is problematic. Andersen et al. (1970) put forward that impact breaking or convection thrusting of the crust releases fractions rich in Fe and Ti. Another explanation is put forward by Engel and Engel (1970) that amounts of titanium present as high as 7% suggest either extreme fractionation of lunar rocks or an unexpected solar abundance of titanium.

However, in connection with this, the high titanium content of terrestrial basalts which indicate that, in exceptional cases, the titanium content of terrestrial basalts can be compared to lunar basalts is of particular interest.

Eu in lunar and corresponding terrestrial rocks

In addition to Ti which is relatively enriched in lunar material in comparison to corresponding terrestrial material, Eu is another element which shows marked differences.

All lunar rare-earth-element distribution patterns resemble those of terrestrial abyssal sub-alkaline basalt, but with Eu depleted by about 60% in all lunar sample compared to the adjacent rare-earth elements (Roman, 1970).

In order to understand the relatively high depletion of Eu in lunar rocks in comparison to its adjacent rare earths perhaps the geochemical studies of Goldschmidt (1954) might be useful "For europium we have ample evidence from our investigations that in even a slightly reducing environment the reduction of the element to the devalent state has taken place."

This distinguishing feature of europium has, as a result

its depletion in lunar basalts, where an environment of high reduction prevailed, in comparison to the adjacent rare-earth elements (Fig. 604). It is thus possible that the depletion of Eu by about 60% in all lunar samples compared to the adjacent rare earth elements is due to its reduction from trivalent to divalent under the high reduction conditions which prevailed on the moon.

In contrast, the distribution and occurrence of rare earths (the yttrium group) in analysed terrestrial basalts shows that the geochemical coherency between the rare-earth elements is preserved and Eu is not depleted in comparison to its adjacent rare earth elements.

Furthermore, Roman et al. (1970) suggest that plagioclase, which readily incorporates divalent Eu and also accepts to a lesser degree the trivalent REE might have played a role in the relative depletion of Eu in lunar basaltic rocks.

Geochemical comparisons of lunar (Apollo 11) samples with terrestrial equivalents

Some remarks regarding the geochemical comparison between terrestrial igneous rocks (basalts) and equivalent lunar samples from the Sea of Tranquility (Apollo 11 Lunar Conference, 30 January 1970):

"The crystalline rocks, which have typical igneous textures, range from very-fine-grained vesicular rocks to vaggy medium-grained equigranular rocks. The most common minerals are pyroxene, often highly zoned with iron-rich rims, plagioclase, ilmenite, olivine and cristobalite. Three new minerals occur in igneous rocks. They are pyrox-manganite (a triclinic pyroxene-like mineral), ferropseudobrookite, and a chromium titanium spinel. Free metallic iron and troilite, both of which are extremely rare on earth are common accessory minerals in the igneous rocks".

"All lunar rocks have unusually high concentrations of titanium, scandium, zirconium, natrium, yttrium, and trivalent rare-earth elements and low concentrations of sodium. One of the most striking features of the igneous rocks is the low abundance of europium relative to the other rare-earth elements. Some of the more volatile elements, for example, bismuth, mercury, zinc, cadmium, thalium, lead, germanium, chlorine and bromine are significantly depleted with respect to their presumed abundance in the primitive solar system".

"In the lunar basalts the high abundance of titanium, the separation of europium from the other rare earth elements and the separation of barium and strontium, suggests that these liquids (lunar igneous rocks) are the end product of an extensive fractional crystallization process".

The above remarks emphasize the most striking differences between the Sea of Tranquility basalts and the corresponding terrestrial equivalent.

However, as the similarity of mineralogical composition and the abundance of intersertal and ophitic intergrowths both in lunar and terrestrial basalts suggest, the lunar basalts (despite their lack of phenocrysts) are comparable to and commensurable with their terrestrial equivalents in the sense that the lunar basalts are mainly produced by melting of the lunar gabbroic crust (mainly due to meteoritic impacts) and the terrestrial basalts by fusion of the earth's gabbroic layer with the participation of mantle and crust (see Chapters 25 and 32).

ILLUSTRATIONS

Fig. 16. Volcanic cones and crater lakes following a fissure fault system in the Bishufto region, south of Addis Ababa, Ethiopian part of the Great Rift Valley. The volcanic cones are mainly of ashes – tuffs and pyroclastics. Also olivine basalts occur in this volcanic line.

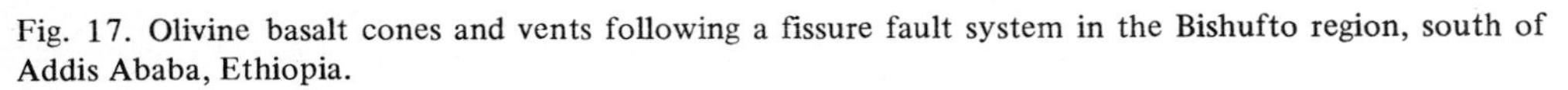

Fig. 17. Olivine basalt cones and vents following a fissure fault system in the Bishufto region, south of Addis Ababa, Ethiopia.

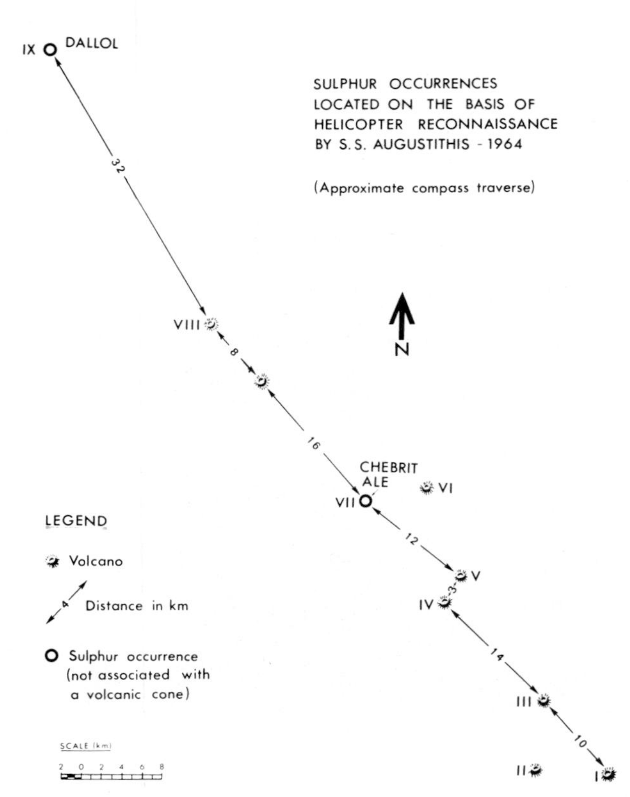

Fig. 18. A sketch diagram based on helicopter reconnaisances by Augustithis (1964) showing the volcanic line Dallol–Lake Jullieta.

Fig. 19. Cone volcanoes occupying a fissure system in the Dallol–Lake Jullieta region of Dankalia, Great Rift Valley, Ethiopian part.
▼

Fig. 20. Cone volcano of the fissure line Dallol–Lake Jullieta. The extrusive material has transversed great thicknesses of evaporites of the Piano del Sale, (salt plane) Dankalia, Great Rift Valley, Ethiopian part. The basaltic magma on its ascent has assimilated components of the evaporite series as a result of which its composition has been altered.

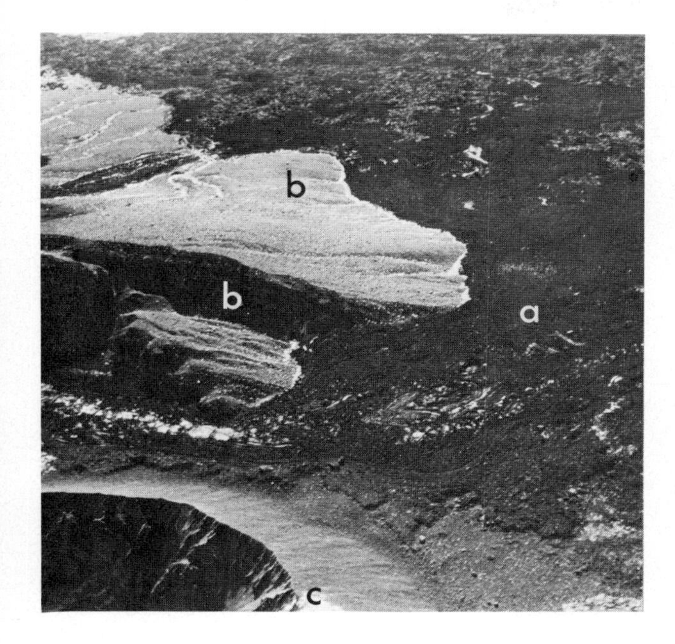

Fig. 21. Desolate landscape of huge quantities of recently out-poured effusives and cone built partly of pyroclastics. The main volcanic cone is partly impregnated by sulphur. The volcanic cone follows a fissure system of the Dallol–Lake Jullieta region of Dankalia, Ethiopian part of the Great Rift Valley.
a = out-poured effusive flow; b = part of the cone built of pyroclastics; c = part of the main volcanic cone impregnated by S.

Fig. 22. Crater and internal secondary cone, as well as an active vent within the crater of a cone volcano, following a fissure system of the Great Rift Valley, Dankalia, Ethiopia.
The internal secondary cone is impregnated by S.

Fig. 23. An active volcano of the Great Rift Valley, Dankalia Ethiopia. The active volcano is situated along a fissure system in the region of Dallol–Lake Jullieta.

Fig. 24. Step-faulting with fissure basaltic flows. Recent basic flows from N.N.W. normal; faults at the S.W. foot of Mt. Borale Ale (Central Range) by Tazieff (1970).

Fig. 25. Fissure basaltic eruption associated with normal faults in the Dankalia central range graben (Rift). By Tazieff (1970).

Fig. 26. Basalts at the margin of the Great Rift Valley at Debra-Sina, Ethiopia. Step-faulting and tectonic terracing of olivine basalts.

Fig. 27. Step-faulting and tectonic terracing of basalts on the Ethiopian Plateau margin of the Great Rift Valley.

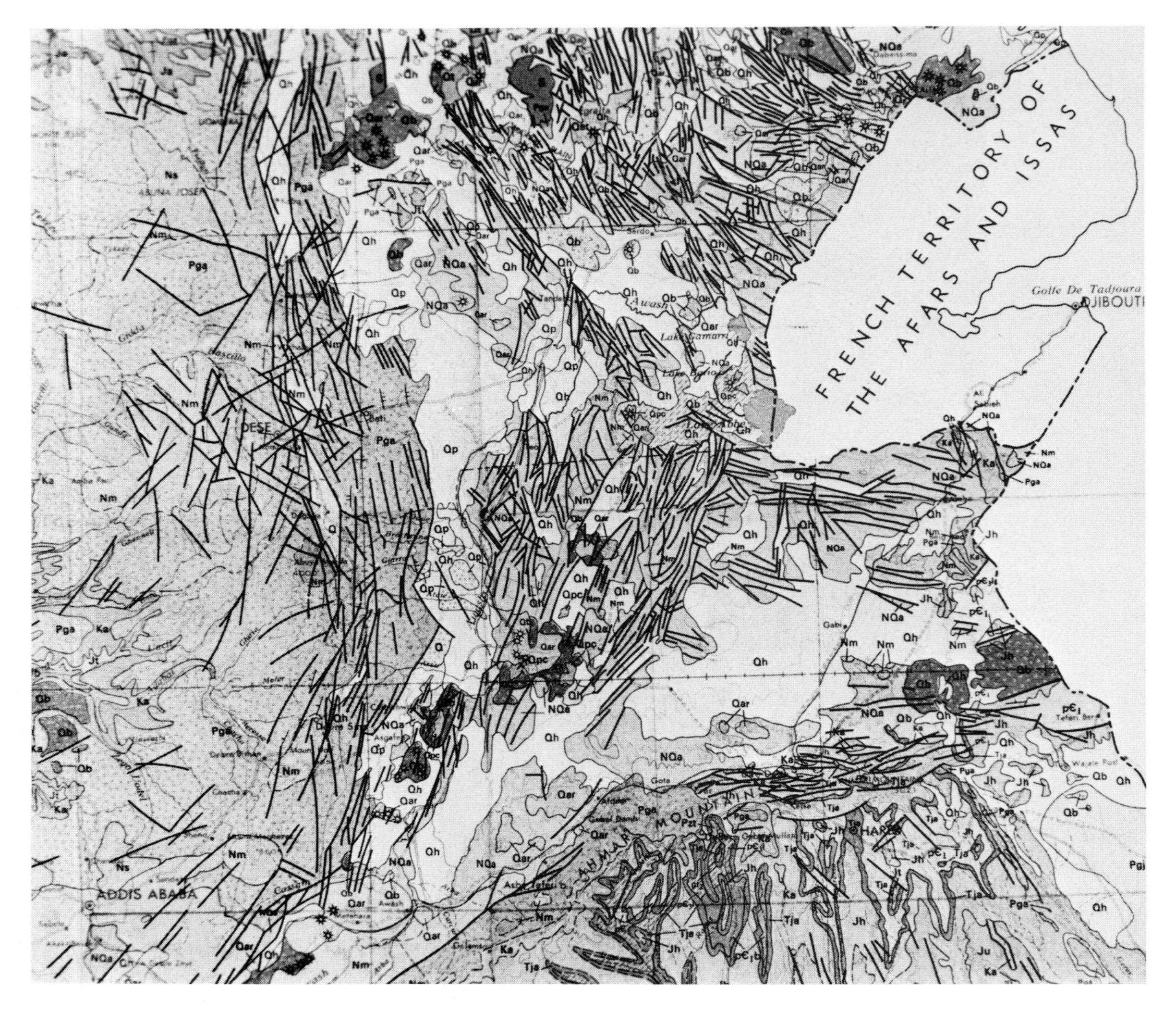

Fig. 28a. A sketch map (based on the geological map of Ethiopia) showing the main fault systems prevailing in the Ethiopian Rift Valley.

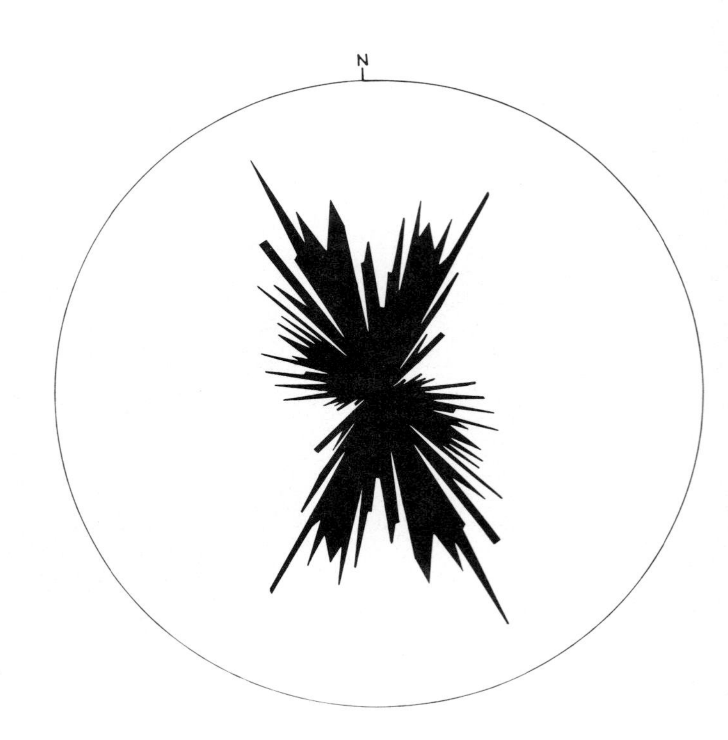

Fig. 28b. A diagram showing statistically the main fault systems of Fig. 28a. (By Dr. E. Boskos.)

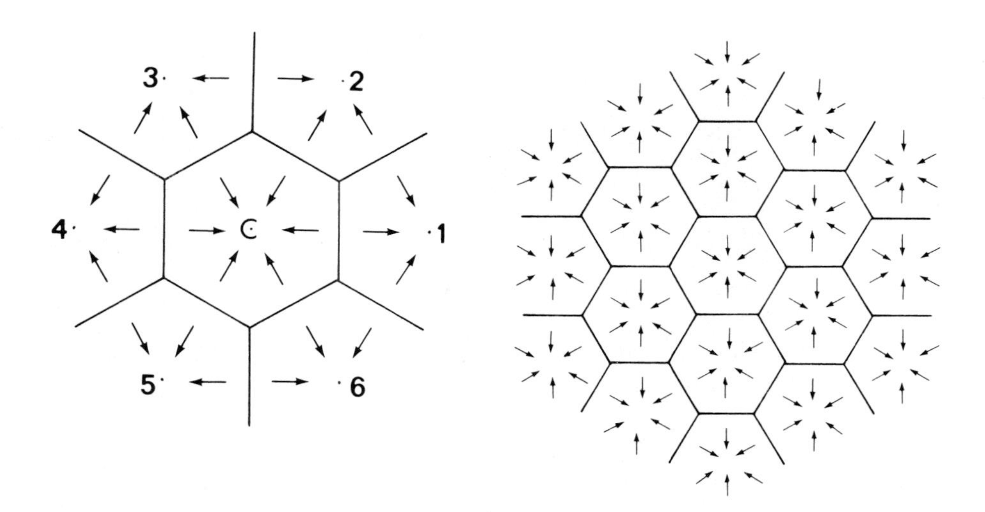

Fig. 29. The formation of an ideal hexagonal pattern of joints by uniform contraction towards evenly spaces centres (Holmes, 1965).

Fig. 30. Greenland's sward. A three-sided columnar basalt from East Greenland (actually a four-sided column). Width of column = 5 cm. (Sample, courtesy of Prof. I. Papagiorgakis.)

Fig. 31. Basalt flow, top and bottom cellular basalt, middle columnar basalt. Isle of Staffa, Argyll. (Photo: H.M. Geological Survey, Great Britain.)

Fig. 32. Basaltic plug showing columnar structure. Devil's Tower, Wyoming, U.S.A. (Photo: T. Nichols)

Fig. 33. Ropy lava (Pahoehoe). Isla Fernandia, Galapagos Islands, Ecquador. (Photo: T. Nichols.)

◀ Fig. 34. Blocky lava. Matahara, Ethiopia.

Fig. 35. Blocky lava. Matahara, Ethiopia.

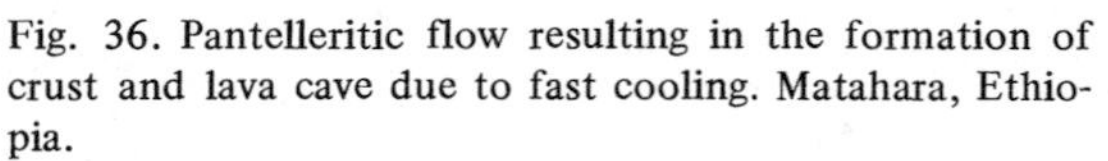

Fig. 36. Pantelleritic flow resulting in the formation of crust and lava cave due to fast cooling. Matahara, Ethiopia.

▼

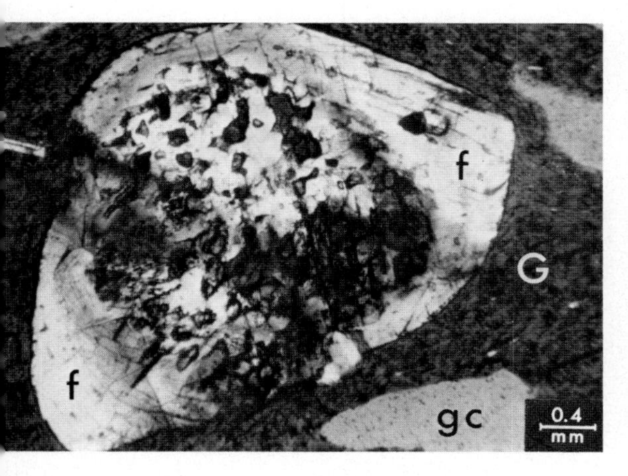

4. Plagioclase collomorph tecoblastoid with gel-relic structures diffuse" twinning, as a phenocryst exercising influence on the of the gas cavity. f = plagioclase collomorph tecoblastoid; G = groundmass; gc = gas cavity.
hyric basalt. Vulcano Islands of Lipari, Tyrrhenian Sea, Italy. rossed nicols.

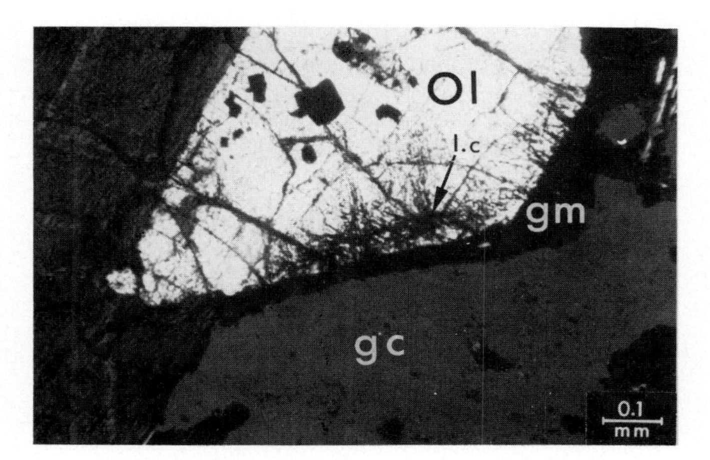

Fig. 55. Olivine phenocryst affected by gas recoil effects, interconnected with gas escape from gas cavities. Ol = olivine; gm = groundmass wall of gas cavities; I.c. = intracrystalline diffusion of gas fronts due to recoil effect; gc = gas cavity.
Olivine basalt. Volcanic line, south of Addis Ababa, Ethiopia. With almost crossed nicols.

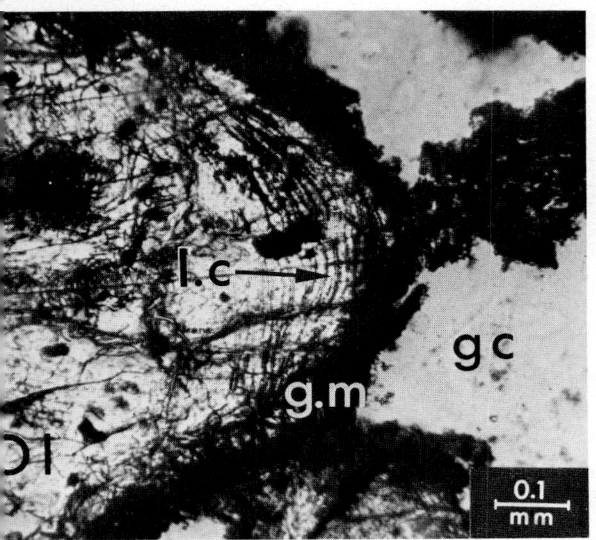

◀ Fig. 56. Olivine phenocrysts affected by gas-recoil effects, interconnected with gas escape from the gas cavities. Ol = olivine phenocryst; g.m = groundmass wall of gas cavities; I.c = intracrystalline diffusion fronts due to recoil effects of gas escape; gc = gas cavities.
Olivine basalt. Volcanic line, south of Addis Ababa, Ethiopia. Without crossed nicols.

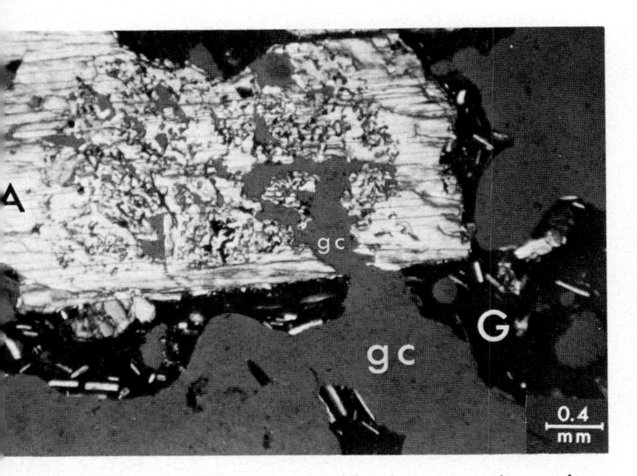

7. Gas cavity extending into an augitic phenocryst. A = augite; as cavity; G = groundmass.
e basalt. Kamchatka, U.S.S.R. With half-crossed nicols.

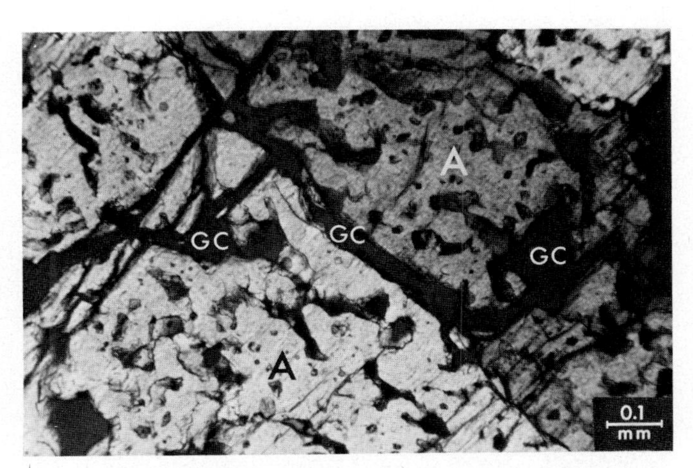

Fig. 58. An intricate system of "channels" in an augitic phenocryst. The system of voids (channels) actually represents extension of cellular structures in an adjacent augitic phenocryst. A = augite; GC = a system of gas-cavity extensions into the augite.
Augite basalt. Kamchatka, U.S.S.R. With half-crossed nicols.

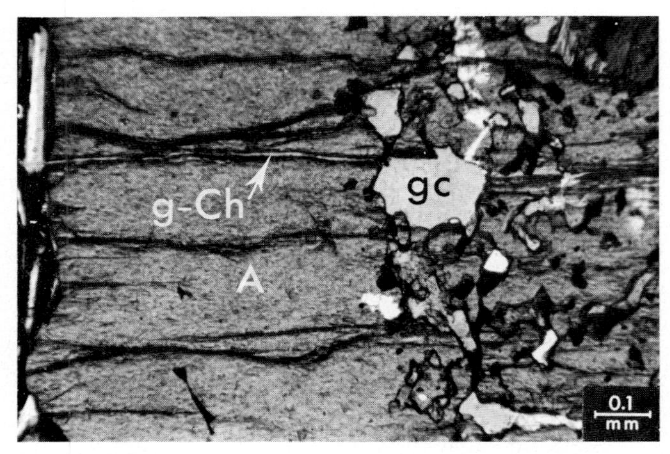

Fig. 59. Cellular structure in an augite phenocryst. Also a fine channel of the gas-cavity system transversing the pyroxene. A = augite; gc = gas cavity; g-Ch = gas-cavity channel transversing the pyroxene.
Augite basalt. Kamchatka, U.S.S.R. With half-crossed nicols.

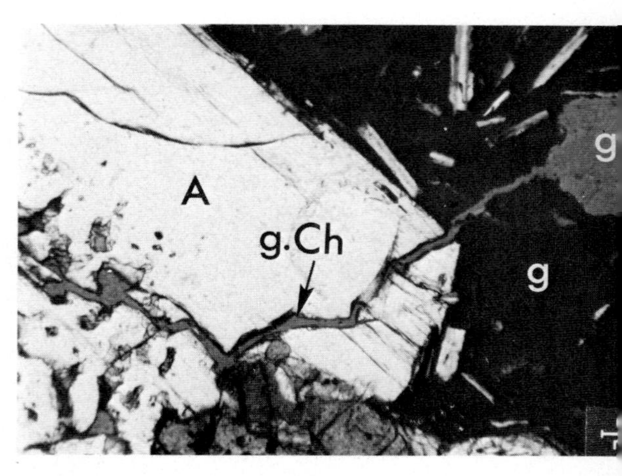

Fig. 60. A gas cavity in the basaltic groundmass with an exte attaining a channel form and branching within an adjacent a phenocryst. A = augite; gc = gas cavity; g.Ch = gas-cavity exte channel; g = basaltic groundmass.
Augite basalt. Kamchatka, U.S.S.R. With half-crossed nicols.

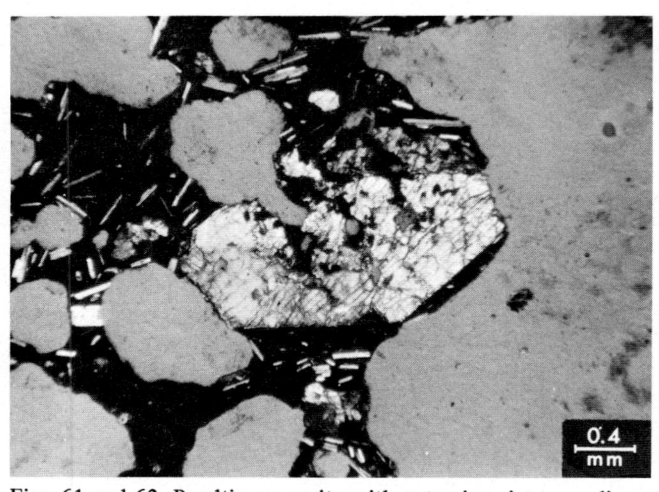

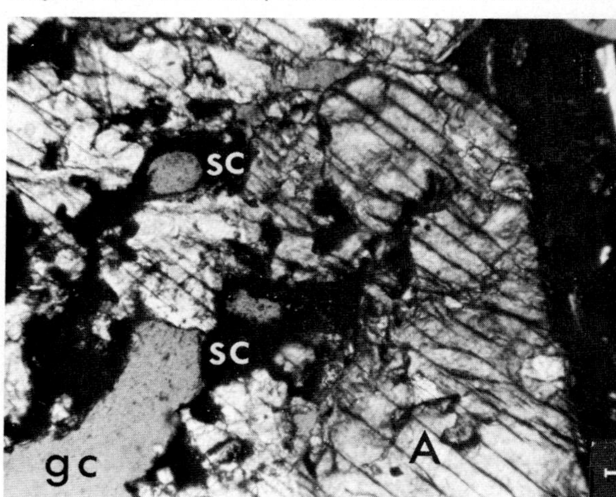

Figs. 61 and 62. Basaltic gas cavity with extensions into an adjacent augitic phenocryst. Due to gas escape, a scoria effect has been produced in the augite adjacent to the gas-cavities extension within the pyroxene. A = augite, gc = gas cavity, sc = scoria due to gas escape within the augitic phenocryst.
Augite basalt. Kamchatka, U.S.S.R. With half-crossed nicols.

Fig. 63. Cluster structure of augitic phenocrysts and magnetite. The augite crystals are radiating out from a centre. py = pyroxene (augites); black = magnetite.
Augite-rich basalt. Selale Mt. Region, Ethiopian Plateau. With crossed nicols.

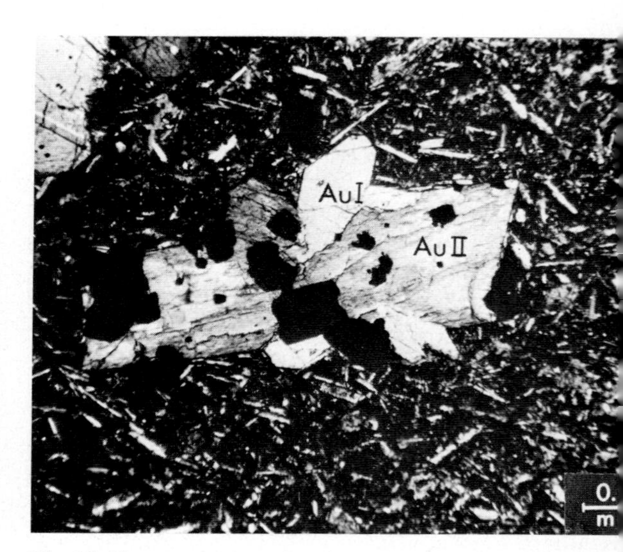

Fig. 64. Cluster structure of augites and magnetites. Augite (I) i perpendicular "orientation" direction, intergrown with augi Au = Augite I; Au II = Augite II; black = magnetite.
Augite-rich basalt. Selale Mt. region, Ethiopian Plateau. With cr nicols.

Fig. 65. Cluster structure consisting of two augites (indicating hourglass structures). A = augite; O = olivine; g = basaltic groundmass.
Augite-rich basalt. East Siberia, U.S.S.R. With crossed nicols.

Fig. 66. Cluster structure of augites. Large augite-twinned enclosing magnetite, with zoned and twinned augite in perpendicular intergrowth with the large augite. Au-I = large twinned augite; Au-II = augite in perpendicular intergrowth with Au-I.
Augite-rich basalt. Selale Mt. region, Ethiopian Plateau. With crossed nicols.

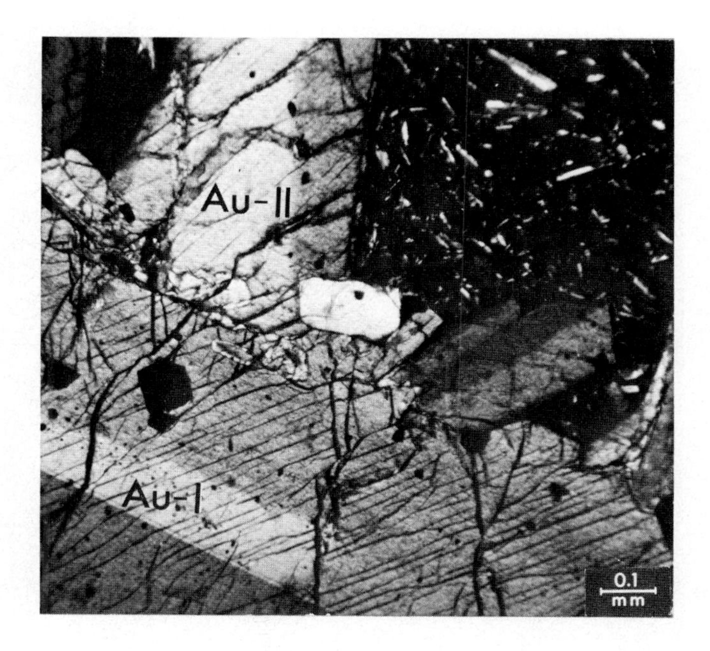

Figs. 67 and 68. Cluster structure consisting of olivine and pyroxene phenocrysts with intrapenetrations of sub-phenocrystalline plagioclase laths. PO = olivine phenocrysts; p = pyroxene; pl = plagioclase sub-phenocrysts.
Olivine basalt. Mid-Atlantic Ridge. With crossed nicols.

Fig. 69. Olivine megablasts following a rupture(?) zone in an olivine fels, with tremolite. Along the rupture zone talc is also formed.
Prata South Chiavenna, Alps. (scale shown by knife).

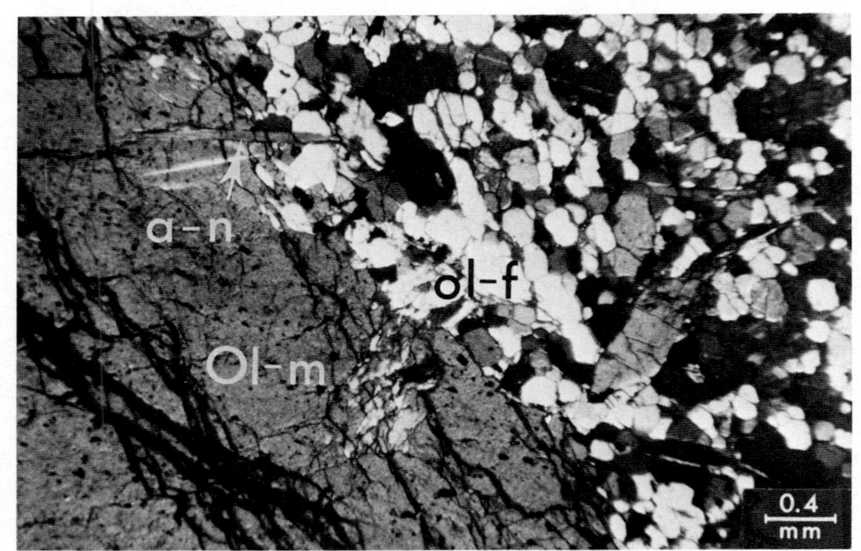

Fig. 70. Megablastic olivine in contact with micro-granoblastic granular olivine fels. The megablastic olivines are formed as post-kinematic blastesis along regenerated rupture planes of the olivine-fels. Amphibole nematoblasts often develop in the olivine-fels and in cases transverse the olivine-fels–olivine megablast boundary. Ol-m = olivine megablast; ol-f = micrograno-blastic granular olivine-fels; a-n = amphibole nematoblast (tremolite) crossing the olivine-fels–olivine megablast boundary.
Prata South Chiavenna, Alps. With crossed nicols.

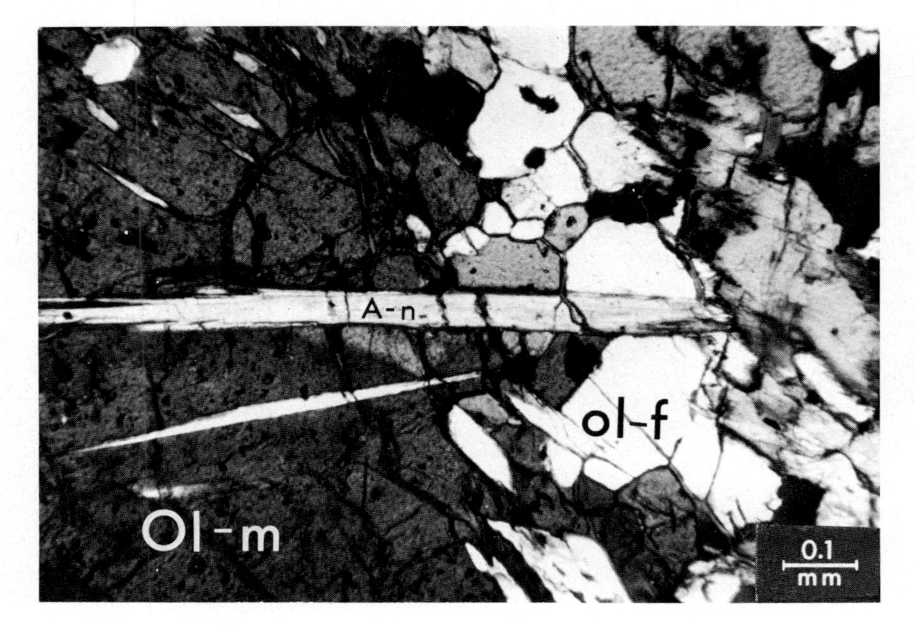

Fig. 71. Amphibole nematoblast (velonoblast), due to its blastogenic force of crystallization, transverses the boundary olivine-fels–olivine megablast. Ol-m = olivine megablast; ol-f = microgranoblastic "granlar" olivine-fels; A=n = amphibole nematoblast (velonoblast) – due to its blastogenic force of crystallization penetrates through both the olivine-fels and the olivine megablast.
Prata South Chiavenna, Alps. With crossed nicols.

Fig. 72. Folded Triassic marbles metasomatically affected by solution fronts. The metasomatic solutions have, in addition to their penetration, selectively followed bands of the folded Triassic marble or anticlinal crests. m = folded Triassic marbles; m-A = metasomatically replaced anticlinal crest; m-b = marble bands, selectively replaced by metasomatic solutions.
Malga Trivena, Val di Breguzzo, Adamello, Alps.

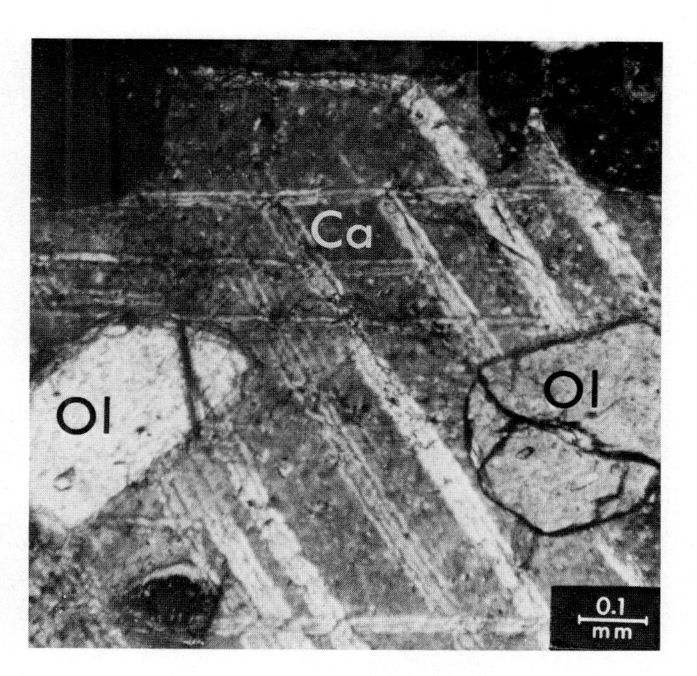

Fig. 73. Granoblastic olivine in re-crystallized calcite (coarse-grained marble). Ol = granoblastic olivine; Ca = calcite twinned.
Metasomatically affected marbles. Malga Trivena, Val di Breguzzo, Adamello, Alps. With crossed nicols.

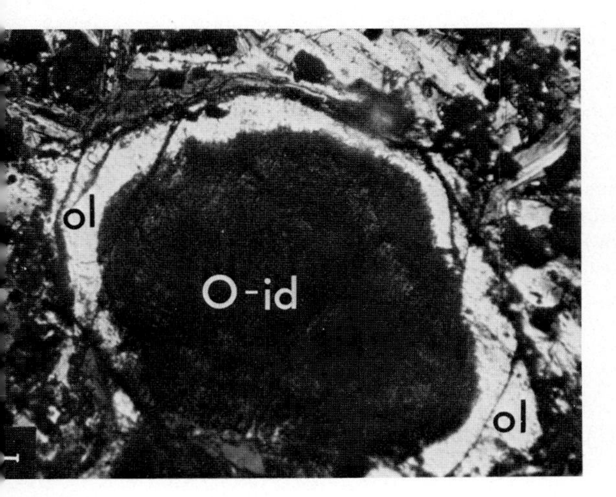

Fig. 74. First-generation olivine completely changed into iddingsite. An overgrowth of second generation of olivine is exhibited. O-id = first-generation olivine changed to iddingsite; ol = olivine overgrowth.
Olivine basalt. Ankober, Ethiopia. With crossed nicols.

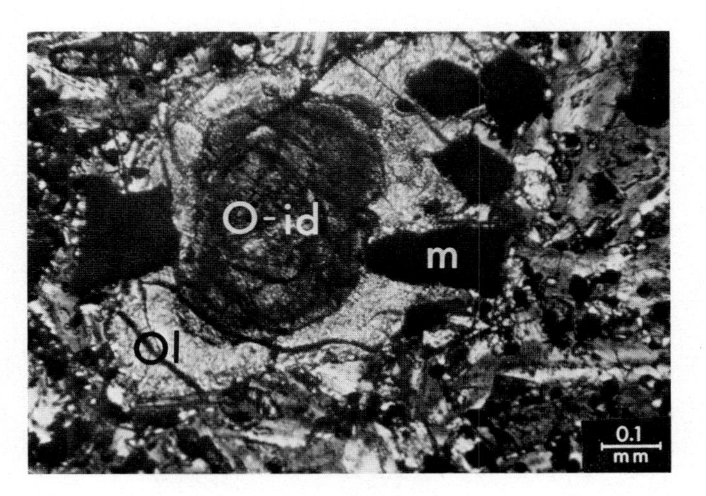

Fig. 75. First generation of olivine almost completely changed to iddingsite. An overgrowth of olivine partly enclosing magnetite crystal grains of the groundmass. O-id = first-generation olivine changed to iddingsite, Ol = olivine overgrowth, m = magnetite crystal grains of the groundmass.
Olivine basalt. Ankober, Ethiopia. With crossed nicols.

Fig. 76. General view of olivine basalt with tecoblastic idiomorphic plagioclase phenocrysts. Rounded and magmatically corroded olivine phenocryst is also shown. F = plagioclase tecoblastic phenocrysts; Au = augite phenocrysts, Ol = rounded olivine phenocrysts most probably representing a mantle xenocryst.
Olivine basalt. Khidane Meheret S. Michele H. Dodde/Addis Ababa. With crossed nicols.

Fig. 77. Olivine phenocryst in basaltic groundmass. The olivine phenocryst shows lamellar stress twinning due to tectonic influences.
Olivine basalt. Adama (Nazaret), Ethiopia. With crossed nicols.

Fig. 78. Twinned olivine in basalt, a and b twinned lamellae differently orientated. Arrow (a) shows corrosion of the olivine phenocryst by the groundmass.
Olivine basalt. Adama (Nazaret), Ethiophia. With crossed nicols.

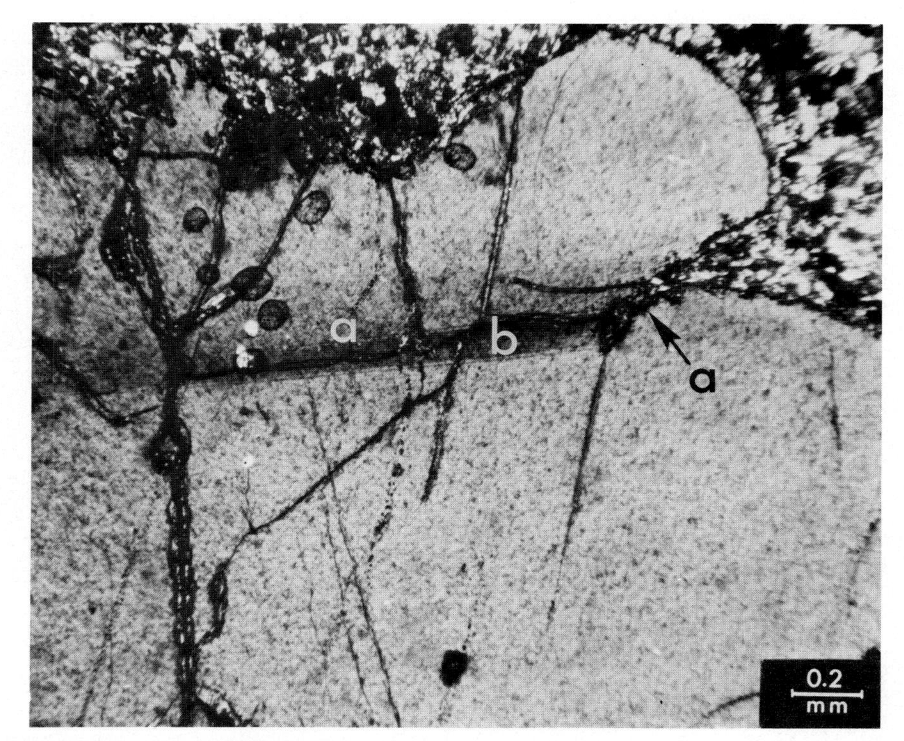

Fig. 79. Olivine xenocryst in basalt (representing an olivine nodule) surrounded by an isotropic alteration margin. The olivine shows a complex pattern of deformation twin lamellae. Ol = olivine with deformation twin lamellae, Is = isotropic margin of olivine alteration, bg = basaltic groundmass.
Olivine basalt. Yubdo, W. Ethiopia. With crossed nicols.

Fig. 80. Co-existence of rounded olivine xenocryst, most probably of mantle derivation, and idiomorphic second-phase olivine phenocryst in picritic basalt. r-ol = rounded olivine phenocryst (xenocryst); i-ol = idiomorphic olivine phenocryst.
Picritic basalt. Agia Marina, Syliatou, Cyprus. With crossed nicols.

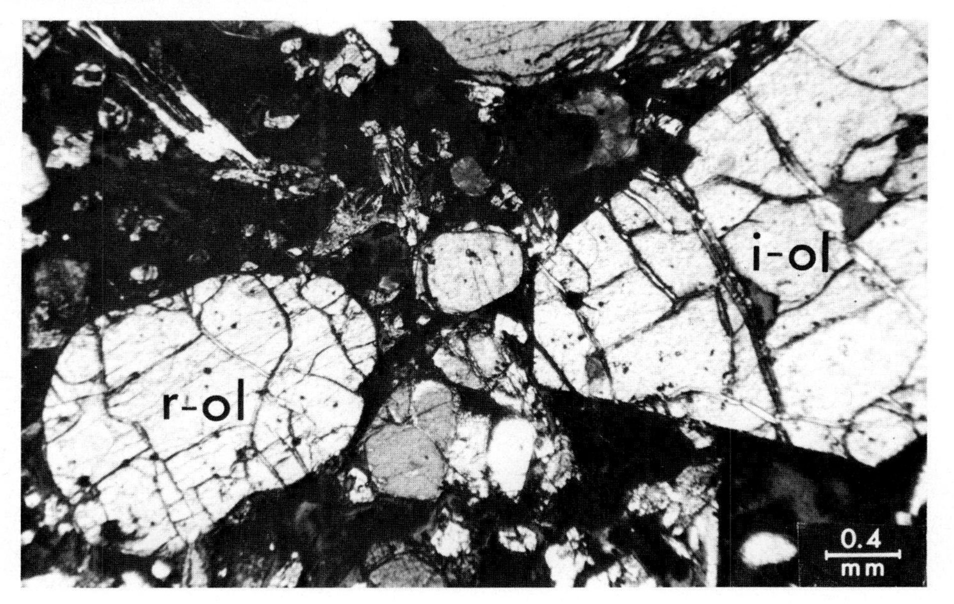

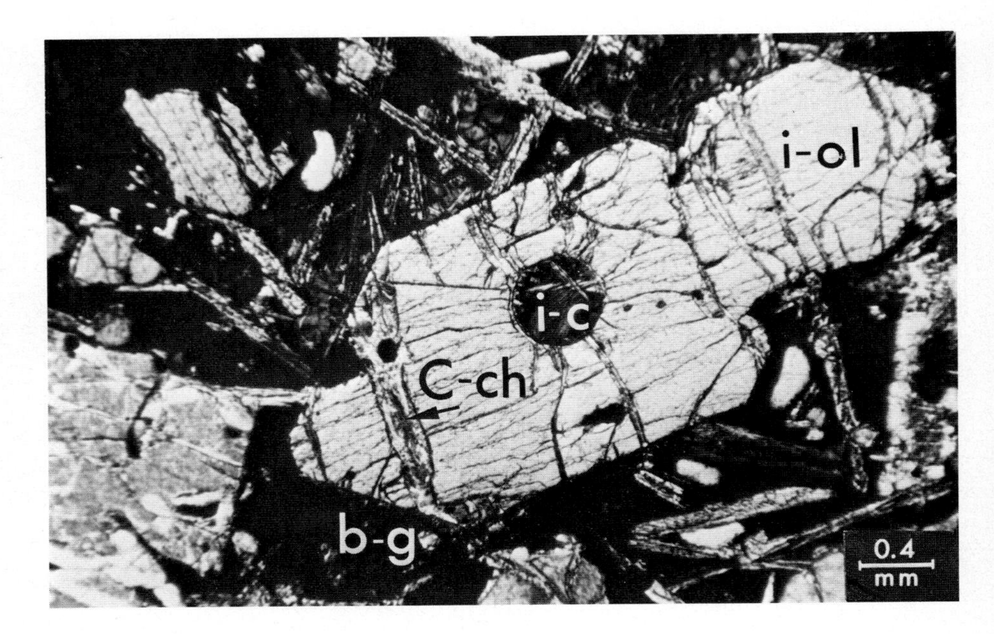

Fig. 81. Idiomorphic olivine phenocryst in picritic basalt. A rounded intracrystalline cavity of the phenocryst is filled with chlorite, also chloritic "channels" transverse the phenocryst. i-ol = idiomorphic olivine phenocryst of the second phase (generation); C-ch = chloritic channels transversing the olivine phenocryst; i-c = rounded cavity in the olivine phenocryst, b-g = picritic basalt groundmass.
Picritic basalt. Agia Marina, Syliatou, Cyprus. With crossed nicols.

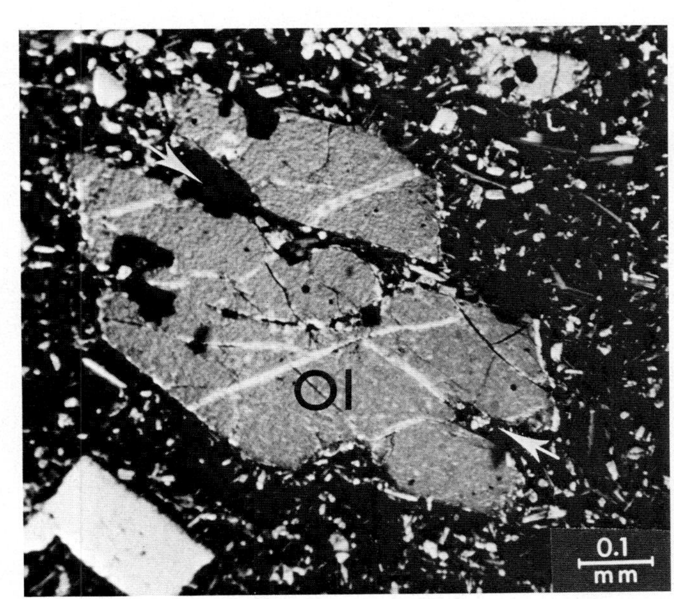

Fig. 82. Idiomorphic olivines in olivine bombs containing basalts. The idiomorphic olivine shows magmatic corrosion and groundmass extensions in the olivine phenocrysts. Ol = olivine phenocrysts; arrow shows groundmass extension in corroded enclaves of the olivine.
Olivine "boms" in basalt. Lekempti, W. Ethiopia. With crossed nicols.

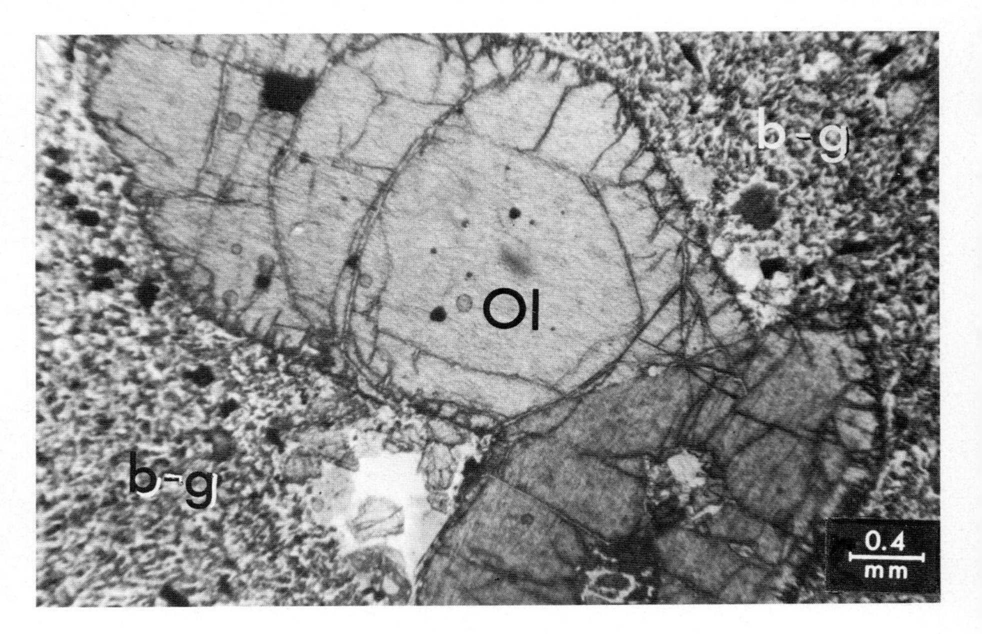

Fig. 83. Olivine phenocryst of the second olivine generation with a common "junction face" comparable to a twin plane. Ol = idiomorphic olivine phenocrysts; b-g = basaltic groundmass.
Olivine basalt. Deccan, India. Without crossed nicols.

Fig. 84. Idiomorphic olivine phenocryst. Ol = idiomorphic olivine; b-g = basaltic groundmass; g-c = gas cavity.
Olivine basalt. Akaki volcanic line, south of Addis Ababa, Ethiopia. With half-crossed nicols.

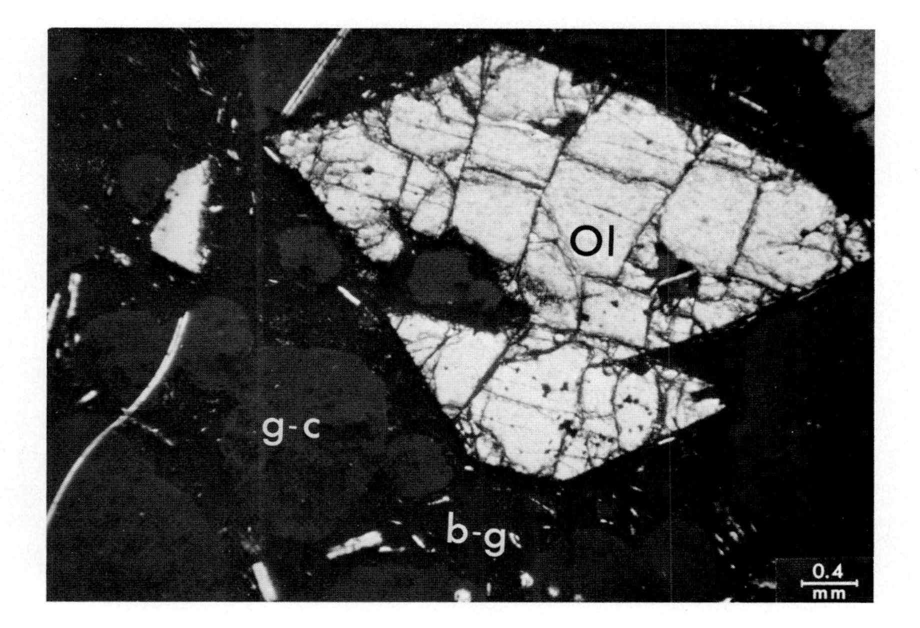

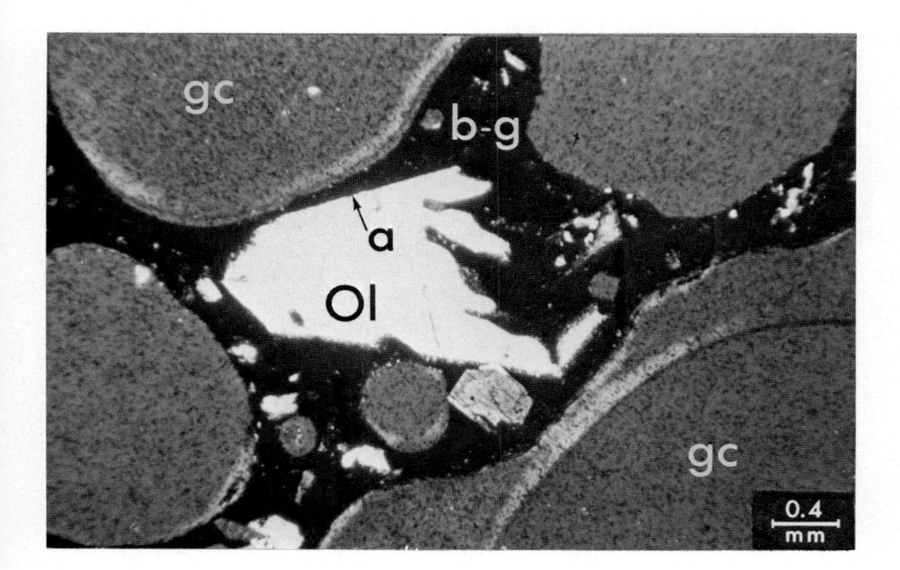

Fig. 85. Magmatically corroded olivine phenocryst. It should be pointed out that due to magmatic corrosion, no intermediate or new mineral phases have been produced. Ol = magmatically corroded olivine phenocryst, one of the crystal faces is unaffected by corrosion (arrow a); b-g = basaltic groundmass; gc = gas cavity.
Olivine basalt. Akaki volcanic line, south of Addis Ababa, Ethiopia. With half-crossed nicols.

Fig. 86. "Equi-granular" olivine texture of the olivine bombs in the basalt. Melt-diffusion paths, transversing olivine grains.
Olivine bombs in basalt. Eifel, Germany. With crossed nicols.

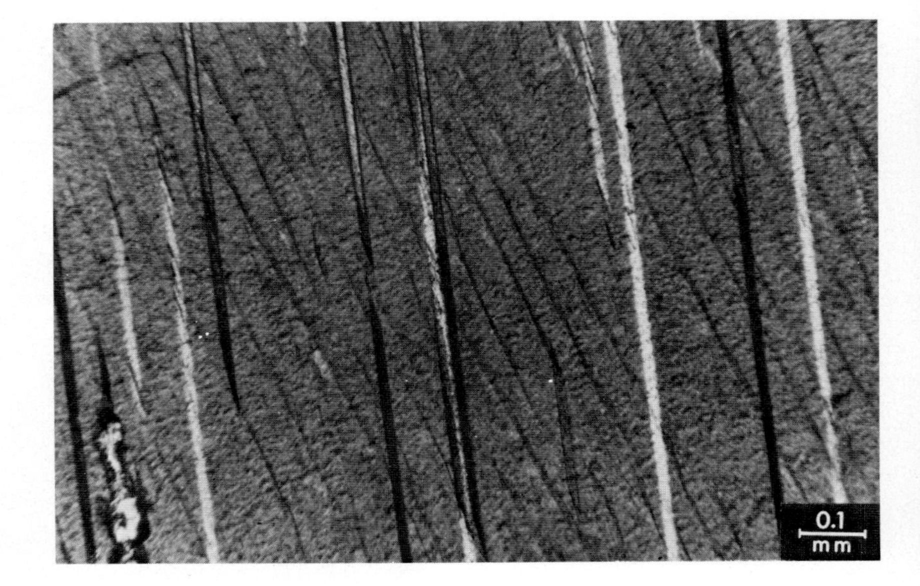

Fig. 87. Bronzite with polysynthetic twinning, the twin lamellae are at an angle to the cleavage.
Jato, Lekempti, W. Ethiopia. With crossed nicols.

Fig. 88. Bronzite of mantle fragment in basalt indicating a rounded fracture pattern; probably due to tectonic mantle deformation. The rounded structure is optically differently orientated.
Olivine bomb in basalt. Finkenberg near Bonn-Beuel, Rhineland, Germany. Without crossed nicols.

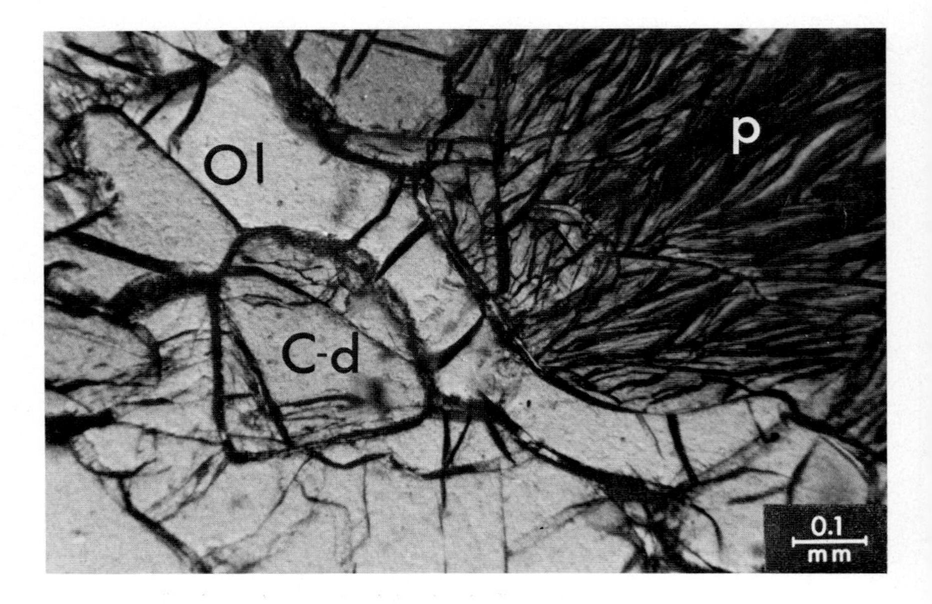

Fig. 89. Chrome-diopside (green in colour) with forsterite and with pyroxene showing stress-cleavage (cleavage pattern due to pressure). C-d - chrome-diopside; Ol = olivine (forsterite); p = pyroxene.
Olivine bomb in basalt. Gerona, Spain. With crossed nicols.

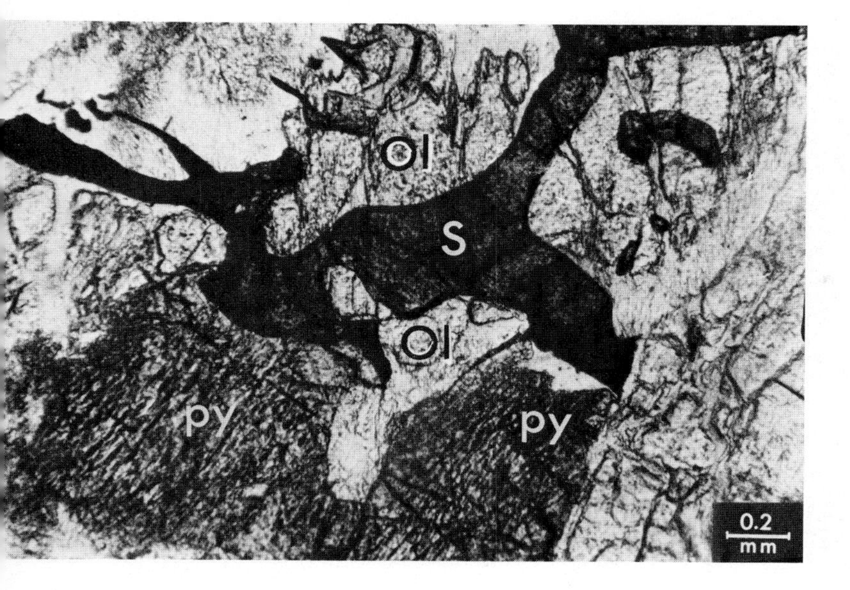

Fig. 90. Olivine and pyroxenes in intergrowth with "graphic" spinel (skeleton spinel crystals) from a mantle diapir. Ol = olivine; py = pyroxene, S = spinel.
Mantle diapir in orogen. Near Konitsa, Greece. Without crossed nicols.

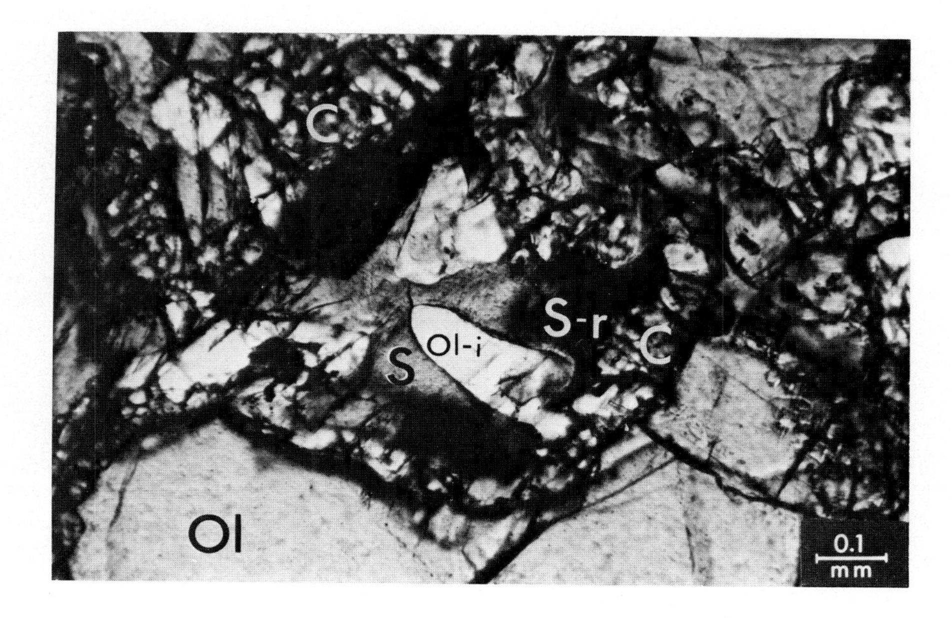

Fig. 91. Spinel engulfing olivine parts and simultaneously showing a reaction margin with the corona structure. Ol = olivine; S = spinel; S-r = spinel with reaction margin with corona structure; C = corona reaction structure; Ol-i = olivine included or engulfed by the spinel.
Olivine bomb in basalt. Jato, Lekempti, W. Ethiopia. With crossed nicols.

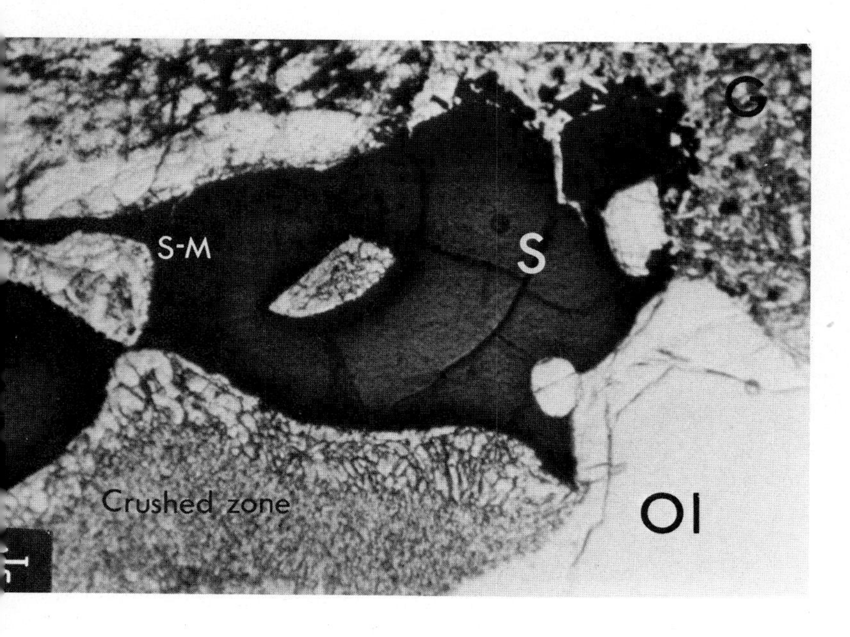

Fig. 92. Spinel (with marginal changes into magnetite) in contact with the basaltic groundmass by which it is affected and changed into magnetite and in intergrowth with forsterite. Also it partly encloses and engulfs parts of the forsterite. In addition, a "crushed zone" is developed marginally to the spinel. G = basaltic groundmass; S = spinel; S-M = spinel marginally changed into magnetite; Ol = olivines partly engulfed by the spinel.
Olivine bomb in basalt. Jato, Lekempti, W. Ethiopia. Without crossed nicols.

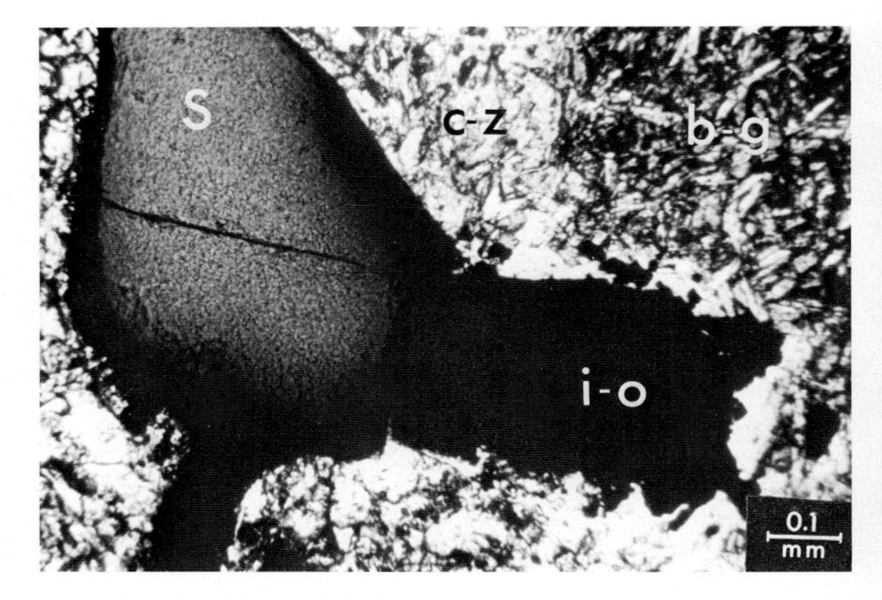

Fig. 93. Spinel surrounded by an olivine–pyroxene crushed zone with marginal transitions to opaque irion oxides. S = spinel; i-o = iron oxides as opaque margins of the spinel, due to thermal influence by the basaltic groundmass; c-z = crushed zone of pyroxene and olivine surrounding the spinel; b-g = basaltic groundmass.
Olivine bombs (mantle fragments) in basalt. Jato, Lekempti, W. Ethiopia. Without crossed nicols.

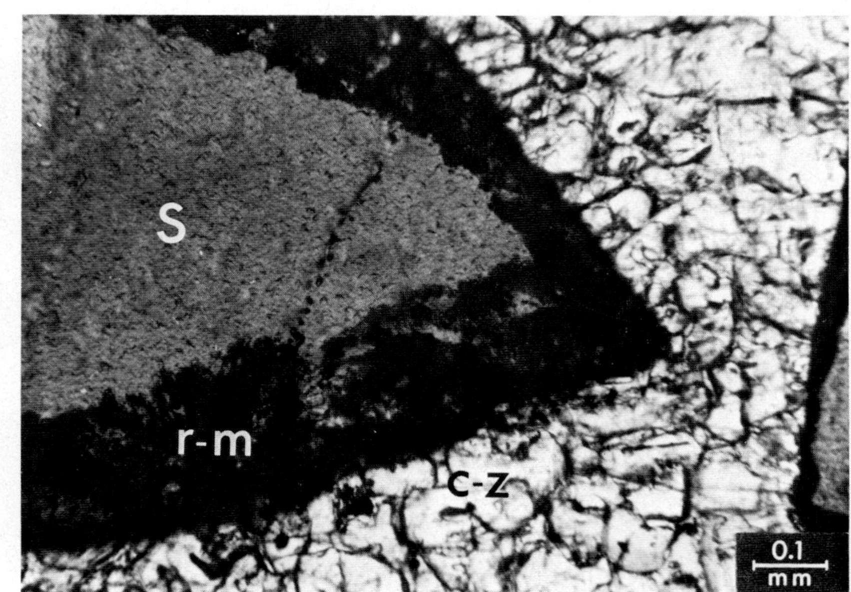

Fig. 94. Spinel with its reaction margin in contact with the reaction corona structure (crushed zone). The spinel reaction margin is produced due to intracrystalline infiltration of "melts" in the spinel. S = spinel; r-m = reaction margin of the spinel in contact with the corona structure; c-z = crushed zone (corona reaction structure).
Olivine bombs in basalt. Jato, Lekempti, W. Ethiopia. Without crossed nicols.

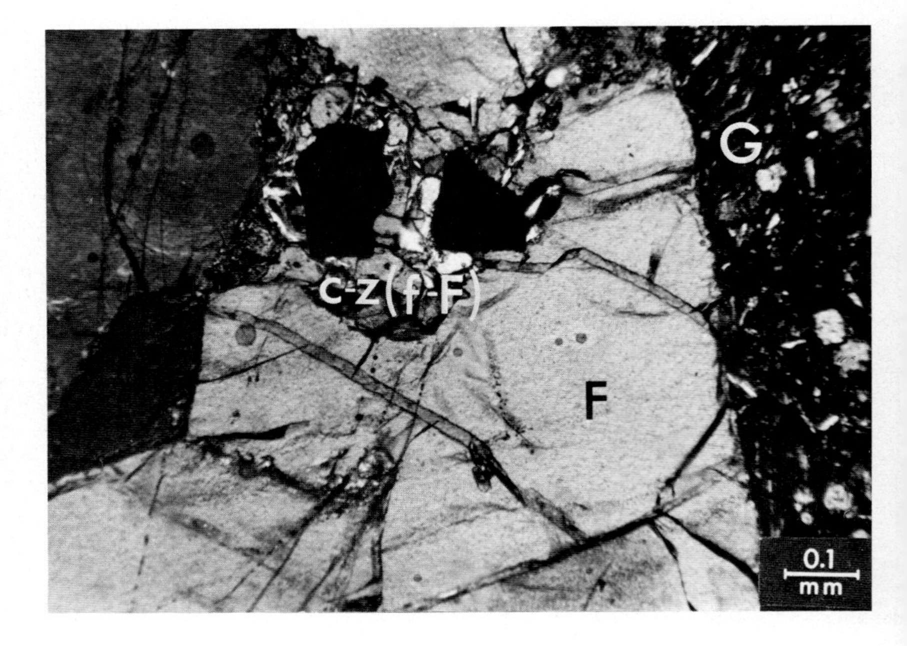

Fig. 95. Magnetite in intergrowth with forsterite, component of olivine bombs in basalts. F = forsterite; black magnetite; G = groundmass; f-F = fine-grained forsterite surrounding the magnetite forming a corona structure (possibly due to basaltic groundmass infiltration "intergranular" between the forsterite and the magnetite); c-z = crushed zone around the magnetite.
Olivine basalt with olivine nodules. Steinwitzhügel, Glatzer circle, Sudetes. With crossed nicols.

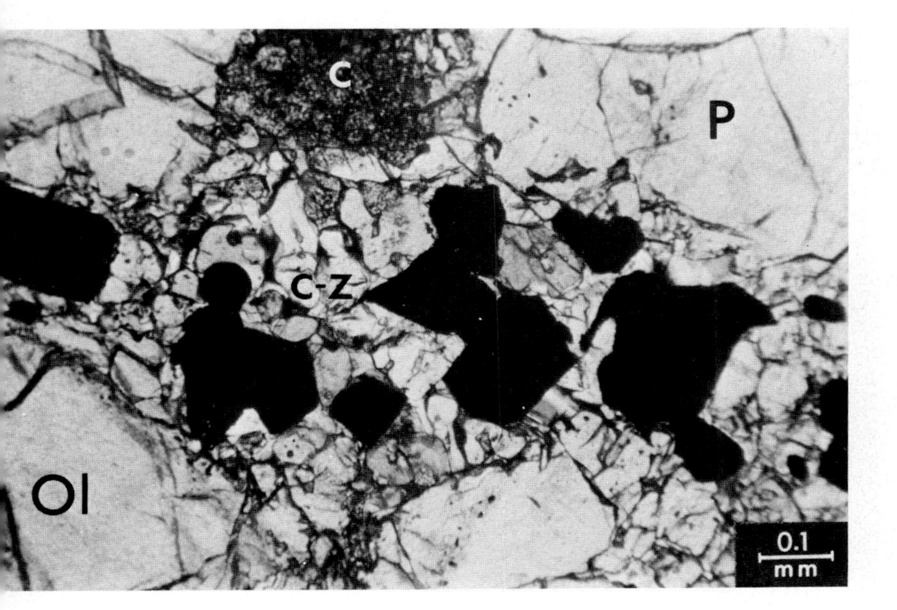

Fig. 96. Comparable to Fig. 95, showing the intergrowth of the magnetite with the reaction corona structure of the forsterite. Ol = olivine, black = magnetite; c = corona structure due to infiltration of "basaltic melts" intergranular between olivines, or olivines and pyroxenes or between the silicates and the magnetite; P = pyroxene; c-z = crushed zone around the magnetite.
Olivine basalt with olivine nodules. Steinwitzhügel, Glatzer circle, Sudetes. With crossed nicols.

Fig. 97. Pyroxene and forsterite in graphic intergrowth with chrome-spinel. A crushed zone is also shown to be associated with the spinels. p = pyroxene; f = forsterite; s = spinels; c-z = crushed zone. Olivine bomb in basalt. Lanzarote, Canary Islands. With crossed nicols.

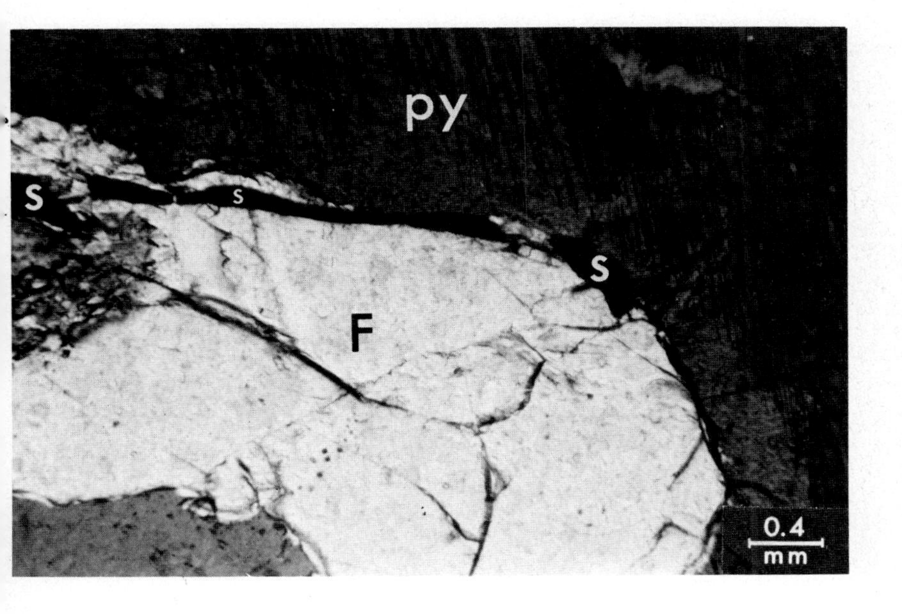

Fig. 98. Forsterite and pyroxene components of mantle-derived olivine nodules with chrome spinel in intergrowth with the olivine and partly following the boundary forsterite–pyroxene. F = forsterite; py = pyroxene; S = chrome-spinel.
Olivine bomb in basalt. Lanzarote, Canary Islands. With crossed nicols.

Fig. 99. Rounded pyroxene (bronzite with characteristic cleavage pattern, as inclusion of olivine). Ol = olivine; p = pyroxene (bronzite).
Olivine bolmb in basalt. Lanzarote, Canary Islands. With crossed nicols.

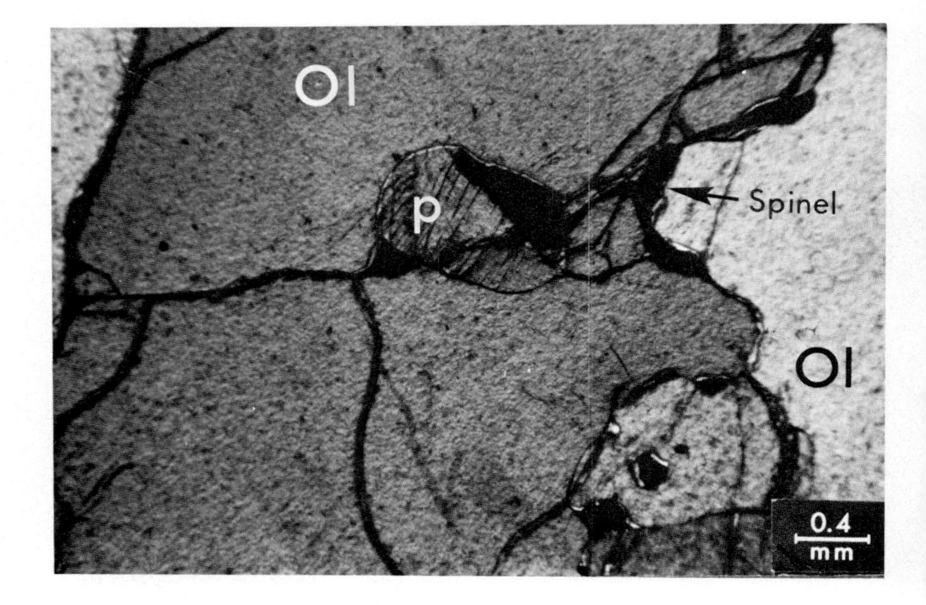

Fig. 100. Synantectic reaction intergrowths of pyroxene (bronzite) in contact with olivine. Extension of the bronzite protrudes into forsterite. B = pyroxene (bronzite); Ol = olivine (forsterite); py-ex = pyroxene extensions into the forsterite as synantectic reaction intergrowths.
Olivine bomb in basalt. Lanzarote, Canary Islands. With crossed nicols.

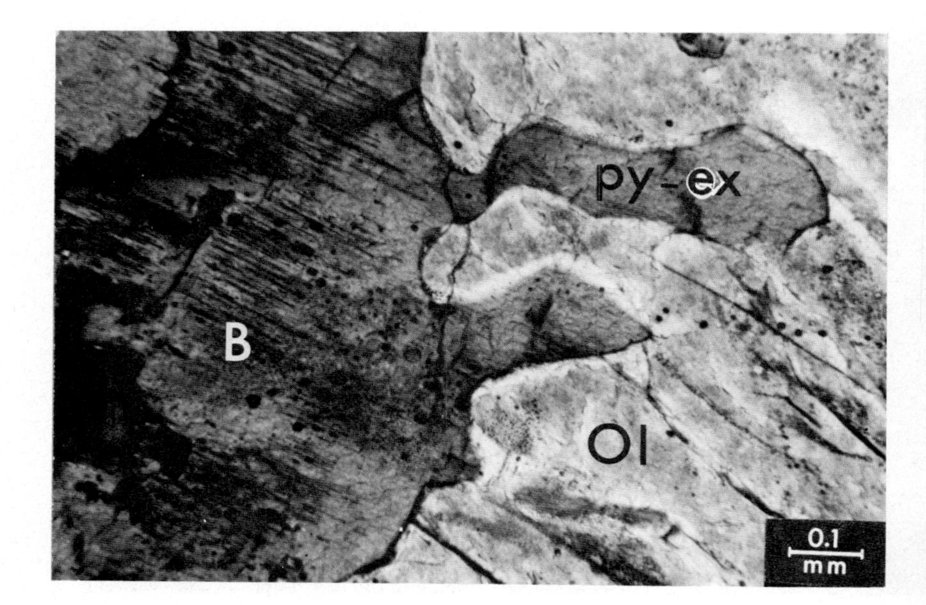

Fig. 101. Deformation lamellae of forsterite. Along the weakness plane of the tectono-twinning, pyroxene (bronzite) extension protrudes into the olivine. Ol = olivine; Tec-T = tectono-polytwinning (deformation lamellae) of the olivines; ext-p = extension of the pyroxene into the olivine along the planes of tectono-twinning.
Olivine bombs in basalt. Lanzarote, Canary Islands. With crossed nicols.

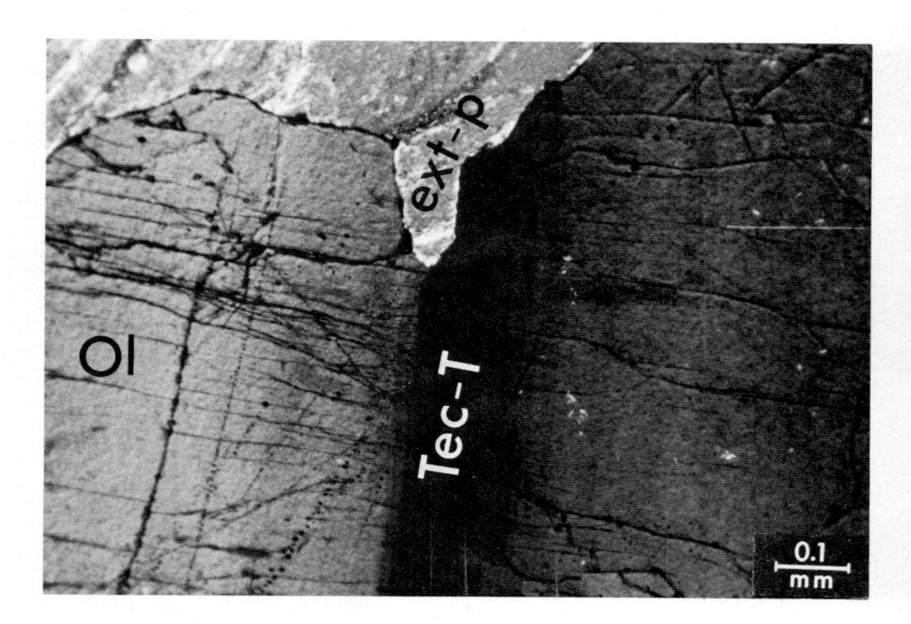

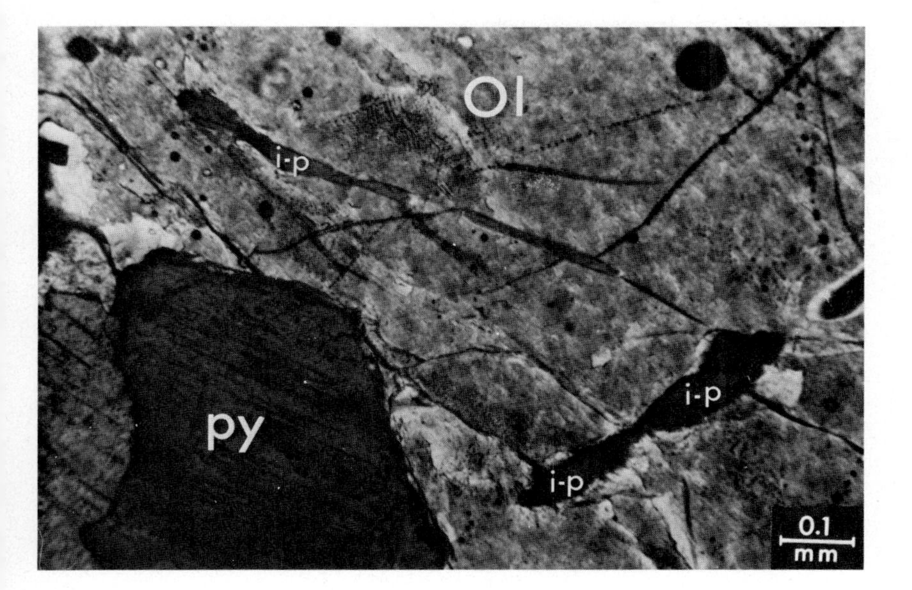

Fig. 102. Pyroxene (bronzite) in contact with olivine. Interleptonic infiltration of pyroxene exists in the olivine, often following crystal cracks. Ol = olivine; py = pyroxene; i-p = interleptonic "infiltrations of pyroxenes into the forsterite". Olivine bomb in basalt. Lanzarote, Canary Islands. With crossed nicols.

Fig. 103. Deformed forsterite with pyroxene following the interleptonic cracks of the olivine. Ol = olivine; py = pyroxene following cracks of the olivine. Olivine bomb in basalt. Lanzarote, Canary Islands. With crossed nicols.

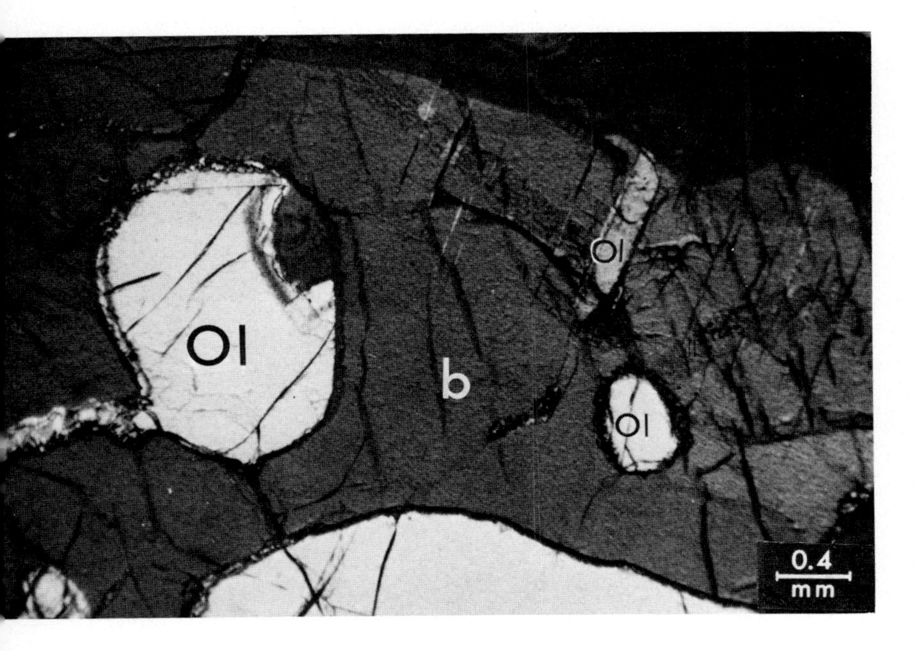

Fig. 104. Rounded forsterite in bronzite and lamellar forsterite orientated parallel to the bronzites' cleavage (and fine twinning). Ol = olivine; b = bronzite.
Olivine bomb in basalt. Jato, Lekempti, W. Ethiopia. With crossed nicols.

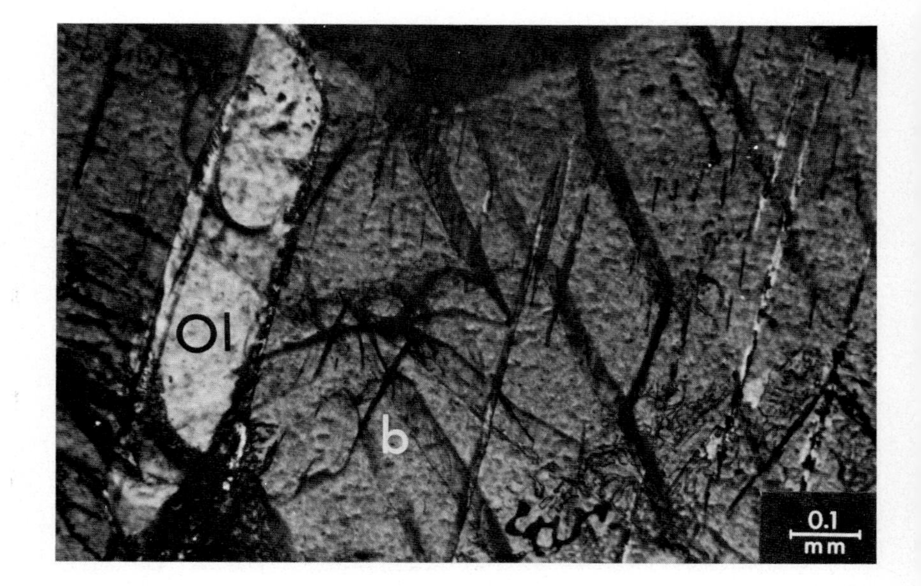

Fig. 105. Detail of Fig. 104. Olivine lamella orientated parallel to the cleavage and twinning of the bronzite host. Ol = olivine; b = bronzite.
Olivine bomb in basalt. Jato, Lekempti, W. Ethiopia. With crossed nicols.

Fig. 106. Pyroxene (bronzite) in contact with and partly corroded and rounded by another crystalline pyroxenic phase (diopside) which protrudes and extends into the bronzite following its cleavage direction. B = bronzite; d = diopside following cleavage penetrability of the host bronzite; arrow (a) shows extensions of the diopside following the interleptonic cleavage spaces of the bronzite.
Olivine bomb in basalt. Lanzarote, Canary Islands. With crossed nicols.

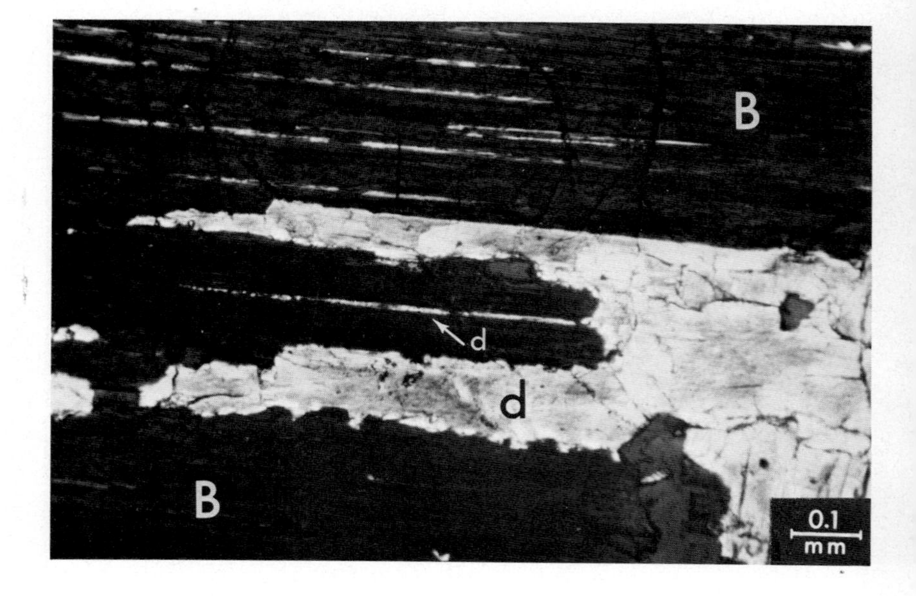

Fig. 107. Detail of Fig. 106. The diopsidic phase extends along the interleptonic cleavage spaces of the bronzite host. B = bronzite; d = diopside extending along the cleavage of the host bronzite.
Olivine bomb in basalt. Lanzarote, Canary Islands. With crossed nicols.

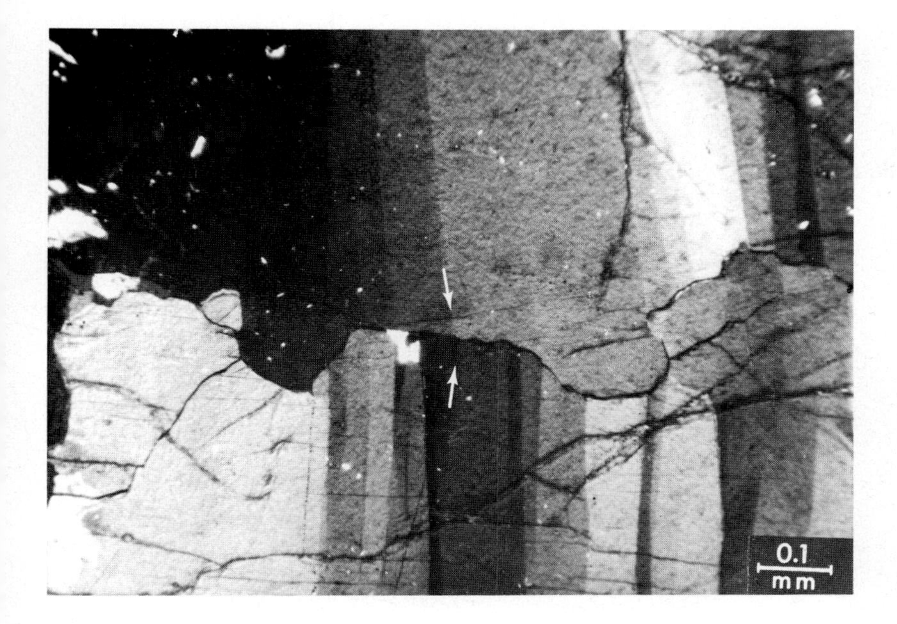

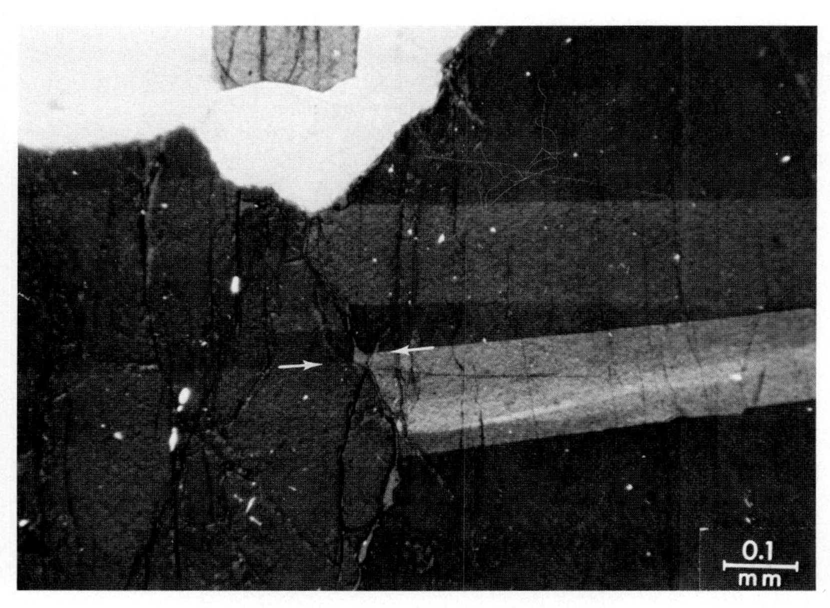

Figs. 119 and 120. Two olivine individual grains in which subparallel tectono-poly-twinning has developed. The twinning in both individuals is similarly orientated and it seems to cross the olivine grain boundaries. Arrows show twin lamellae crossing the crystal grain boundaries.
Olivine bomb in basalt. Lanzarote, Canary Islands. With crossed nicols.

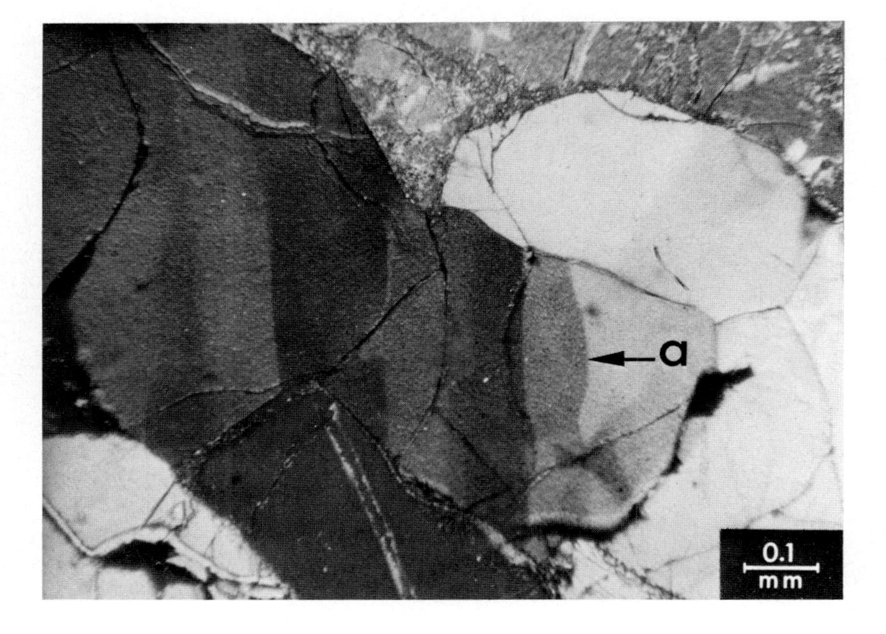

Fig. 121. Deformation lamellae of olivine. Arrow (a) shows the curved outline of the olivine deformation lamellae.
Olivine bomb in basalt (mantle fragment in basalt). Jato, Lekempti, W. Ethiopia. With crossed nicols.

Fig. 122. Olivine nodule (forsterite bomb in basalt), partly rounded and with a reaction margin with the basaltic groundmass. The forsterite shows an impressive and complex in orientation deformation polysynthetic twinning. The forsterite nodules represent xenocrysts picked up by the basalt from the dunite which is partly covered by the olivine-nodules containing basalt. T-ol = tectonically twinned (stressed) olivine nodule; black margin surrounding the olivine is a reaction of the olivine nodule and the basalt; g = basaltic groundmass.
Olivine nodules containing basalt. Yubdo, Wollaga, W. Ethiopia. With crossed nicols.

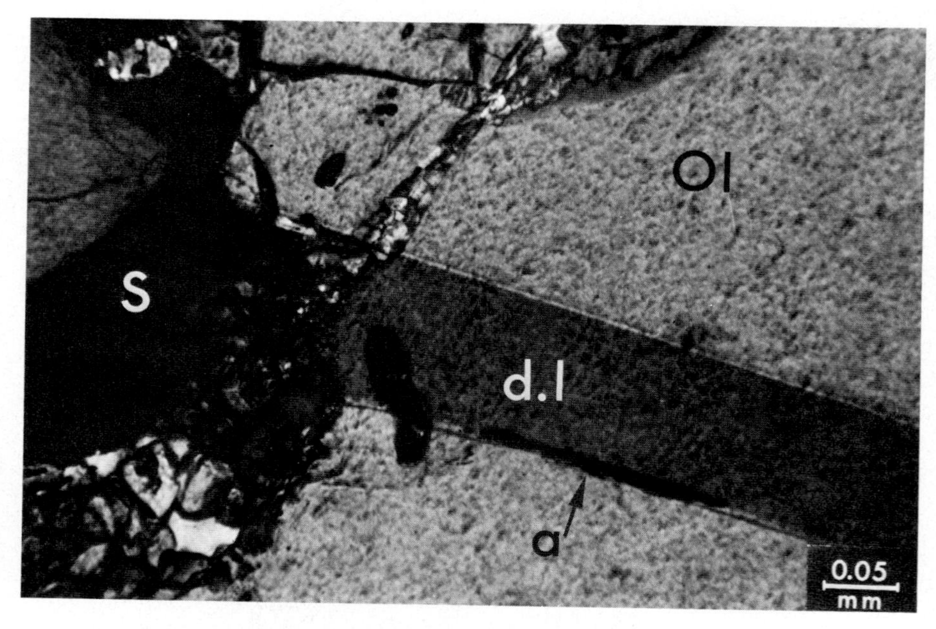

Fig. 123. Deformation lamella of olivine-bomb forsterite. Spinel (see arrow "a") marginal to the deformation lamella of the olivine. Ol = olivine (forsterite) d.l = deformation lamella, S = spinel. Olivine bomb in basalt. Lanzarote, Canary Islands. With crossed nicols.

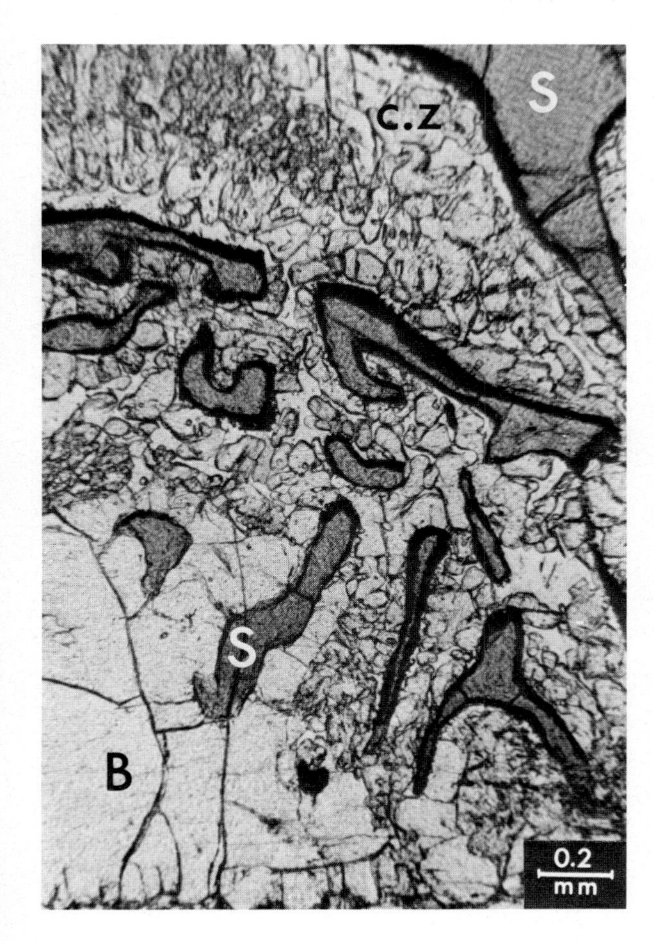

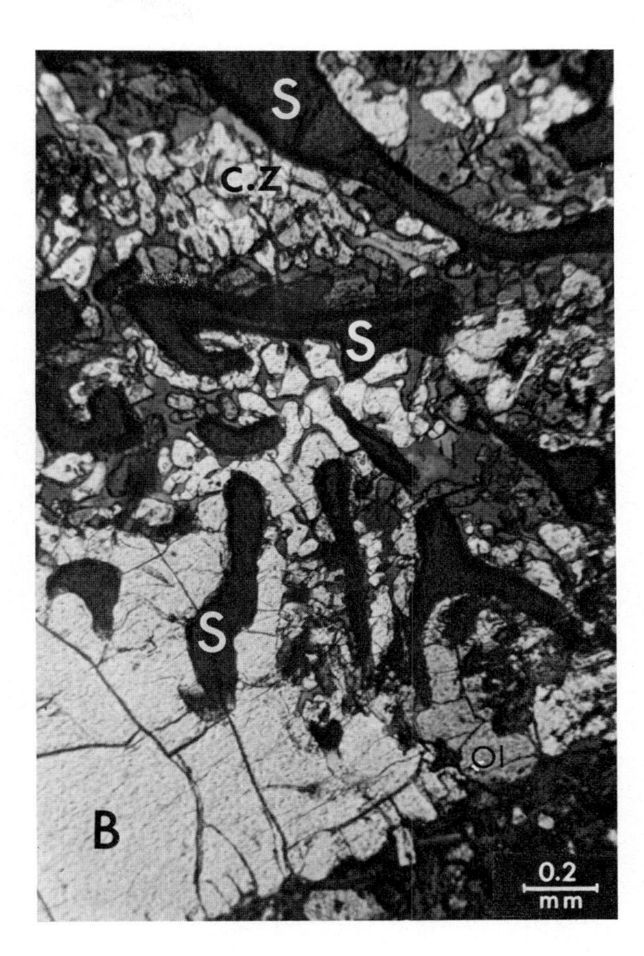

Fig. 124 and 125. Spinels in graphic intergrowth with bronzite and forsterite. The spinels represent crystal skeletons. In almost all the cases where there is an intergrowth of spinel, chrome-spinels, or magnetites in graphic intergrowth or in granular intergrowth with the pyroxenes and olivines of the mantle derived nodules, there is a "crushed zone" of olivines and pyroxenes in close association with the spinel-group minerals. Also see comparable and commensurable crushed zones in Figs. 126, 127 and 133.
The crushed zone is often invaded by "corona" reaction infiltrations which can be traced to the basaltic material invading the bomb and which often take advantage of the crushed zone round the spinel-group minerals, which are pressure-resistant and their formation often is indicative of high pressures. B = bronzite; S = spinels; c.z = crushed zone; Ol = olivine.
Jato, Lekempti, W. Ethiopia. Without crossed nicols and with crossed nicols.

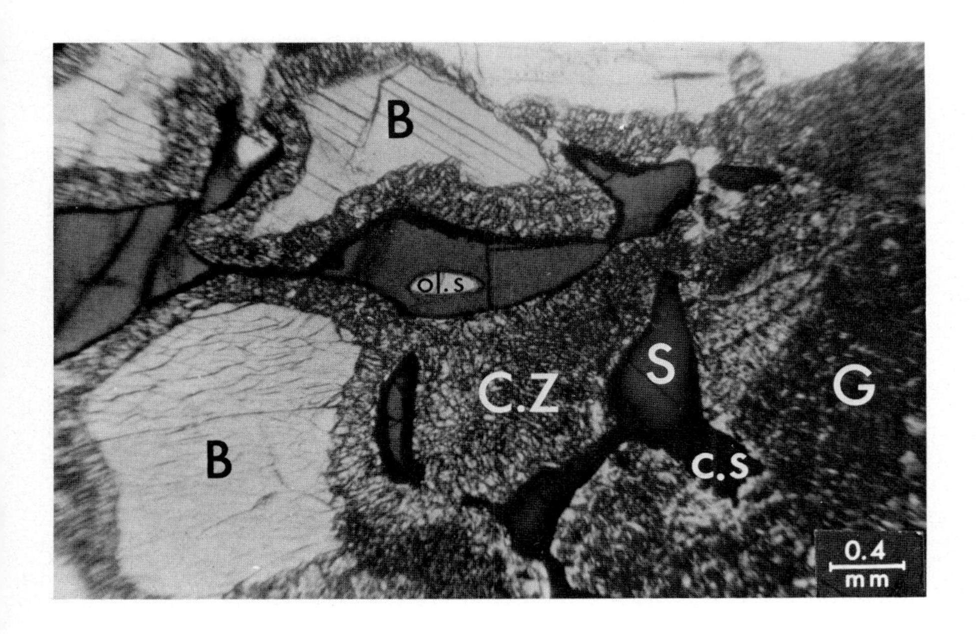

Fig. 126. Bronzite in intergrowth with graphic spinel corona structure between bronzite and spinel is shown. The spinel, in contact with the basaltic groundmass, is magmatically affected and changed into ferro-spinel ("thermal" reaction corona). B = bronzite, S = spinel in graphic intergrowth with bronzite; G = basaltic groundmass; c.s = "thermal" reaction corona of spinel with the basaltic groundmass; C-Z = crushed zone (see Fig. 93); ol-s = uncrushed olivine included and protected in the spinel.
Olivine bomb in basalt. Jato, Lekempti, W. Ethiopia. Without crossed nicols.

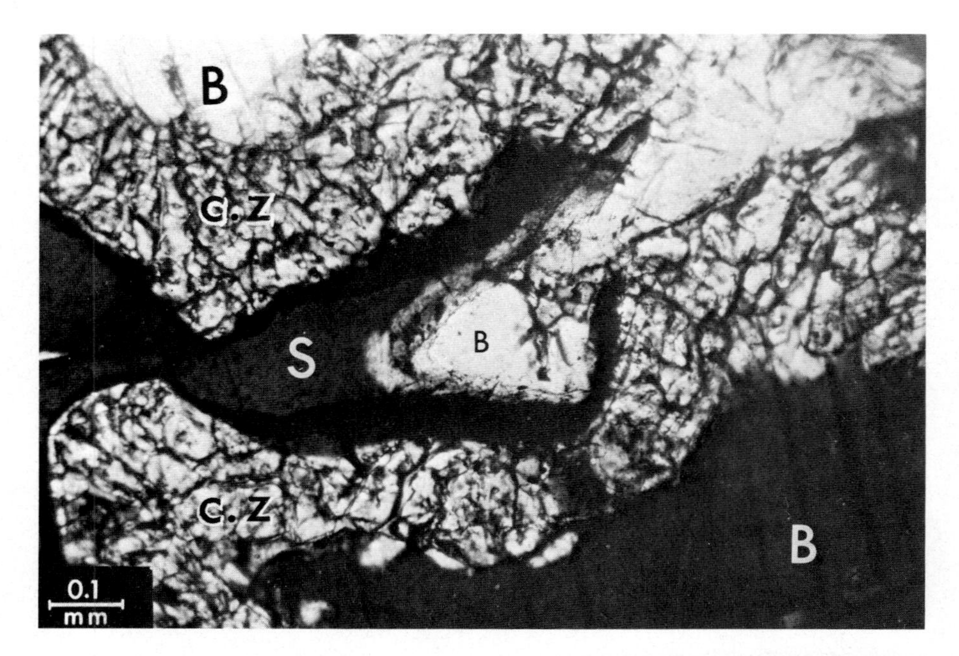

Fig. 127. Detail showing the intergrowth of the graphic spinel with the bronzite and with the corona-structure bronzite–spinel. S = spinel in graphic intergrowth with bronzite and with the reaction corona. B = bronzite (also in extinction orientation); C.Z = crushed zone and corona structure (see Fig. 124).
Olivine bomb in basalt. Jato, Lekempti, W. Ethiopia. With crossed nicols.

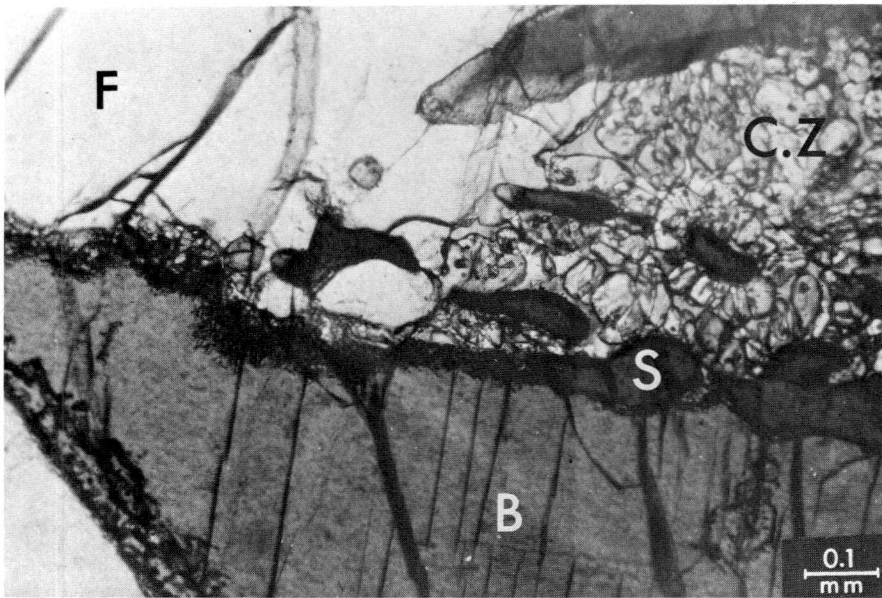

Fig. 128. Graphic spinel in intergrowth with bronzite and forsterite. F = forsterite; B = bronzite; S = spinel; C.Z = crushed zone. Olivine bomb in basalt. Jato, Lekempti, W. Ethiopia. With crossed nicols.

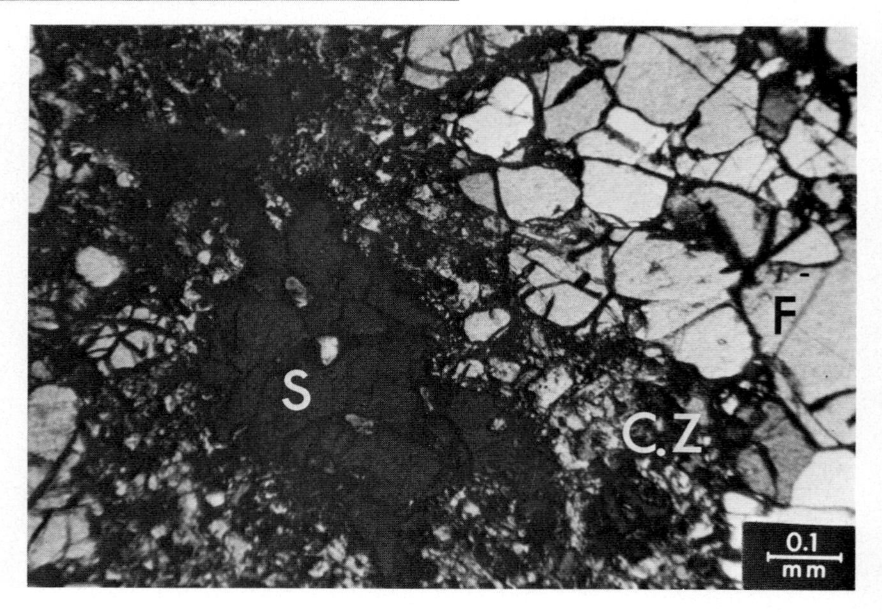

Fig. 129. Forsterite with corona structure and with spinel in intergrowth with it. F = forsterite grains, components of olivine bomb in basalt, S = spinel, C.Z = crushed zone (see Fig. 127). Alkali basalt with xenolith, dunitic in composition, actually mantle fragment.
Finkenberg near Bonn-Beuel, Rhineland, Germany. With half-crossed nicols.

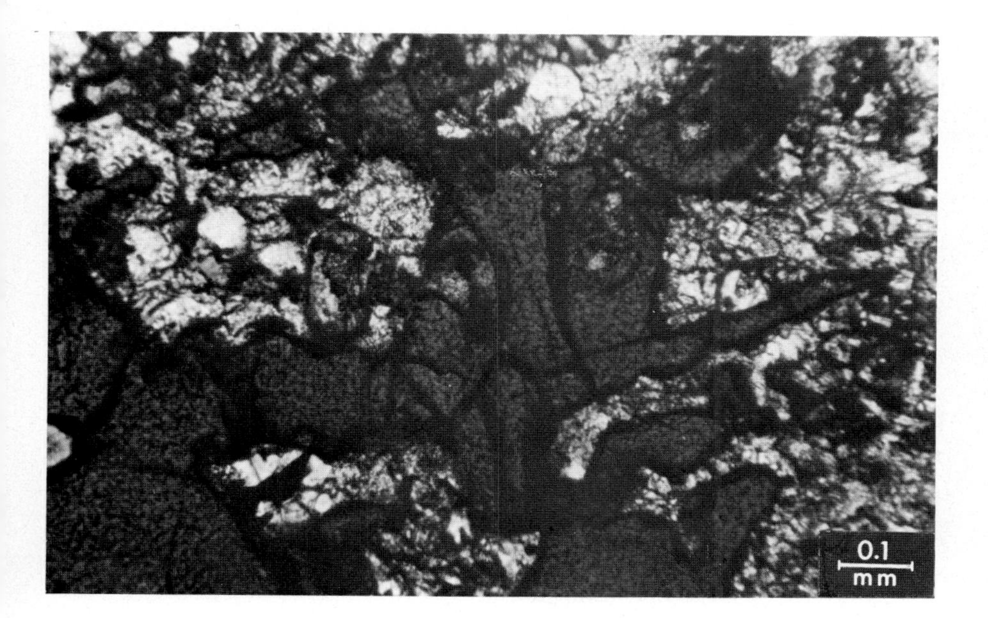

Fig. 130. Spinels in "graphic" intergrowth with a crushed zone consisting of olivines and pyroxenes. Often these crushed zones are invaded by the basaltic melts forming corona-reaction intergrowths. Alkali basalt with xenolith, dunitic in composition, actually mantle fragments.
Finkenberg near Bonn-Beuel, Rhineland, Germany. With half-crossed nicols.

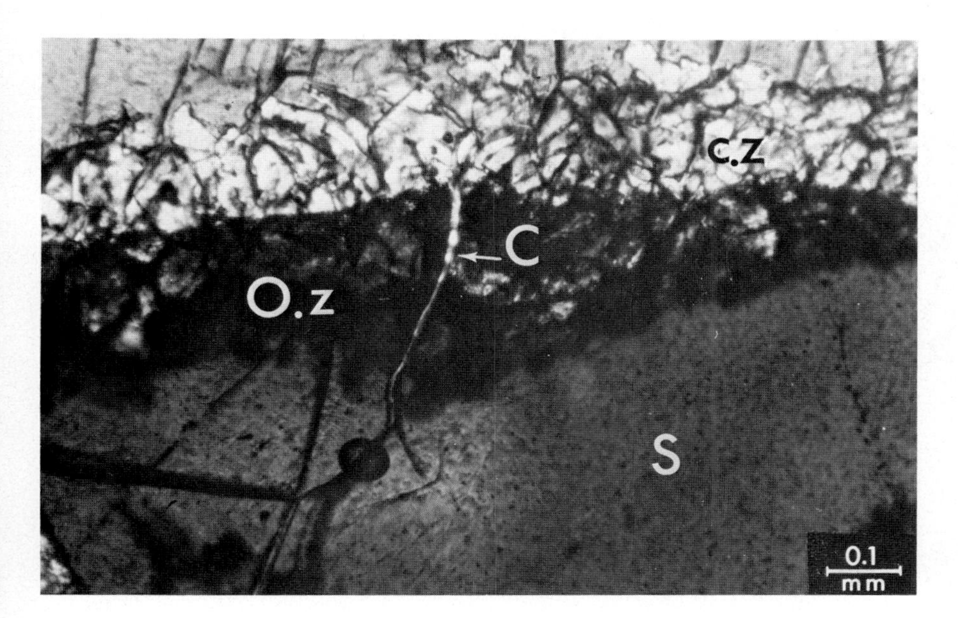

Fig. 131. Spinel with oxidation margin due to diffusion and with fine extension of the crushed zone, mobilized into cracks of the spinel. Mantle fragment in basalt. S = spinel; O.z = oxidized zone of the spinel; c.z = crushed zone; C = crack of spinel with the crushed zone mobilized into it.
Olivine bomb in basalt. Jato, Lekempti, W. Ethiopia. Without crossed nicols.

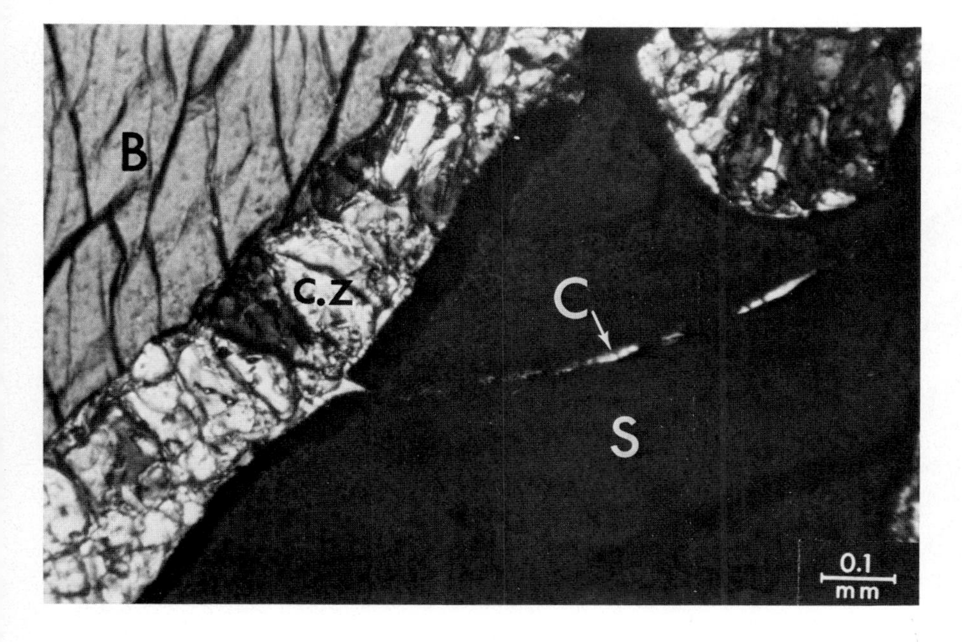

Fig. 132. Spinel with an oxidized margin and with a crack into which the crushed zone has been mobilized. S = spinel; c.z = crushed zone; B = bronzite; C = crack of the spinel into which the crushed zone has been mobilized.
Olivine bomb in basalt. Jato, Lekempti, W. Ethiopia. Without crossed nicols.

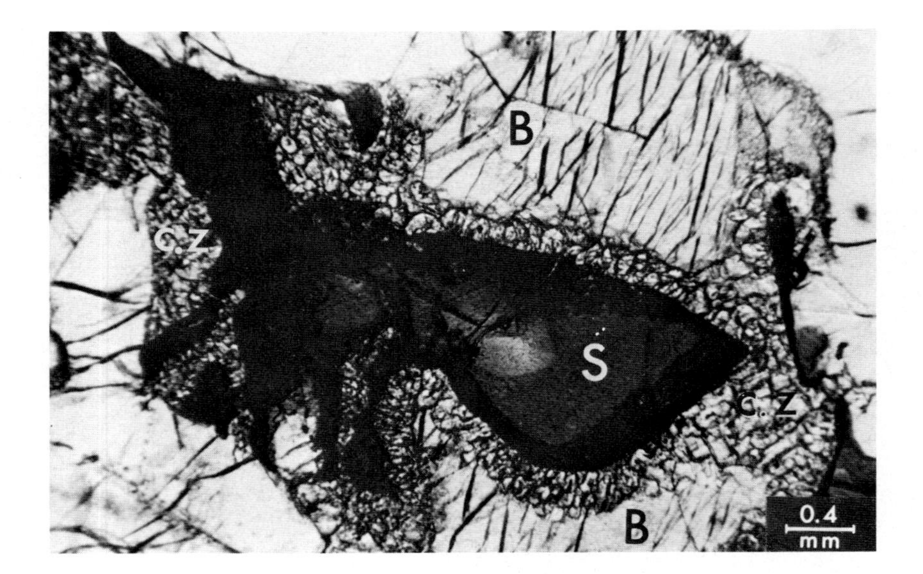

Fig. 133. Spinel with an oxidized zone surrounded by a crushed zone formed due to mantle deformation. Also adjacent pyroxene (bronzite) with a cleavage pattern due to mantle deformation. S = spinel; c.z = crushed zone; B = bronzite with cleavage pattern due to mantle deformation.
Olivine bomb (mantle fragment) in basalt. Jato, Lekempti, W. Ethiopia. Without crossed nicols.

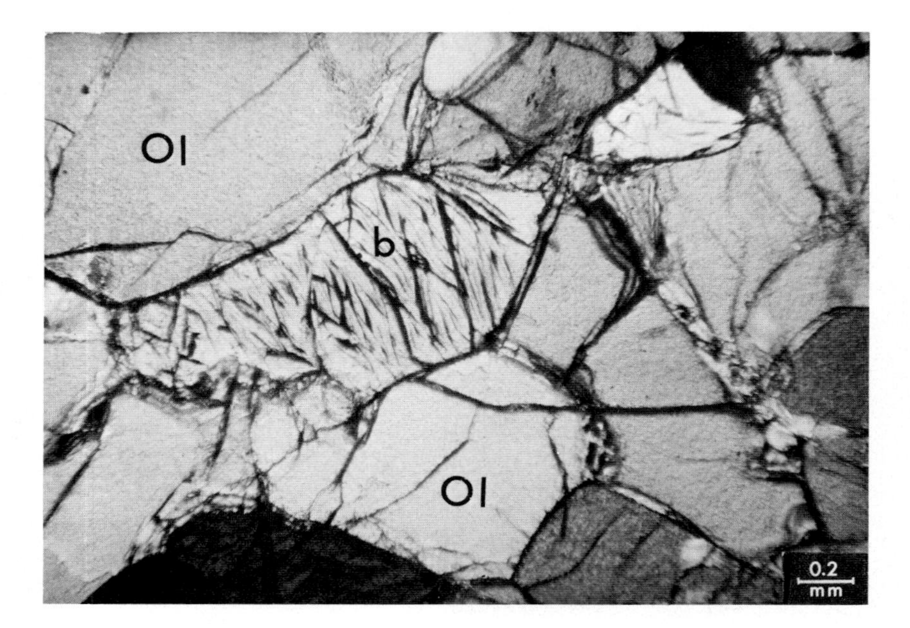

Fig. 134. Equigranular olivine pyroxene texture of mantle fragment in basalt (olivine bomb) with bronzite showing a cleavage pattern due to mantle deformation. Ol = olivine; b = bronzite (with cleavage pattern due to mantle deformation).
Olivine bombs in basalt. Eifel, Germany. With crossed nicols.

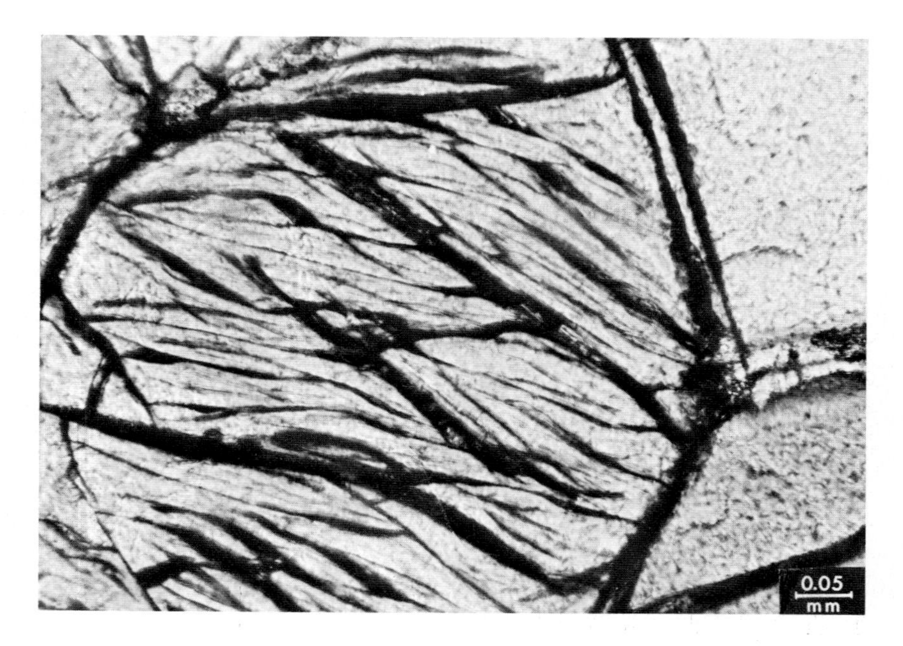

Fig. 135. Detail of Fig. 134. Pyroxene (bronzite) with a cleavage pattern due to mantle deformation.
Eifel, Germany. With crossed nicols.

Fig. 136. An olivine bomb, rounded and ovoidal in shape, partly enclosed in basalt. Jato, Lekempti, W. Ethiopia. Hand specimen, half natural size.

Fig. 137. Volcanic bomb having as a nucleus a mantle fragment (olivine bomb) consisting of forsterite, pyroxene spinels. Half of natural size. Lanzarote, Canary Islands.

Fig. 138. Olivine xenocrysts (olivine nodule) in basalt. The olivine phenocryst shows a complex pattern of deformation lamellae due to mantle tectonics, and an isotropic (synisotropic) margin in reaction with the basaltic groundmass.
Olivine nodule in basalt. Yubdo, Wollaga, W. Ethiopia. With crossed nicols.

Fig. 139. Detail of Fig. 138. A rounded olivine crystal (representing an olivine nodule) surrounded by an isotropic alteration margin. The olivine shows a complex twinning.
Olivine nodule in basalt. Yubdo, Wollaga, W. Ethiopia. With crossed nicols.

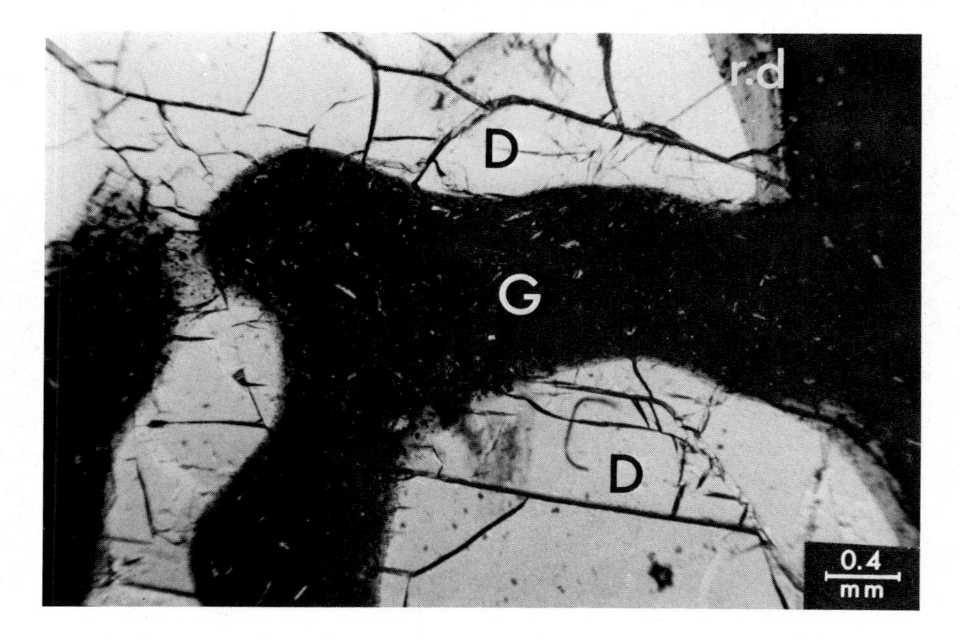

Fig. 140. Large diopside, corroded and invaded by the basaltic groundmass. D = diopside; G = groundmass invading the pyroxene; r.d = reaction margin of the pyroxene with the basaltic groundmass.
Olivine and pyroxene nodules in basalt. Yubdo, Wollaga, W. Ethiopia. Without crossed nicols.

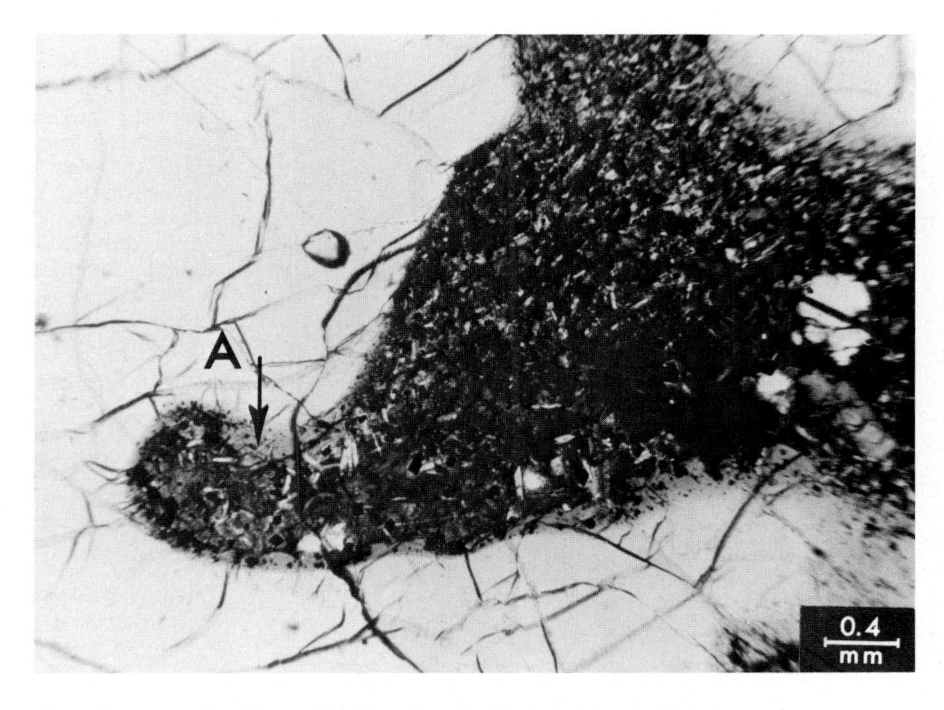

Fig. 141. Large diopside, corroded and attacked by the groundmass of the basalt. Arrow (A) marks isotropic pyroxene part due to alteration (this part is shown in greater detail in Fig. 142). Olivine and pyroxene nodules in basalt. Yubdo, Wollaga, W. Ethiopia. Without crossed nicols.

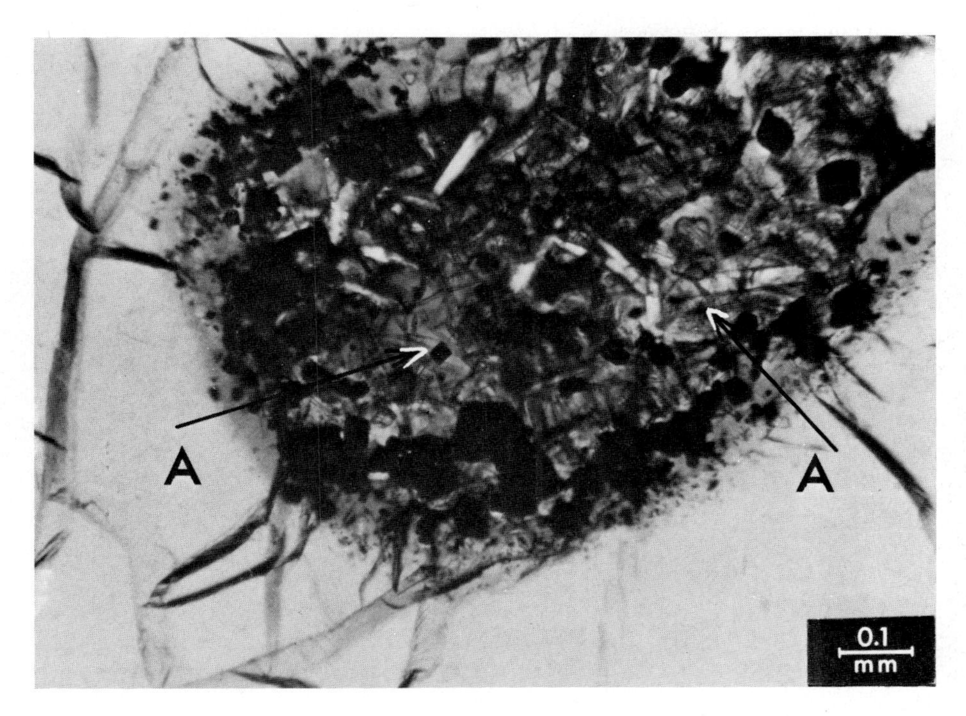

Fig. 142. Pyroxene part (diopside), changed into an isotropic substance (marked by arrow "A"); small magnetite cubes and feldspars are also present as part of the invading basaltic groundmass.
Olivine and pyroxene nodules in basalt, Yubdo, Wollaga, W. Ethiopia. Without crossed nicols.

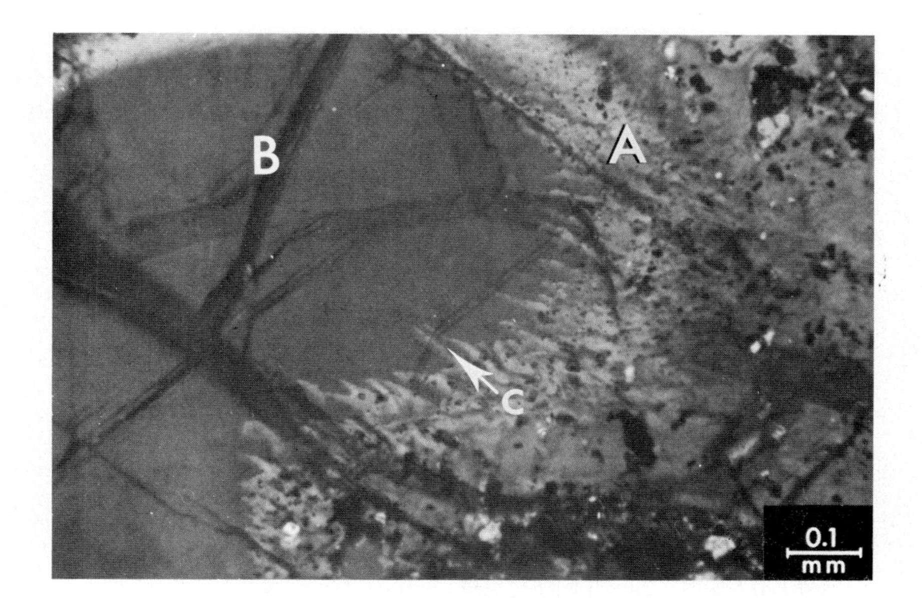

Fig. 143. Corrosion and alteration of pyroxene by basaltic groundmass. A = altered pyroxene (these altered pyroxene margins are anisotropic); B = unaltered diopside (pyroxene); C = fine canals representing the alteration front advancing into the unaltered pyroxene. Olivine and pyroxene nodules in basalt. Yubdo, Wollaga, W. Ethiopia. With crossed nicols.

Fig. 144. A path of basaltic melt penetrating into diopside, a nodule xenocryst in basalt. The basaltic "melt path" on consolidation shows a myrmekitic symplectic pattern with the diopside.
Olivine and pyroxene nodules in basalt. Yubdo, Wollaga, W. Ethiopia. With crossed nicols.

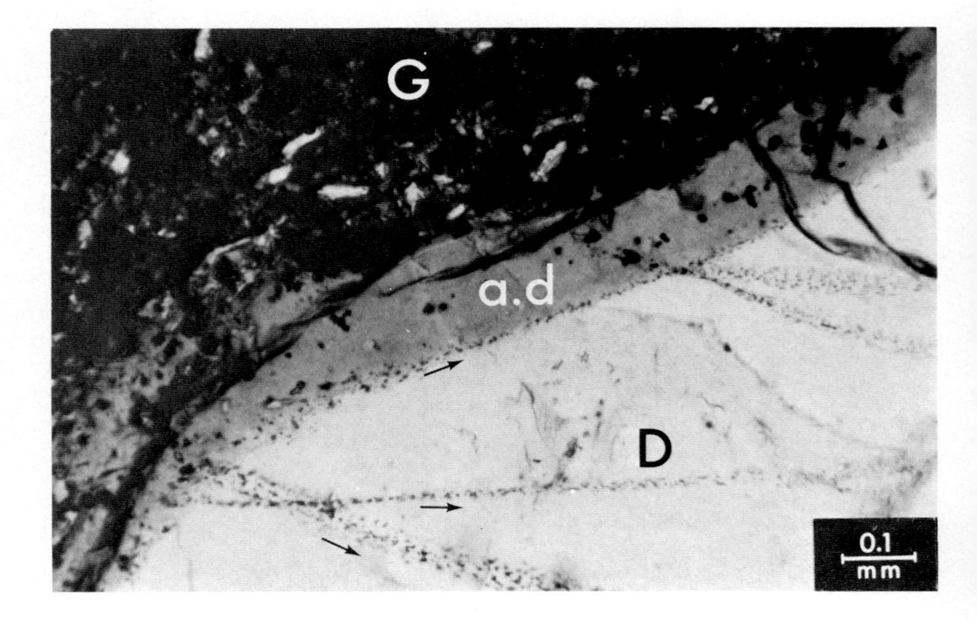

Fig. 145. Diopside xenocryst of an olivine–pyroxene nodule in basalt corroded and with a reaction margin, invaded by basaltic melts forming paths, either following the margin of affected – unaffected diopside or penetrating into the pyroxene. G = basaltic groundmass, D = diopside, a.d = affected diopsidic margin. Arrows show infiltration melt paths.
Olivine and pyroxene nodules in basalts. Yubdo, Wollaga, W. Ethiopia. Without crossed nicols.

Fig. 146. Myrmekite-like bodies representing symplectic intergrowth of the "infiltrations" of basaltic melts into olivines or pyroxenes.
Olivine bomb in basalt. Eifel, Germany. Without crossed nicols.

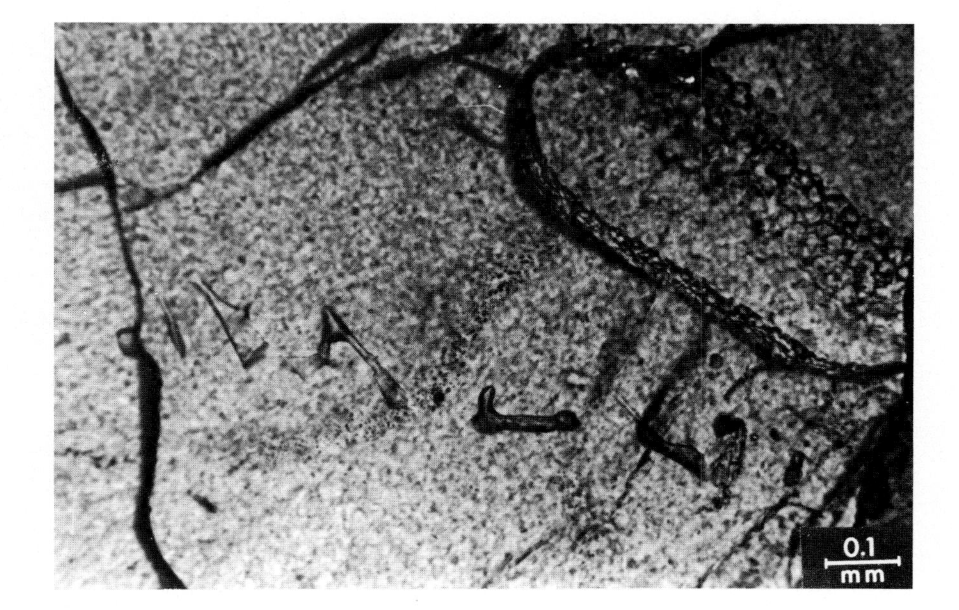

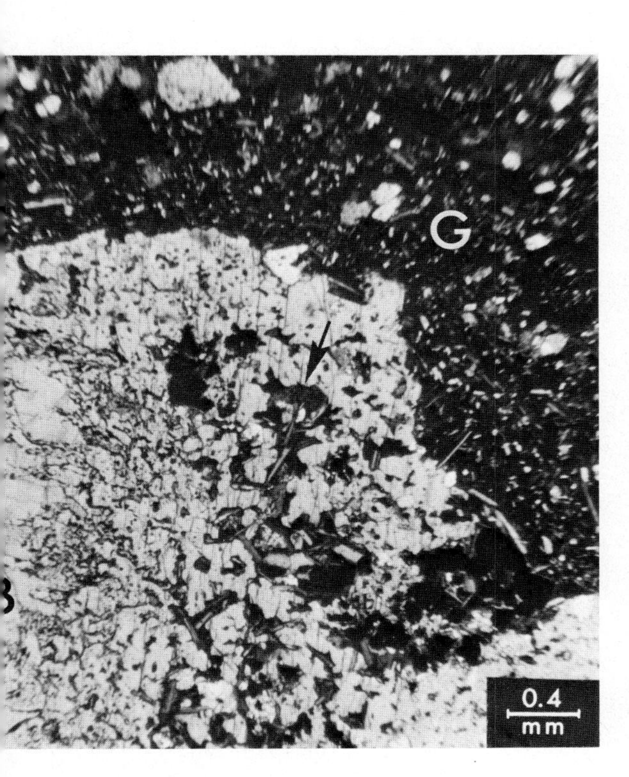

Fig. 147. Bronzite of the olivine–pyroxene bombs in basalt, shows magmatic corrosion and infiltration into the bronzite of basaltic melt which consolidated in groundmass components. B = bronzite; G = groundmass; arrow marks groundmass infiltrations in the bronzite.
Olivine bombs (mantle fragments) in basalt. Lekempti, W. Ethiopia. With crossed nicols.

Fig. 148. Detail of Fig. 147, indicating groundmass components (mainly basaltic groundmass plagioclases) as infiltration of subsequently consolidated basaltic melts.
Olivine bombs in basalt. Jato, Lekempti, W. Ethiopia. Olivine bombs (mantle fragments) in basalt. With crossed nicols.

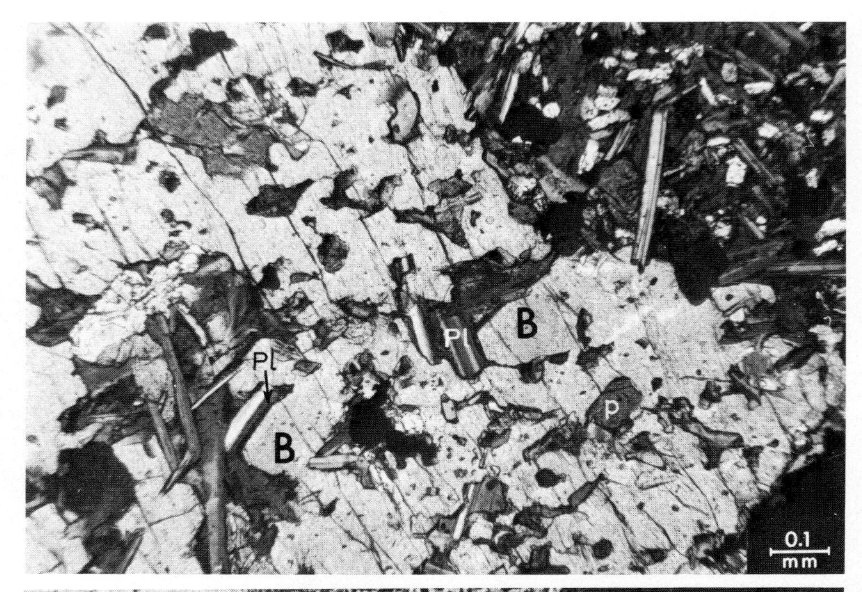

Fig. 149. A typical picture of the reaction: bomb components – basaltic melt. Bronzite initially of "olivine bombs" now in the basaltic groundmass, corroded and invaded by the basaltic melt which on subsequent crystallization resulted in plagioclases and other basaltic components (pyroxenes) in the bronzite. B = bronzite; Pl = plagioclase; p = pyroxene.
Mantle fragments in basalt. Jato, Lekempti, W. Ethiopia. With crossed nicols.

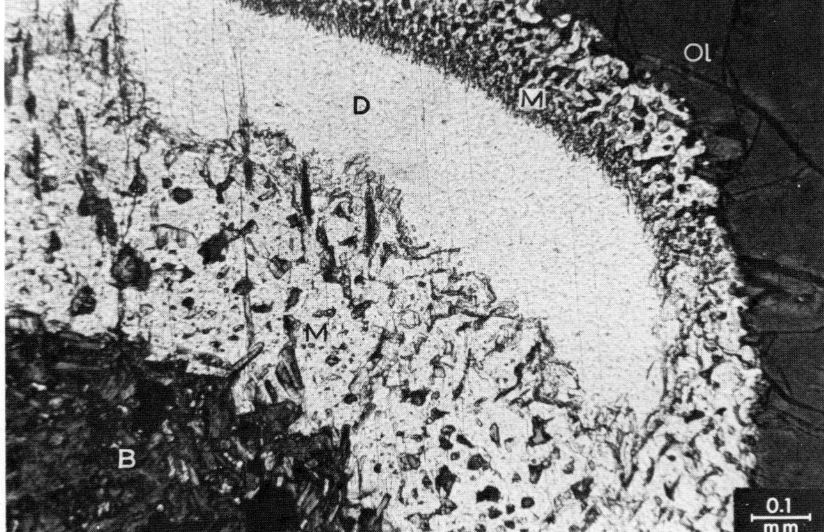

Fig. 150. Bronzite and forsterite of the "olivine bombs" in contact with basalt. As a result of the magmatic corrosion and infiltration, a "myrmekite" type of fringe is formed marginal to the bronzite. The myrmekite consists of basaltic components (e.g. plagioclases) which have been formed by infiltration along directions of penetrability (i.e. cleavage) of the host bronzite. D = bronzite; M = myrmekitic fringe, Ol = olivine; B = basalt.
Mantle fragments in basalt. Jato, Lekempti, W. Ethiopia. With crossed nicols.

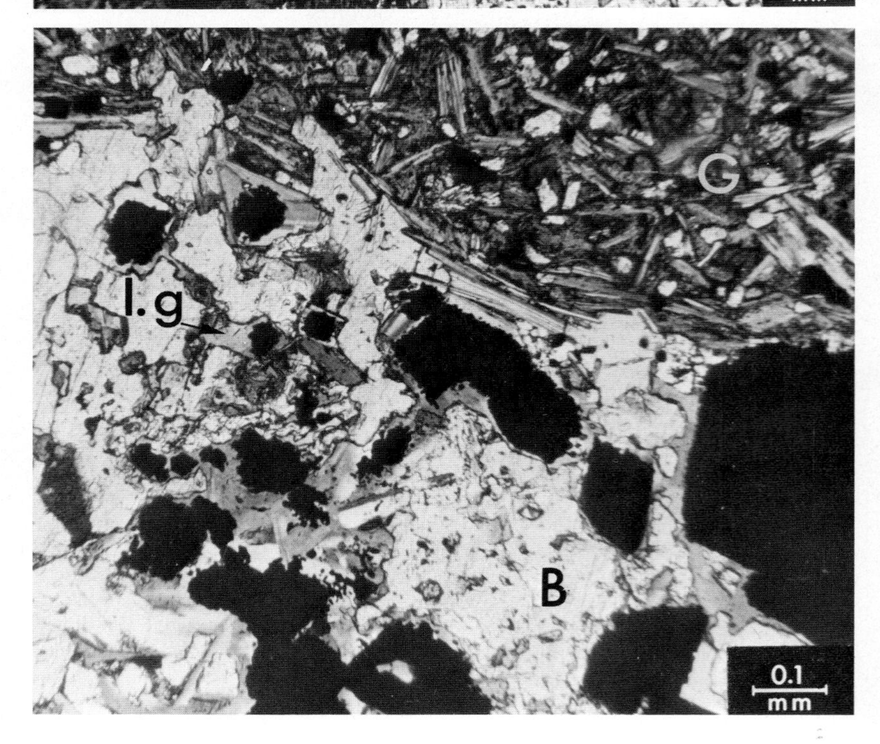

Fig. 151. Bronzite with magnetite inclusions invaded by melts which subsequently crystallized into groundmass plagioclases, the feldspars often formed round the magnetite inclusions of the bronzite. B = bronzite, black = magnetite; G = groundmass; I.g = basaltic melts subsequently consolidated into plagioclases – infiltrations into the bronzite.
Olivine bombs in basalt. Jato, Lekempti, W. Ethiopia. With crossed nicols.

Fig. 152. Bronzite with a corona reaction margin consisting of olivine crystal grains. As a result of the complex reaction with the basaltic melt (or forsterite), a corona structure is produced, a prolongation of which extends into the bronzite. D = bronzite; Ol = olivine crystal grains of the corona; C = corona structure; My-I = myrmekitic intracrystalline infiltrations of melts into the bronzite.
Olivine bomb in basalt. Jato, Lekempti, W. Ethiopia. With crossed nicols.

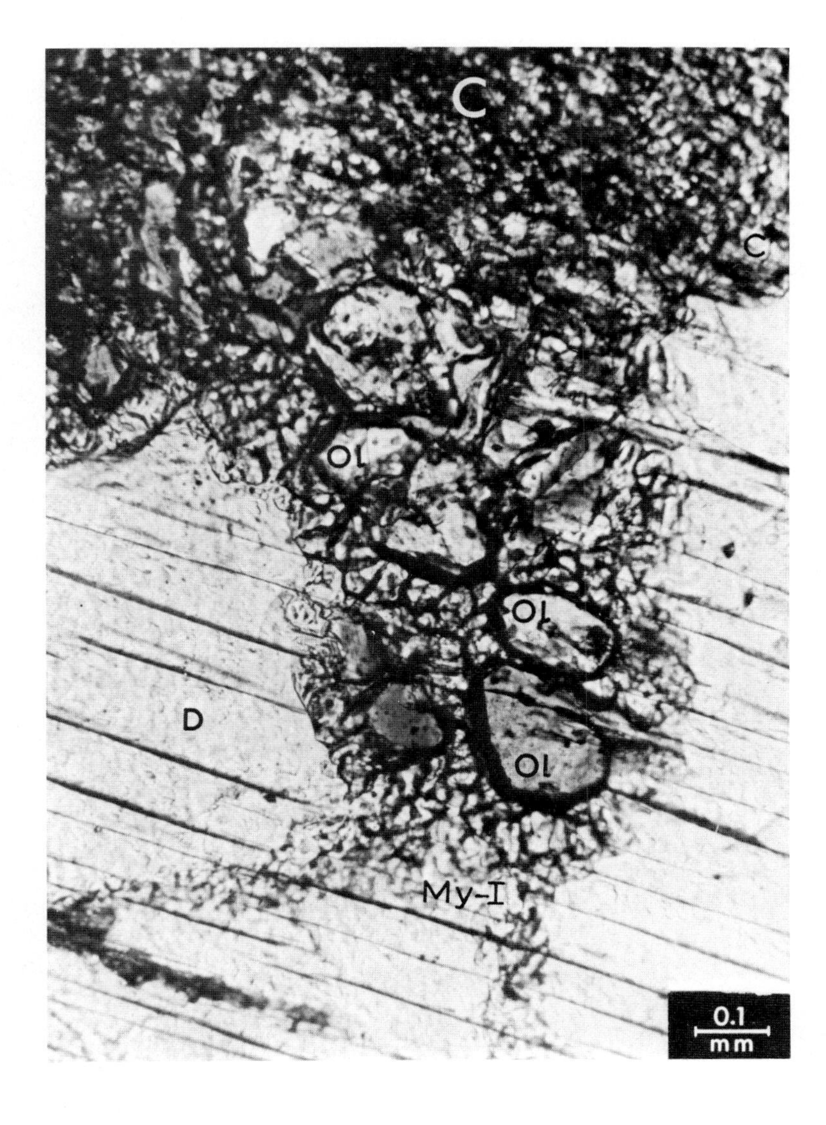

Fig. 153. Olivine with a veinlet showing differences in interference colours and partly occupied by olivine agglutubations of olivine regrowths due to topometasomatic mobilizations. Ol = olivine, r.g = agglutinations of olivine regrowth.
Olivine bomb in basalt. Jato, Lekempti, W. Ethiopia. With crossed nicols.

Fig. 154. Metasomatic agglutinations consisting of magnesium silicates formed due to melt infiltration into the forsterite. Ol = forsterite, M.ag = Magnesium silicate agglutinations.
Olivine bomb in basalt (mantle fragments in basalt). Jato, Lekempti, W. Ethiopia. With crossed nicols.

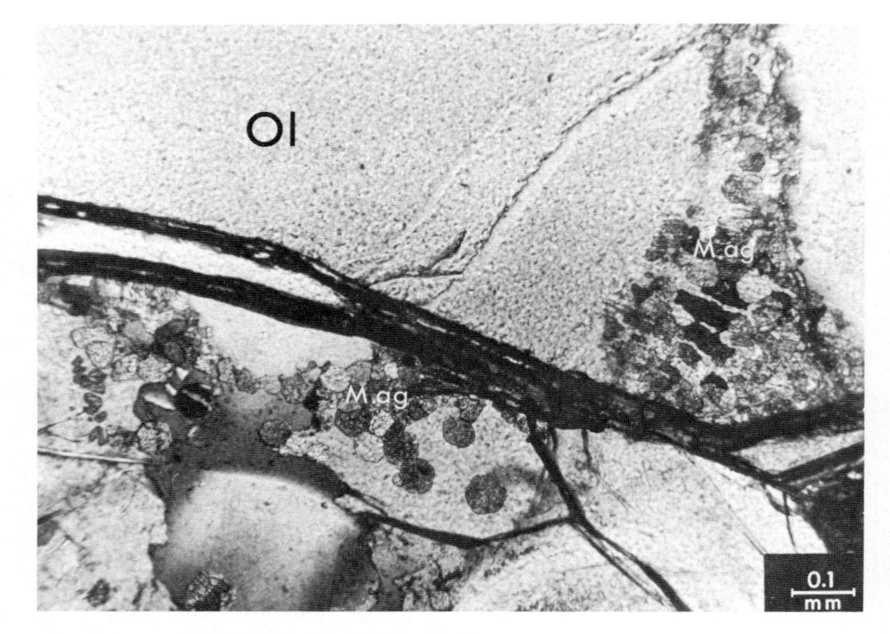

Fig. 155. Metasomatic agglutinations consisting of magnesium silicates formed due to melt infiltration into the forsterite. Ol = forsterite, M.ag = magnesium silicate agglutinations into the forsterite. Mantle fragments in basalt. Jato, Lekempti, W. Ethiopia. With crossed nicols.

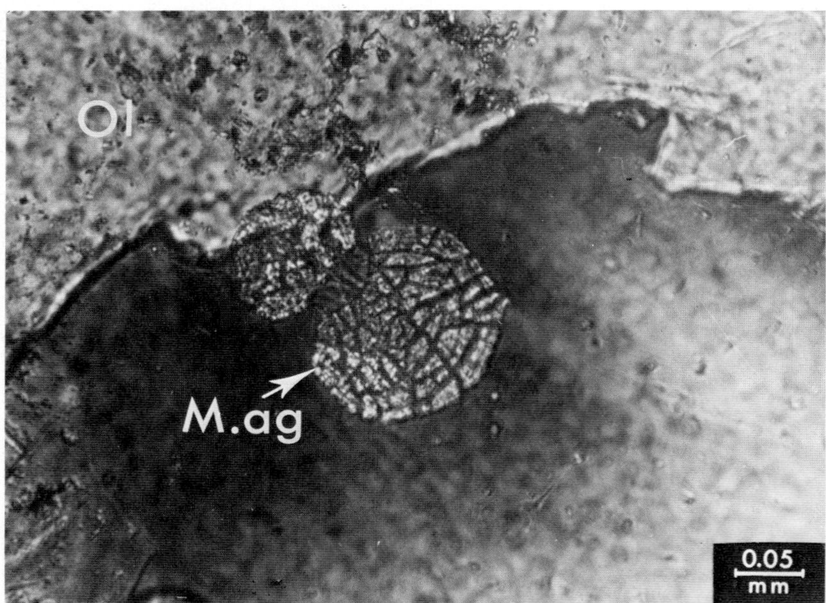

Fig. 156. Rounded chromite grain (xenocryst) in basalt. The chromite has a margin of magnetite. Ch = chromite; S = silicate in the basalt. White scattered spots (crystals) are basaltic magnetite.
Olivine bomb containing basalt. Polished section with one nicol. Yubdo, Wollaga, W. Ethiopia.

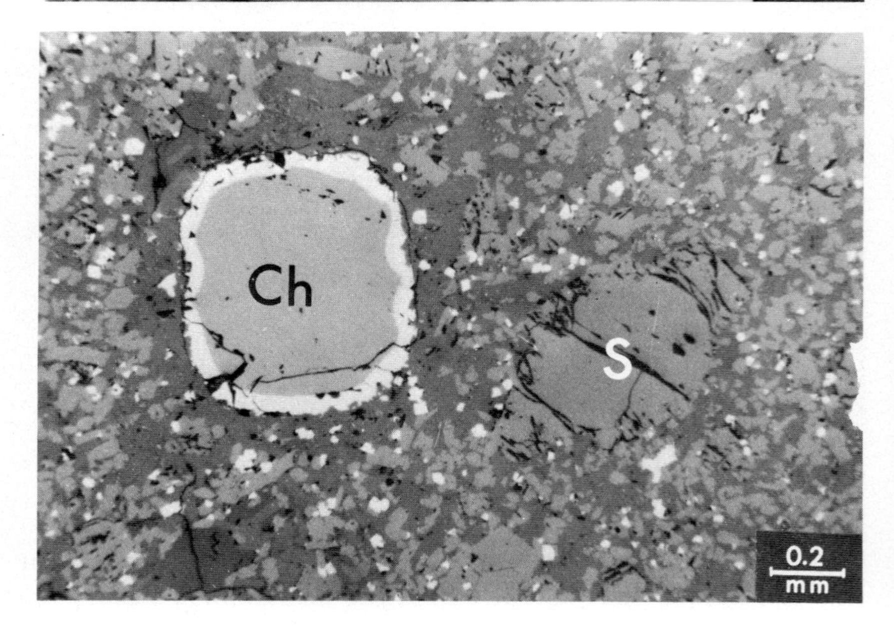

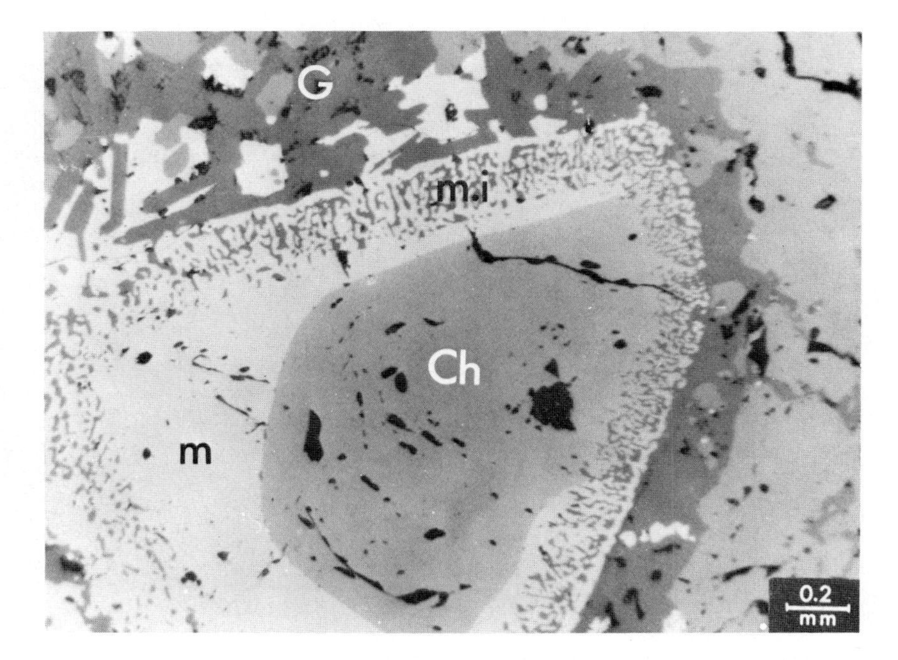

Fig. 157. Chrome-spinel xenocrysts in the groundmass of olivine basalt. The spinel xenocryst has a magnetite margin. Both the magnetite margin and the chrome-spinel show "myrmekitic intergrowths" as a result of the synantetic reaction of chrome-spinel–magnetite with the basaltic groundmass which has corroded and infiltrated in the metallic minerals. Ch = chrome-spinel; m = magnetite margin; m-i = myrmekitic intergrowth of basaltic melts and chrome-spinel magnetite; G = basaltic groundmass.
Mantle fragments in basalt. Polished section. Jato, Lekempti, W. Ethiopia. With one nicol.

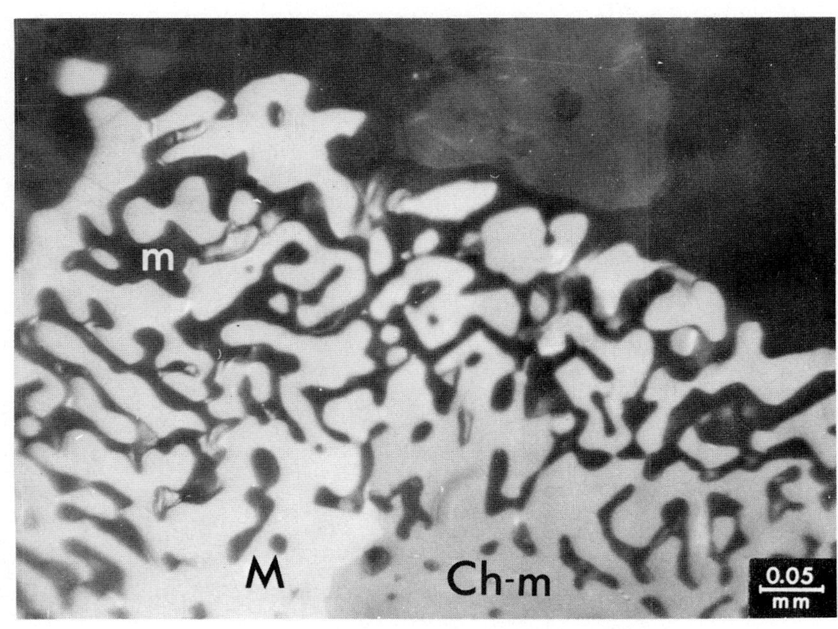

Fig. 158. Chrome-spinel with magnetite (components of the olivine-bombs) both myrmekitised by later basaltic melt. Olivine bomb in basalt. Ch-m = chrome-spinel; M = magnetite; m = "basalt melt" myrmekitic intergrowth with the chrome-spinel and the magnetite.
Mantle fragments in basalt. Jato, Lekempti, W. Ethiopia. Polished section, oil-immersion. With one nicol.

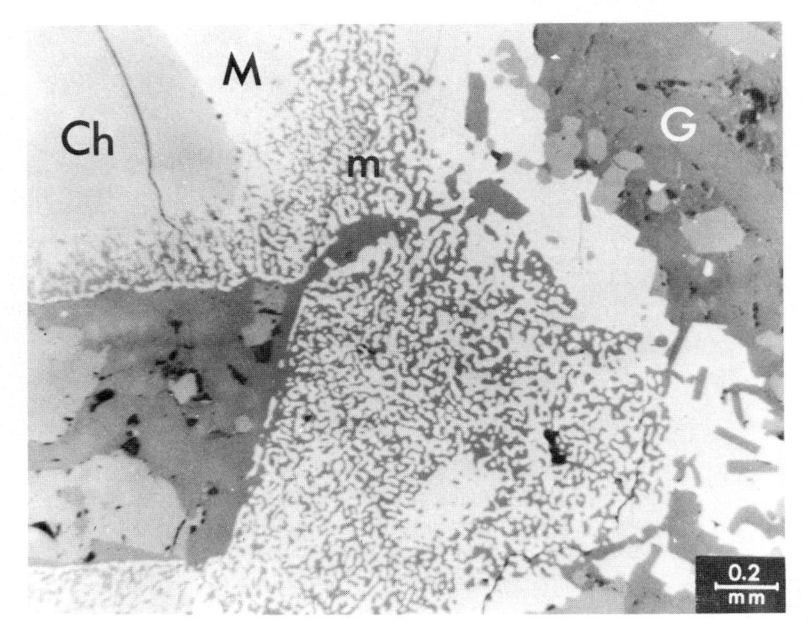

Fig. 159. Chrome-spinel with a margin of magnetite in contact with olivine crystal grain, components of olivine bomb in basalt. The basaltic melts show myrmekitic intergrowths with the magnetite and the chrome-spinel. G = basaltic groundmass; Ch = chrome-spinel; M = magnetite margin; m = myrmekitic intergrowths of basaltic melts with chrome-spinel and magnetite.
Olivine bomb in basalt. Jato, Lekempti, W. Ethiopia. Oil-immersion with one nicol.

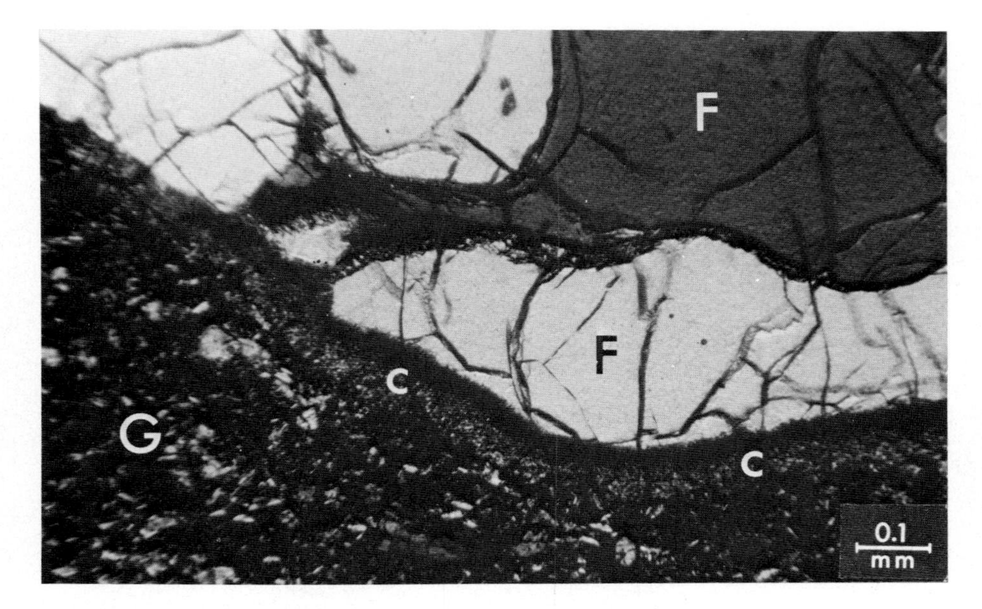

Fig. 160. Forsterite (part of olivine bomb) in contact with basaltic groundmass. Corona reaction structures are formed between the olivine xenocrysts and the groundmass intragranular between the forsterite grains. F = forsterite; G = groundmass; c = corona reaction structures. Olivine basalt (with olivine nodules). Steinwitzhügel, Glatzer circle, Sudetes. With crossed nicols.

Fig. 161. Reaction corona structure between olivine crystal (of the mantle fragment in basalt) and the basaltic groundmass. Ol = olivine (forsterite, mantle fragment); b.g = basaltic groundmass; c = corona reaction structure, consisting of perpendicularly arranged prismatic pyroxenes. Olivine basalt (with olivine nodules) Steinwitzhügel, Glatzer circle, Sudetes. With crossed nicols.

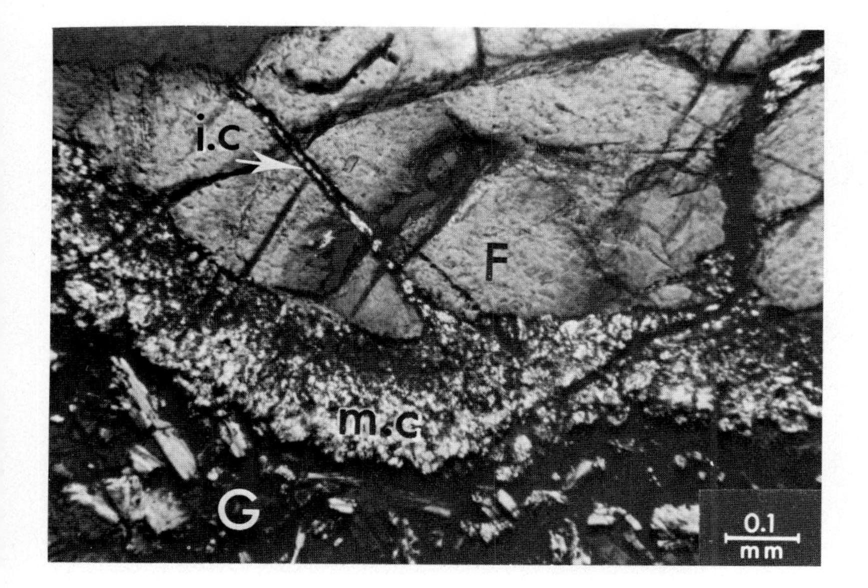

Fig. 162. Forsterite olivine bomb component with a reaction corona structure with the basaltic groundmass. An extension of the corona is protruding from the corona margin into the forsterite. F = for forsterite; m.c =marginal corona; i.c = intracrystalline corona extension; G = groundmass.
Olivine basalt (with olivine nodules) Steinwitzhügel, Glatzer circle, Sudetes. With crossed nicols.

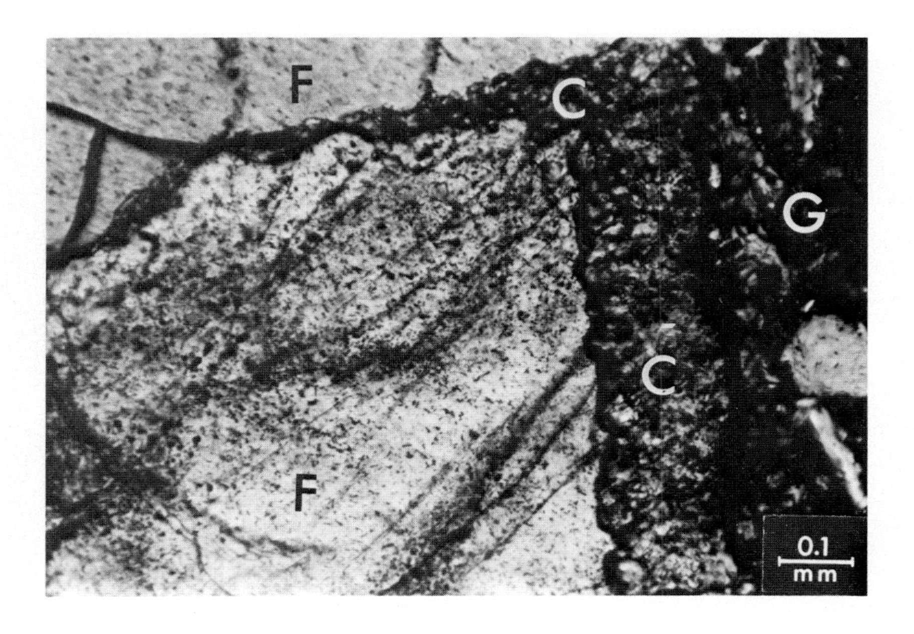

Fig. 163. Forsterites, olivine bomb components with a reaction corona structure and with an extension of it protruding between the intergranular of the olivines. Also fine magmatic (basaltic groundmass) as fine intracrystalline infiltration in the forsterite. F = forsterite, C = corona structures; G = groundmass.
Olivine basalt (with olivine nodules) Steinwitzhügel, Glatzer circle, Sudetes. With crossed nicols.

Fig. 164. A corona reaction structure intergranular between two olivines and consisting of fine laths of plagioclase perpendicular to the olivine grain margins. Ol = olivines; pl = plagioclase lath perpendicular to the olivine grain. Jato, Lekempti, W. Ethiopia. With crossed nicols.

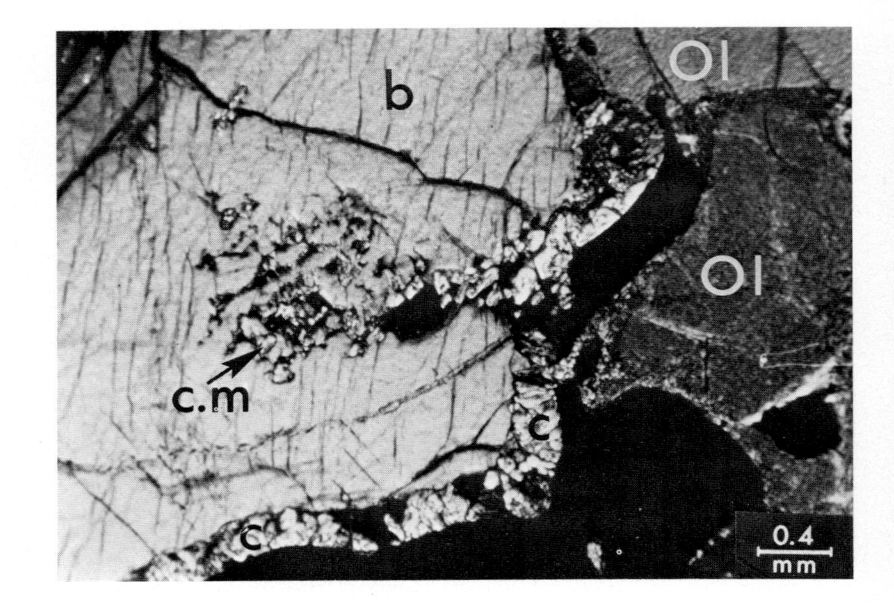

Fig. 165. A corona reaction structure which follows the boundary spinel–bronzite and extends as intracrystalline "infiltration" into the bronzite, attaining myrmekitoid forms. Ol = olivines; black = spinel; b = bronzite, c = corona reaction structure; c.m = myrmekitoid intracrystalline infiltrations of the corona into the bronzite.
Mantle fragments in basalt. Jato, Lekempti, W. Ethiopia. With crossed nicols.

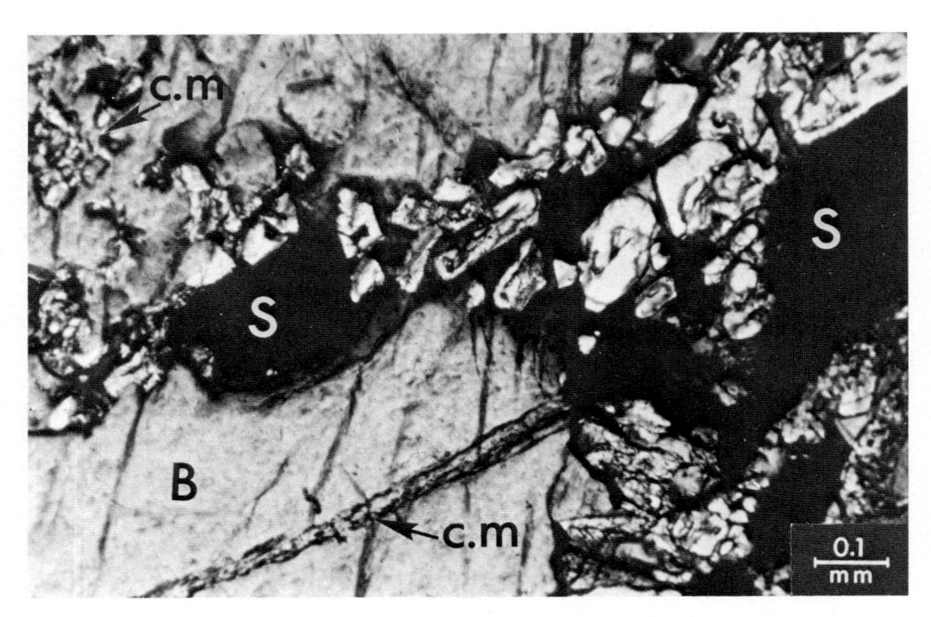

Fig. 166. Detail of Fig. 165. Specifically showing the intracrystalline infiltration of the corona reaction into the bronzite in which it attains myrmekitoid character. S = spinel (black); B = bronzite; c.m = myrmekitoid intracrystalline infiltrations of the "corona reaction" into the bronzite.

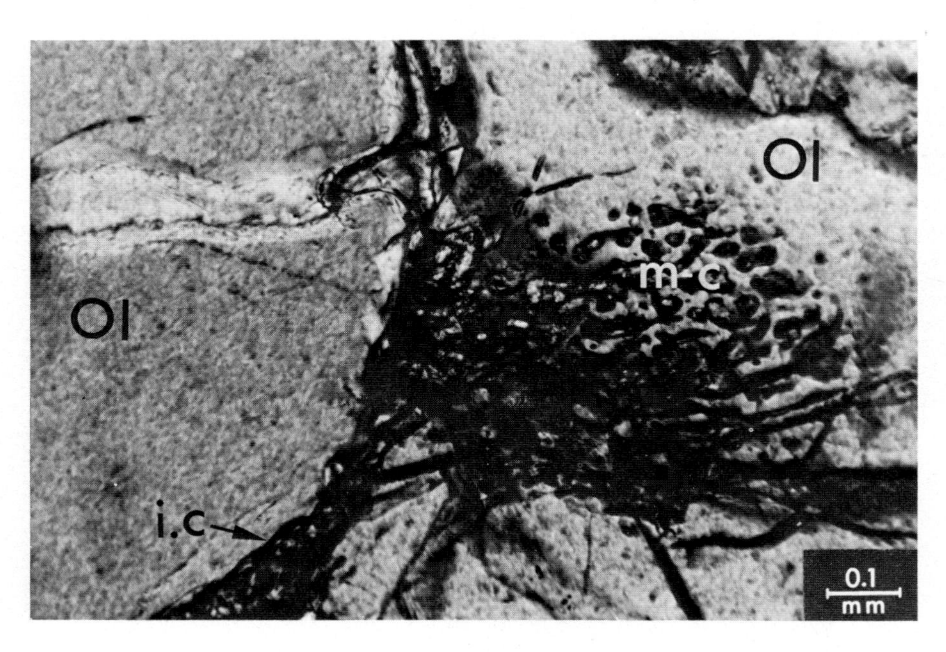

Fig. 167. Intergranular "corona reaction" infiltration between two olivines with myrmekitoid infiltrations extending into one of the forsterites. Ol = olivine; i.c = intergranular "corona reaction" between two olivine grains; m-c = myrmekitic-like extensions "infiltrated" into one of the olivine grains, starting from the intergranular corona-reaction. Jato, Lekempti, W. Ethiopia. With crossed nicols.

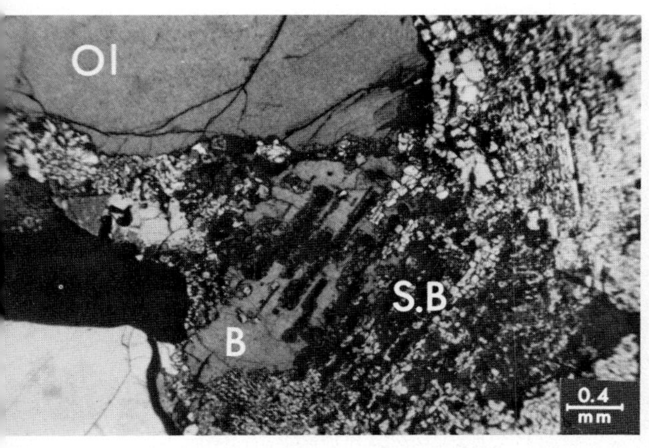

Fig. 168. Olivine (forsterite) and bronzite invaded along crystal penetrability direction by corona infiltrations which, if predominant, can result in a substitution of the bronzite individual by an aggregate of corona infiltrations. Ol = olivine (forsterite); B = bronzite with corona infiltrations, following crystal penetrability directions of the host; S.B = substituted bronzite by the corona infiltrations.
Jato, Lekempti, W. Ethiopia. With crossed nicols.

Fig. 169. Detail of Fig. 168, showing bronzite replaced along crystal penetrability directions by corona reaction infiltrations. The lattice penetrability direction might be at an angle to the host's cleavage.

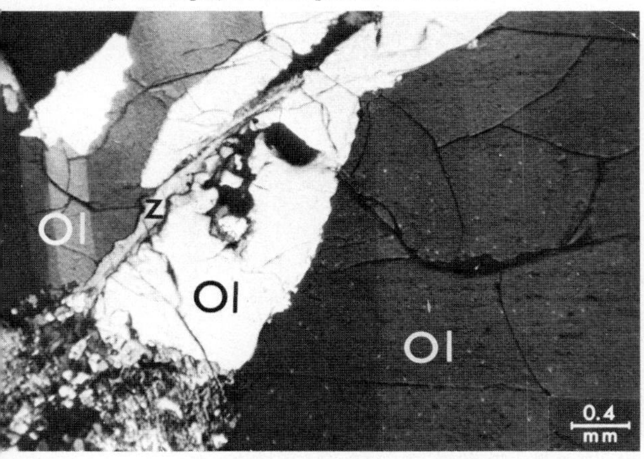

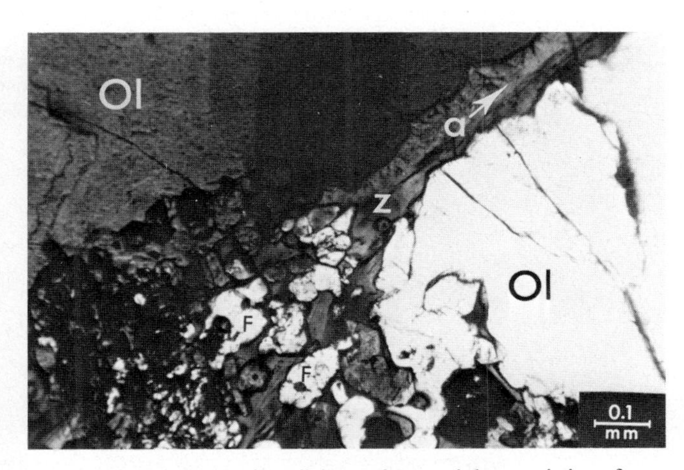

Figs. 170 and 171. Deformed olivines with a crushed zone. Along a crack fissure transgressing olivine grains, a veinlet consisting of zeolites extends into the crushed zone and occupies the interspaces between the fragments of the crushed zone, as shown in Fig. 171 by arrow "a". Ol = deformed olivine; z = veinlet of zeolite occupying fissure of the olivine grains; F = fragments in the crushed zone consisting of olivine.
Jato, Lekempti, W. Ethiopia. With crossed nicols.

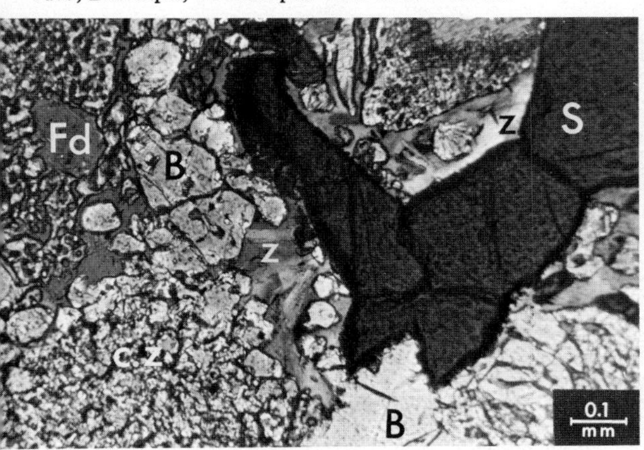

Fig. 172. Graphic spinel in intergrowth with bronzite. The bronzite is infiltrated and replaced by magmatic melt resulting in the formation of feldspars and zeolites. S = spinel; B = bronzite; Fd = feldspars; z = zeolites; Ol = forsterite; c.z = crushed zone.
Jato, Lekempti, W. Ethiopia. With half-crossed nicols.

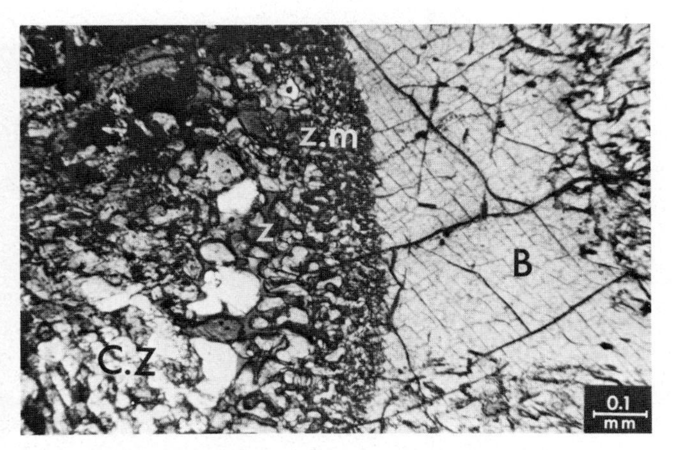

Fig. 173. Bronzite with a crushed zone with zeolites occupying the interspaces between the fragmented pyroxene of the crushed zone. Also zeolites extend as myrmekitic intergrowths into the pyroxene (the pyroxene margin that has been loosened by mantle-tectonics and which is next to the crushed zone). B = bronzite; C.Z = crushed zone marginal to the pyroxene; z = zeolites occupying the interspaces of the fragments of the crushed zone; z.m = zeolites in myrmekitic intergrowth with the pyroxene.
Jato, Lekempti, W. Ethiopia. With crossed nicols.

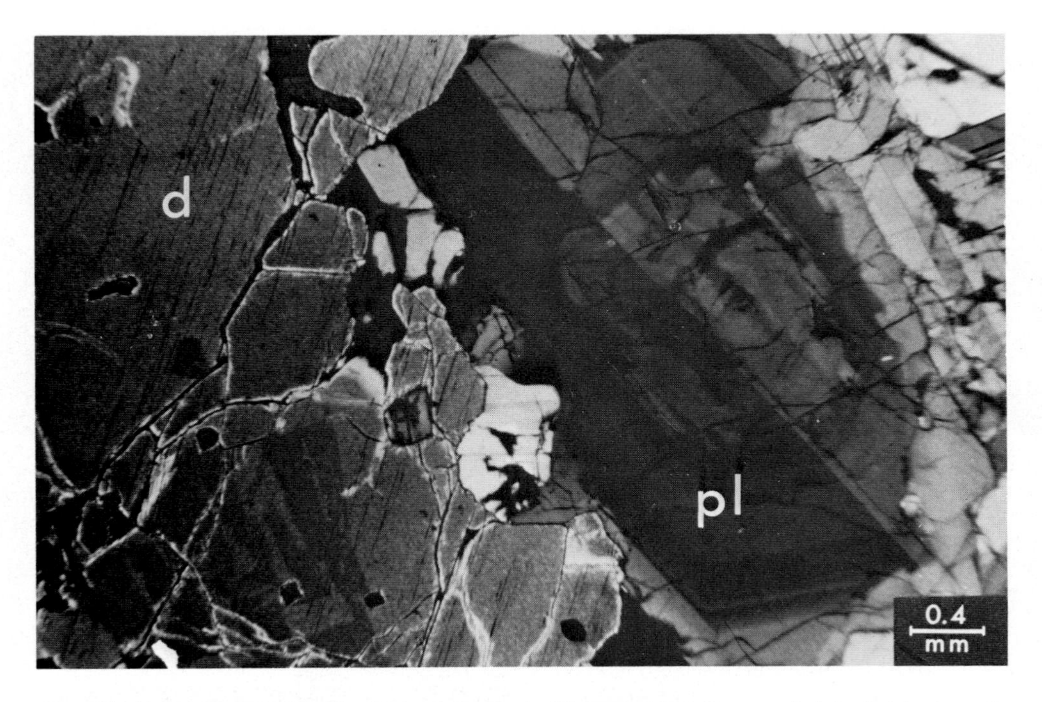

Fig. 174. A zoned plagioclase phenocryst in contact with a pyroxene phenocryst. A typical texture of the holo-phenocrystalline (protolyte) of gabbroic composition and texture is shown in the volcanic rock of the Duncan volcanic line. d = diopside; pl = plagioclase.
Volcanic line, Duncan, Ethiopia. With crossed nicols.

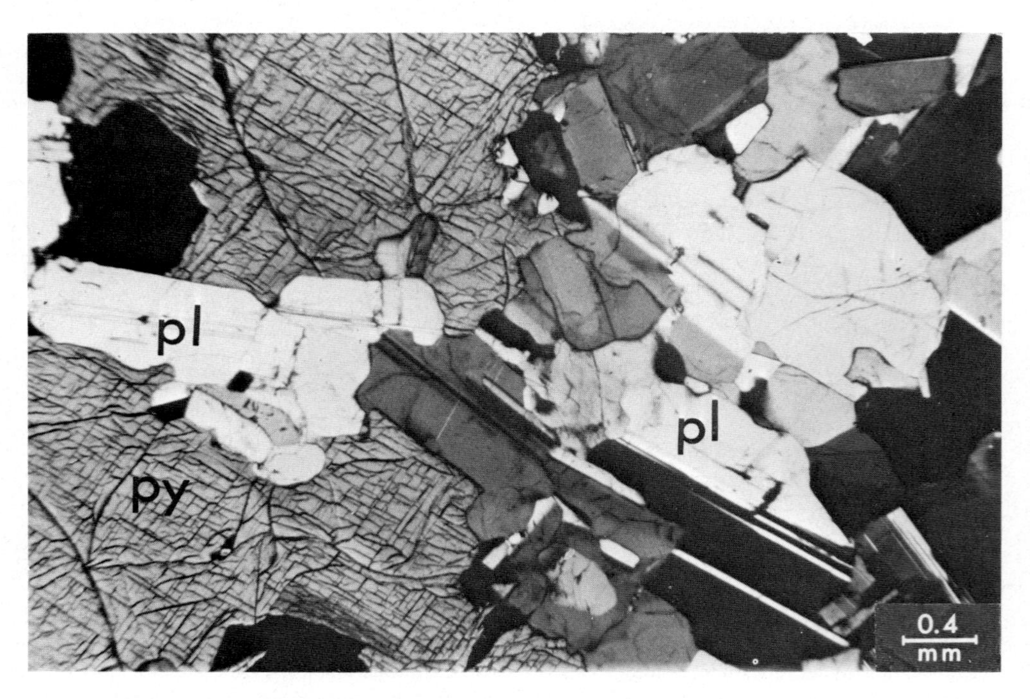

Fig. 175. Intimate gabbroic intergrowth of pyroxene (augite) and plagioclase phenocrysts. The plagioclases are corroded and enclosed by the pyroxene (augite). py = augite pyroxene; pl = plagioclase.
Volcanic line, Duncan, Ethiopia. With crossed nicols.

Fig. 176. Plagioclase in ophitic intergrowth with twinned augite phenocrysts. Also plagioclase phenocryst is shown. P = pyroxene (augite); pl = plagioclase. Duncan volcanic line, Ethiopia. With crossed nicols.

Fig. 177a. Pyroxene in contact with zoned plagioclase. The augite shows undulating extinction due to deformation effects. Similarly, Fig. 177b shows deformation lamellae of an olivine phenocryst of the gabbroic protolyte occurring in the volcanic line of Duncan, Ethiopia. With crossed nicols.

Fig. 177b. Olivine with undulating extinction due to deformation. Holophenocrystalline gabbroic rock (protolyte).
Volcanic line, Duncan, south of Addis Ababa, Ethiopia. With crossed nicols.

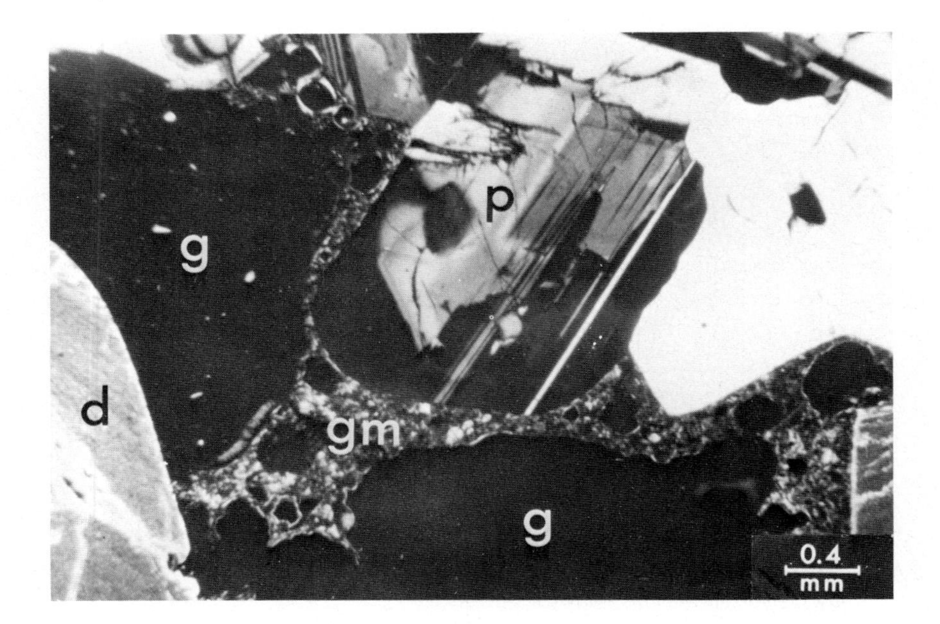

Fig. 178. Plagioclase and pyroxene with rounded margins due to magmatic corrosion. Corroded glass grains are also present. Fine groundmass, intergranular in nature is also seen. d = diopside; p = plagioclase; g = glass; gm = groundmass. Holophenocrystalline volcanic (protolyte). Duncan, south of Addis Ababa, Ethiopia. With crossed nicols.

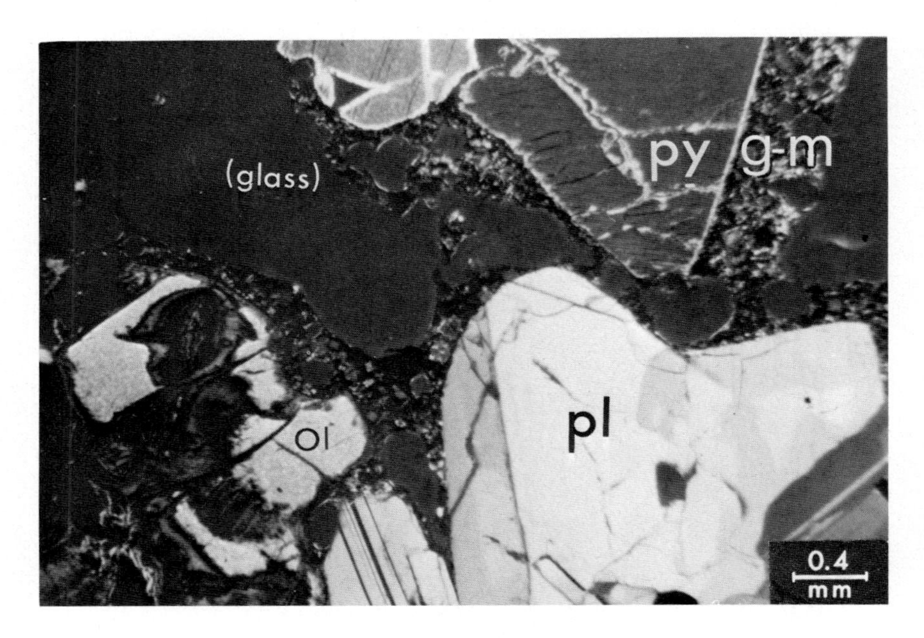

Fig. 179. Plagioclase, olivine and pyroxene phenocrysts, rounded and surrounded by the basaltic groundmass. Ol = olivines; py = pyroxenes; pl = plagioclase; g-m = basaltic groundmass. Holophenocrystalline volcanic (protolyte). Duncan, south of Addis Ababa, Ethiopia. With crossed nicols.

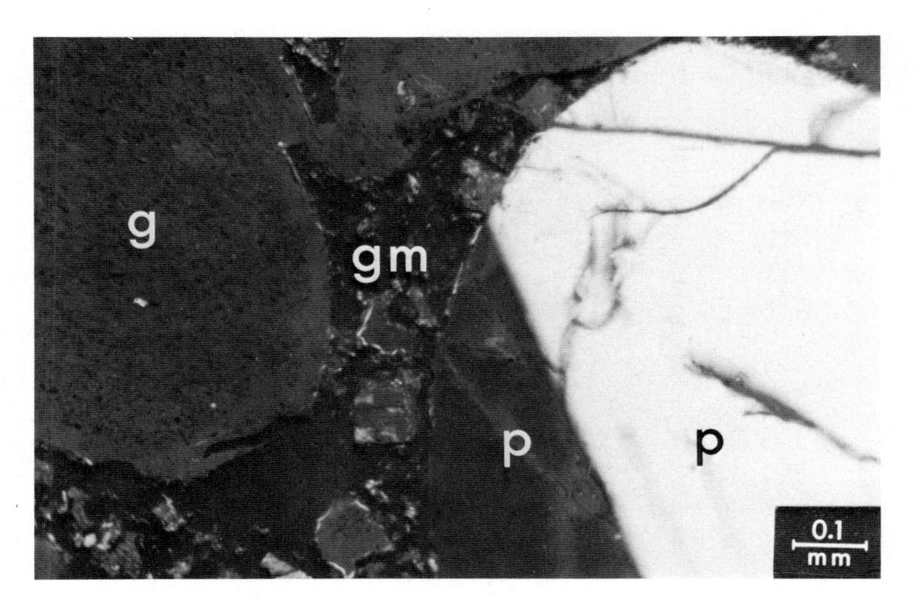

Fig. 180. Corroded (rounded) plagioclase and glass grains in fine crystalline groundmass. p = plagioclase, g = glass, gm = groundmass. Holophenocrystalline volcanic (protolyte).
Duncan, south of Addis Ababa, Ethiopia. With crossed nicols.

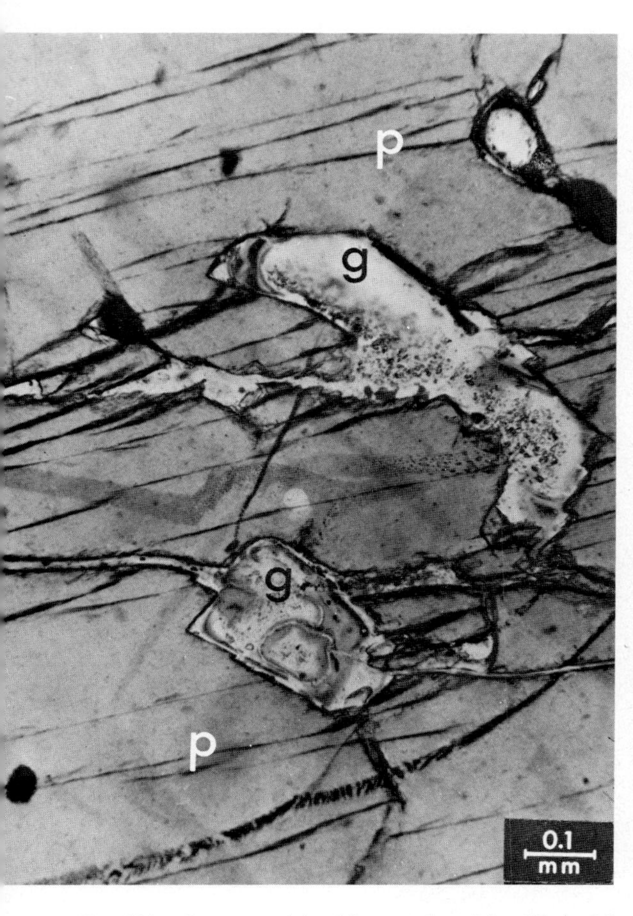

Fig. 181a. Pyroxene (p) with crystal cavities filled with glass (g). Holophenocrystalline volcanic.
Duncan, south of Addis Ababa, Ethiopia. Without crossed nicols.

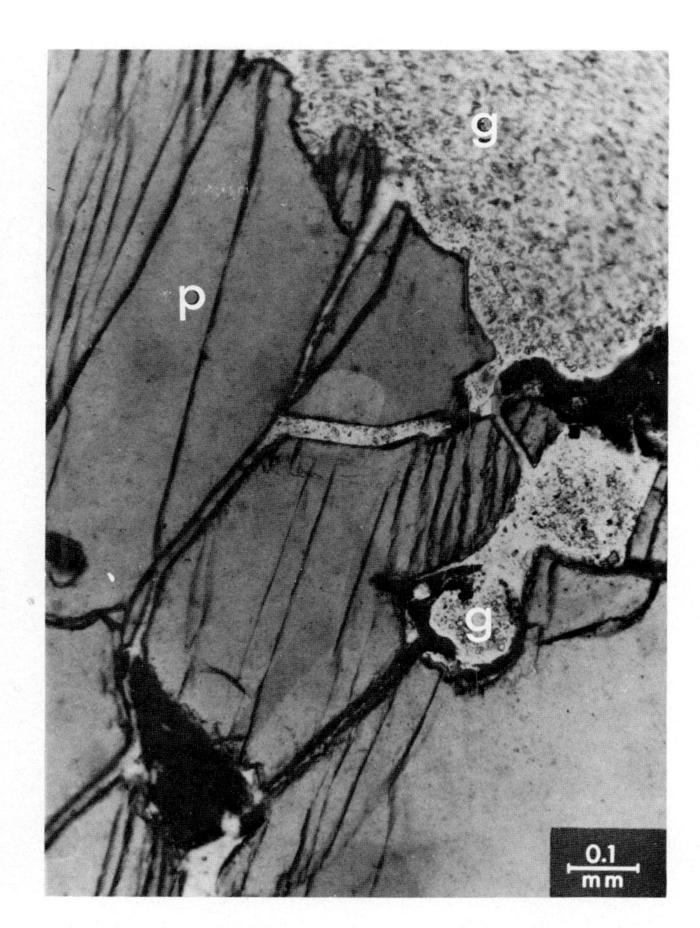

Fig. 181b. Augite with glass (exterior to the pyroxene). Fine canals of glass extend along cracks into the augite. Volcanic line of Duncan, Ethiopia. Holophenocrystalline volcanic (protolyte). p = augite, g = glass.
Duncan, south of Addis Ababa, Ethiopia. Without crossed nicols.

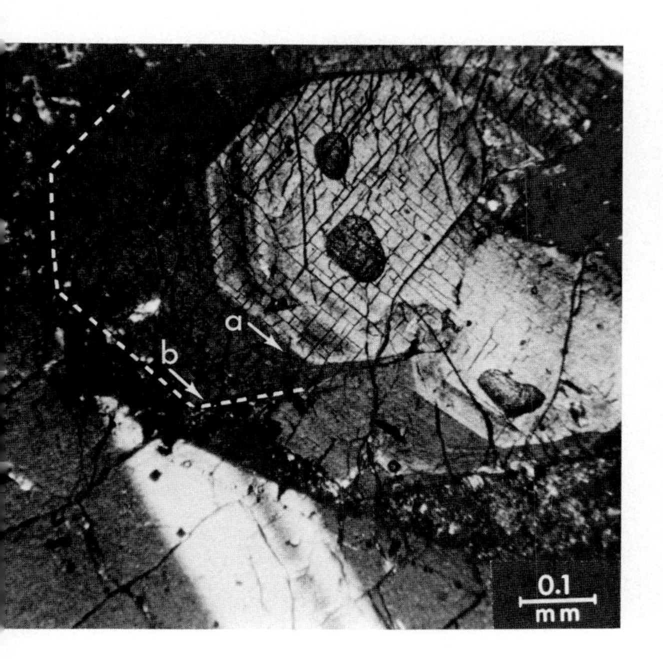

Fig. 182. An augite phenocryst section normal to the c-axis of the pyroxene and indicating cleavage parallel to (110) and ($\bar{1}10$). The zonal growth of the augite corresponds to the external face development of the crystal, as is indicated by arrows "a" and "b".
Selale Mt. region, Ethiopian Plateau. With crossed nicols.

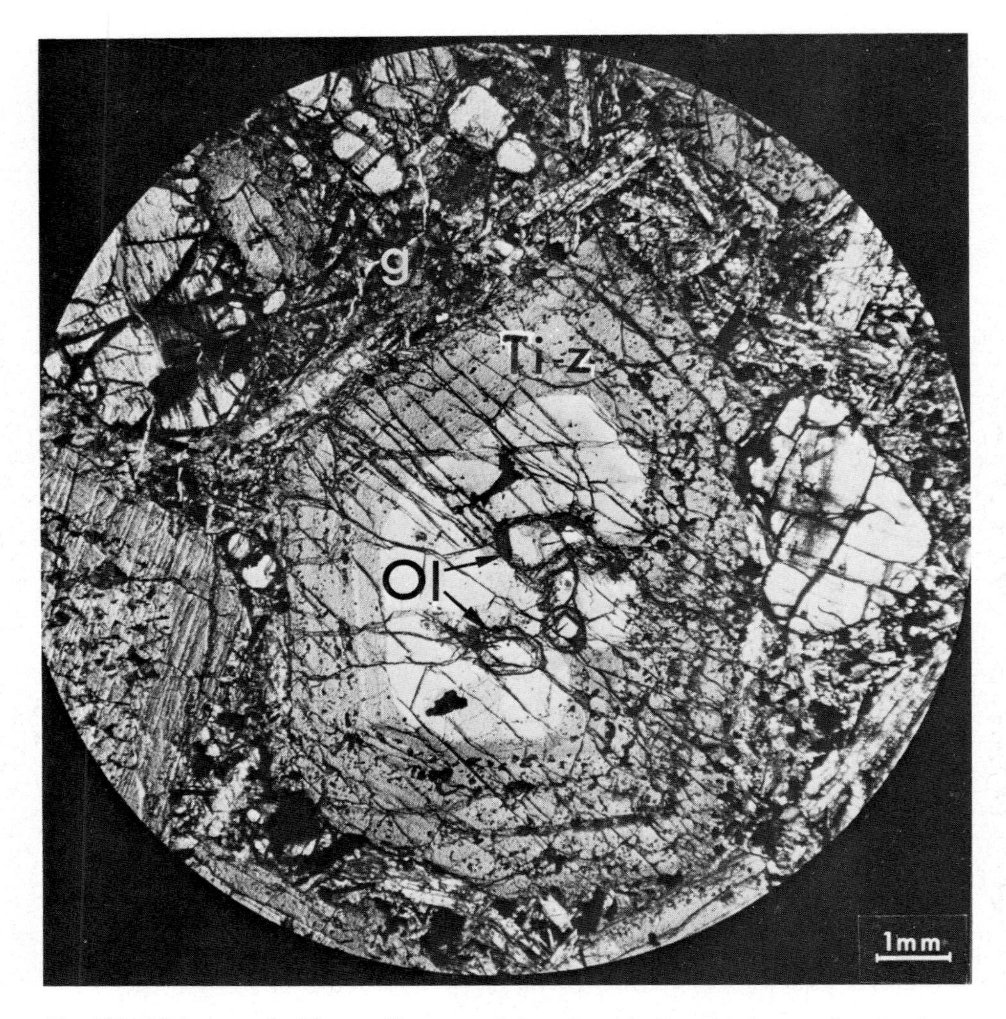

Fig. 183. Olivines, partly idiomorphic or rounded, enclosed in the central zone of augite phenocryst. The external augite zones are rich in Ti pigments, rendering the pyroxene a titano-augite. Ol = olivines included in the central zone of the augite; Ti-Z = external zone of the augite with Ti pigmentation; g = groundmass.
Batie-Dessie, Ethiopia. Without crossed nicols.

Fig. 184. Detail of Fig. 183. Rather idiomorphic olivines included in the central augitic zone.
Batie-Dessie, Ethiopia. With crossed nicols.

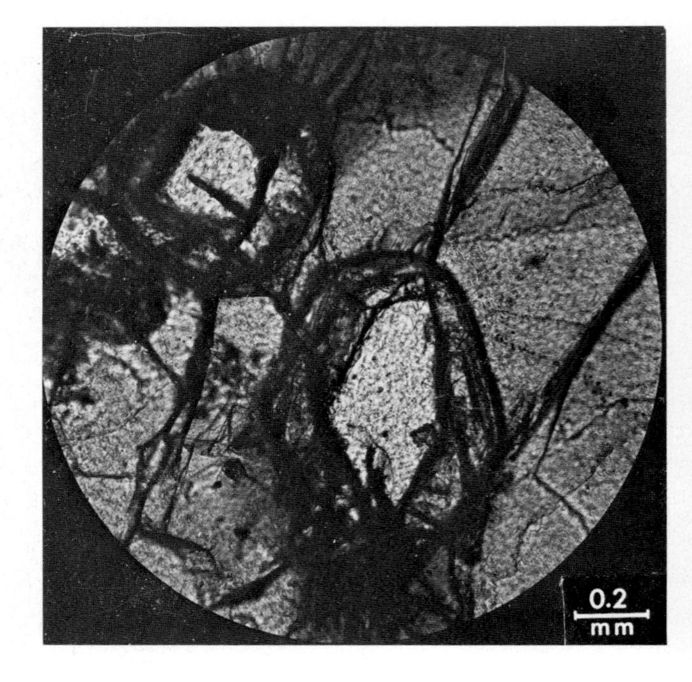

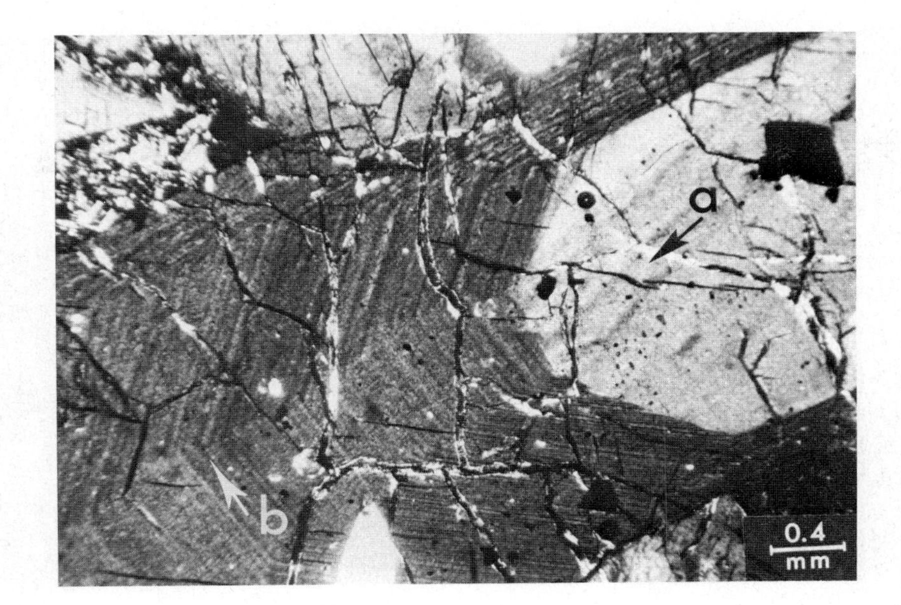

Fig. 185. Zoned titano-augite with central nucleus free from Ti-pigmentation and with fine oscillatory zoning marked by arrow "a". In contrast, the external part exhibits a fine oscillatory zoning and is heavily Ti-pigmented. There is no conformity between the zoning of the central nucleus and the external fine zoning. Arrow "b" shows the external fine oscillatory zoning of the titano-augite phenocryst.
Augite-rich basalt. Selale Mt. region, Ethiopian Plateau. With crossed nicols.

Fig. 186. Zoned augite phenocrysts. The nucleus central part is free from Ti-pigmentation. The external zoning is heavily pigmented. In the case of an adjacent zoned titano-augite, the polysynthetic twinning transverses the fine zoning.
Augite-rich basalt. Selale Mt. region, Ethiopian Plateau. With crossed nicols.

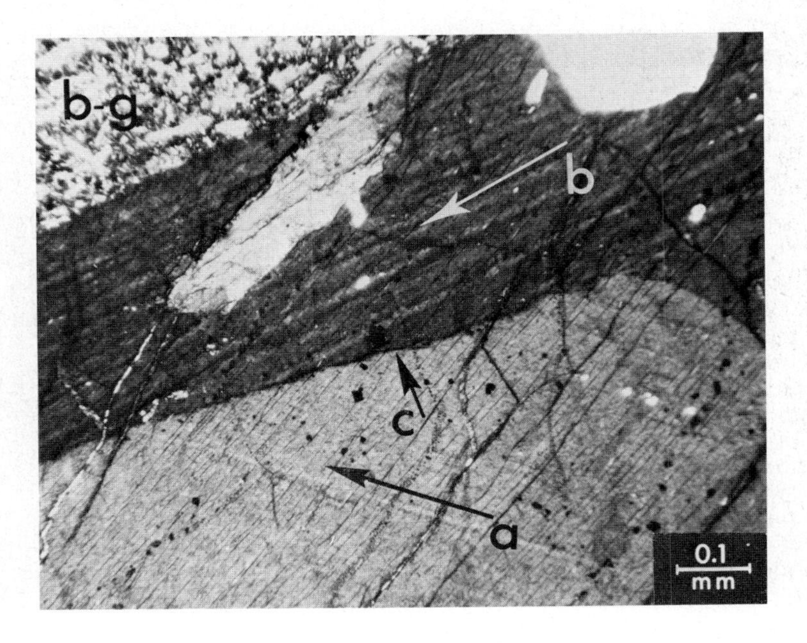

Fig. 187. Zoned titano-augite with a marked unconformity between the central (nucleus) zoning and the external zoning which is heavily pigmented. Arrow "a" shows the zoning of the central nucleus (almost free of Ti pigmentation) and arrow "b" the zoning of the external part. Arrow "c" shows the corroded outline of the central nucleus, b-g (basaltic groundmass).
Augite-rich basalt. Selale Mt. region, Ethiopian Plateau. With crossed nicols.

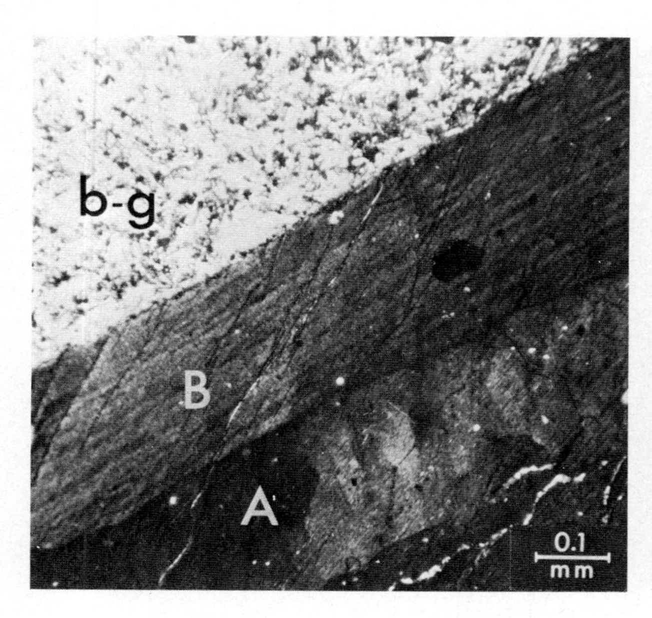

Fig. 188. Zoned titano-augite with the central nucleus showing undulating extinction (deformation effects) and the external part exhibiting fine oscillatory zoning. A = the nucleus part with deformation lamellae; B = the external part of the phenocryst with fine oscillatory zoning; b-g = basaltic groundmass.
Augite-rich basalt. Selale Mt. region, Ethiopian Plateau. With crossed nicols.

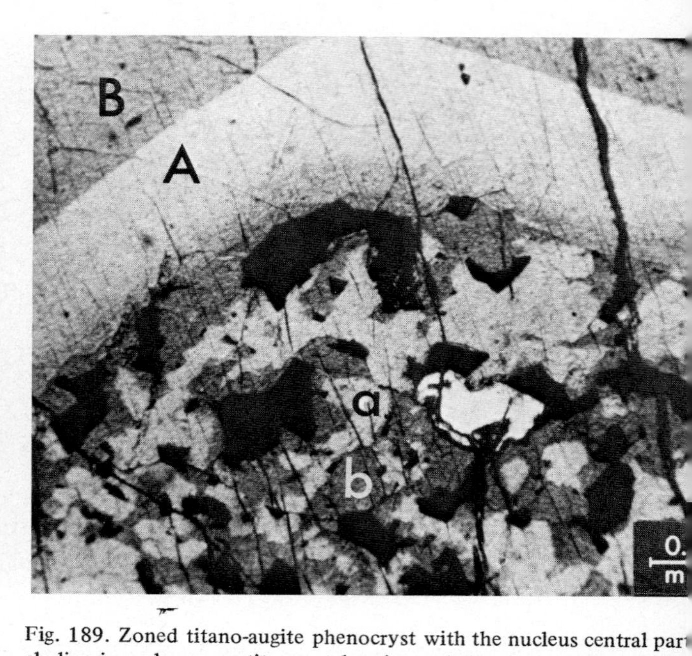

Fig. 189. Zoned titano-augite phenocryst with the nucleus central par cluding irregular magnetite crystal grains and showing a patchy differ of composition (indicated by shadowing). The external zones are free f magnetite and are uniform in composition. Magnetite = black,"a", and showing patchy differences of composition of the central nucleus pa the zoned augite phenocryst. A and B = external zones of titano-au free from magnetite inclusions and uniform in composition.
Augite-rich basalt. Selale Mt. region, Ethiopian Plateau. With cro nicols.

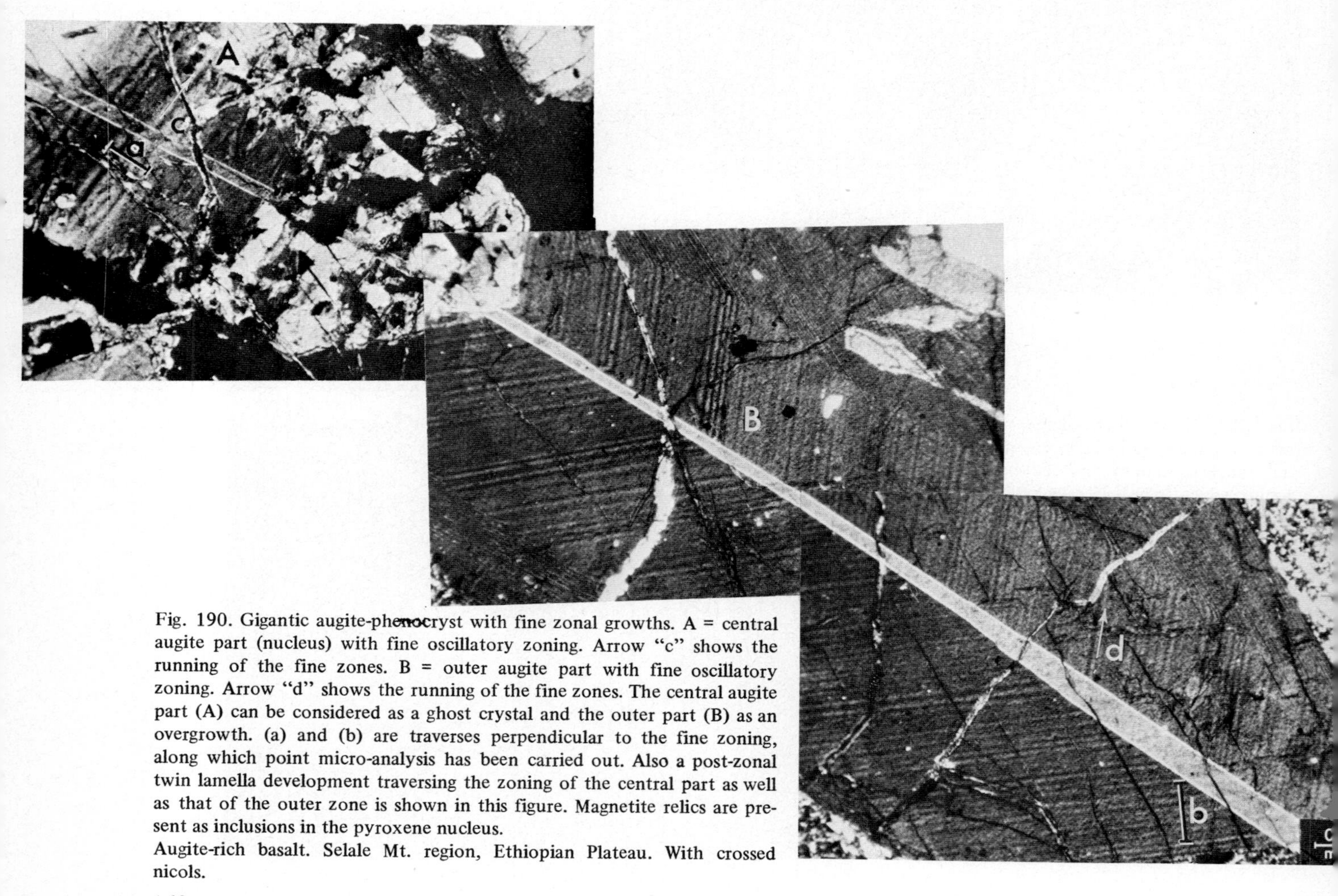

Fig. 190. Gigantic augite-phenocryst with fine zonal growths. A = central augite part (nucleus) with fine oscillatory zoning. Arrow "c" shows the running of the fine zones. B = outer augite part with fine oscillatory zoning. Arrow "d" shows the running of the fine zones. The central augite part (A) can be considered as a ghost crystal and the outer part (B) as an overgrowth. (a) and (b) are traverses perpendicular to the fine zoning, along which point micro-analysis has been carried out. Also a post-zonal twin lamella development traversing the zoning of the central part as well as that of the outer zone is shown in this figure. Magnetite relics are present as inclusions in the pyroxene nucleus.
Augite-rich basalt. Selale Mt. region, Ethiopian Plateau. With crossed nicols.

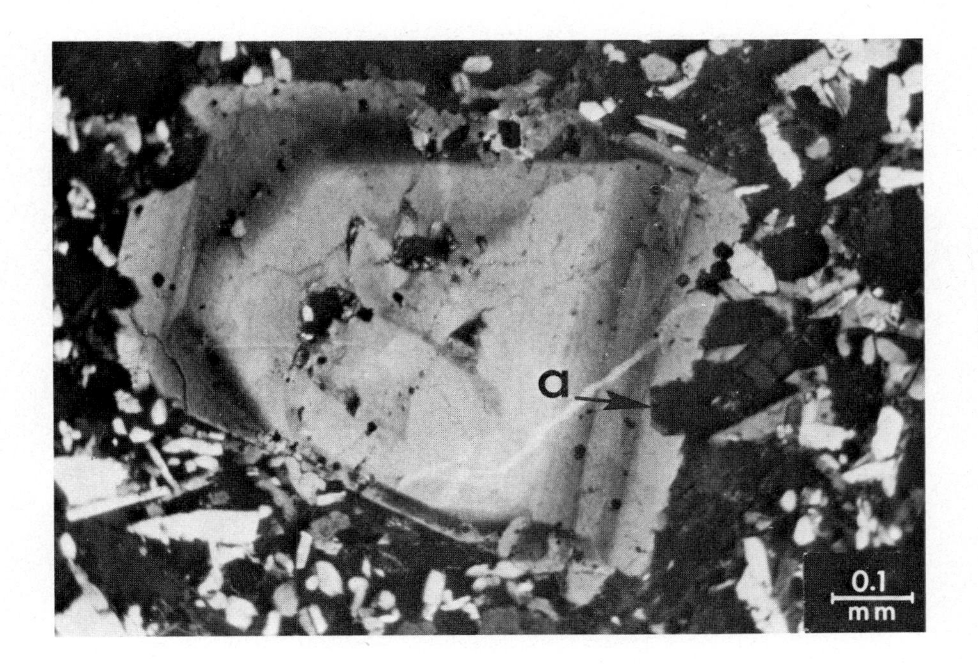

Figs. 191 and 192. Zoned augite phenocryst exhibiting simple zoning and with the external zones partly enclosing and partly engulfing groundmass components. Arrow "a" shows groundmass either enclosed or partly engulfed by the external augite zoning.
Augite-rich basalt. Selale Mt. region, Ethiopian Plateau. With crossed nicols.

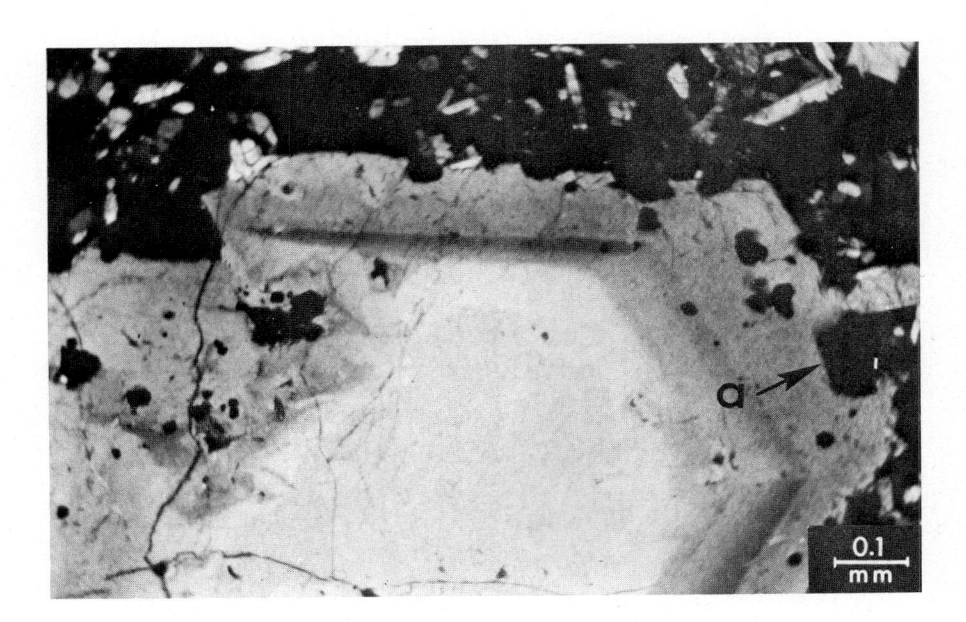

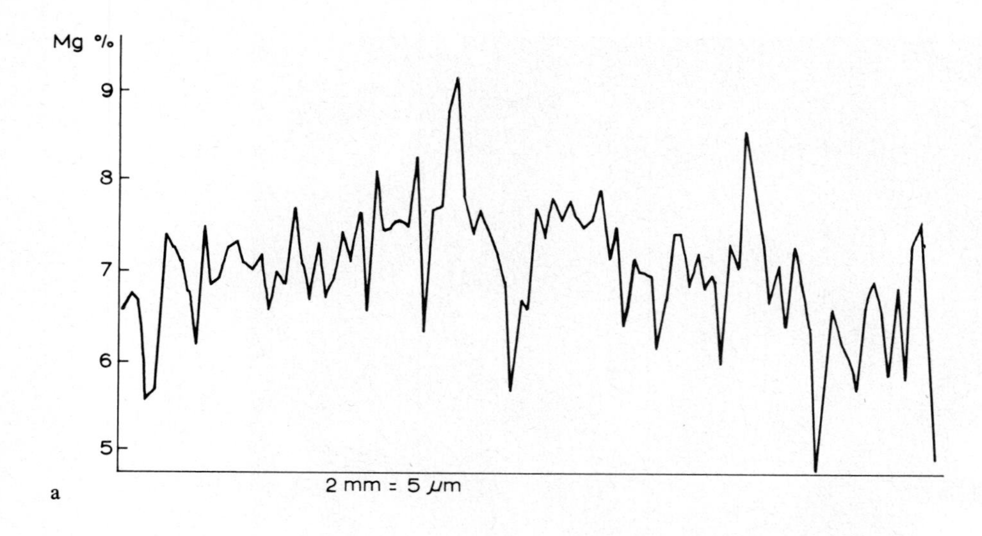

a

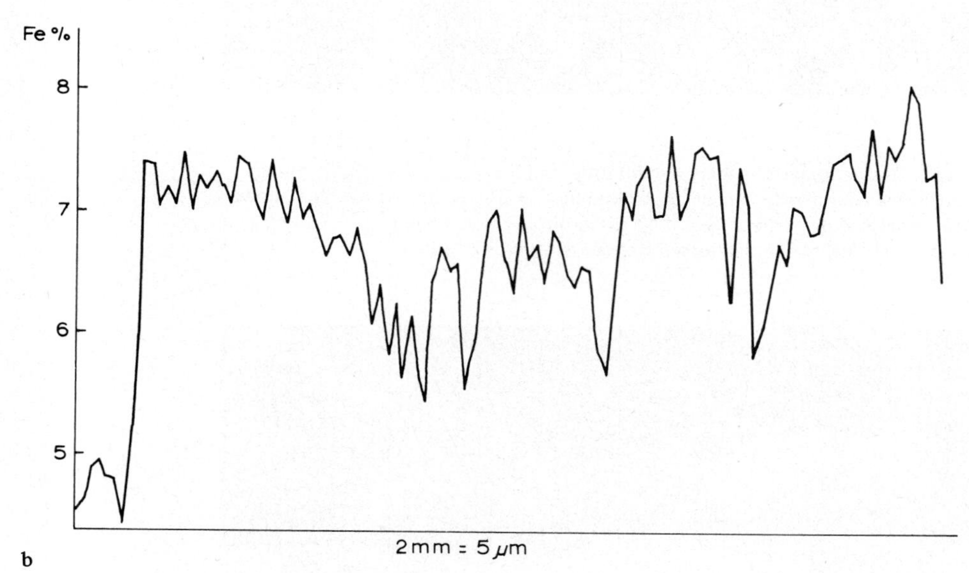

b

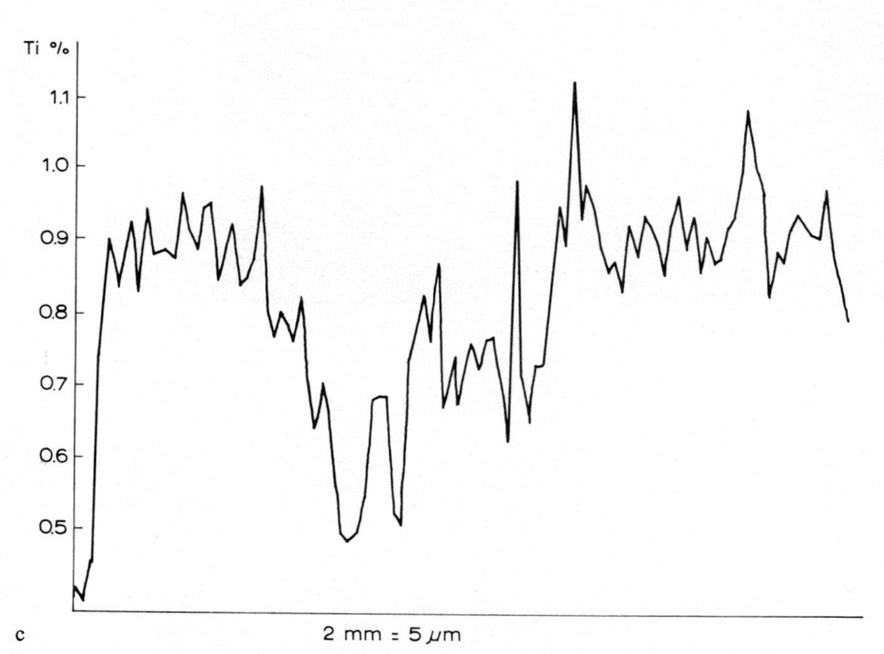

c

Fig. 193. Point microprobe analysis along traverse "a" (Fig. 190) perpendicular to the fine zoning of the central augitic phenocryst part. The Mg and Fe oscillations appear to be reversed and the Ti resembles the Mg in its oscillatory fluctuations.

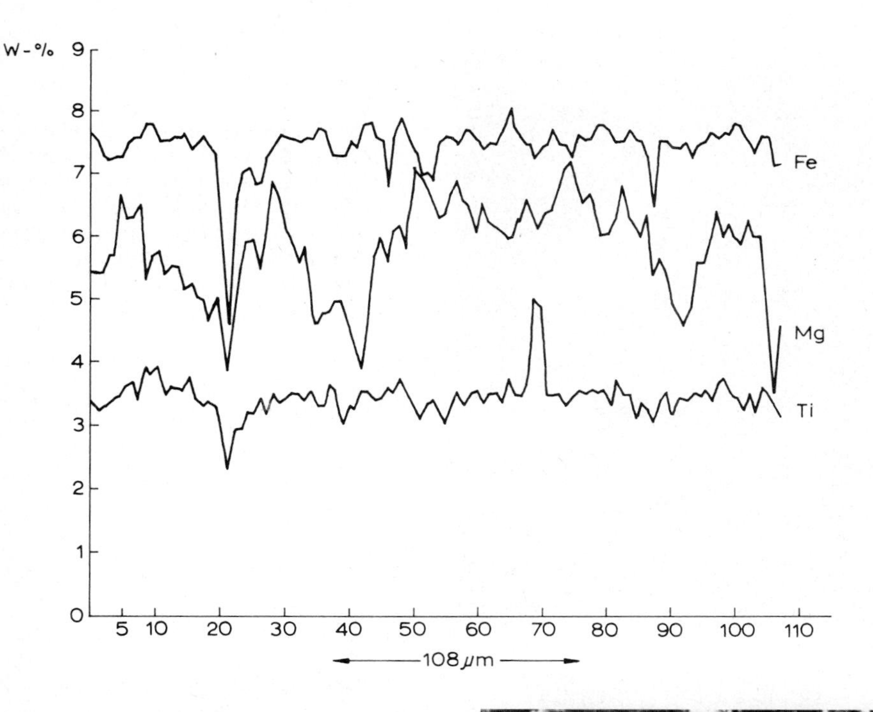

Fig. 194. Point microprobe analysis along traverse "b" (Fig. 190) perpendicular to the fine zoning of the outer augitic phenocryst part. The Mg and Fe oscillations appear to be reversed and the Ti is increased in comparison to the Ti contents of the central augitic part and remains almost constant at about 3.5%, i.e. it does not show oscillatory fluctuations.

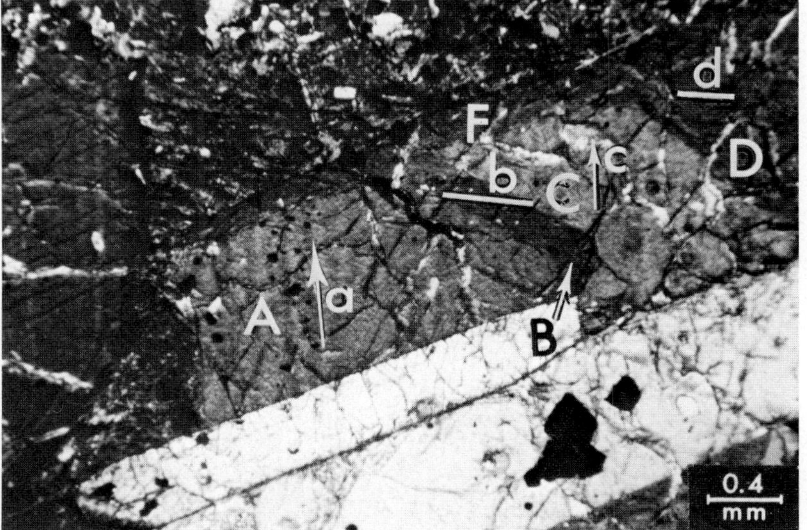

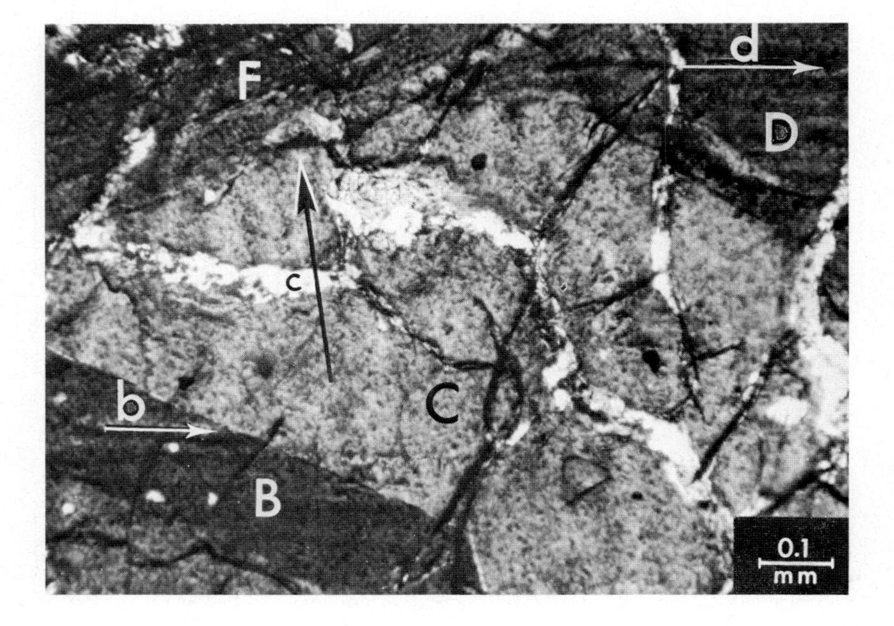

Figs. 195 and 196. Oscillatory-zoned augitic phenocrysts with marked interruptions and crystal growth and overgrowths. The growth phases A and C, and B and D have similarly orientated oscillatory zoning. The direction of the zoning in each phase of growth is marked correspondingly by a,b,c and d. F is an overgrowth and is in continuation with zone D.
Augite-rich basalt. Selale Mt. region, Ethiopian Plateau. With crossed nicols.

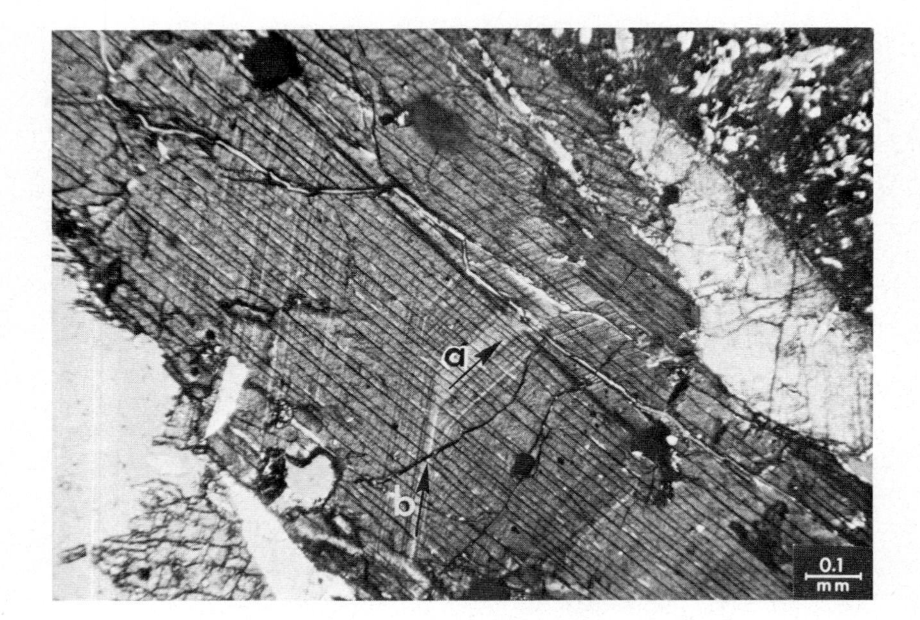

Fig. 197. Oscillatory-zoned augitic phenocryst. The width of the zone marked by arrow "a" is different from the width of the zone marked by arrow "b". The width of the same zone corresponding to different crystal faces is different.
Augite-rich basalt. Selale Mt. region, Ethiopian Plateau. With crossed nicols.

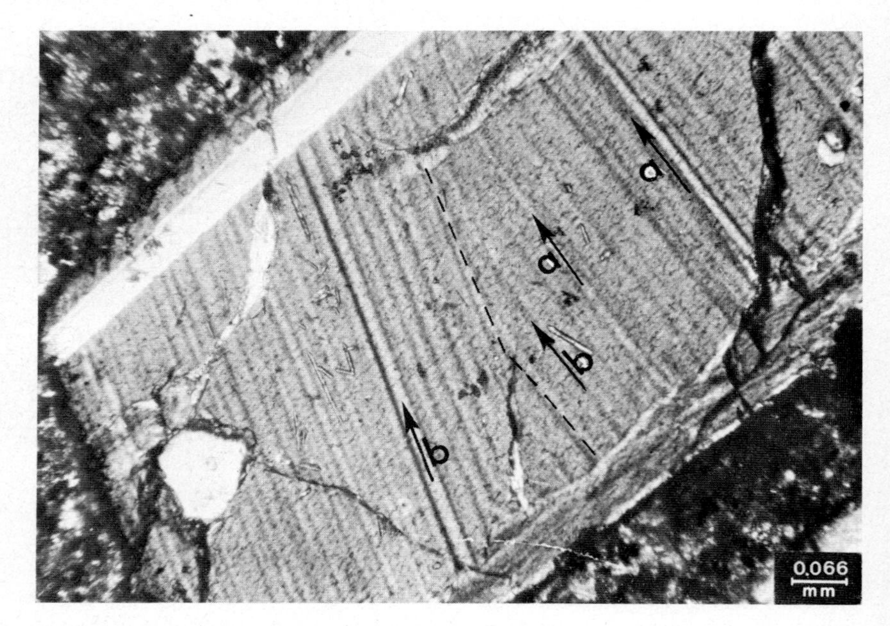

Fig. 198. Oscillatory-zoned augitic phenocryst. The zones marked by arrow "a" are parallel in growth, whereas the zones marked "b" are not parallel due to an interzonal "unconformity" marked by the interrupted line. Augite-rich basalt. Selale Mt. region, Ethiopian Plateau. With crossed nicols.

Fig. 199. Oscillatory-zoned augitic phenocryst with needles of partly resorbed rutile, orientated parallel to the augitic zoning.
Augitic-rich basalt. Selale Mt. region, Ethiopian Plateau. With crossed nicols.

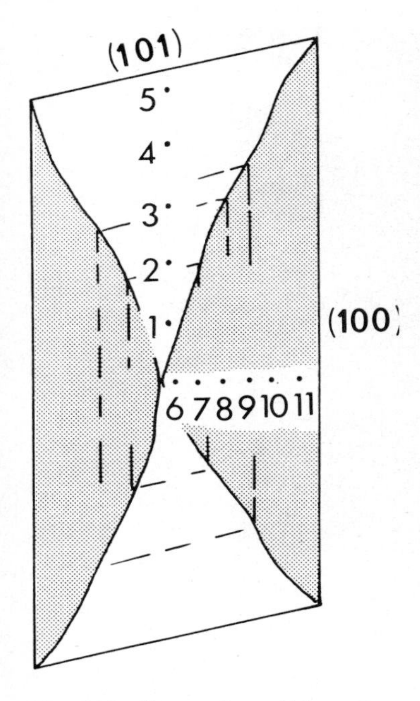

Fig. 200. Shows the position of the point analyses on the "hourglass" titano-augite by G. Schorer (1970).

Fig. 201. Augitic phenocryst with oscillatory zoning and "hourglass structure". The width of the zones of growth determine the shape of the "hourglass structure". There is a reversed orientation in the "sides" of the zoning which corresponds to adjacent crystal faces, thus the "hourglass structure" is produced. Reversal of a part of the fine zones independently of crystal face orientation has taken place, resulting in patches comparable to "hourglass structures".
Augite-rich basalt. Selale Mt. region, Ethiopian Plateau. With crossed nicols.

Figs. 202 and 203. Oscillatory-zoned pyroxene phenocrysts with "hourglass structures" and with a zonal growth "a" which does not show reversal of orientation corresponding to crystal faces. This zone appears as a continuous "belt" traversing the "hourglass sectors" and is wider and does not conform to the general pattern of the oscillatory zoning of the "hourglass structure".
Augite-rich basalt. Selale Mt. region, Ethiopian Plateau. With crossed nicols.

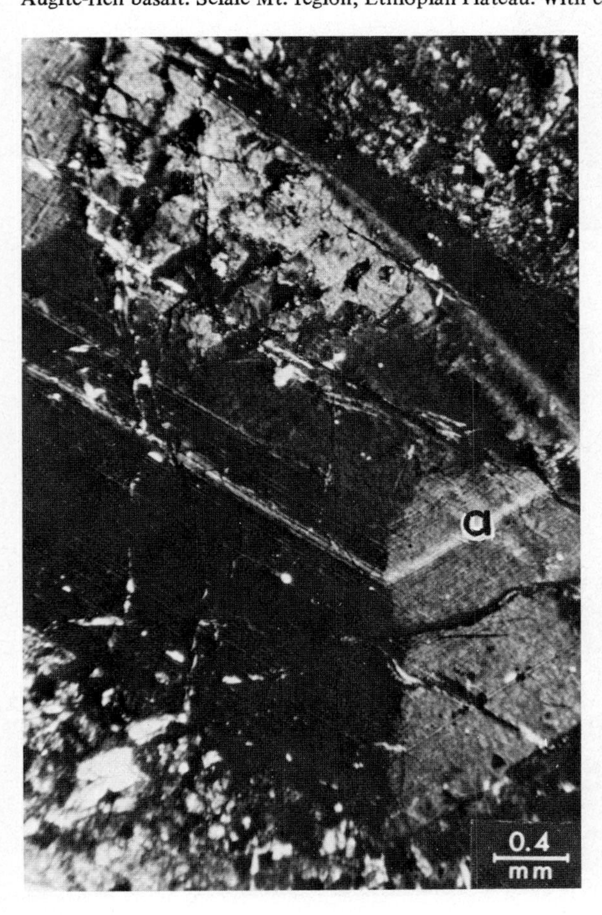

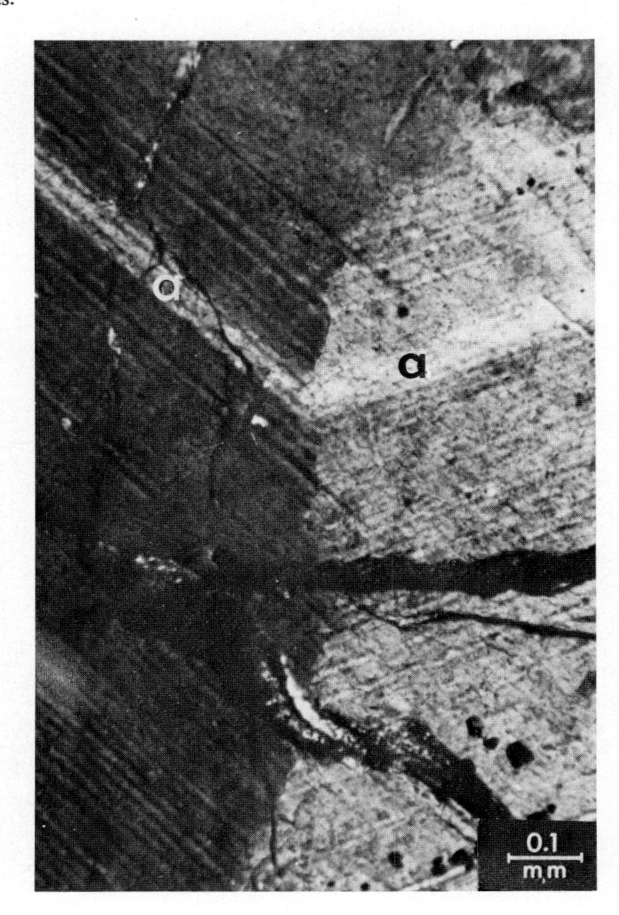

Fig. 204. Sub-phenocrystalline augite exhibiting hourglass structure. A = sub-phenocrystalline augite with hourglass structure, b-g = basaltic groundmass.
Augite-rich basalt. Selale Mt. region, Ethiopian Plateau. With crossed nicols.

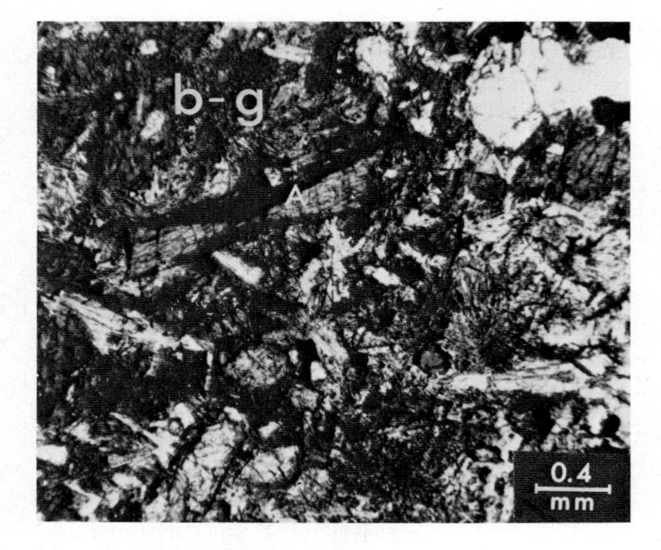

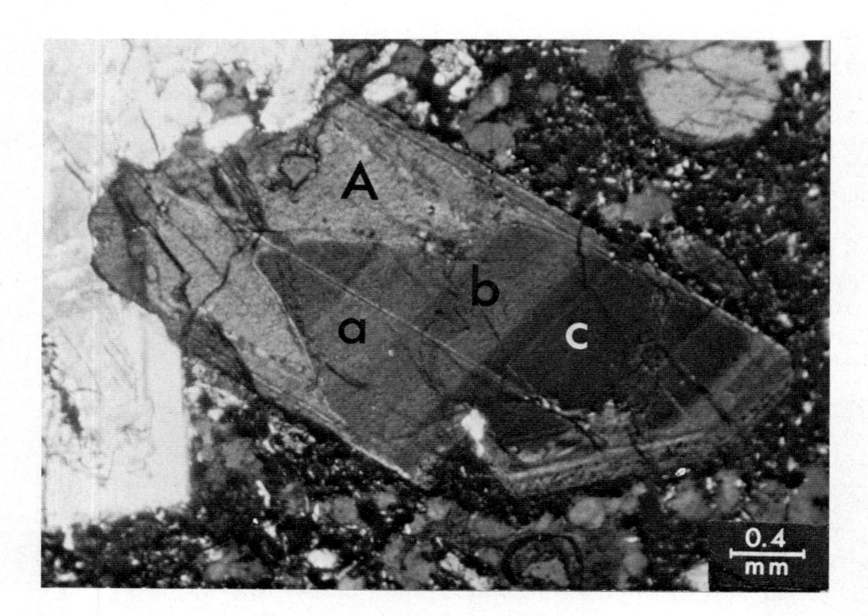

Fig. 205. Augite phenocryst with hourglass structure, exhibiting wide zones of growth sectorally corresponding to the "sectorial structure" of the hourglass. A = augite phenocryst with hourglass structure; a,b,c = wide zones of growth corresponding to the sectorial structure of the hourglass. Augite-rich basalt. Selale Mt. region, Ethiopian Plateau. With crossed nicols.

Fig. 206. Augite phenocryst with fine oscillatory zoning and with multi-sectoral structure (comparable to hourglass) dependent on the orientation of the fine zoning in the respective sectors. a,b,c,d = indicate sectors of the augite phenocryst. Augite-rich basalt. Selale Mt. region, Ethiopian Plateau. With crossed nicols.

Fig. 207. Cross-section of augite phenocryst showing polysynthetic twinning parallel to the (100) of the augite.
Augite-rich basalt. Selale Mt. region, Ethiopian Plateau. With crossed nicols.

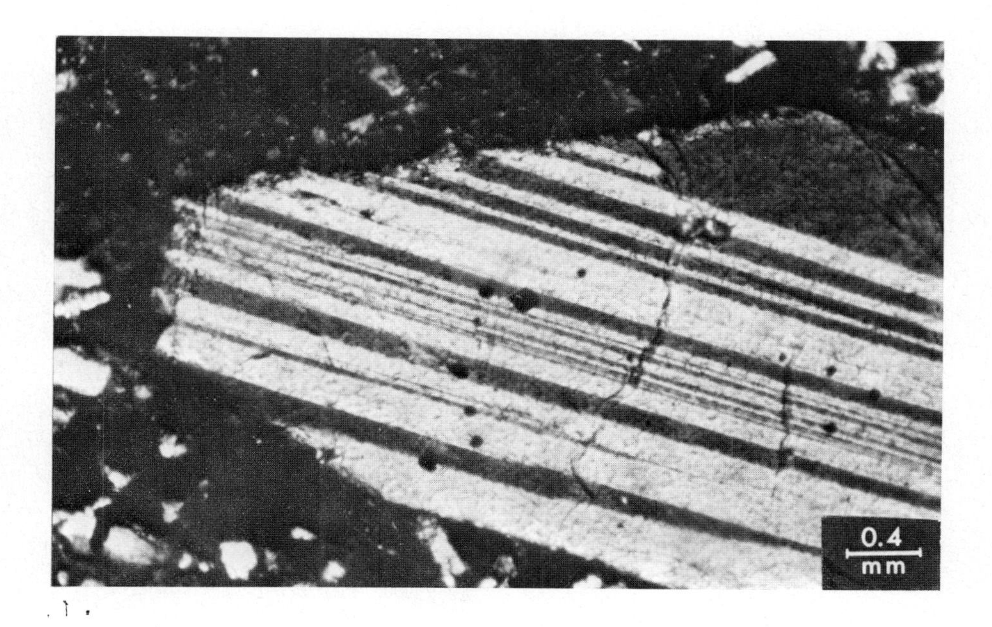

Fig. 208. Cross-section of augite phenocryst exhibiting polysynthetic twinning parallel to the (100).
Augite-rich basalt. Selale Mt. region, Ethiopian Plateau. With crossed nicols.

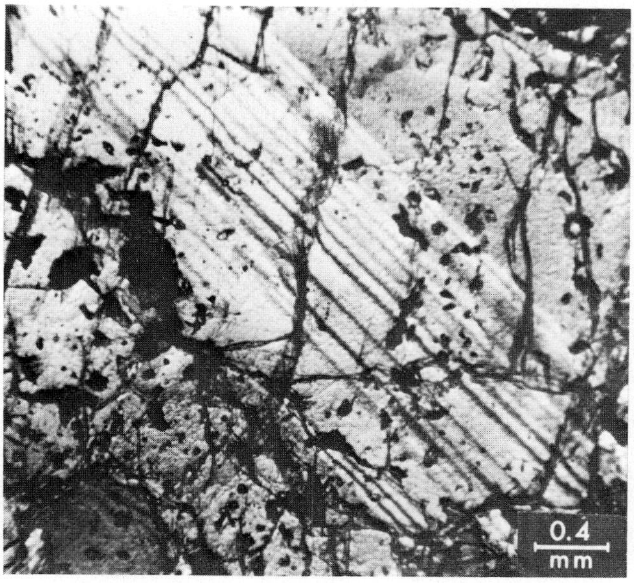

Fig. 209. Augite phenocryst exhibiting fine polysynthetic twinning.
Augite-rich basalt. Selale Mt. region, Ethiopian Plateau. With crossed nicols.

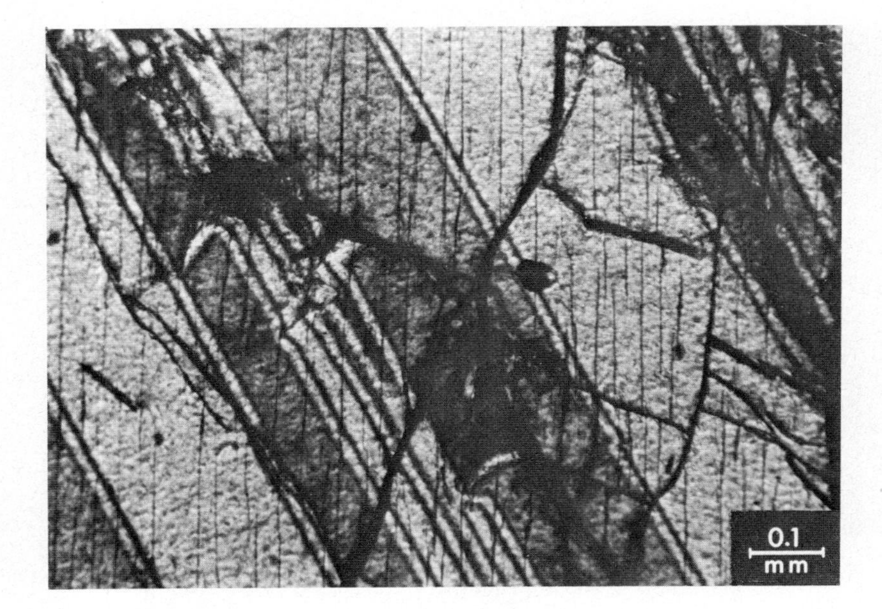

Fig. 210. Augite phenocryst with polysynthetic twinning parallel to (100) and with cleavage traces at an angle of 36° to the twinning.
Augite-rich basalt. Selale Mt. region, Ethiopian Plateau. With crossed nicols.

Fig. 211. Zoned augite phenocryst, with the zoning displaced by twinning. The zig-zag (interrupted line) marks the twinning "plane".
Augite-rich basalt. Selale Mt. region, Ethiopian Plateau. With crossed nicols.

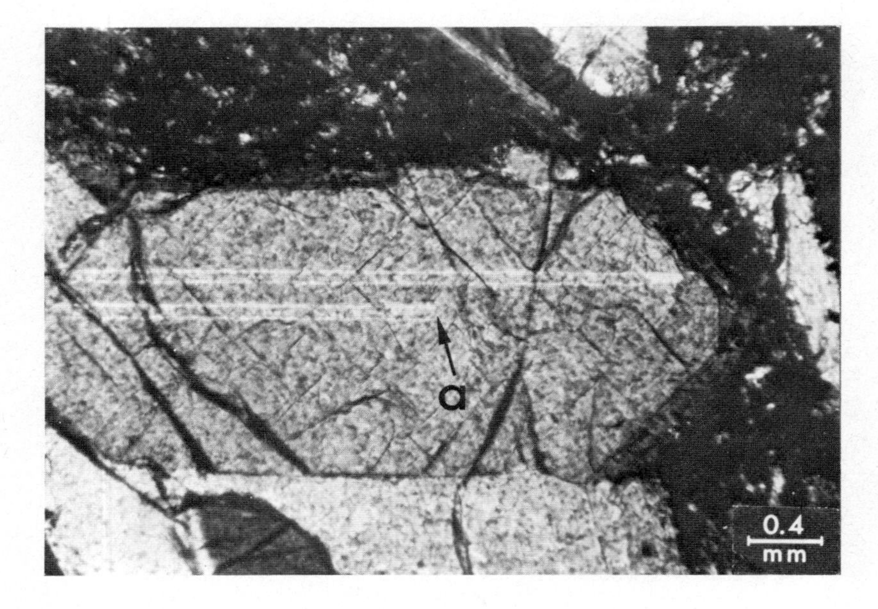

Fig. 212. Augite with polysynthetic twinning. Arrow "a" shows the abrupt termination of one of the lamellae, while the other continues.
Augite-rich basalt. Selale Mt. region, Ethiopian Plateau. With crossed nicols.

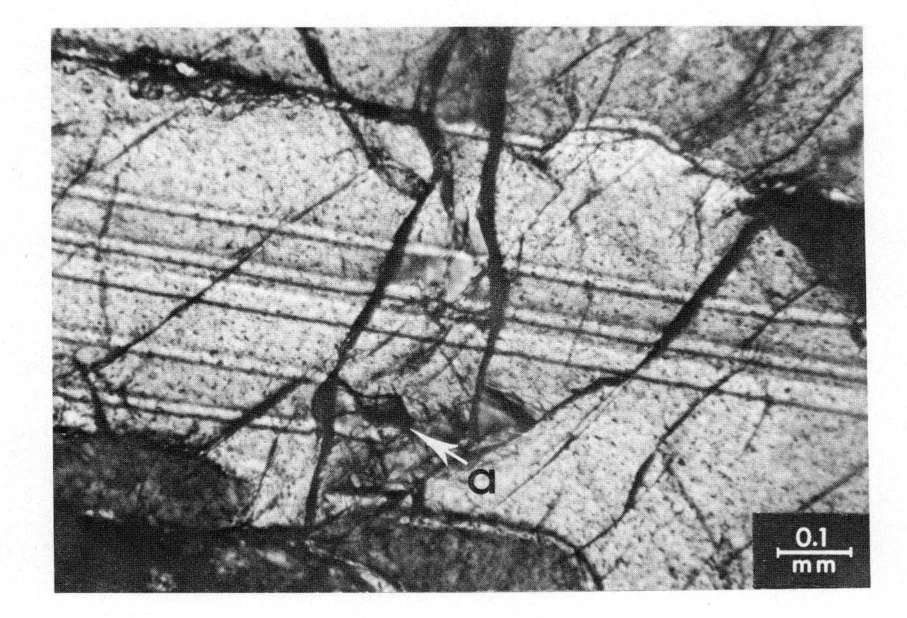

Fig. 213. Augite phenocryst with polysynthetic twin lamellae. One of the lamellae terminates abruptly. Arrow "a" shows the termination of a polysynthetic augitic twin lamella.
Augite-rich basalt. Selale Mt. region, Ethiopian Plateau. With crossed nicols.

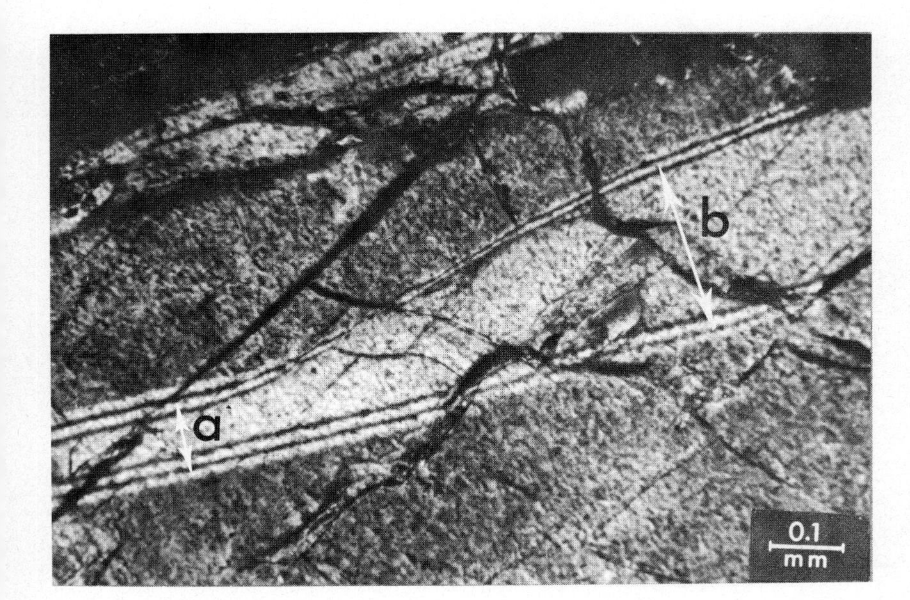

Fig. 214. Augite phenocryst with a twin lamella showing variation in shape, deviating greatly from a regular lamellar shape. Arrows "a" and "b" show the change of width of the augite twin lamella.
Augite-rich basalt. Selale Mt. region, Ethiopian Plateau. With crossed nicols.

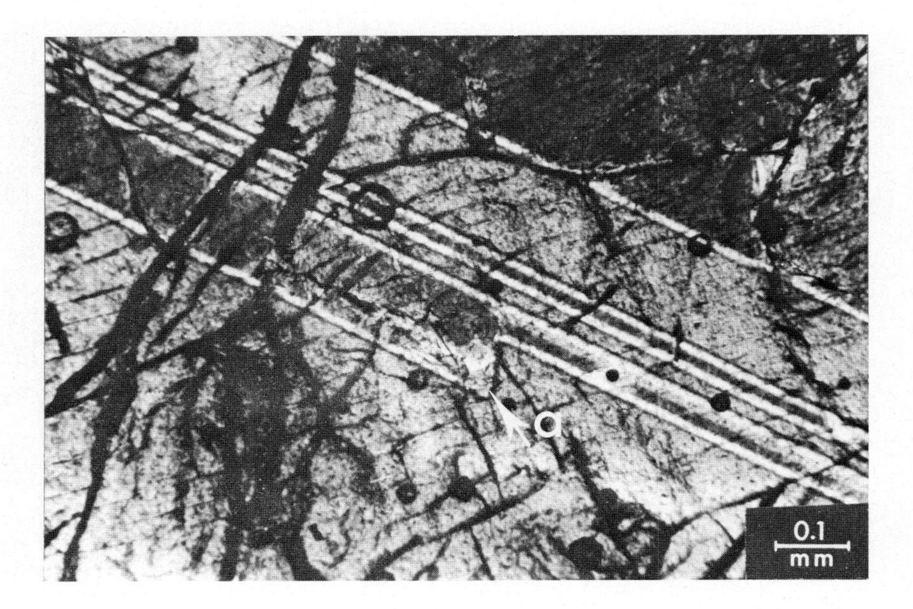

Fig. 215. Augite with polysynthetic twinning. As arrow "a" shows, one of the lamellae abruptly changes width.
Augite-rich basalt. Selale Mt. region, Ethiopian Plateau. With crossed nicols.

Fig. 216. Polysynthetic twinning in augite phenocrysts. Arrow shows a knick in the twinning.
Augite-rich basalt. Selale Mt. region, Ethiopian Plateau. With crossed nicols.

Fig. 217. Zoned augite (arrow "a" shows the zonal crystal growth) transversed by a diagonal twin-lamella showing a knick (see arrow "b").
Augite-rich basalt. Selale Mt. region, Ethiopian Plateau. With crossed nicols.

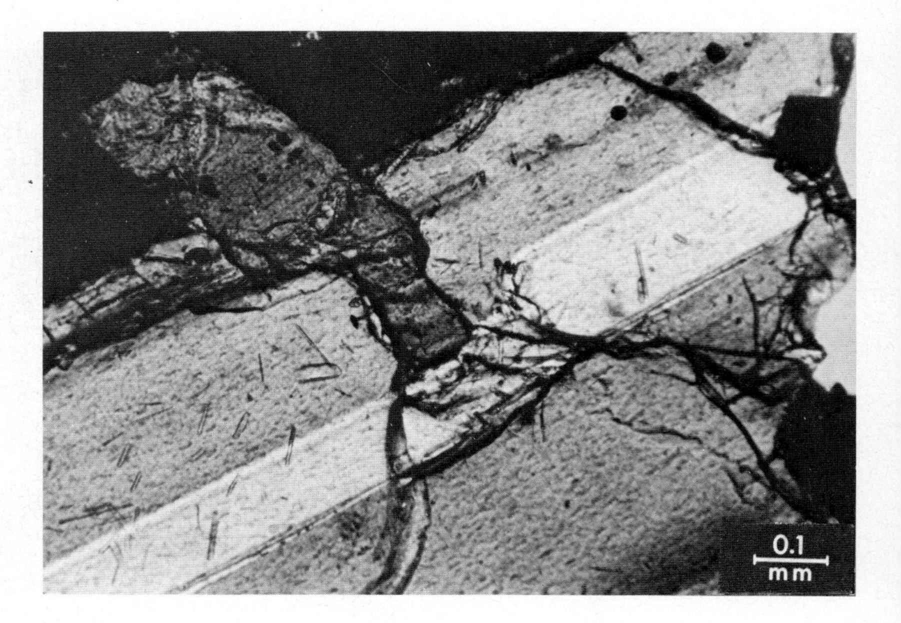

Fig. 218. Twinned augite phenocryst invaded by a smaller augite in a growth direction perpendicular to the phenocryst's twinning. The twin lamella of the phenocryst is bent and crushed.
Augite-rich basalt. Selale Mt. region, Ethiopian Plateau. With crossed nicols.

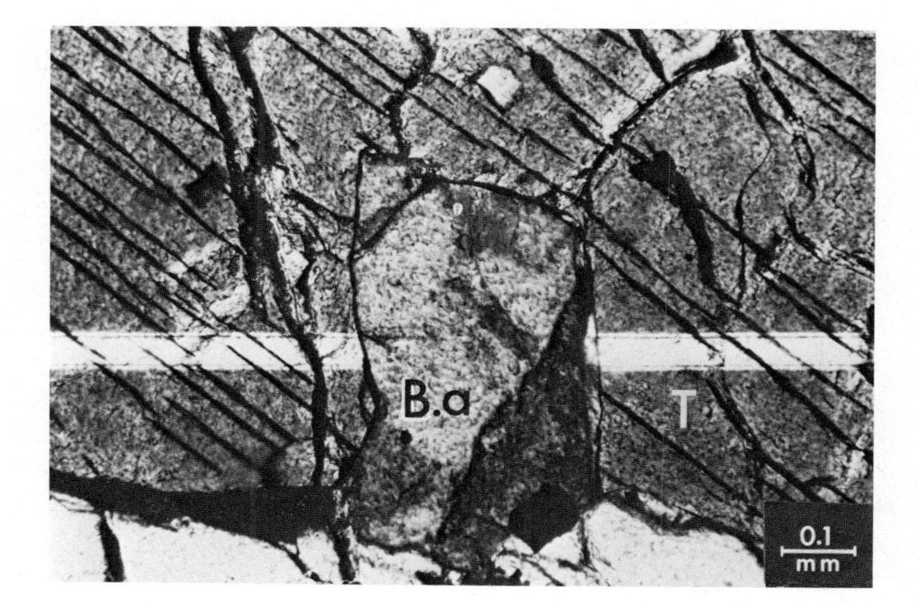

Fig. 219. Twinned augite, with the twin lamella interrupted by an augite block differently orientated. T = twinned augite; B.a = differently orientated augite block interrupting the twin lamella.
Augite-rich basalt. Selale Mt. region, Ethiopian Plateau. With crossed nicols.

Fig. 220. Oscillatory-zoned phenocryst with post-zonal polysynthetic twinning, which affects the zoning and cuases a "knicking", i.e. evidence of the post-zonal development of twinning.
Augite-rich basalt. Selale Mt. region, Ethiopian Plateau. With crossed nicols.

Fig. 221. Twinned augitic pyroxene with the augite individuals having the plane of twinning as a rotation axis. A swallow-tailed pyroxene is thus formed.
Augite-rich basalt. Selale Mt. region, Ethiopian Plateau. With crossed nicols.

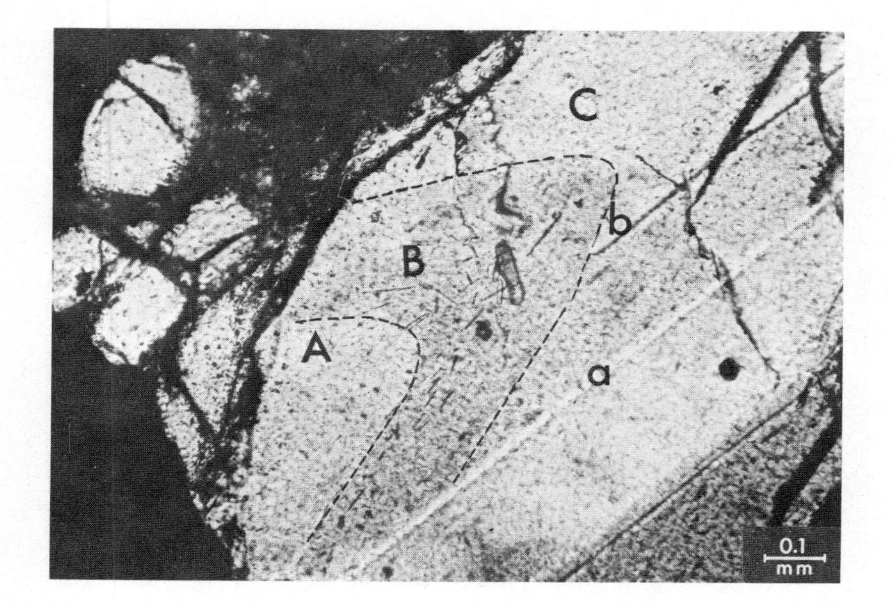

Fig. 222. The relation between zoning and twinning. Twin-lamella (a) traverses the augite zoning, whereas twin lamella (b) is developed only at the outer zone (C) of the zoned pyroxene, the succesion of the zonal growth of augite being A,B and C. In zone B microneedles are abundant, probably rutiles or micro-apatites.
Augite-rich basalt. Selale Mt. region, Ethiopian Plateau. With crossed nicols.

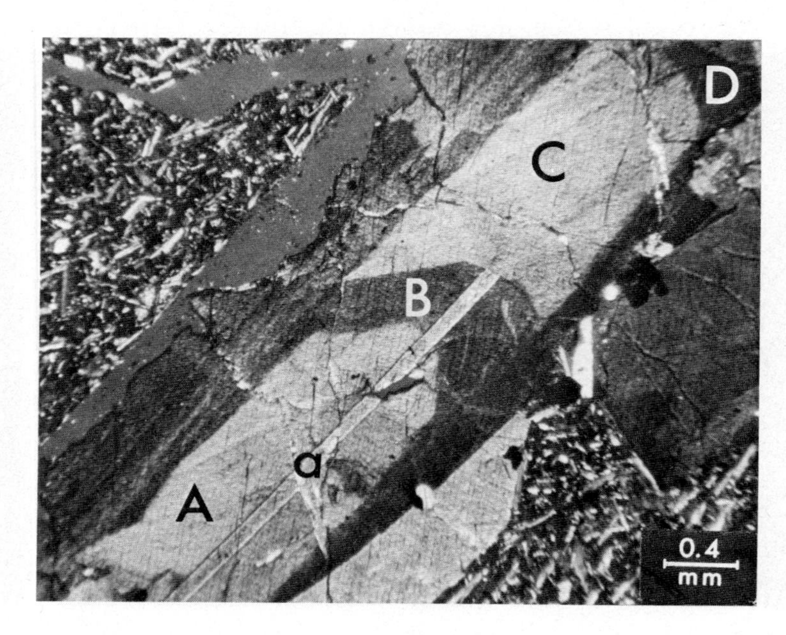

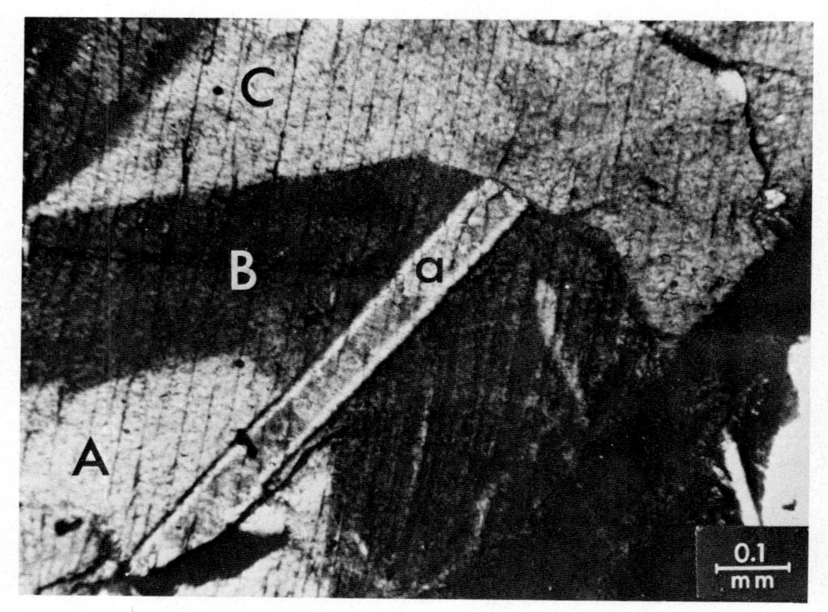

Fig. 223 (a,b). Augite with zonal growths A, B,C and D has a twin lamella "a" traversing zones A and B and is post-zonal in respect to these zones. In contrast, the zones C and D are formed later than the twinning. It is possible that the outline of zone B represents a pronounced interruption in crystal growth.
Augite-rich basalt. Selale Mt. region, Ethiopian Plateau. With crossed nicols.

Fig. 224. Polysynthetically twinned augite with a cleavage pattern. No displacement of the augitic twin lamellae is to be observed. A = polysynthetic twin lamellae; b = cleavage pattern.
Augite-rich basalt. Batie-Dessie, Ethiopia. With crossed nicols.

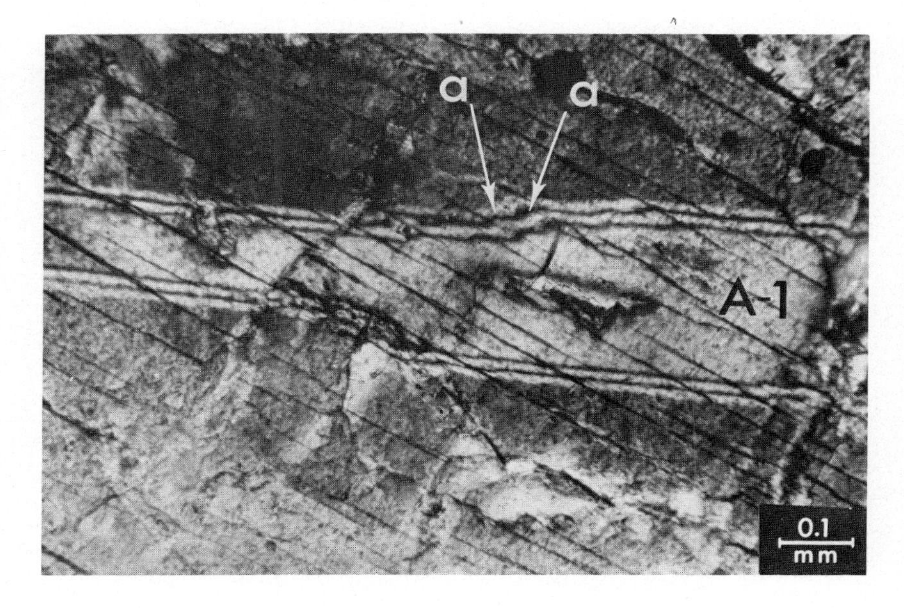

Fig. 225. Augitic twin lamella with margins clearly showing displacements due to the post-twinning crystal cleavage. Arrows "a" show displacement (a type of microfault) of the margins of the twin lamella by the crystal cleavage planes that acted as microfault displacement planes. A-1 = augite twin-lamella.
Augite-rich basalt. Batie-Dessie, Ethiopia. With crossed nicols.

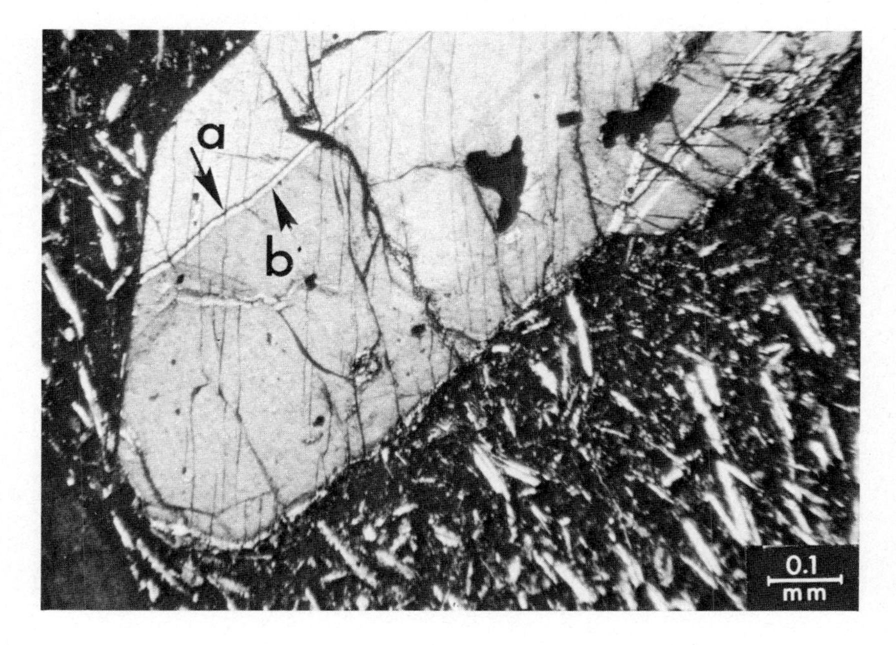

Fig. 226. Shows twinned augite with the twin-planes showing microdisplacement due to the crystal cleavage acting as microfault planes. Arrow "a" shows micro-displacement of the twin planes by cleavage. Arrow "b" marks the twin plane which show displacement by the cleavage micro-faulting.

Augite-rich basalt. Selale Mt. region, Ethiopian Plateau. With crossed nicols.

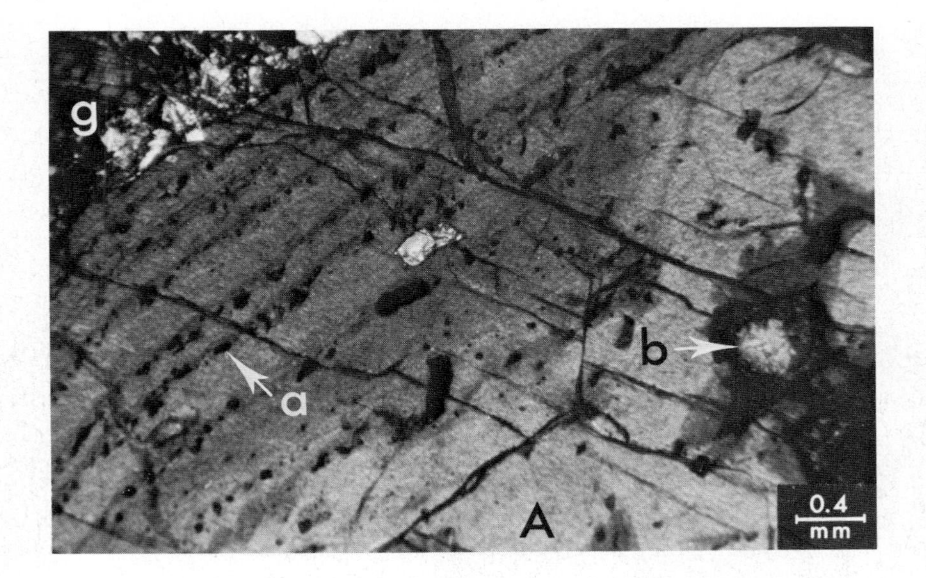

Fig. 227. Gigantic augite phenocryst partly grown after groundmass components which are either zonally incorporated and reduced to pigment size or are enclosed, indicating differences of composition around them. Arrow "a" = groundmass components reduced to pigment size and zonally incorporated; arrow "b" = olivine enclosed in the pyroxene phenocryst causing composition differences around it in the host augite. A = augite phenocryst; g = groundmass.
Olivine–pyroxene-rich basalt. Batie-Dessie, Ethiopia. With crossed nicols.

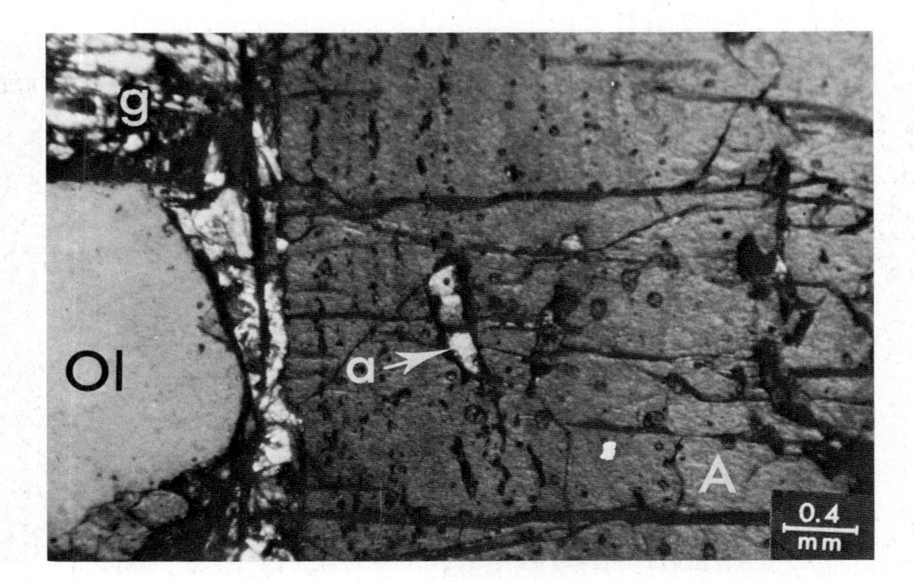

Fig. 228. Pigment-size groundmass and feldspar laths zonally incorporated in the augite phenocryst. Arrow "a" shows groundmass plagioclase zonally incorporated in the augite. A = augite phenocryst; Ol = early olivine crystallization phase; g = groundmass.
Olivine–pyroxene-rich basalt. Batie-Dessie, Ethiopia. With crossed nicols.

Fig. 229. Gigantic zoned pyroxene (tecoblast) with plagioclase laths of the groundmass incorporated parallel to the zoning of the augitic host. py = pyroxene; pl = plagioclase.
Essexite, Crawford, J., Lanarkshire, Scotland. With crossed nicols.

Figs. 230 and 231. Pyroxene phenocryst digested and assimilated along a crack by later tecoblastic plagioclase. The following zones are recognizable in the plagioclase: Zone I = 38–39% An; II = 44% An; III = 31–32% An; IV = 34% An. Zone II is as wide as the pyroxene. It also contains relics of pyroxene not assimilated by the blastic feldspar. In addition, pigment-rests are present, representing elements which could not be incorporated in the lattice of the plagioclase.
Olivine basalt. Khidane Meheret (between Entoto and British Embassy) Addis Ababa, Ethiopia. With crossed nicols.

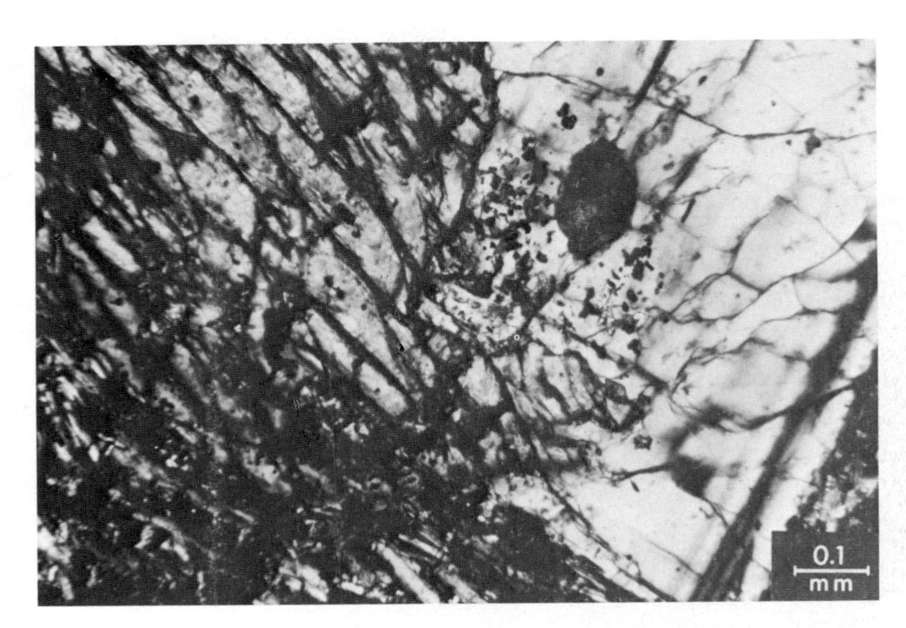

Fig. 232. Plagioclase tecoblast in contact with pyroxene phenocryst. Due to contact reaction, the later feldspar has partly assimilated and partly digested the pyroxene, with parts and relics enclosed in the feldspar.
Olivine basalt. Khidane Meheret (between Entoto and British Embassy) Addis Ababa, Ethiopia. With crossed nicols.

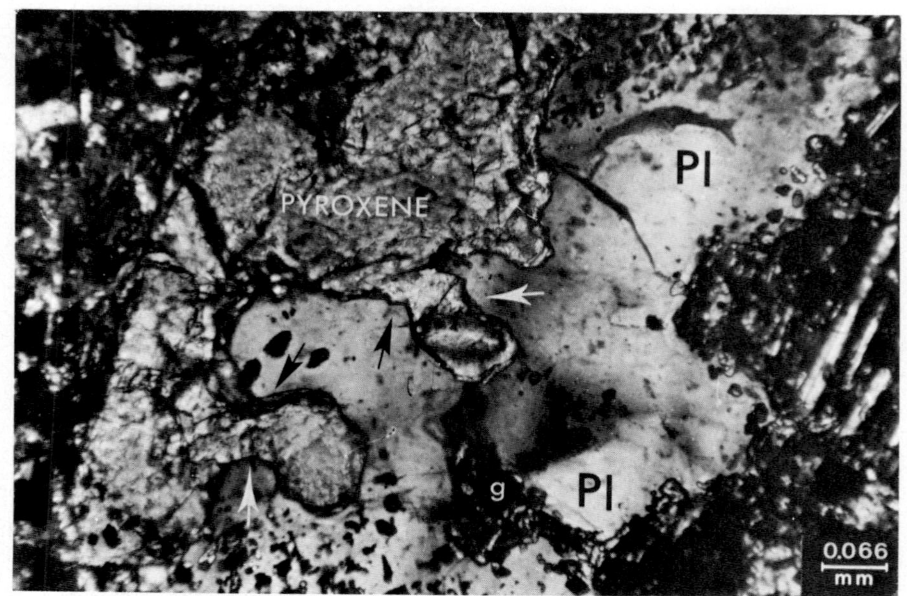

Fig. 233. Plagioclase tecoblast with amoeboidal protruberances partly engulfing and assimilating pyroxene. Also basaltic groundmass is enclosed by the tecoblast. Pl = plagioclase tecoblast; arrows show plagioclase engulfing the pyroxene; g = groundmass enclosed in the tecoblastic phenocryst.
Olivine basalt. Khidane Meheret (between Entoto and British Embassy), Addis Ababa, Ethiopia. With crossed nicols.

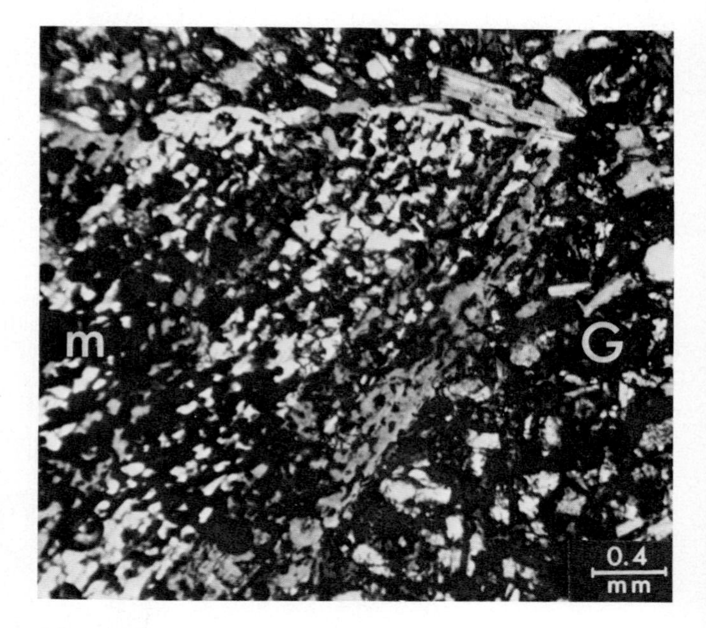

Fig. 234. Groundmass components and particularly early crystallized magnetite components, enclosed partly corroded and assimilated by a plagioclase tecoblast, showing poikiloblastic pattern. G = basaltic groundmass; m = magnetites, components of the groundmass enclosed in the tecoblast.
Khidane Meheret (between Entoto and British Embassy), Addis Ababa, Ethiopia.

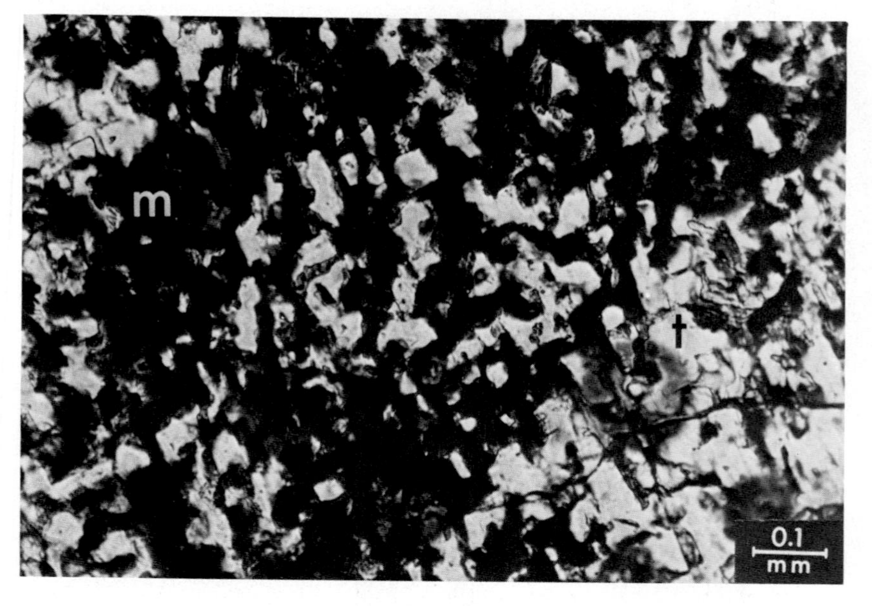

Fig. 235. Detail of Fig. 234, showing magnetite-rich basaltic groundmass, corroded and engulfed by the poikiloblastic tecoblast. m = magnetite groundmass components corroded and enclosed by the tecoblast, t = plagioclase tecoblast.
Olivine basalt. Khidane Meheret (between Entoto and British Embassy), Addis Ababa, Ethiopia. With crossed nicols.

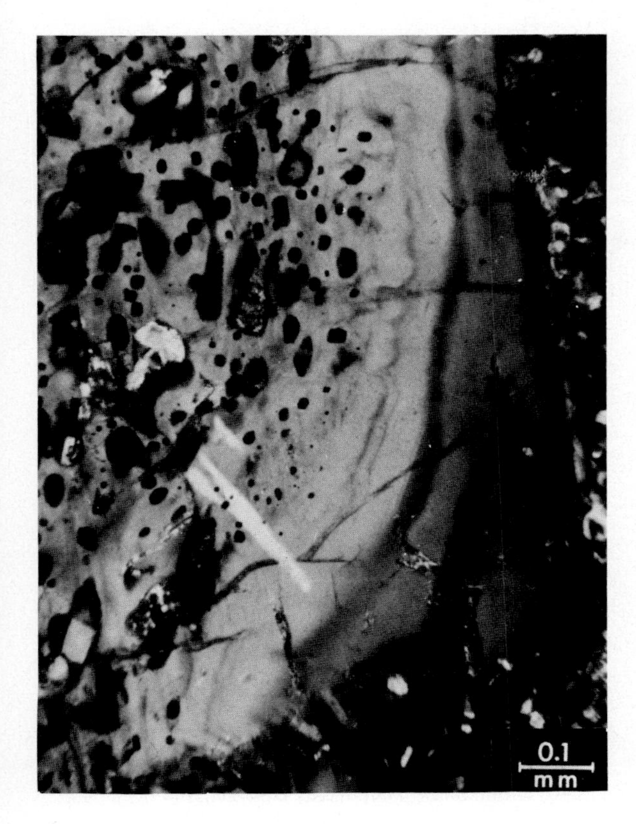

Fig. 236. Tecoblastic plagioclase with zoned structure due to variation in composition. A great many groundmass components (with various corrosion appearances) are enclosed in the tecoblast.
Olivine basalt. Khidane Meheret (between Entoto and British Embassy), Addis Ababa, Ethiopia. With crossed nicols.

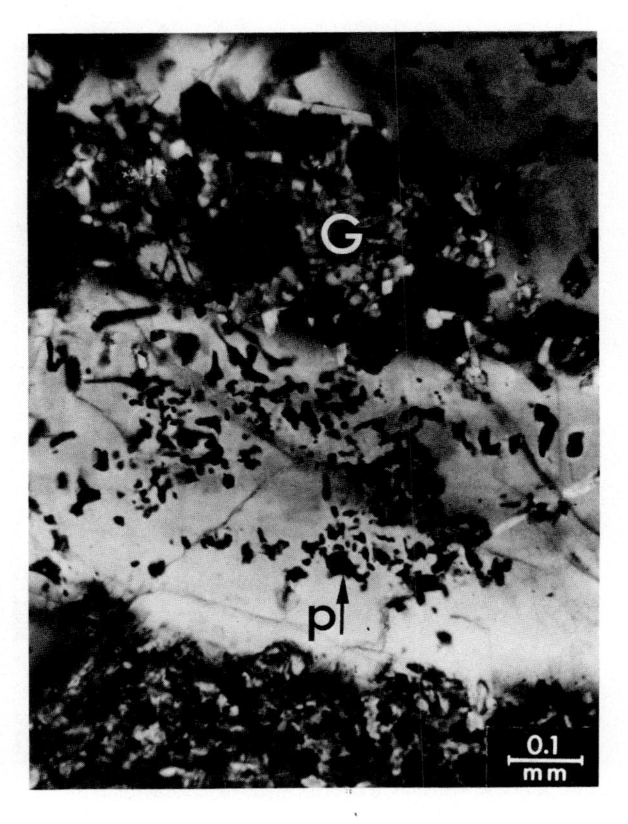

Fig. 237. Groundmass components (unaffected) enclosed in the plagioclase; rest-pigments also enclosed represent relics of assimilated groundmass. The tecoblastic zoning is influenced in its course by the pre-existing groundmass enclosed in the tecoblast. G = groundmass enclosed by tecoblast; p = pigment rests in tecoblast; black = magnetite components.
Olivine basalt. Khidane Meheret (between Entoto and British Embassy), Addis Ababa, Ethiopia. With crossed nicols.

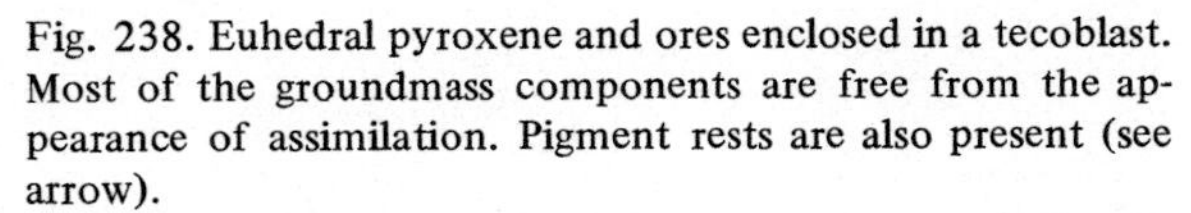

Fig. 238. Euhedral pyroxene and ores enclosed in a tecoblast. Most of the groundmass components are free from the appearance of assimilation. Pigment rests are also present (see arrow).
Olivine basalt. Khidane Meheret (between Entoto and British Embassy), Addis Ababa, Ethiopia. With crossed nicols.

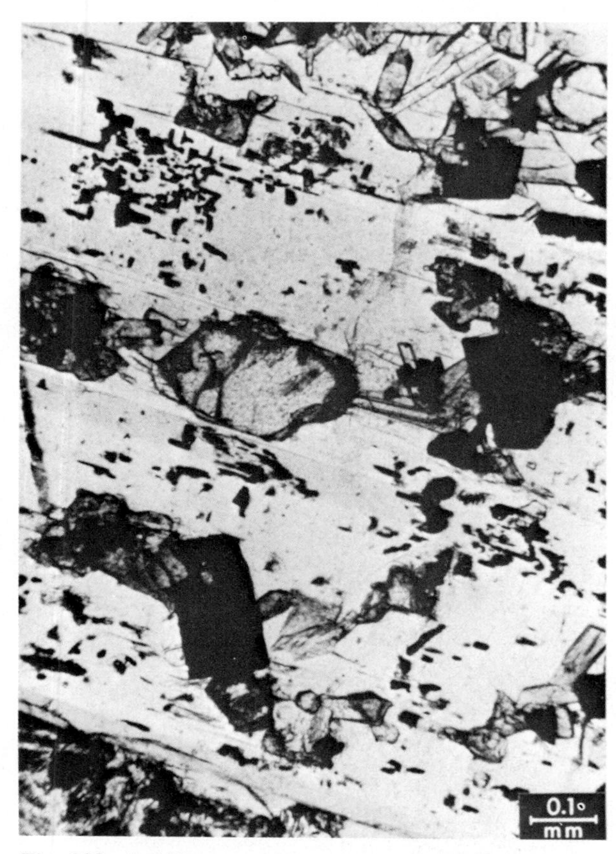

Fig. 239. A feldspar tecoblast with inclusions of corroded and rounded groundmass. (Groundmass components are shown in greater detail in Fig. 240). Olivine is included in the tecoblast.
Olivine basalt. Khidane Meheret (between Entoto and British Embassy), Addis Ababa, Ethiopia. Without crossed nicols.

Fig. 240. Groundmass components corroded and assimilated by the tecoblast. (See also Fig. 239.) Pigments following cleavage.
Olivine basalt. Khidane Meheret (between Entoto and British Embassy), Addis Ababa, Ethiopia. Without crossed nicols.

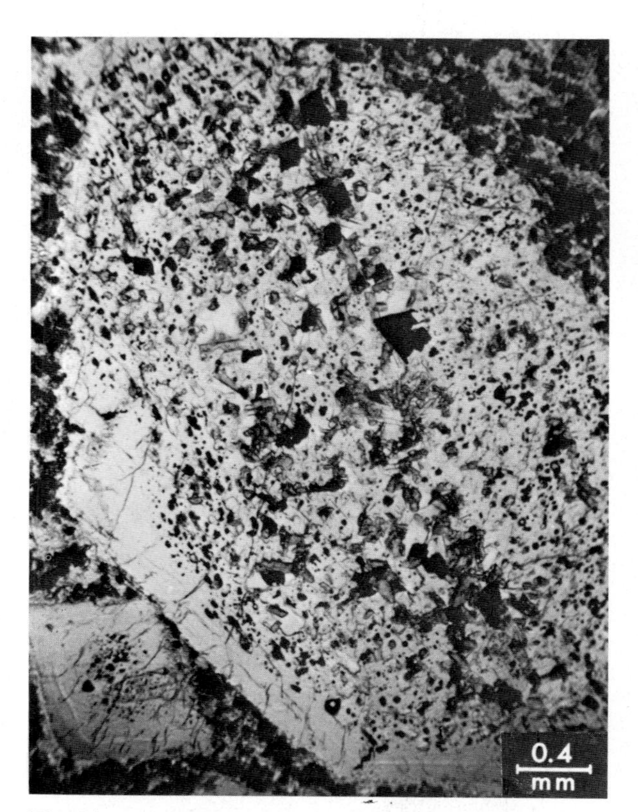

Fig. 241. A general view of a tecoblast plagioclase full of inclusions consisting of components of the groundmass. Figs. 242, 243 and 244 show the components of the groundmass in greater detail.
Olivine basalt. Khidane Meheret (between Entoto and British Embassy), Addis Ababa, Ethiopia. Without crossed nicols.

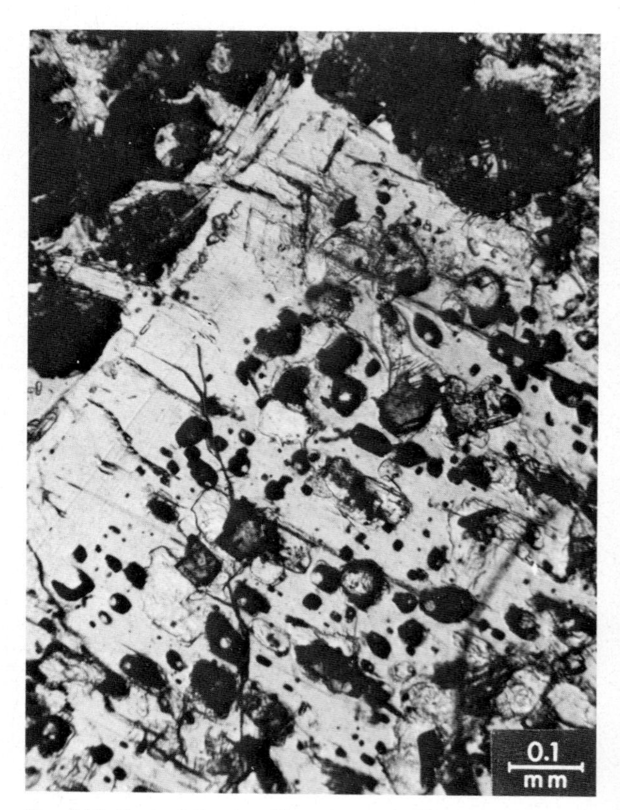

Fig. 242. Part of the tecoblast shown in Fig. 241. Corroded and assimilated groundmass components in the tecoblast. In contrast, the groundmass exterior to the tecoblast consists of large and non-corroded mineral components.
Olivine basalt. Khidane Meheret (between Entoto and British Embassy), Addis Ababa, Ethiopia. Without crossed nicols.

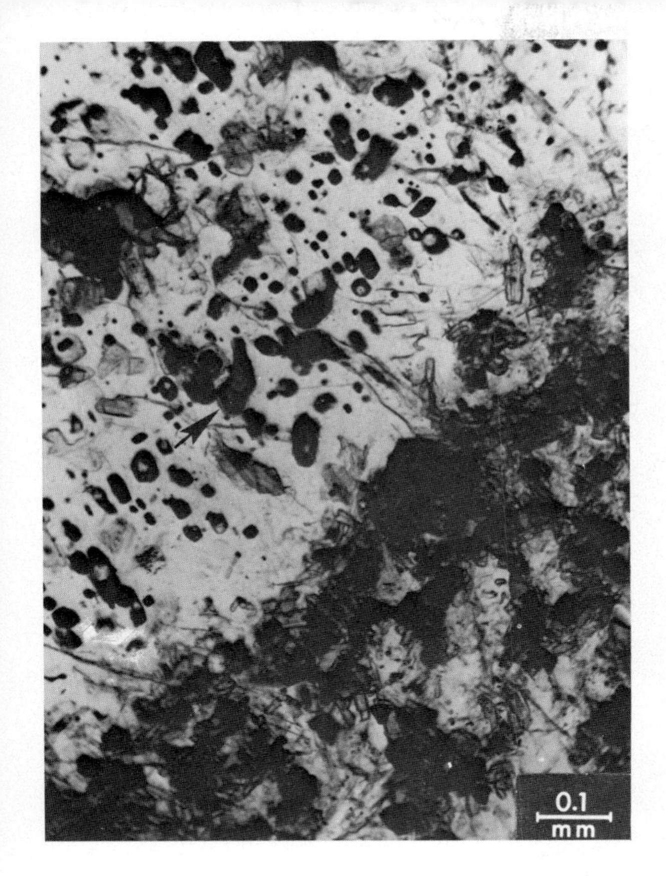

Fig. 243. A feldspar tecoblast with inclusions of corroded and rounded groundmass. (Arrow marks groundmass components shown in greater detail in Fig. 244). In comparison, the components of the groundmass exterior to the tecoblast are larger and apparently free from corrosion.
Olivine basalt. Khidane Meheret (between Entoto and British Embassy), Addis Ababa, Ethiopia. Without crossed nicols.

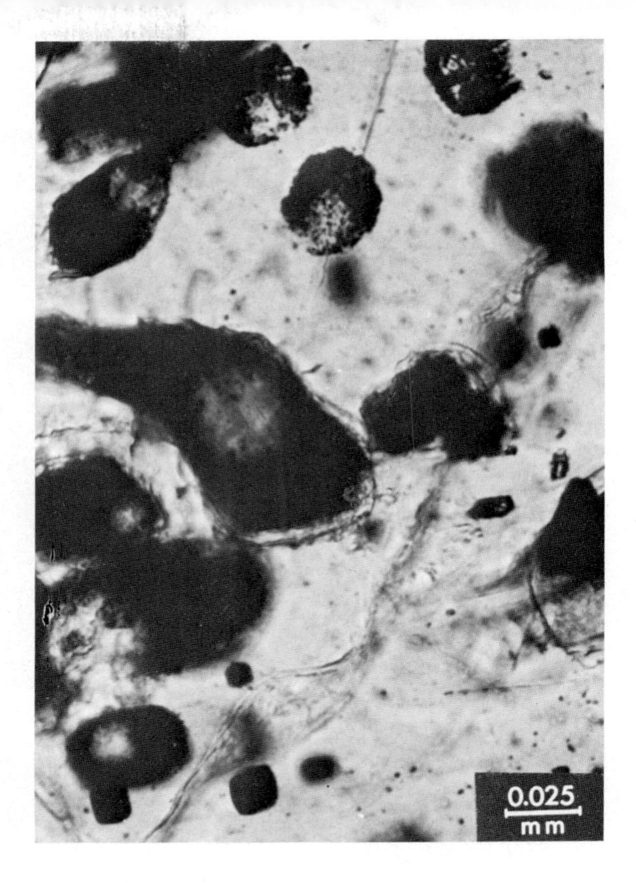

Fig. 244. Groundmass components, corroded and assimilated by the tecoblast (see also Figs. 236 and 239).
Olivine basalt. Khidane Meheret (between Entoto and British Embassy), Addis Ababa, Ethiopia. Without crossed nicols.

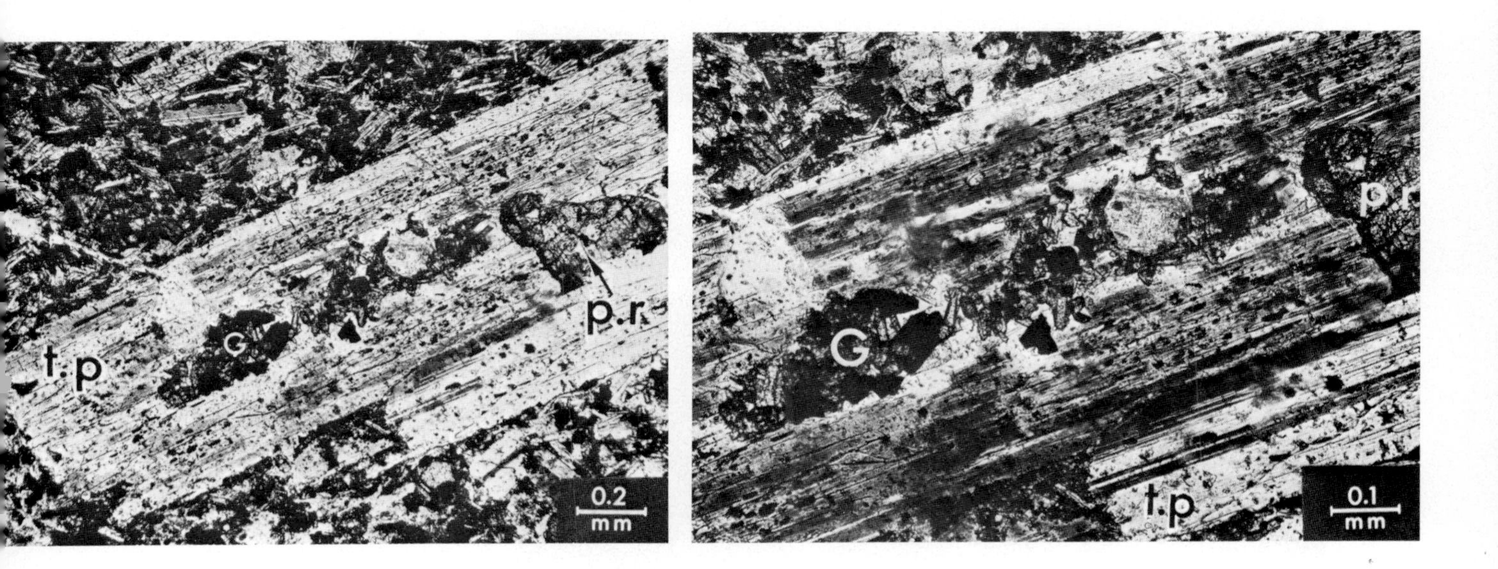

Figs. 245 and 246. Groundmass components enclosed in a plagioclase tecoblast. Pigment rests are also present as inclusions. A large corroded pyroxene of the groundmass is also in the tecoblast. G = groundmass enclosed in the tecoblast; p.r = corroded pyroxene enclosed by the tecoblast; t.p = tecoblastic plagioclase.
Olivine basalt. Khidane Meheret (between Entoto and British Embassy), Addis Ababa, Ethiopia. With crossed nicols.

Fig. 247. Plagioclase tecoblast enclosing and corroding pyroxene and groundmass which is reduced to pigment size and is partly zonally incorporated in the tecoblast. Pl = plagioclase tecoblast; py = pyroxene enclosed and corroded by the plagioclase tecoblast; P = groundmass reduced to pigment size; g = basaltic groundmass.
Olivine basalt. Khidane Meheret (between Entoto and British Embassy), Addis Ababa, Ethiopia. With crossed nicols.

Fig. 248. Plagioclase tecoblast with groundmass components enclosed and corroded, partly reduced to pigments (representing elements that cannot be incorporated by the tecoblastic lattice).
Olivine basalt. Khidane Meheret (between Entoto and British Embassy), Addis Ababa, Ethiopia. With crossed nicols.

(For Figs. 249 and 250 see p.189)

Fig. 251. Groundmass "extending" into a tecoblast (exceptional case). Also, parts of the groundmass are enclosed in the tecoblast.
Olivine basalt. Khidane Meheret (between Entoto and British Embassy), Addis Ababa, Ethiopia. With crossed nicols.

Fig. 249. Groundmass components enclosed in tecoblast feldspar. Often the groundmass is reduced to pigment forms. Surrounding the pigments and the groundmass components there are marked reaction margins in the tecoblast. The arrow marks a reaction zone around an individual pigment.
Olivine basalt. Khidane Meheret (between Entoto and British Embassy), Addis Ababa, Ethiopia. With crossed nicols.

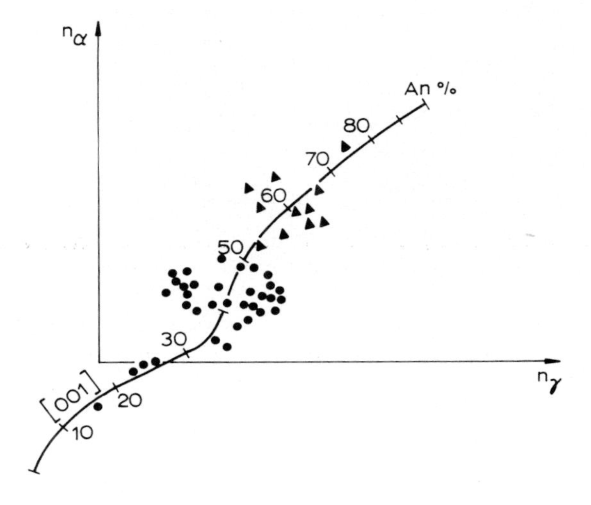

Fig. 252. The anorthite content of the groundmass (small) plagioclase is shown by triangles. The composition of the large feldspar tecoblasts (phenocrysts) is indicated by points.
Olivine basalt. Khidane Meheret (between Entoto and British Embassy), Addis Ababa, Ethiopia.

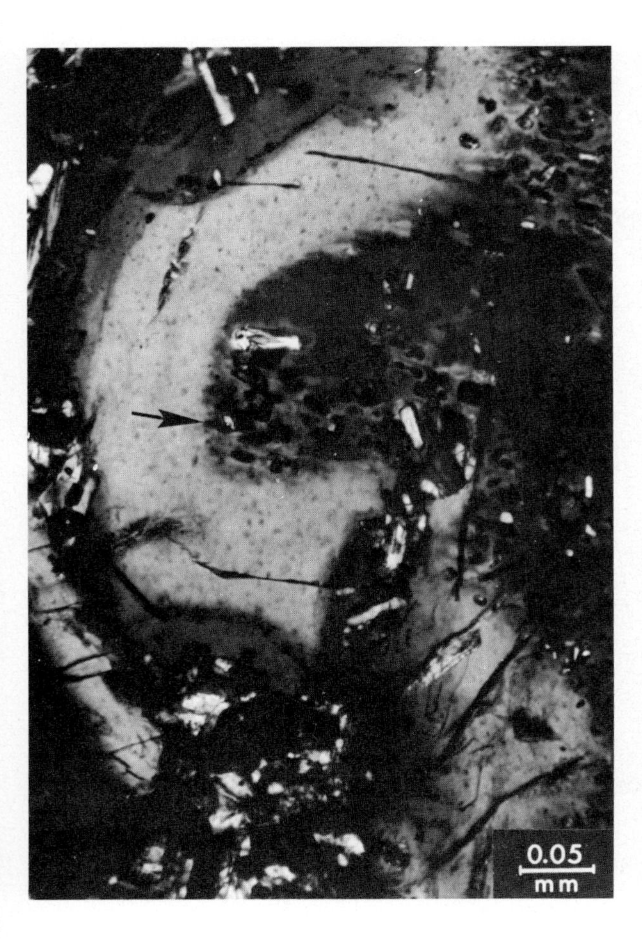

Fig. 250. A tecoblast plagioclase with pigment rests following a zonal distribution in the feldspar. It is interesting to note the coincidence of the feldspar "zone" with the pigment distribution. Arrow shows plagioclase zoning and pigment rests restricted in the internal zone of the plagioclase.
Olivine basalt. Khidane Meheret (between Entoto and British Embassy), Addis Ababa, Ethiopia. With crossed nicols.

Fig. 253 (a and b). A tecoblast with an irregular type of zoning. The zones of the feldspar have the following An-contents: Zone I = 25% An; II = 15% An; III = 24% An; IV = 25–27% An. Zone I contains pigment-rests. Also parallel orientated cavities are present in the tecoblast.
Olivine basalt. Khidane Meheret (between Entoto and British Embassy), Addis Ababa, Ethiopia. With crossed nicols.

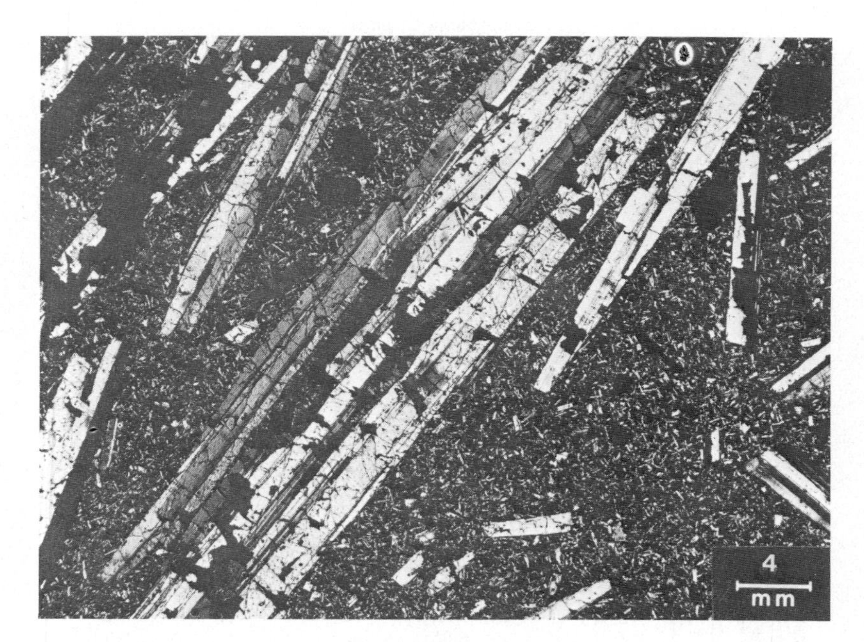

Fig. 254. Large plates of plagioclase phenocrysts in basalt groundmass.
Olivine basalt. Debra-Sina (margin of the Ethiopian part of the Great Rift Valley). With crossed nicols.

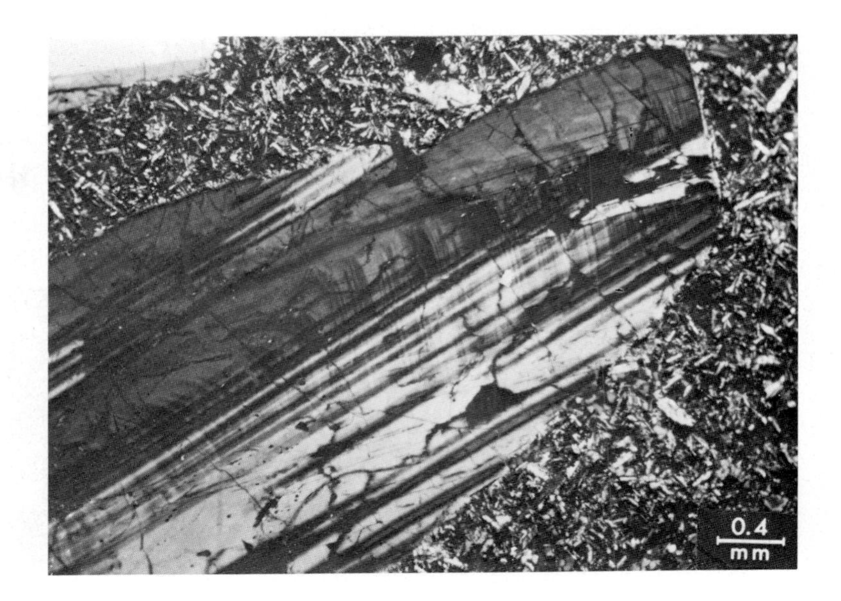

Fig. 255. Plagioclase with polysynthetic twinning varying from fine to broad lamellae. Olivine basalt. Debra-Sina, Ethiopia.

Fig. 256. Plagioclase phenocryst with a great number of fine zones (exceeding 100). A resorption twinning can be seen.
Olivine basalt. Debra-Sina (margin of the Ethiopian part of the Great Rift Valley). With crossed nicols.

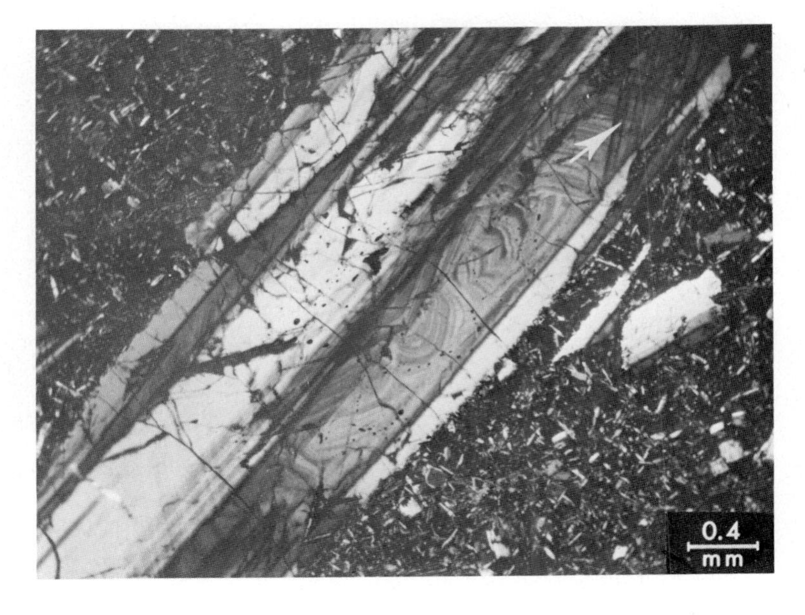

Fig. 257. Complex zonal growth with fine twinning (arrow marks parquet structure).
Olivine basalt. Debra-Sina (margin of the Ethipian part of the Great Rift Valley). With crossed nicols.

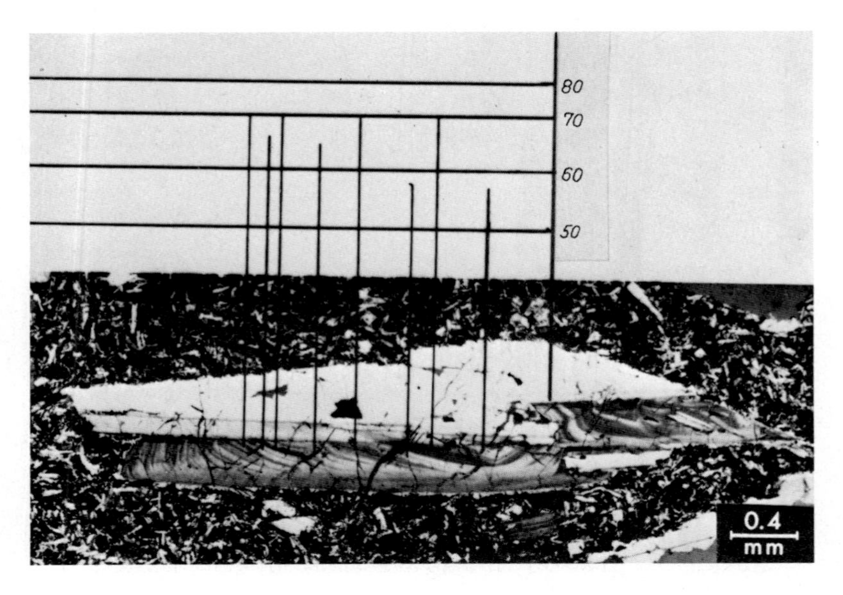

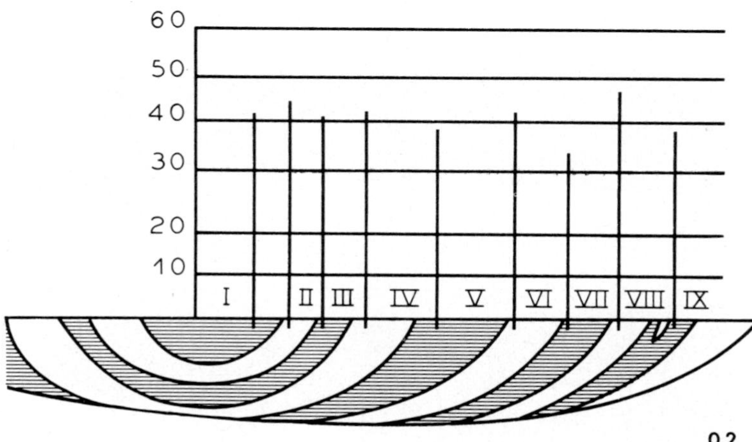

Figs. 258 and 259. Plagioclase phenocryst, with oscillatory zoning from the central zone outwards. (The composition of the first zone, Fig. 258, was not determined). The determinations of Ab–An% are diagrammatically represented. The vertical lines show the An% on the basis of M. Reinhard's Table No. 2, Federoff-Nikitin Stereogram (Wandering der Flächenpole bei fixen optischen Vektoren), curve for 001.
Olivine basalt. Debra-Sina (margin of Ethiopian part of the Great Rift Valley). With crossed nicols.

Fig. 260. Plagioclase phenocryst with zoning. The oscillatory fine zoning of the plagioclase phenocryst resembles colloform pattern.
Olivine basalt. Debra-Sina (margin of Ethiopian part of the Great Rift Valley). With crossed nicols.

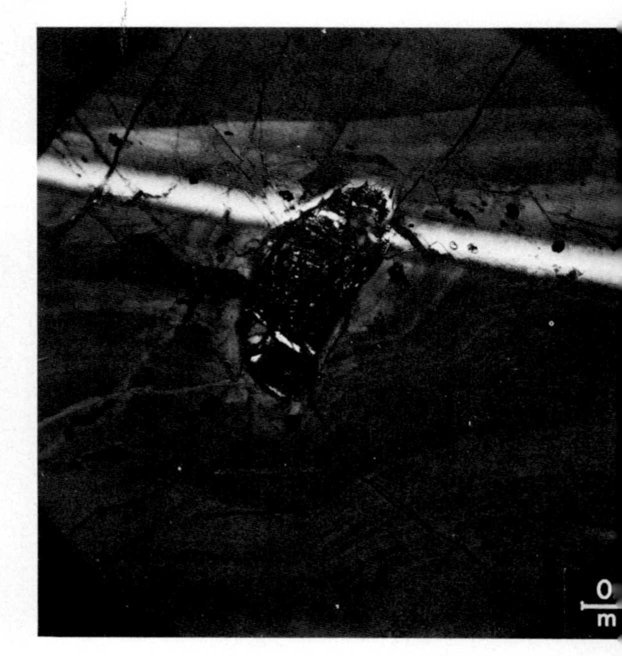

Fig. 261. Groundmass components included in the plagioclase feldspar and surrounded by fine zoning resembling colloform patterns.
Olivine basalt. Debra-Sina (margin of Ethiopian part of the Great Rift Valley). With crossed nicols.

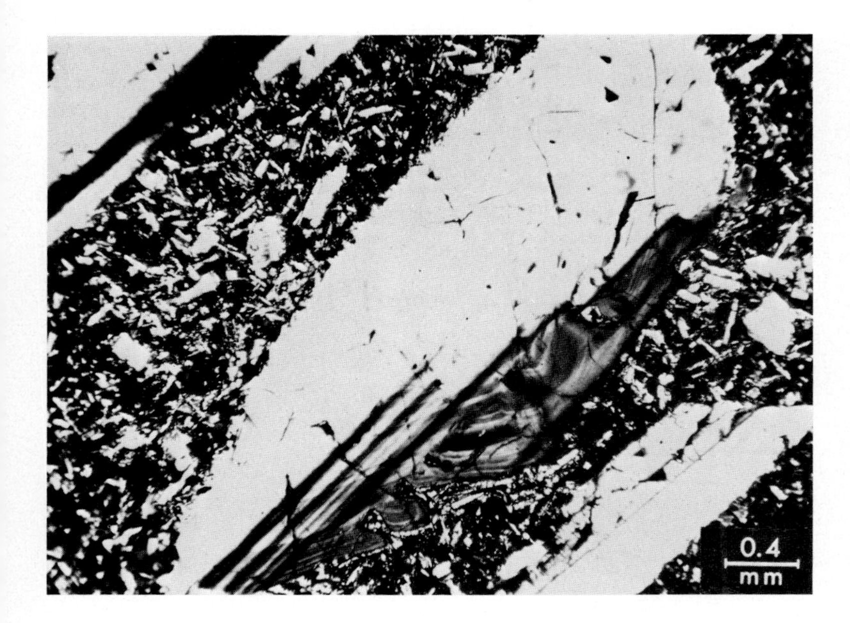

Figs. 262 and 263. Groundmass influencing the "running" of the plagioclase zones. Arrow "a" shows the curving of the plagioclase zones influenced by the existence of the groundmass.
Olivine basalt. Debra-Sina (margin of Ethiopian part of the Great Rift Valley). With crossed nicols.

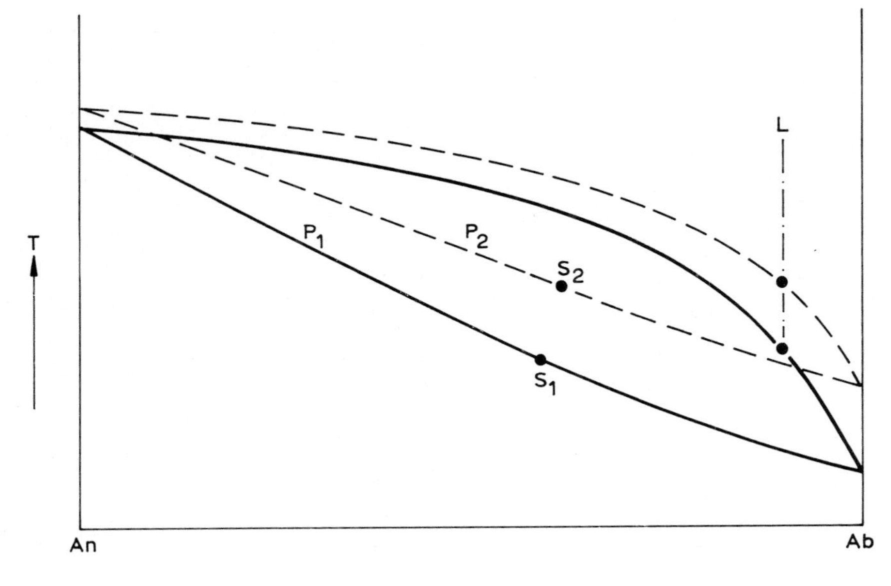

Fig. 264. Hypothetical representation of the possible effect of pressure upon the plagioclase equilibria. Solid curves represent the liquidus–solidus at atmospheric pressure (P_1). Broken curves represent the liquidus–solidus at high pressure (P_2). L is the unchanging composition of the liquid phase. With increasing pressure to P_2, the solid phase in equilibrium would change from S_1 to more sodic composition S_2. (Diagram from Carr, 1954.)

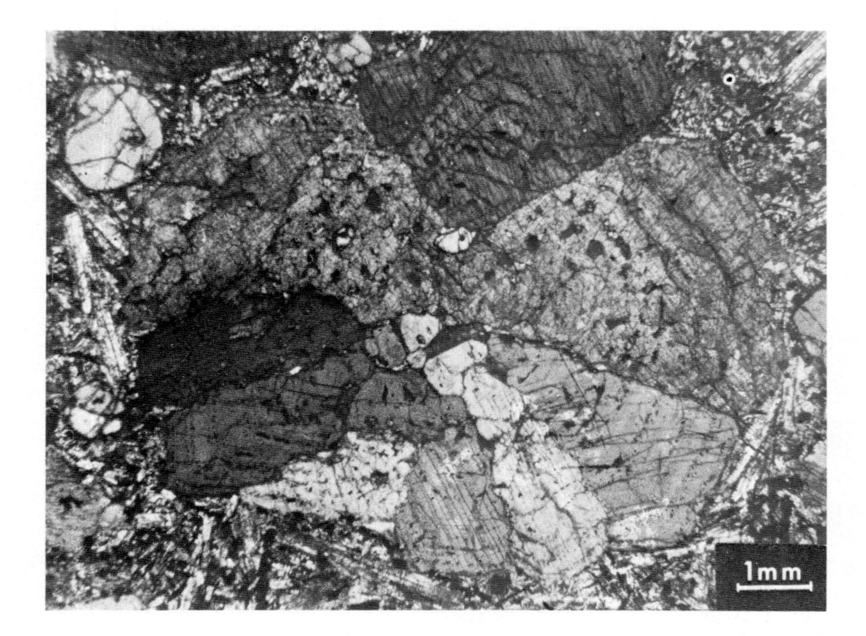

Fig. 265. Zoned and fractured augite phenocryst. The fractured segments are rotated as is indicated by the cleavage pattern. The augite phenocryst is zoned.
Augite-rich basalt. Dessie-Batie, Ethiopia.

Fig. 266. Part of large zoned and fractured augite phenocryst. The fracture segments are rotated, as the differences in the orientation of the cleavage pattern of adjacent segments indicate. a,b and c = radially arranged segments of the raptured phenocrysts; arrows d and e show the pattern of cleavages in adjacent segments and their different orientations; arrows z = zoning of the phenocrysts.
Augite-rich basaltic rock. Batie-Dessie, Ethiopia. With crossed nicols.

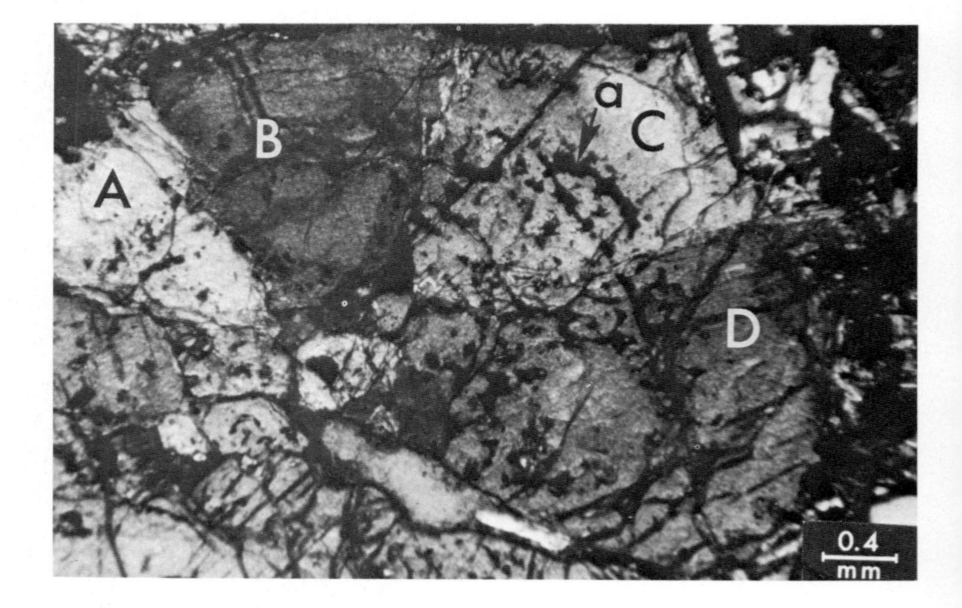

Fig. 267. Augite phenocryst consisting of radiating segments. A zonal distribution of inclusion transverses the radially arranged segments. A,B, C,D = radiating segments; arrow "a" shows the zonal inclusions transversing the segment.
Augite-rich basaltic rock. Batie-Dessie, Ethiopia. With crossed nicols.

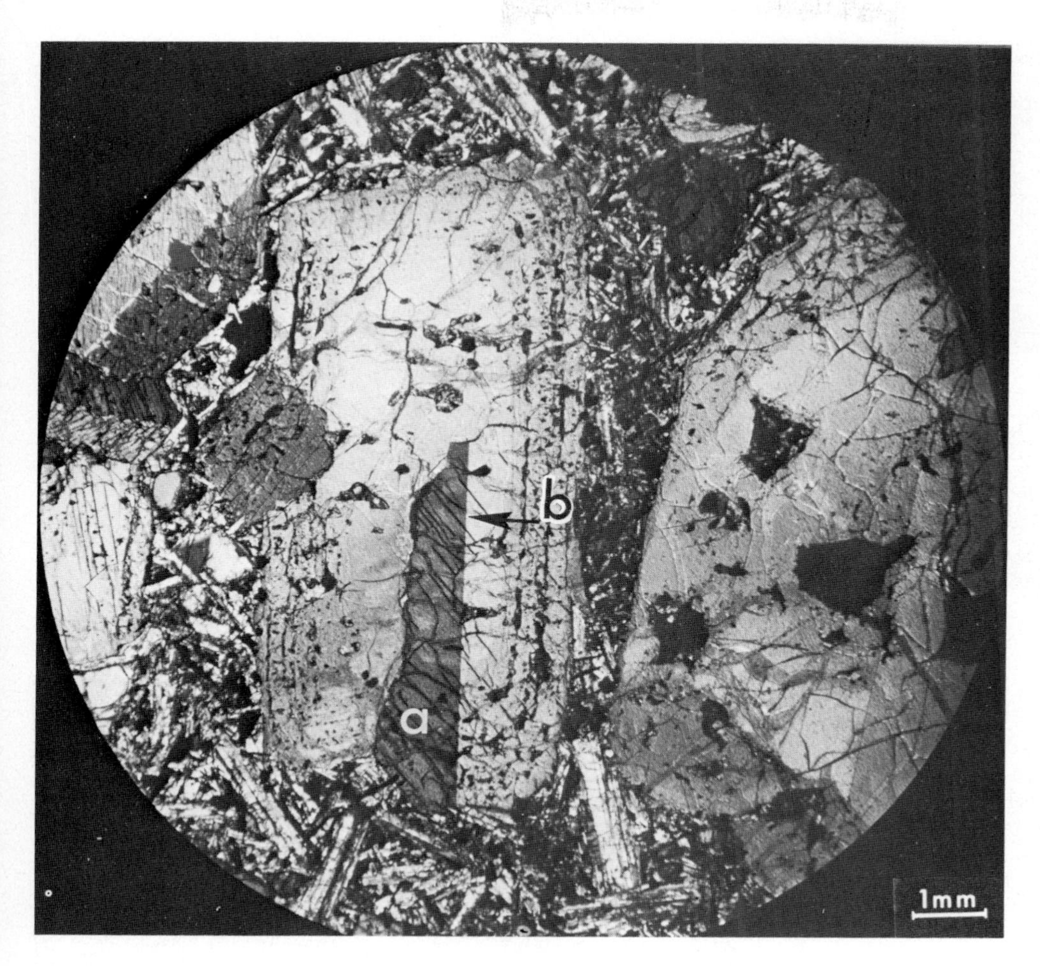

Fig. 268. Titano-augite phenocryst with a "rotated block", marked "a". Arrow "b" shows that the rotated block is delimited by a twin plane.
Augite-rich basaltic rock. Batie-Dessie, Ethiopia. With crossed nicols.

Fig. 269. Displaced block within twinned augite phenocryst. A = displaced block within augite phenocryst; B = polysynthetic twin lamellae of augite phenocryst.
Augite-rich basalt. Selale Mt. region, Ethiopian Plateau. With crossed nicols.

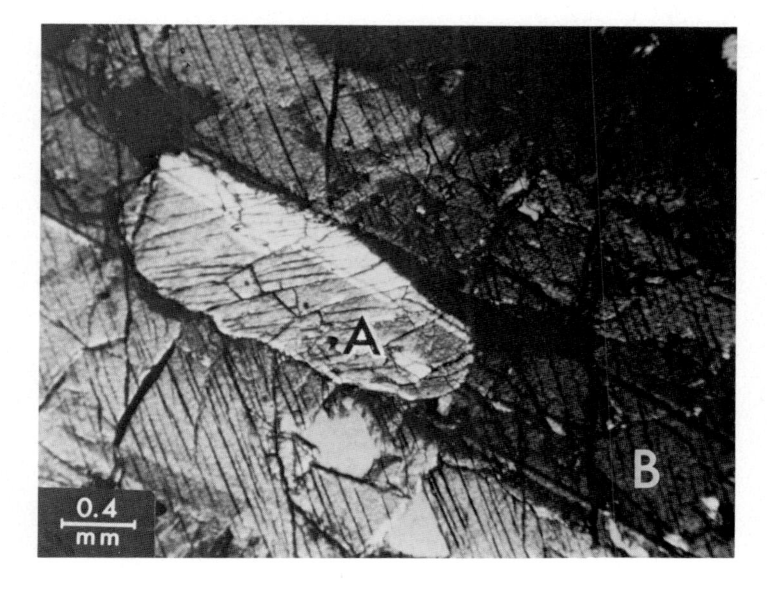

Fig. 270. Detail of augite displaced block within augite phenocryst. Twin lamella of augite block in turn displaced by cleavage set "a" and cleavage set "b". Augite-rich basalt. Selale Mt. region, Ethiopian Plateau. With crossed nicols.

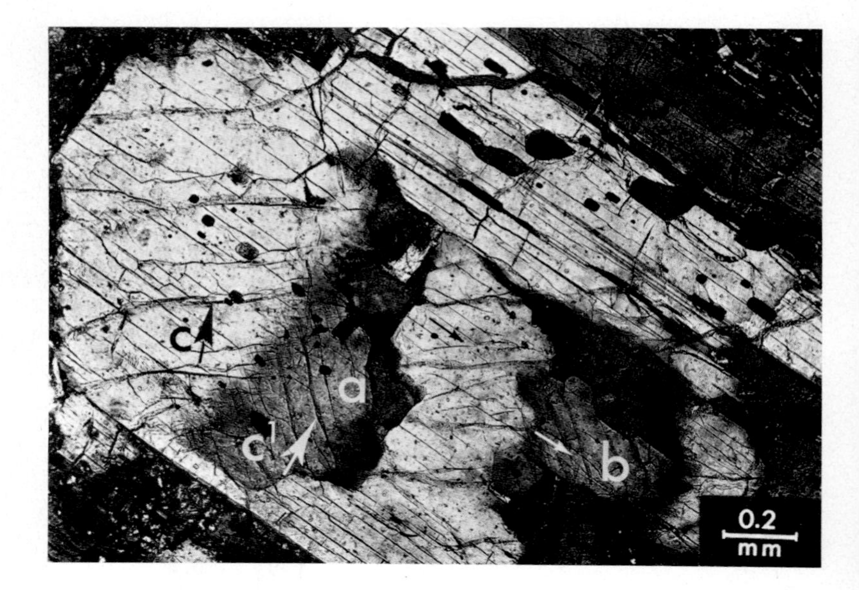

Fig. 271. Plagioclase phenocryst with parts of it showing block distortion. a and b = plagioclase parts which are actually blocks rotated within the plagioclase phenocryst; c and c^1 show the difference in orientation of cracks of the phenocryst and of its "block" parts. The cleavage orientation remained the same in the rotated blocks and in the phenocrysts outside and is marked by arrows.
Olivine basalt. Khidane Meheret (between Entoto and British Embassy), Addis Ababa, Ethiopia. With crossed nicols.

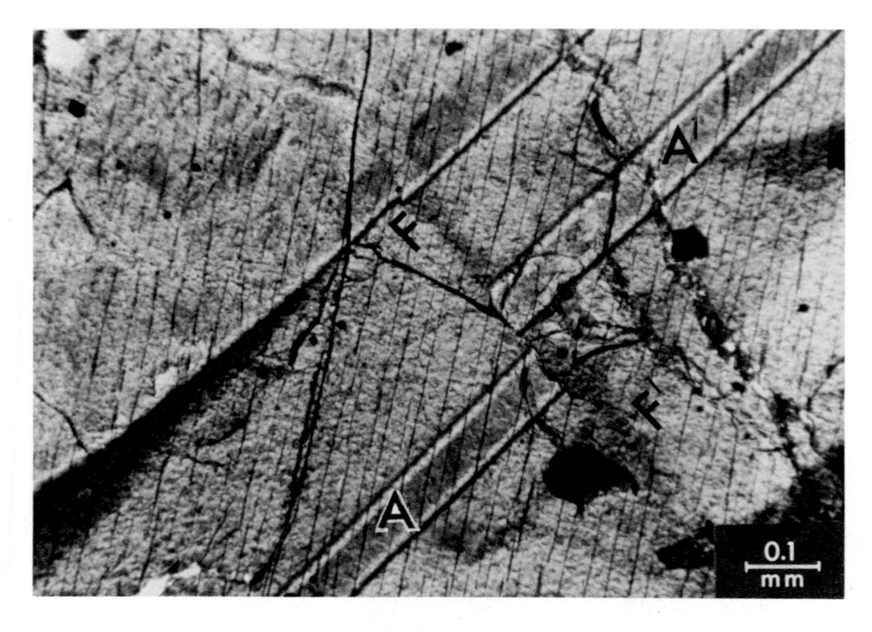

Fig. 272. Augite phenocryst with lamella showing displacement along a microfault fracture. A and A^1 = twin lamella displaced along microfault ($F-F^1$). The microfault (microfractured) is healed.
Augite-rich basalt. Selale Mt. region, Ethiopian Plateau. With crossed nicols.

Fig. 273. Zoned augite phenocrysts with block displacement, as is shown by displacement of the augite zoning. z = zoning of augite phenocryst; f = microfaults and block displacement of the augite phenocryst.
Basalt. Pelm, Eifel, Germany. With crossed nicols.

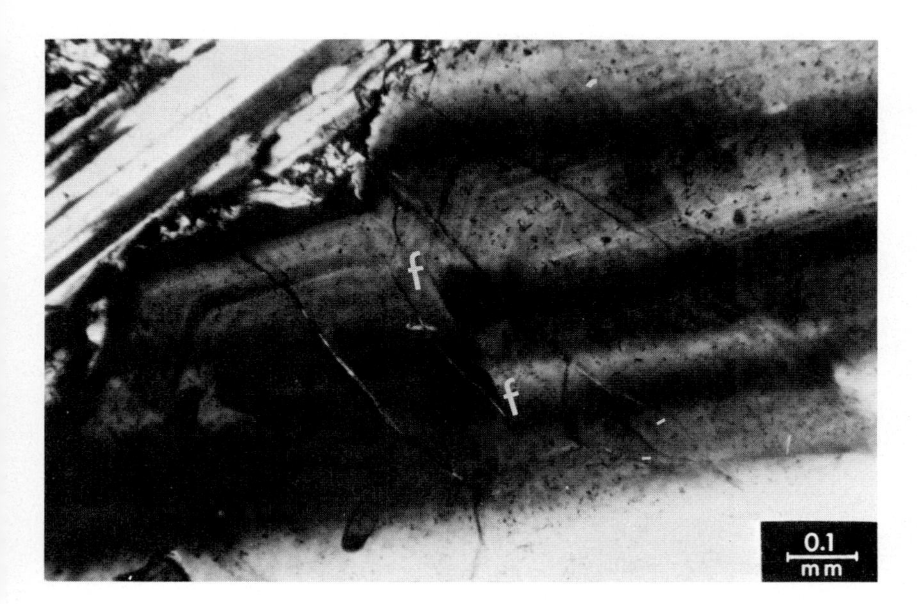

Fig. 274. Zoned plagioclase phenocryst with block displacement along fracture (microfaults). f–f = microfault cracks along which the zoned plagioclase is displaced.
Hyalobasalt. Ustica, Sicily, Italy. With crossed nicols.

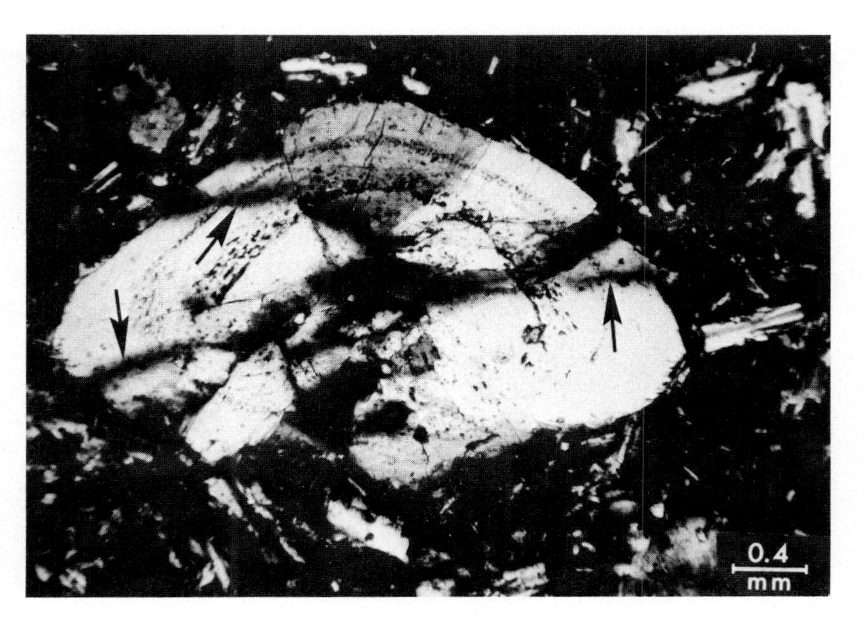

Fig. 275. Colloform plagioclase tecoblastoid in leucite-containing basaltic rock. The zonal gel-structure of the colloform tecoblastoid is indicated by inclusion-rich zones. Arrows show "shadow-like" plagioclase polysynthetic twinning.
Basaltoid-rock leucite–tephrite. Travolata Rome, Italy. With crossed nicols.

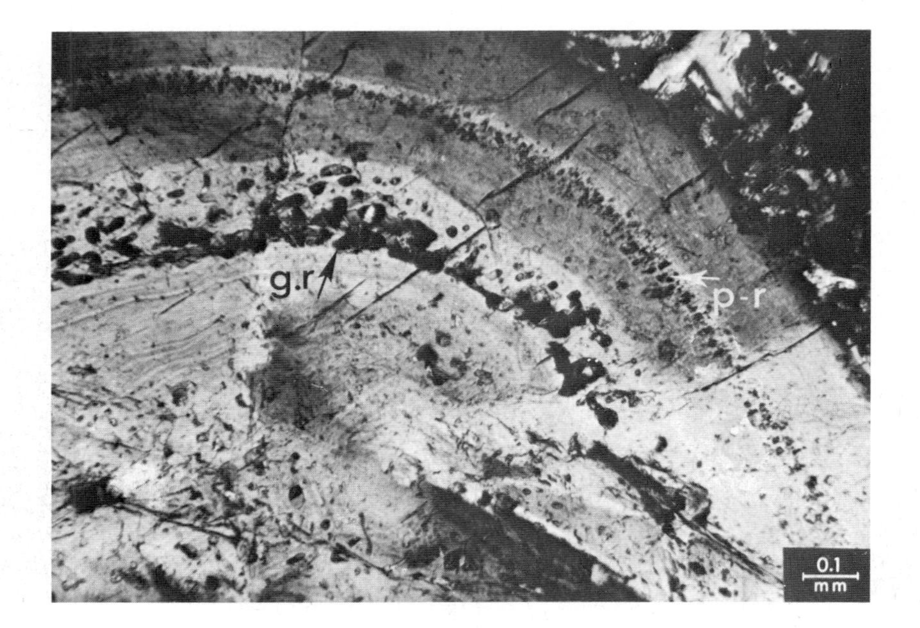

Fig. 276. Colloform plagioclase tecoblast. Gel-zonal growths with "pigment" rests or groundmass rests following certain gel-zones. p-r = pigment rests following gel-zonal growths; g.r = groundmass rests following gel-zonal growths.
Basaltoid-rock leucite–tephrite. Travolata, Rome, Italy. With crossed nicols.

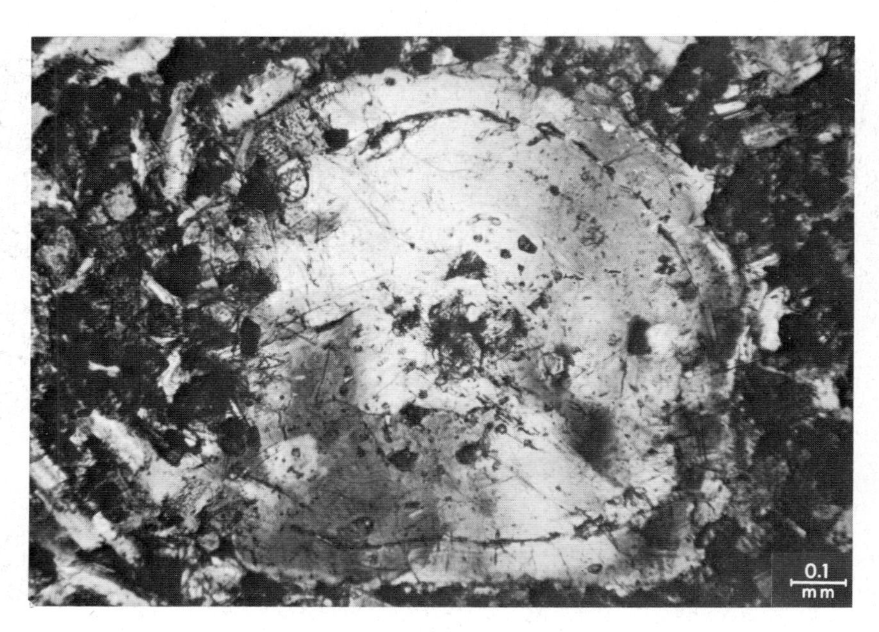

Fig. 277. Colloform plagioclase tecoblastoid with "diffused" polysynthetic twinning and with a distinct gel-zonal growth which is marked by pigment rests.
Basaltoid-rock leucite–tehprite. Travolata, Rome, Italy. With crossed nicols.

Fig. 278. Colloform – general view. Plagioclase poly-colloform tecoblastoid.
Tephritic basalt (Etnaite Lava 1886). Nicolosi, Atna, Sicily, Italy. With crossed nicols.

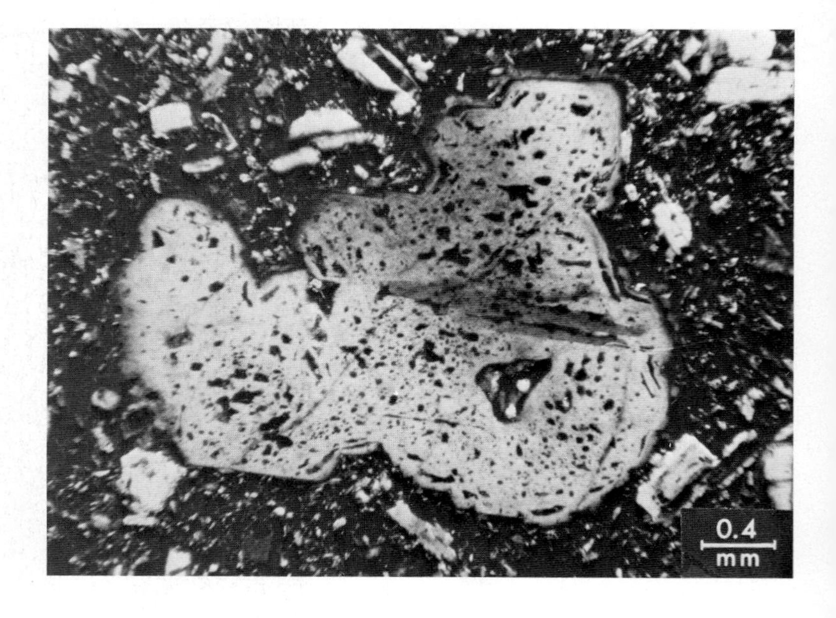

Fig. 279. Detail of plagioclase poly-colloform tecoblastoid. The poly-colloform structure is to be understood as topo-crystalline gel-zonal development round enclosed groundmass pockets (pyroxene aggregates often act as nuclei for the development of the poly-colloform structure). Arrows "a" show gel-zonal growths of the polycolloform plagioclase tecoblastoid; p = pyroxene, enclosed by the poly-colloform tecoblastoid; z = zeolites, associated with the pyroxene "nuclei".
Tephritic basalt (Etnaite Lava 1886). Nicolosi, Atna, Sicily, Italy. With crossed nicols.

Fig. 280. Plagioclase collocryst consisting of fine zoning which from the centre to the margin tends to become more like colloform zonal banding.
Vesicular basalt. Chaffée County, Colorado, U.S.A. With crossed nicols.

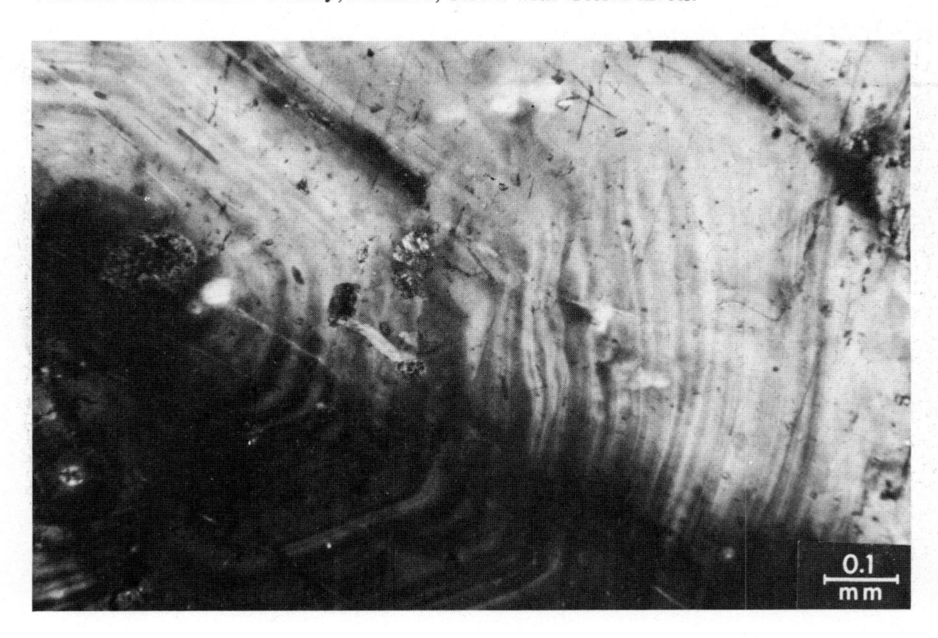

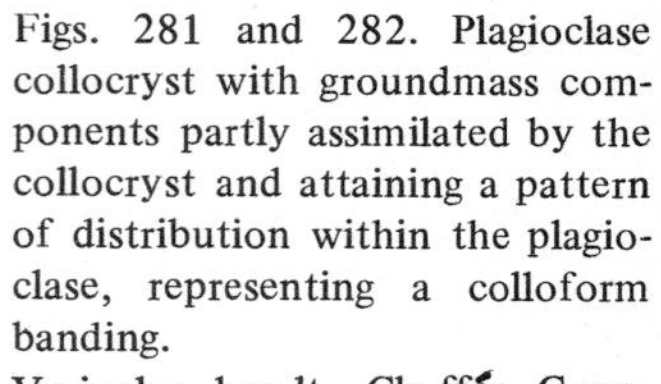

Figs. 281 and 282. Plagioclase collocryst with groundmass components partly assimilated by the collocryst and attaining a pattern of distribution within the plagioclase, representing a colloform banding.
Vesicular basalt. Chaffée County, Colorado, U.S.A. With crossed nicols.

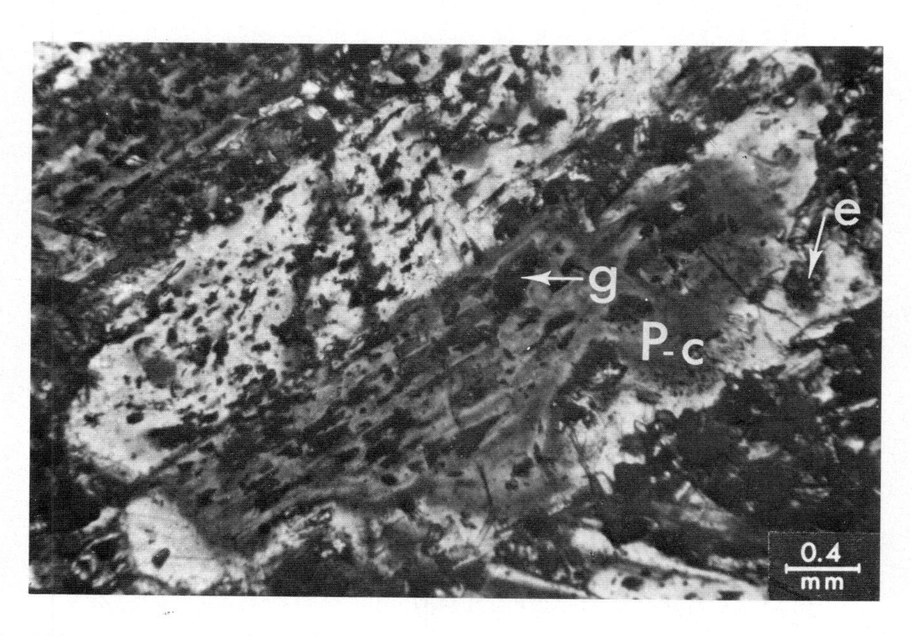

Fig. 283. Large colloform tecoblastoid with the colloform structures forming an intricate pattern with groundmass rests and components interwoven with the intricate colloform structure. P-c = plagioclase colloform tecoblastoid; g = groundmass or groundmass relics within the colloform tecoblastoid; arrow "e" shows groundmass embayment in the colloform blastoid (collocrysts).
Tephritic basalt (Etnaite Lava 1886). Nicolosi, Atna, Sicily, Italy. With crossed nicols.

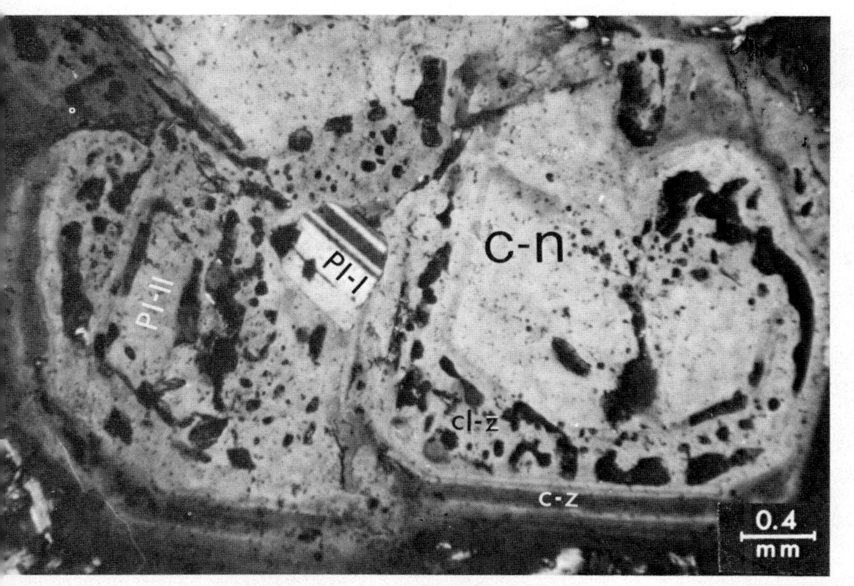

Fig. 284. Plagioclase generation I, enclosed by a second-generation (plagioclase II) or plagioclase phenocryst (tecoblastoid). The interrupted line shows the crystalline outline of the central "nucleus" of the plagioclase tecoblastoid. The crystalline nucleus (c-n) is followed by a colloform in pattern zone (cl-z) which, in turn, is followed again by a typical crystal zoning (c-z). Pl-I = plagioclase generation I; Pl-II = plagioclase generation II; arrow "a" shows resorbed outline of the crystalline nucleus (c-n).
Tephritic basalt (Etnaite Lava 1886), Nicolosi, Atna, Sicily, Italy. With crossed nicols.

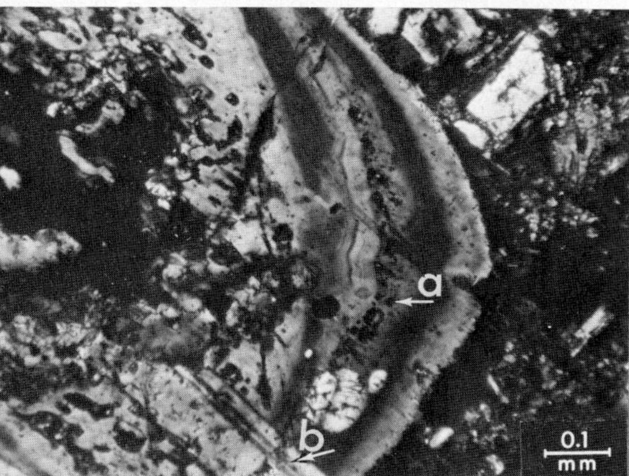

Fig. 285. Colloform plagioclase tecoblastic phenocrysts with groundmass assimilation rests which, in their topocrystalline distribution within the phenocryst, can either be in the central part of the phenocryst or following colloform zoning. Arrow "a" shows pigment-rests following gel-zoning of the plagioclase phenocryst. Arrow "b" shows post-zonal twinning.
Tephritic basalt (Etnaite Lava 1886). Nicolosi, Atna, Sicily, Italy. With crossed nicols.

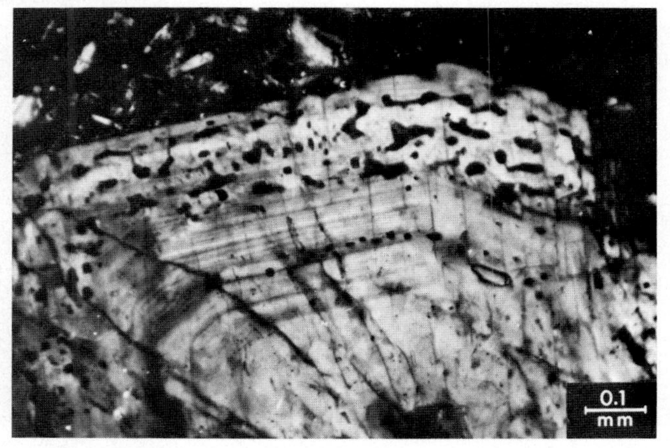

Fig. 286. A collocryst with zoning and shape closer to a crystalloid than to a typical collocryst spheroidal structure.
Chaffée County, Colorado, U.S.A. With crossed nicols.

Fig. 287. Zoned plagioclase plates. Complex (even colloform in appearance) zonal patterns are often observed in plagioclase phenocrysts.
Basalt. Adama, Ethiopia. With crossed nicols.

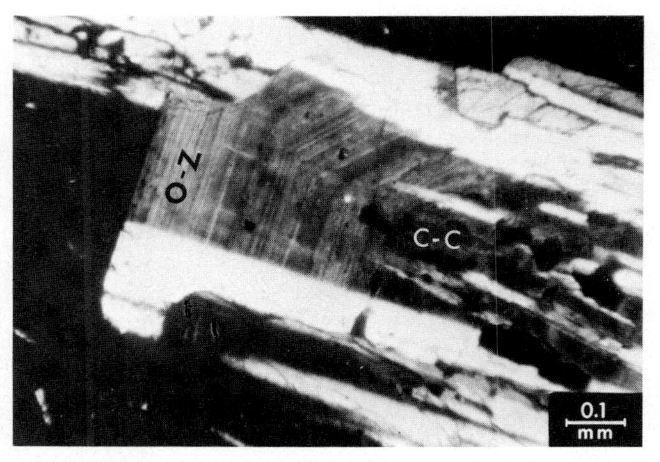

Fig. 288. Zoned plagioclase with "colloform pattern" of the central zonal part which eventually changes into zonal growths following crystal faces. C-C = central zonal part of colloform tecoblast; o-z = outer zones of plagioclase phenocryst – parallel to crystal faces.
Vitrophyric basalt. Vulcano, Islands of Lipari, Tyrrhenian Sea, Italy. With crossed nicols.

Fig. 289. General view of an aggregate of radiating plagioclases (with characteristic polysynthetic twinning) originally colloform tecoblastoid in origin. The initial gel-pattern of the plagioclase aggregates is obvious and is particularly indicated by the colloform zonal growths and their pigment-rest inclusions. Arrow "a" shows detail shown in Fig. 290.
Basaltic rock type – leucite – tephrite. Travolata, Rome, Italy. With crossed nicols.

Fig. 290. Detail of Fig. 289 showing gel-zonal growth of the plagioclase colloform tecoblastoids with pigment rests following the gel-growth zonal pattern. Basaltic rock type – leucite–tephrite. Travolata, Rome, Italy. With crossed nicols.

Figs. 291 and 292. Differently orientated plagioclase colloform tecoblastoid with crystalline and gel-zonal growths. The different orientation positions show the polysynthetic twinning. Arrow "a" shows the crystal-zonal growth and arrow "b" the colloform gel-zonal growths.
Tephritic basalt (Etnaite Lava 1886). Nicolosi, Atna, Sicily, Italy. With crossed nicols.

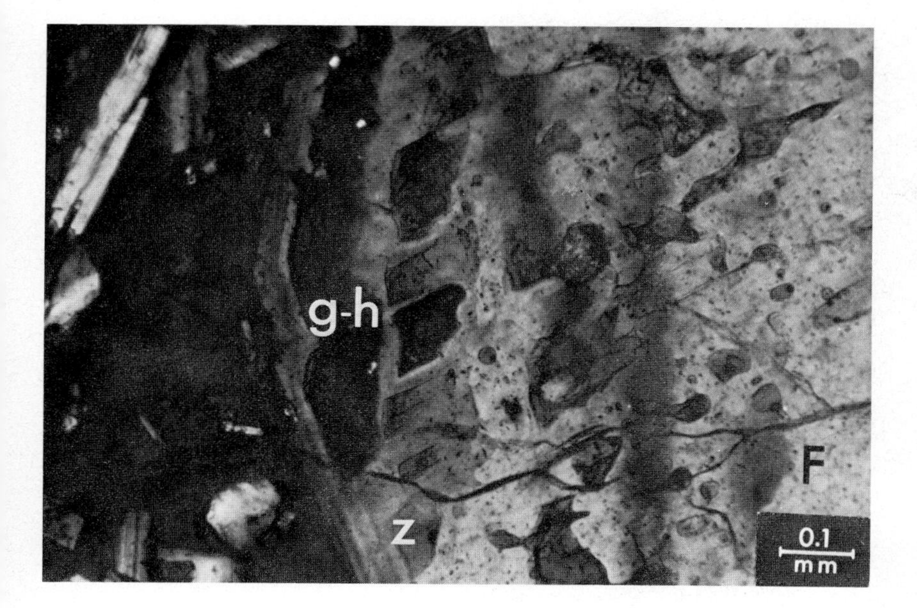

Fig. 304. Plagioclase phenocryst in a glassy groundmass. Hyaloids are interzonal and are transition relics of a transmutation from hyalo to crystal. F = plagioclase phenocrysts; g-h = gel hyaloids; z = zoned plagioclase zoning.
Hyalobasalt. Ustica, Sicily, Italy. With half-crossed nicols.

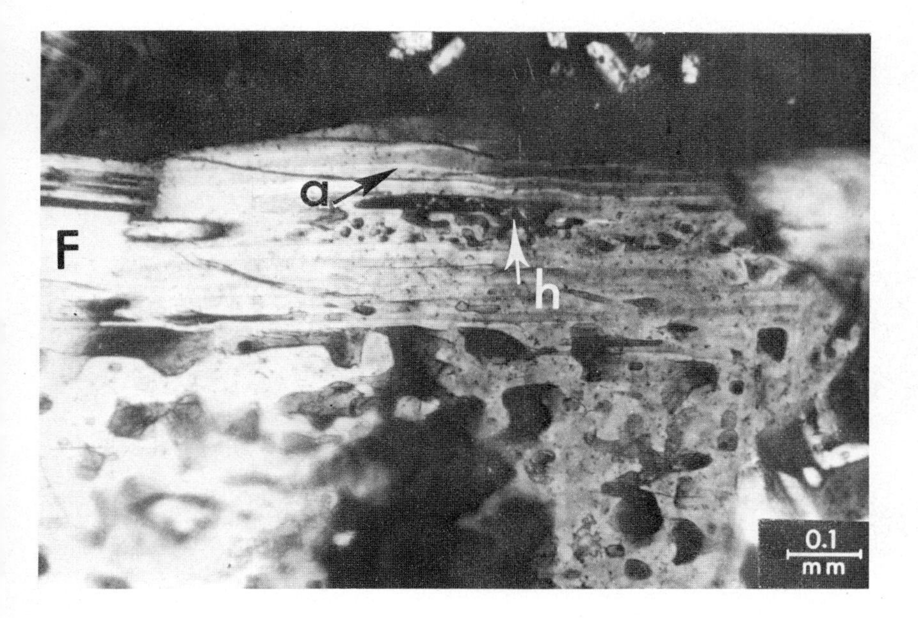

Fig. 305. Plagioclase phenocrysts in glassy groundmass. Hyaloids interzonally within the feldspar. Arrow "a" shows curving of the phenocryst zoning because of the existence of the interzonal hyaloids. F = plagioclase phenocrysts; h = interronal hyaloids. Again the hyaloids are a transitory phase in the transmutation gel hyaloids to crystalline feldspars.
Hyalobasalt, Ustica, Sicily, Italy.
With half-crossed nicols.

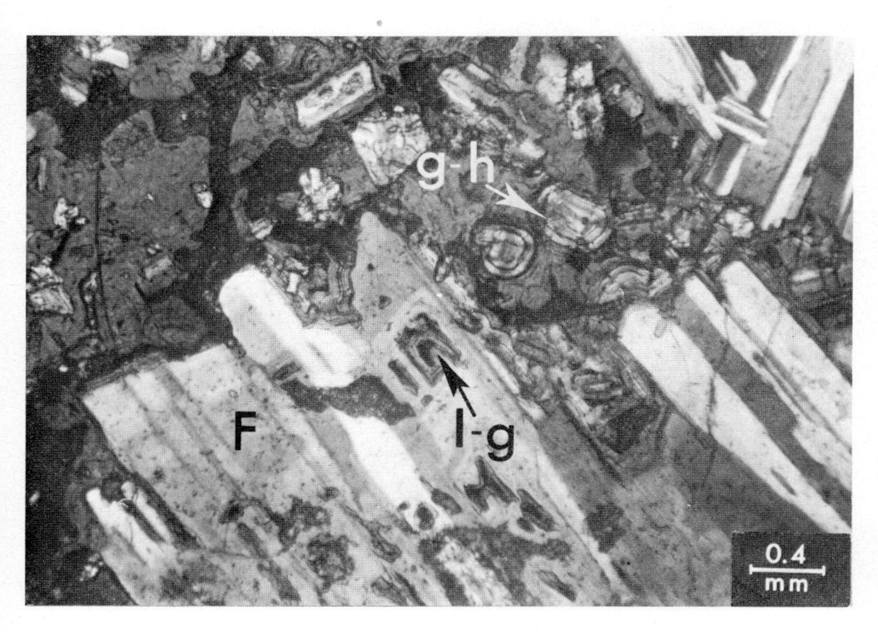

Fig. 306. Plagioclase phenocryst with gel hyaloids in the phenocrysts and gel hyaloids in the glassy groundmass. F = plagioclase phenocrysts; I-g = internal gel hyaloids in the plagioclases; g-h = gel hyaloids in the glassy groundmass.
Hyalobasalt. Ustica, Sicily, Italy. With half-crossed nicols.

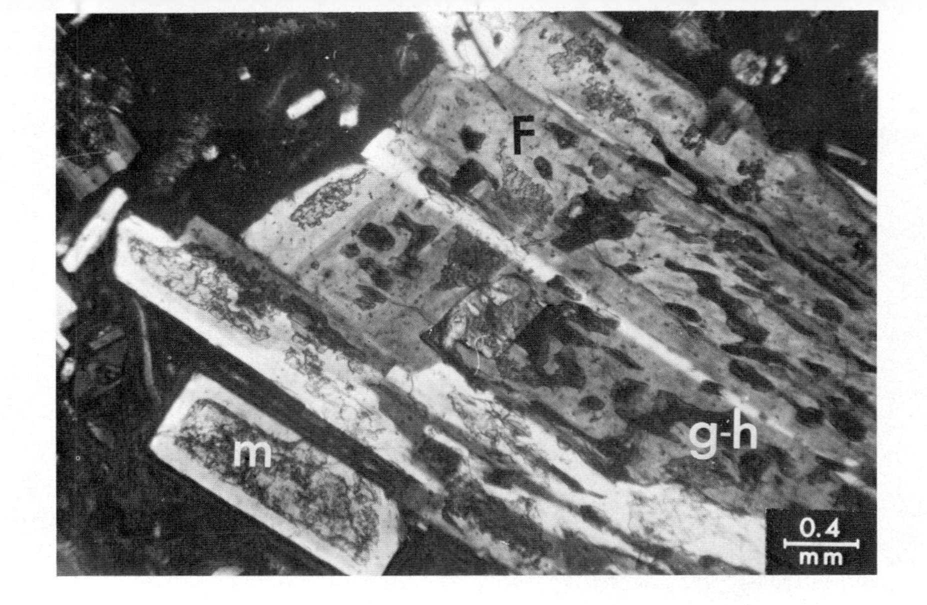

Fig. 307. Plagioclase phenocrysts with gel hyaloids in the phenocryst. Mesostatic phases of gel hyaloids to feldspar are also present in the phenocryst. F = plagioclase phenocryst; g-h = gel-hyaloids; m = mesostatic phase.
Hyalobasalt. Ustica, Sicily, Italy. With half-crossed nicols.

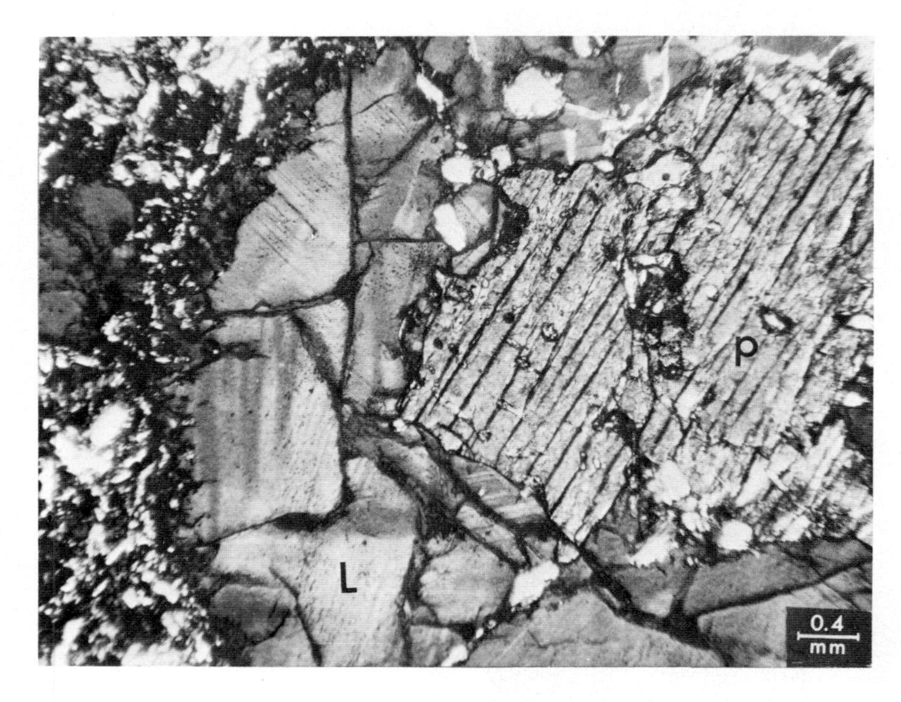

Figs. 308 and 309. Large leucite tecoblastic phenocryst entirely enclosing corroded augite phenocryst in its central part. L = leucite tecobastic phenocryst; p = augitic pyroxene enclosed by the leucite.
Basaltic rock type, leucite – tephrite, Travolata, Rome, Italy. With crossed nicols.

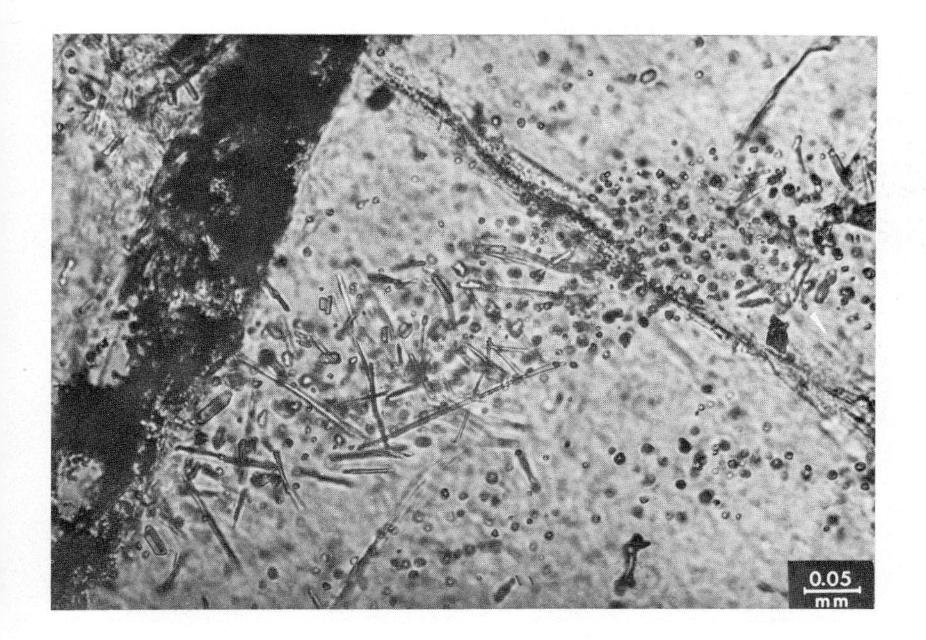

Fig. 310. The zonal growth of leucites is marked by fine needles of apatite which form "bands" in zonal arrangement within the leucite phenocryst.
Basaltic rock type, leucite – tephrite. Travolata, Rome, Italy.

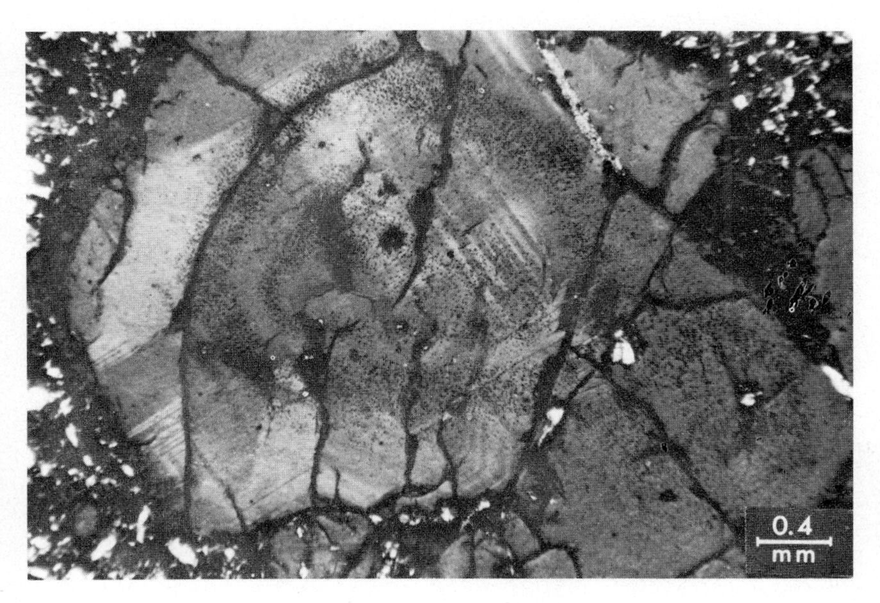

Fig. 311. Leucite phenocryst with packet-twinning and with needles of apatite in the zonal distribution within the leucite, i.e. the fine needles of apatite follow the zonal growth of the leucite.
Basaltic rock type, leucite – tephrite. Travolata, Rome, Italy.

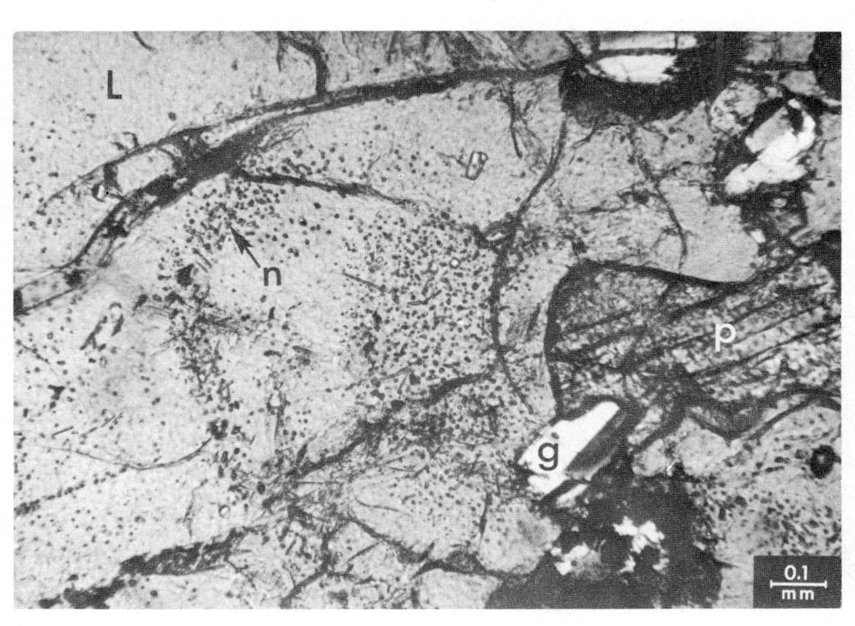

Fig. 312. Rounded and corroded pyroxene and plagioclase components surrounded by zonal leucite. The zonal structure of the leucite is indicated by fine apatite needles (fine dots). The rounded pyroxene and corroded pyroxene outline is in accordance with the leucite zoning. L = leucite; p = corroded and rounded pyroxene; g = groundmass plagioclases; n = needles of apatite in a zonal arrangement in the leucite.
Basaltic rock type, leucitic – tephrite. Travolata, Rome, Italy.

Fig. 313. Leucite phenocryst with packet polysynthetic twinning (block polysynthetic twinning). The zonal growth of the leucite crystal growth is marked by fine needles of apatite. Interrupted lines "a" and "b" show the zonal growth of the central leucite; interrupted line "c" shows the zonal growths of the outer leucite. There is an unconformity between the arrangement of the internal zoning and the external zoning of the leucite. Arrow "d" shows a rounded outline of the leucite due to magmatic corrosion, i.e. the crystallization of the groundmass was partly contemporaneous (see Fig. 316) and partly post leucitic.
Basaltic rock type, leucite – tephrite. Travolata, Rome, Italy.

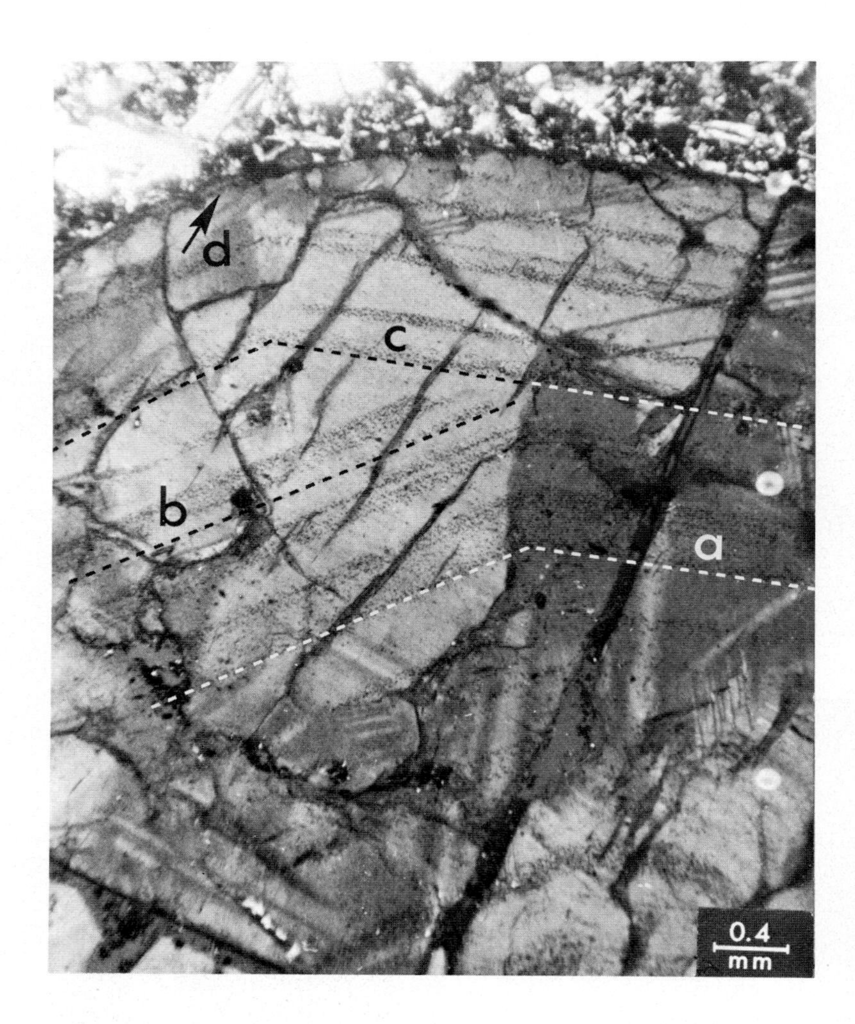

Fig. 314. Olivine and pyroxene groundmass components included in the leucite phenocryst.
Basaltic rock type, leucite – tephrite. Travolata, Rome, Italy.

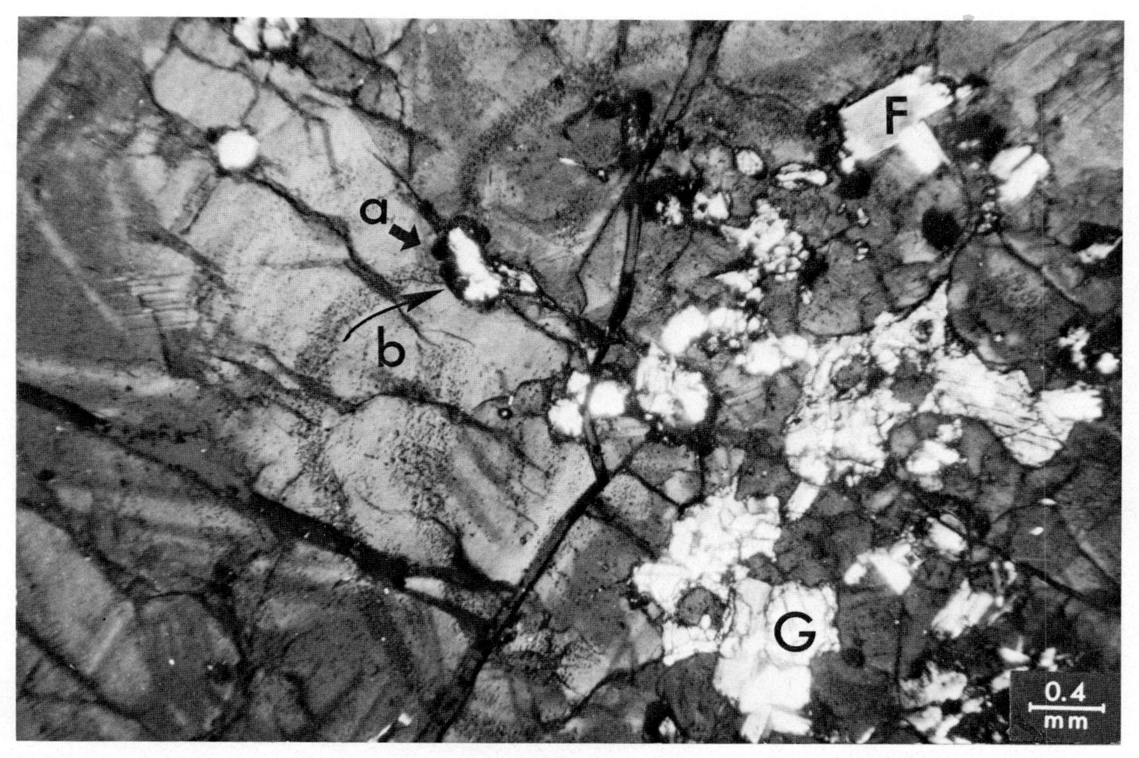

Fig. 315. Pyroxene and groundmass plagioclase enclosed by the leucite tecoblast. The zonal growth of the leucite tecoblast is shown by bands of fine needsles of apatite following the leucite zoning. The groundmass components were either enclosed in the central part of the leucite or as the zonal growth of the phenocryst developed; they existed already and had interferred with the zonal development, as is shown by arrow "a". Plagioclase crystal grain of the groundmass is zonally incorporated by the growth of leucite and has caused a local disturbance and a curving of the zonal bands, see arrow "b". G = groundmass components enclosed in the leucite; F = interzonally incorporated feldspar.
Basaltic rock type – leucite – tephrite. Travolata, Rome, Italy.

Fig. 316. Detail of Fig. 315. Plagioclase and pyroxene components of the groundmass, interzonally incorporated by the leucite phenocryst. Arrows show the curving of the zonal bands (marked by fine needles of apatite) and thus pointing out a pre-leucite crystallization for the feldspar (f) and pyroxene (p) crystal grains that interferred in the development of the zonal growth of the leucite.
Basaltic rock type – leucitic – tephrite. Travolata, Rome, Italy.

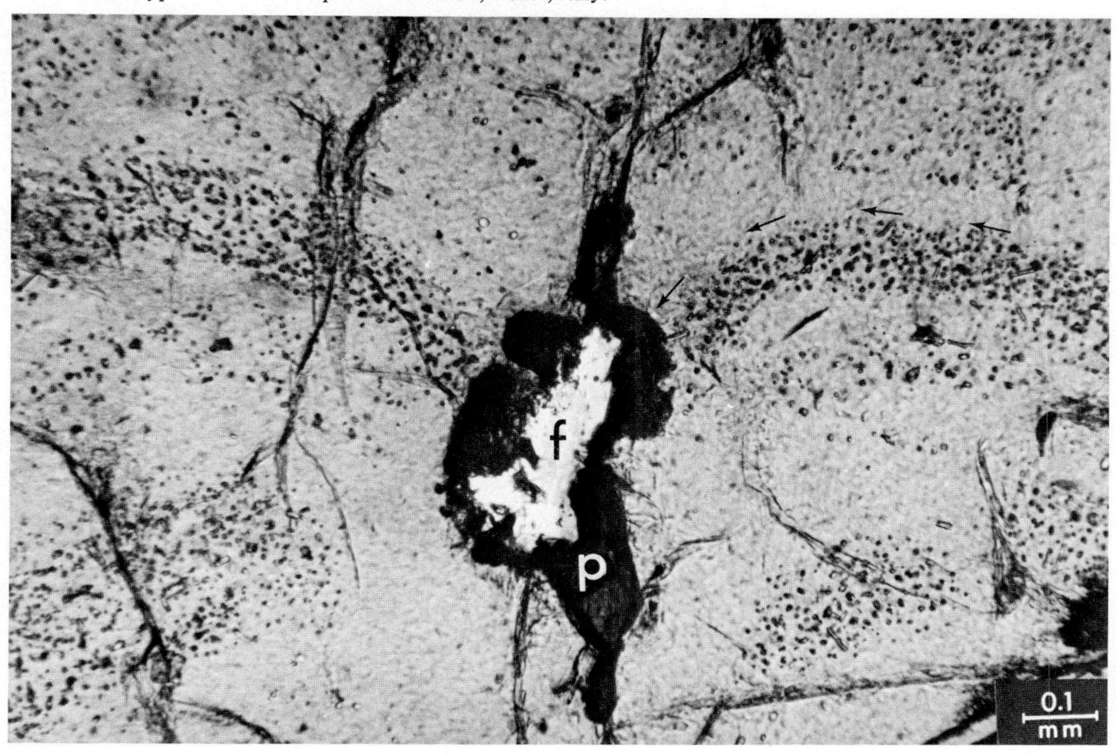

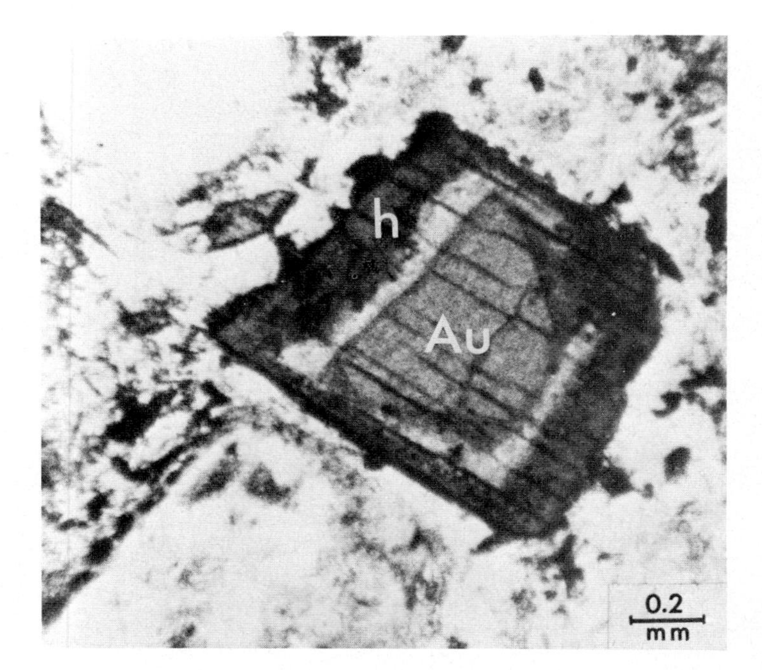

Fig. 317. Augite surrounded by hornblende due to uralitisation. Au = augite; h = hornblende. Nosean nepheline, leucitophyre. Laacher See, Rieden, Germany. Without crossed nicols.

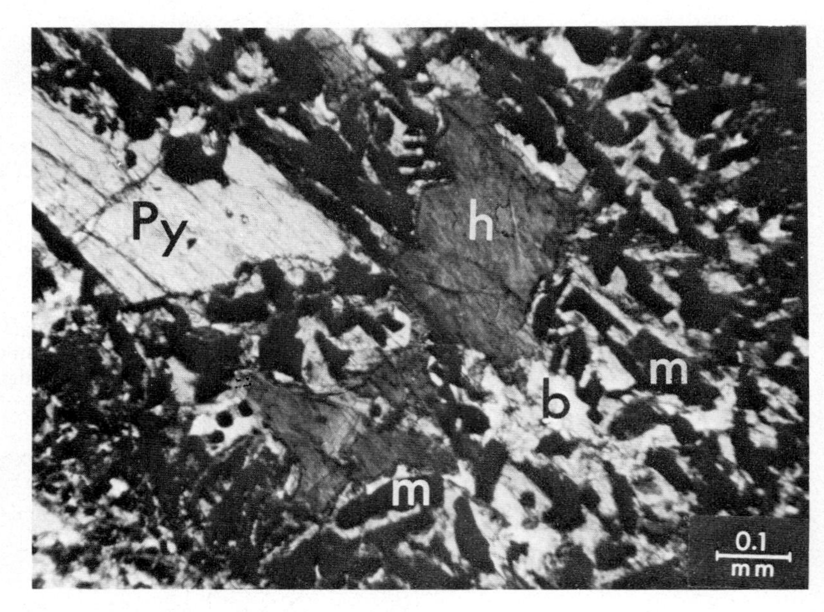

Fig. 318. Pyroxene in intergrowth with brown pleochroic hornblende and with biotite surrounding this complex intergrowth structure; there is also a concentration of fine groundmass magnetite. Py = pyroxene; h = hornblende (brown, pleochroic); b = biotite; m = fine magnetite.
Minderberg near Litz on Rhine River, Germany. With crossed nicols.

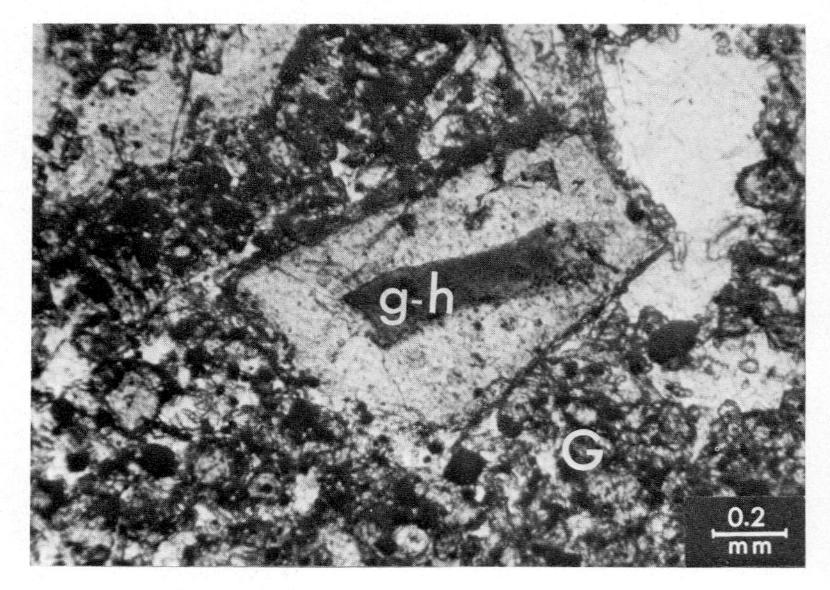

Fig. 319. Green hornblende in pyroxene. g-h = green hornblende; G = basaltic groundmass. Alkali basalt. Maxain near Selters, Westerwald, Germany. Without crossed nicols.

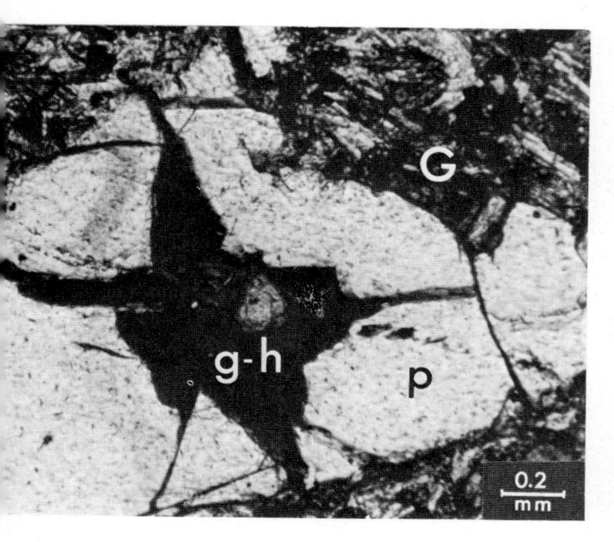

Fig. 320. Green hornblende included in pyroxene. g-h = green hornblende; p = pyroxene; G = groundmass.
Basalt. Steinl, Kleinhahn, Bohemia, Czechoslovakia. With crossed nicols.

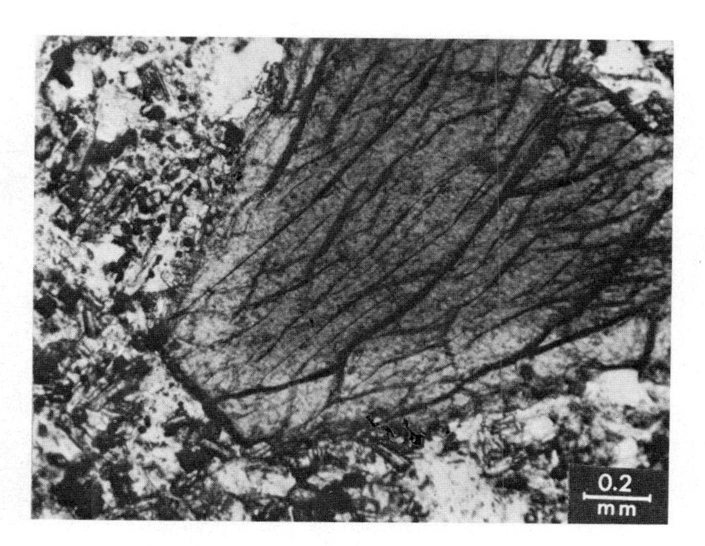

Fig. 321. Idioblastic hornblende phenocryst in basalt.
Nephelinite. Pulberbuk Oberbergen, Kaiserstuhl, Baden, Germany. Without crossed nicols.

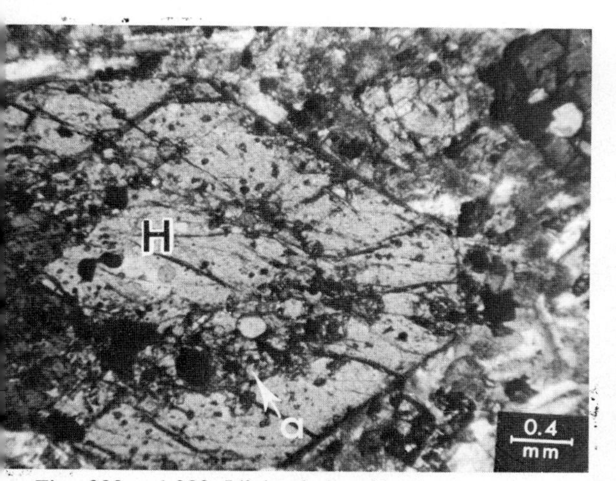

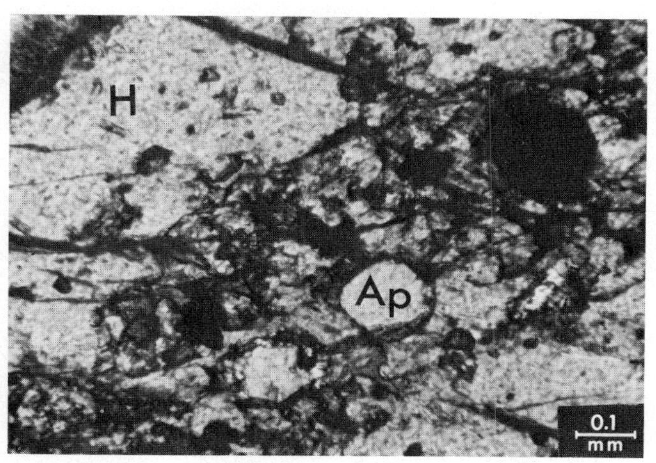

Figs. 322 and 323. Idiobastic hornblende phenocryst in basalt with groundmass component and accessory subidiomorphic apatite zonally included in the amphibole. H = hornblende idioblast. Arrow "a" shows groundmass, zonally included; Ap = sub-idiomorphic apatite.
Tephritic basalt (Etnaite Lava 1886). Nicolosi, Atna, Sicily, Italy. With crossed nicols.

Fig. 324. Poikiloblastic hornblende including iron oxides of the groundmass. h = hornblende poikiloblast; black = magnetite.
Olivine-basalt porphyry. Boulder County, Colorado, U.S.A. Without crossed nicols.

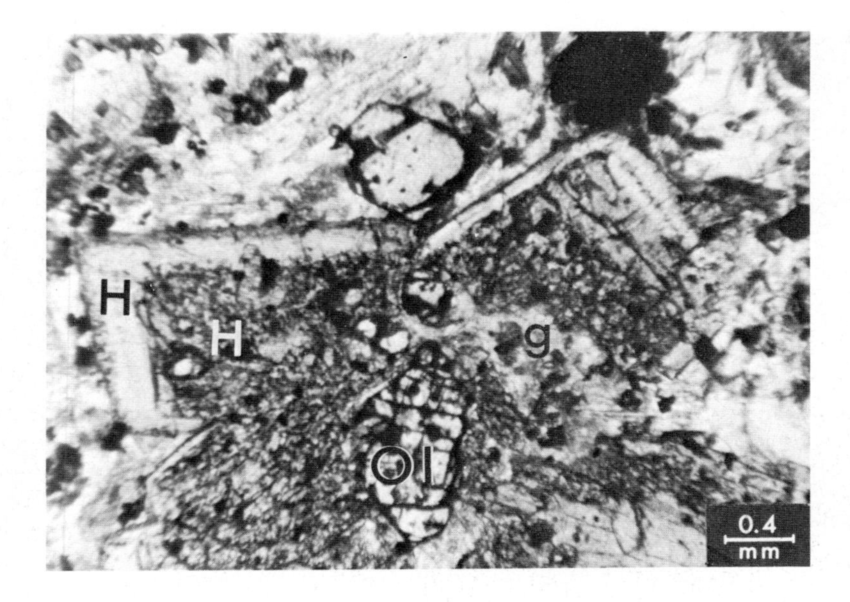

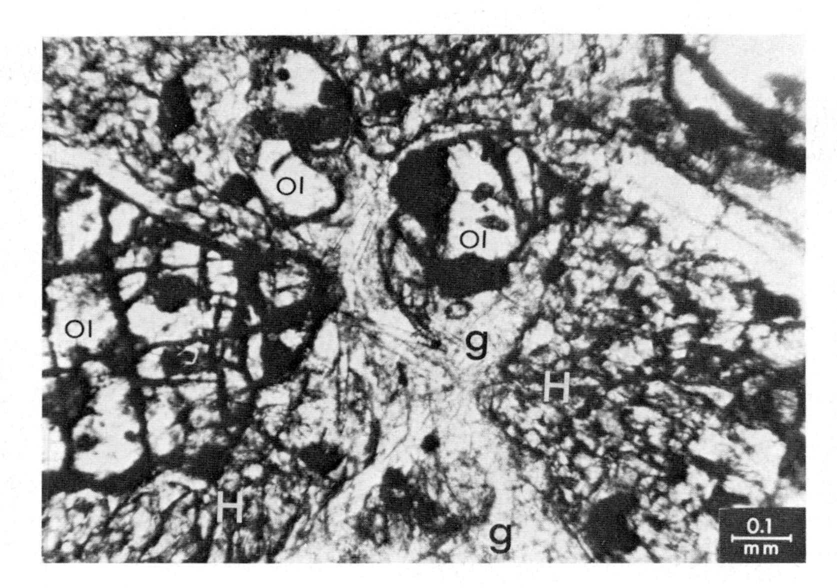

Fig. 325 and 326. Hornblende crystalloblasts (radiating) enclosing early olivine and basaltic groundmass. H = hornblende crystalloblast; Ol = olivine; g = groundmass enclosed in hornblende.
Olivine-basalt porphyry. Boulder County, Colorado, U.S.A. Without crossed nicols.

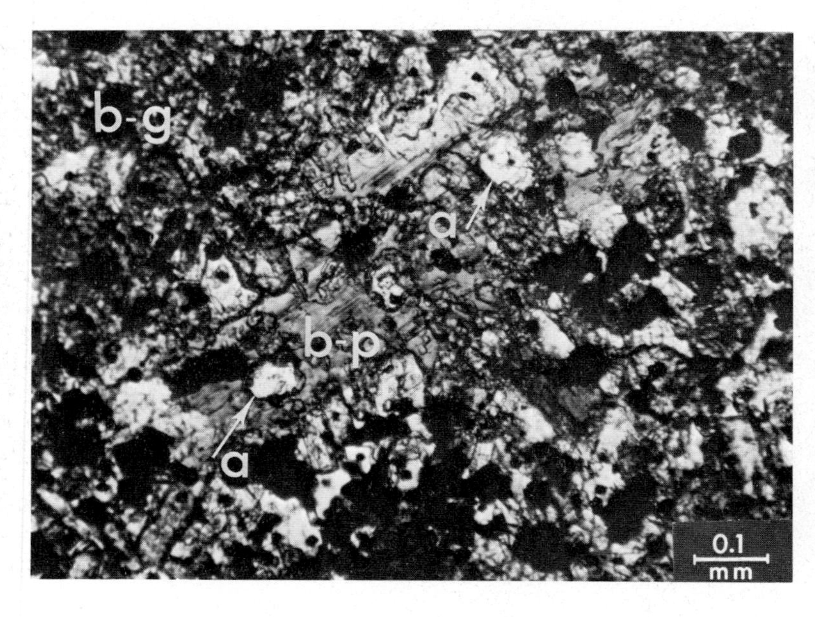

Fig. 327. Biotite poikiloblast enclosing components of the groundmass. b-p = biotite poikiloblast; b-g = basaltic groundmass. Arrow "a" shows basaltic groundmass enclosed in the biotite.
Basalt. Schlüsselberg, Bohemia, Czechoslovakia. With crossed nicols.

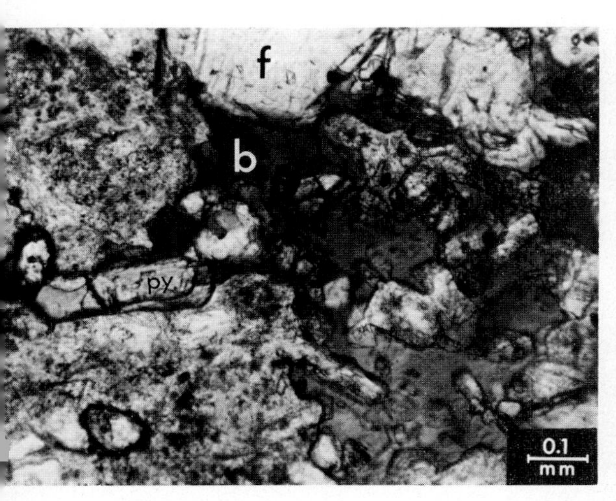

Fig. 328. Biotite poikiloblast occupying the intergranular space between the feldspars and enclosing groundmass components. f = feldspars; b = biotite poikiloblast; py = pyroxene.
Olivine basalt. Jefferson County, Colorado, U.S.A. With crossed nicols.

Fig. 329. Basaltic biotite occupying the intergranular between pyroxene and feldspars of the basaltic groundmass. bi = biotite; py = pyroxene; f = feldspar.
Olivine basalt. Jefferson County, Colorado, U.S.A. With crossed nicols.

Fig. 330. Idiomorphic basaltic biotite laths partly surrounding groundmass pyroxene. py = pyroxene; bi = idiomorphic basaltic biotite laths.
Olivine basalt. Jefferson County, Colorado, U.S.A. With crossed nicols.

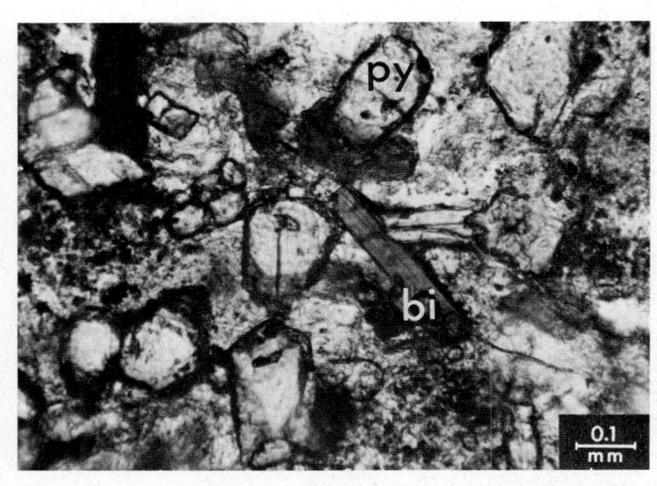

Fig. 331. Idiomorphic basaltic biotite laths with basaltic groundmass and sub-idiomorphic pyroxenes. bi = biotite laths; py = pyroxene.
Olivine basalt. Jefferson County, Colorado, U.S.A. With crossed nicols.

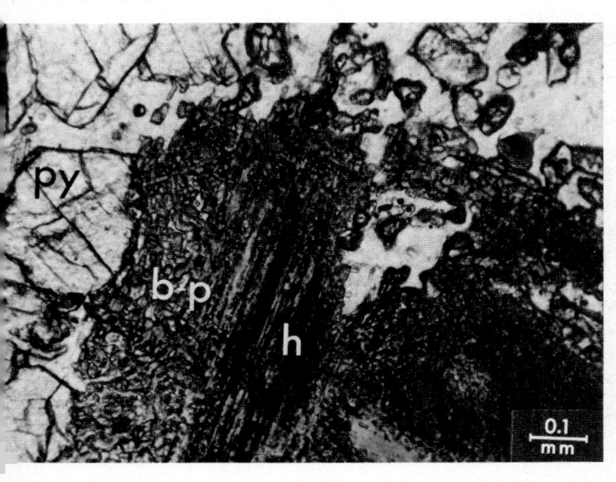

Fig. 332. Basaltic biotite in symplectic intergrowth with pyroxene and hornblende. b-p = symplectic intergrowth of biotite with pyroxene; py = pyroxene; h = hornblende.
Basalt columns. Minderberg near Litz on Rhine River, Germany. With crossed nicols.

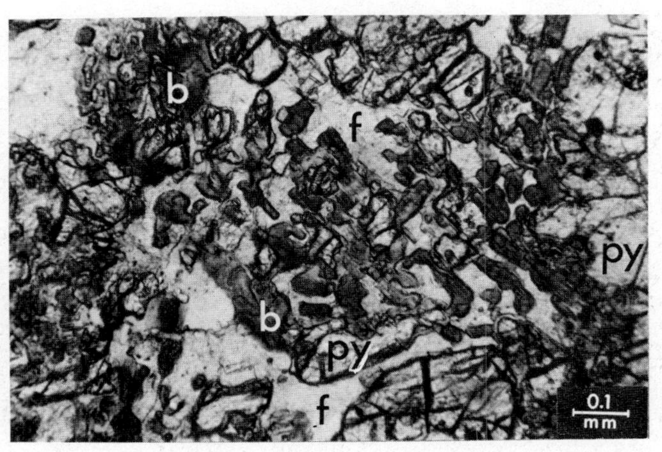

Fig. 333. Biotite in symplectic intergrowth with pyroxene. b = biotite; py = pyroxene; f = plagioclase.
Basalt columns. Minderberg near Litz on Rhine River, Germany. With crossed nicols.

Fig. 334. Chloritic cavity filling with neocrystallization of mica. n-m = neocrystallization of mica; ch-c = chloritic mass filling cavity; g-b = basaltic groundmass.
Doleritic basalt. Jaljuba, U.S.S.R. With half-crossed nicols.

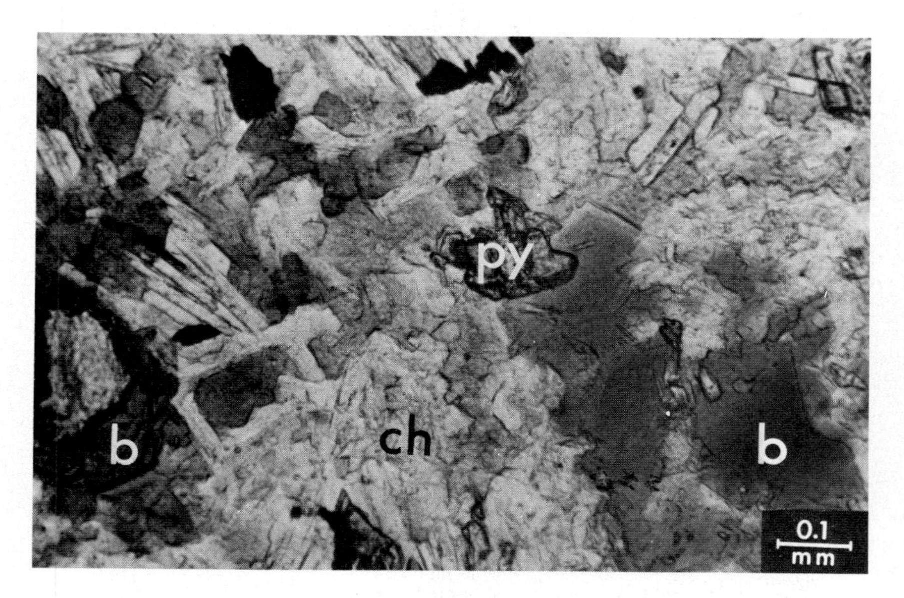

Fig. 335. Biotite with chlorite and partly enclosing groundmass pyroxenes. b = biotite; py = pyroxene; ch = chlorite.
Olivine basalt. Jefferson County, Colorado, U.S.A. Without crossed nicols.

Fig. 336. Magnetite "skeletons" attaining eucrystalline forms in symplectic intergrowth with quartz and phlogopite. M = magnetite; q = quartz; ph = phlogopite; L = leucite.
Leucite basalt. Totenkappel b. Meiches, Vogelsberg, Germany.

Fig. 337. Basaltic titanomagnetite (actually magnetite with an increased Ti content). t = titanomagnetite. The titanomagnetite shows alteration to maghemite.
Olivine basalt. Debra-Sina, Ethiopia. Oil-immersion with one nicol.

Fig. 338. Zoned augite with broad polysynthetic lamellae, with magnetite idiomorphic crystal grains, partly enclosed in the phenocryst and partly in the groundmass. A = twinned and zoned augite phenocrysts; M = idiomorphic phenocrysts; G = basaltic groundmass.
Augite-rich basalt. Selale Mt. region, Ethiopian Plateau. With crossed nicols.

Fig. 339. Ophitic magnetite-plagioclase laths, symplectic intergrowths.
Basalt. Floor of Atlantic Ocean. With crossed nicols.

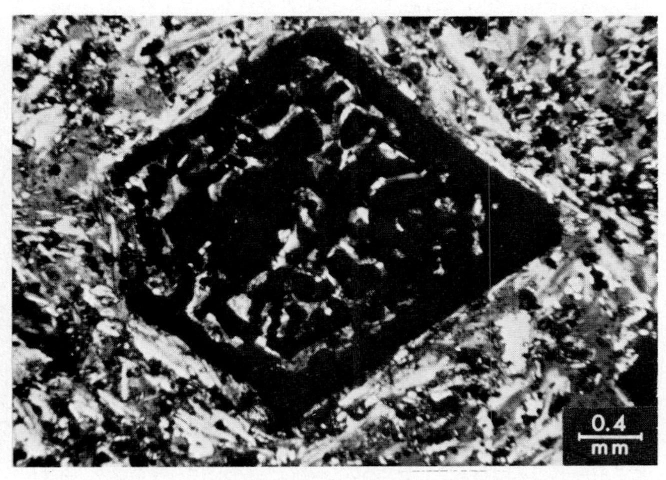

Fig. 340. Idiomorphic magnetite with groundmass components enclosed by the host. The groundmass is an intracrystalline infiltration (see Fig. 341) or enclosed by teco-poikiloblast.
Spotted basalt. Canary Islands. With crossed nicols.

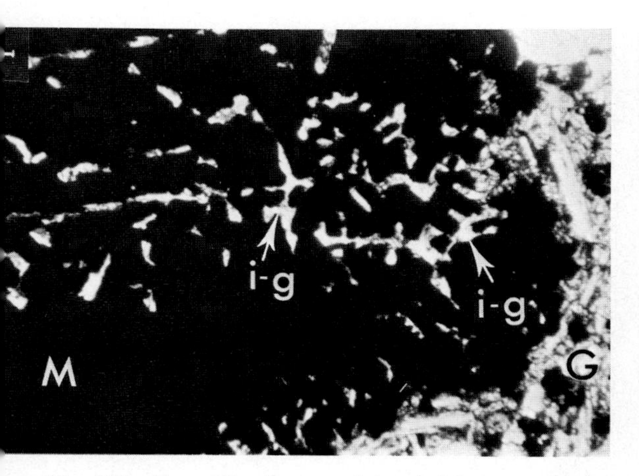

Fig. 341. Magnetite phenocryst invaded by groundmass components. M = magnetite phenocryst; G = groundmass components; i-g = infiltration of groundmass in the magnetite phenocryst.
Basalt. Las Planas, Gerona Prov., Spain. With crossed nicols.

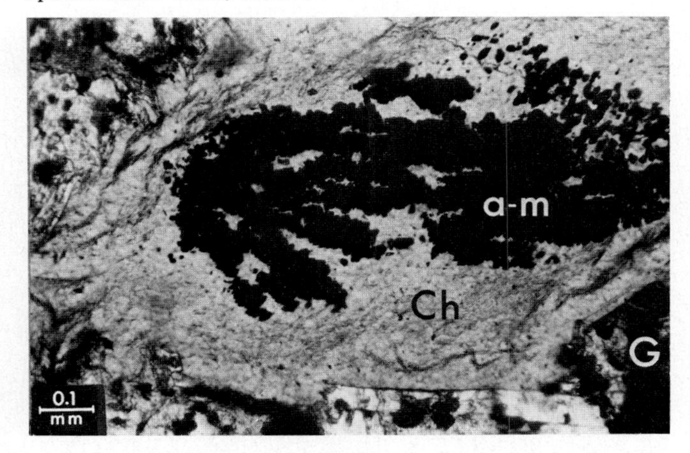

Fig. 342. Chlorite mass in basaltic cavity with magnetite crystal-grain agglutination. Ch = chloritic mass in basaltic groundmass; a-m = agglutination of magnetite crystal grains within the chloritic mass occupying groundmass cavity; G = basaltic groundmass.
Basalt dolerite. Jaljaba, U.S.S.R. Without crossed nicols.

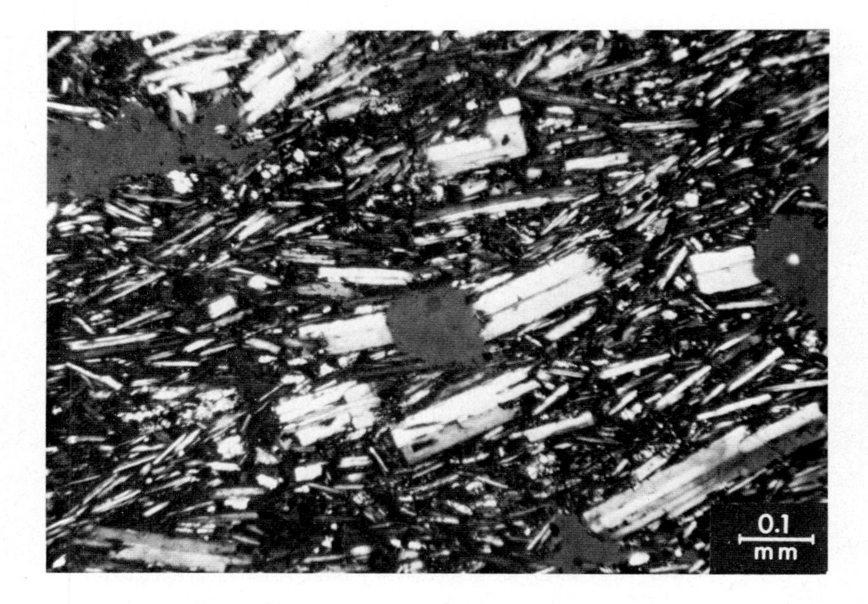

Fig. 343. Fine and sub-prismatic plagioclase groundmass laths showing flow orientation. Small gas-cavities are also shown.
Basalt. Oregon, U.S.A. With half crossed nicols.

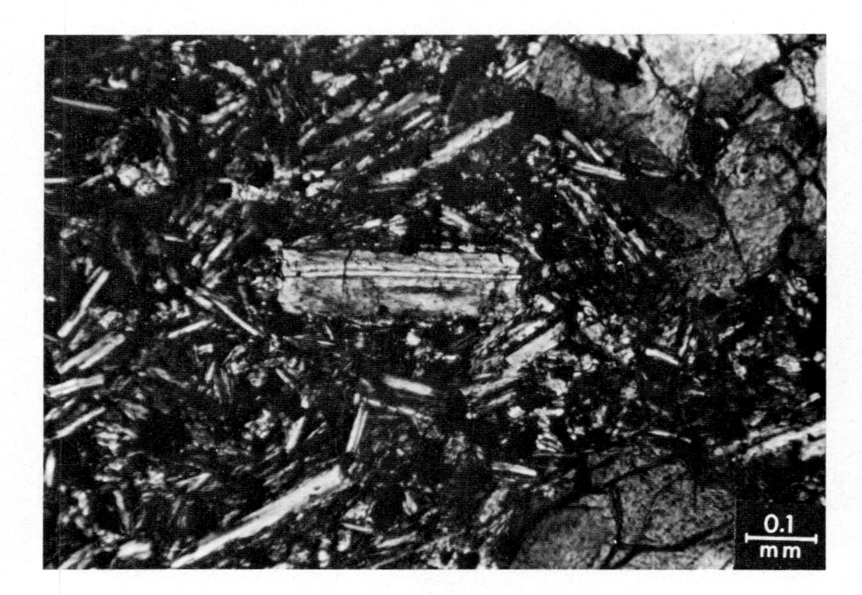

Fig. 344. Basaltic groundmass showing a flow texture with polysynthetically twinned augite also orientated in accordance with the flow structure of the groundmass feldspars.
Basalt. Nidda, Hessen, Germany. With crossed nicols.

Fig. 345. Pyroxene phenocrysts with feldspar laths with a flow orientation texture parallel to the margins of the pyroxene phenocrysts.
Basalt. Nidda, Hessen, Germany. With crossed nicols.

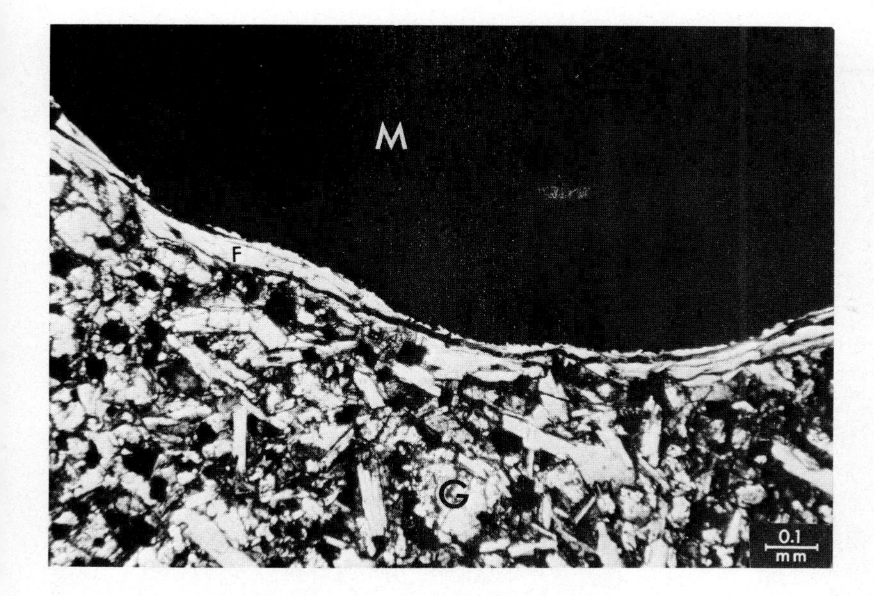

Fig. 346. Magnetite component of the olivine bomb in basalt with fluidal plagioclases following the magnetite outlines with the basaltic groundmass. M = magnetite; G = basaltic groundmass; F = feldspars elongated along the contact with magnetite and their shape adapted to the outline of the magnetite contact with the groundmass.
Olivine bomb in basalt. Jato, Lekempti, W. Ethiopia. With crossed nicols.

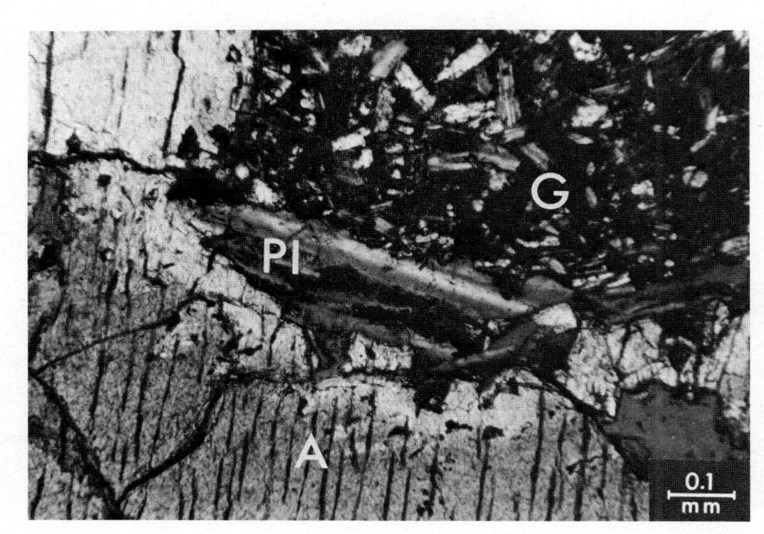

Fig. 347. Basaltic groundmass in contact with pyroxene which shows alteration reaction margins with the basalt. A basaltic plagioclase is shown to be in intergrowth with the reaction margin of the augite. A = augite; G = basaltic groundmass; Pl = plagioclase in intergrowth with the augite.
Olivine pyroxene basalt. Addis Ababa, Rift Valley, Ethiopia. With crossed nicols.

Fig. 348. Sub-idiomorphic and rounded early crystallized olivine, partly engulfed or enclosed by phenocrystalline augite exhibiting hourglass structure. Ol = olivine; py = pyroxene; b-g = basaltic groundmass.
Augite-rich basalt. East Siberia, U.S.S.R. With crossed nicols.

Fig. 349. Early olivine crystallization rounded and surrounded by augite phenocrysts. Ol = olivine; py = pyroxene; b-g = basaltic groundmass.
Augite-rich basalt. East Siberia, U.S.S.R. With crossed nicols.

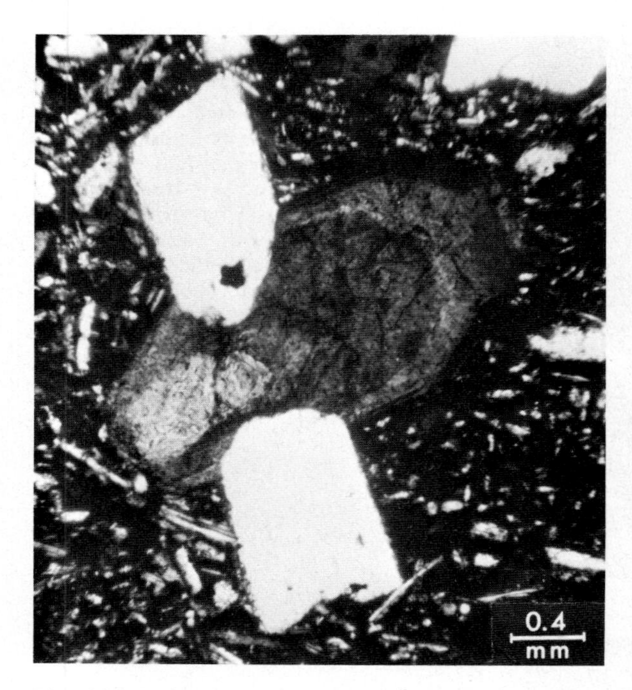

Figs. 350 and 351. Different orientations of an interpenetration growth of two augite phenocrysts exhibiting hourglass structure. Augite-rich basalt. East Siberia, U.S.S.R. With crossed nicols.

Fig. 352. A zonal sub-phenocrystalline augite perpendicularly grown and penetrating into a larger augite phenocryst exhibiting hourglass structure.
Augite-rich basalt. East Siberia, U.S.S.R. With crossed nicols.

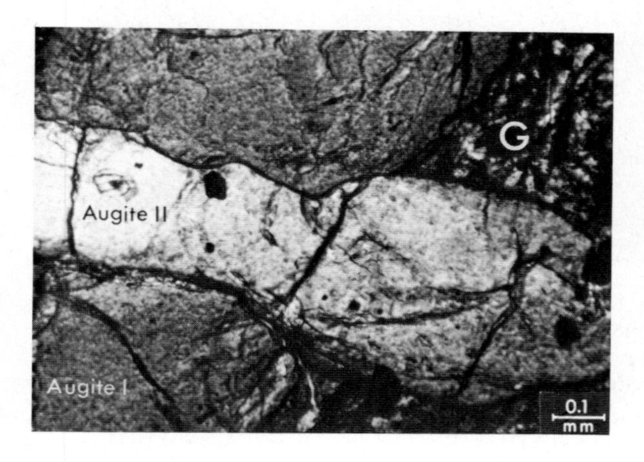

Fig. 353. Large, zoned augite phenocryst (Augite I) is invaded along a crack, perpendicular to the length of augite I by a younger augite (augite II). G = basaltic groundmass.
Augite-rich basalt. Selale Mt. region, Ethiopian Plateau. With crossed nicols.

Fig. 354. Sub-prismatic polysynthetically twinned plagioclase laths exhibiting an intersertal structure and associated with groundmass pyroxenes. Pl = prismatic polysynthetically twinned plagioclase laths; py = pyroxenes.
Plateau basalt. Ethiopian plateau. With crossed nicols.

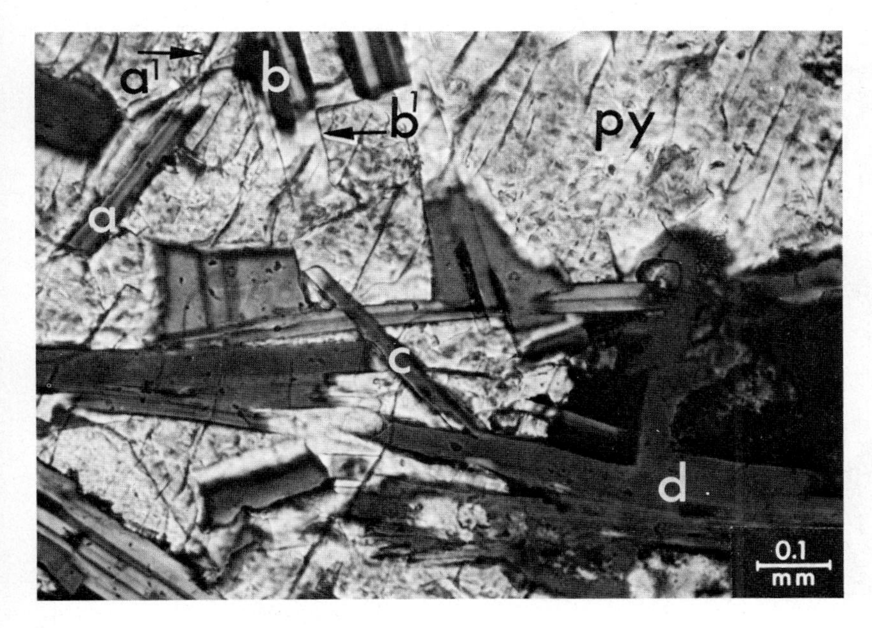

Fig. 355. Ophitic plagioclase–pyroxene intergrowth texture. Plagioclase laths extending into a pyroxene phenocryst; clearly the plagioclases either follow cracks, cleavage directions or weak lattice spaces within the pyroxene; e.g., plagioclase marked "a" follows a crack in the system of cracks marked "a1" and plagioclase marked "b" follows a cleavage direction marked "b1". The plagioclase lath marked "c" penetrates an already-formed plagioclase lath. The plagioclase laths marked "d" extend from the margin of the pyroxene into its interior. The association of feldspar laths with cleavage cracks and other penetrability directions of the host pyroxene clearly support a post-pyroxene formation of the plagioclase. It is interesting to note that the plagioclase lath marked "c" not only extends into the pyroxene but also cuts across another lath. Plagioclase "c" must have had an enormous cyrstallization force. py = pyroxene phenocryst.
Doleritic basalt. Karroo series. South Africa.

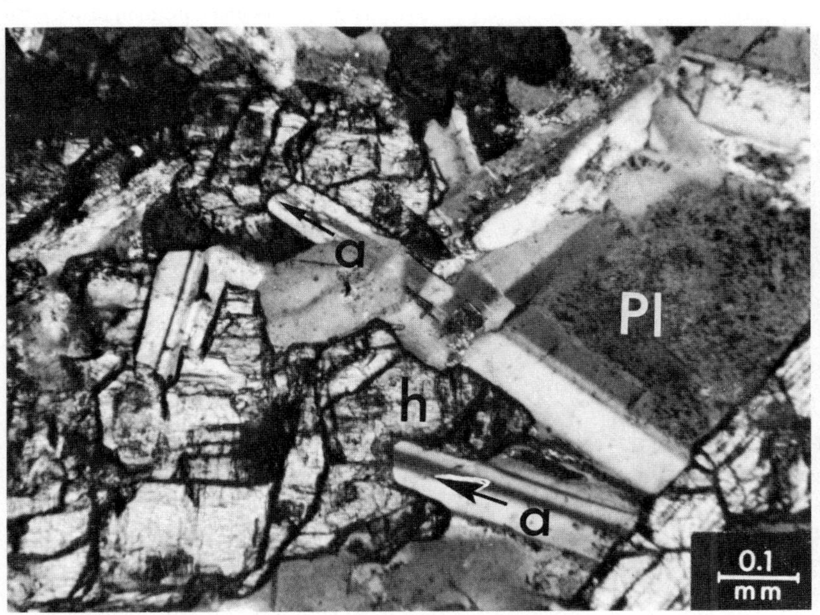

Fig. 356. Plagioclase, partly engulfing and in ophitic intergrowth with basaltic hornblende. Plagioclase laths have penetrated into the amphibole. Pl = plagioclase; h = hornblende. Arrow "a" shows plagioclase laths penetrating into the amphibole.
Hornblende basalt. Chaffee County, Colorado, U.S.A. With crossed nicols.

Fig. 357. Magnetite-plagioclase ophitic intergrowth. Magnetite (black) penetrated by plagioclase laths.
Mid-Atlantic Ridge. With crossed nicols.

Fig 358 (a and b). "Intersertal" ophitic texture of plagioclases basaltic groundmass in intergrowth with olivine phenocrysts. Ol = olivine phenocryst; Pl = plagioclase groundmass in intergrowth with the olivine phenocrysts.
Olivine basalt, ophitic. Säsebühl, Dransfeld, Rhön, Germany. With crossed nicols.

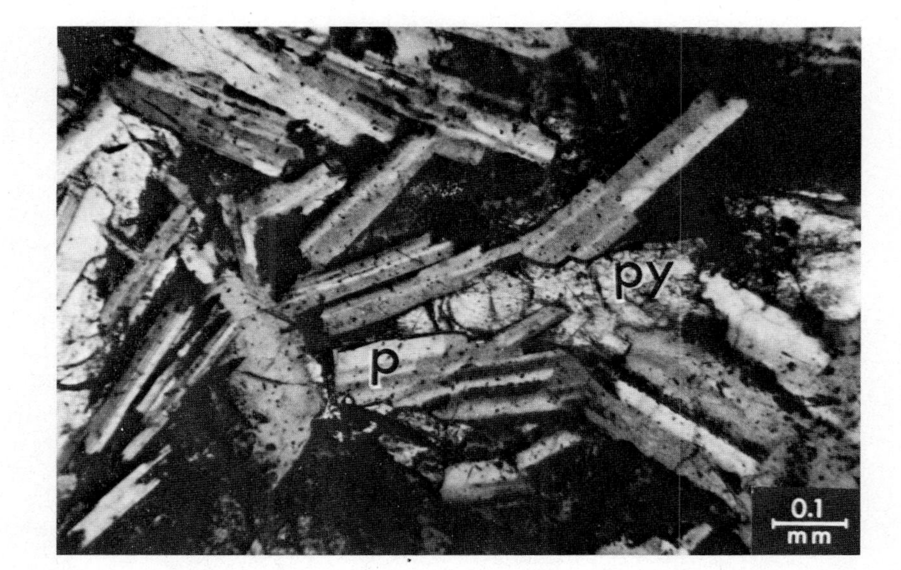

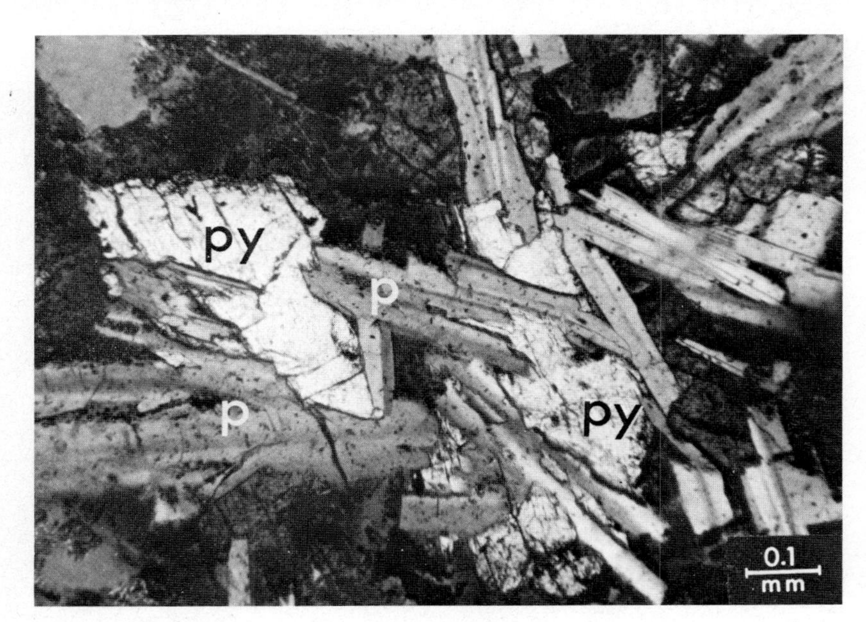

Figs. 359 and 360. Ophitic plagioclase–pyroxene intergrowth texture. Intergrowth patterns of plagioclase and pyroxene. Admittedly, such textural patterns are difficult to decipher, since it is difficult to know which extends into which. However, in the same thin sections, other textural associations often exist which clearly show that the plagioclase laths are following cracks and cleavage directions of the host-pyroxene. p = plagioclase laths; py = pyroxene.
Compact basalt. Deccan Plateau, India. With crossed nicols.

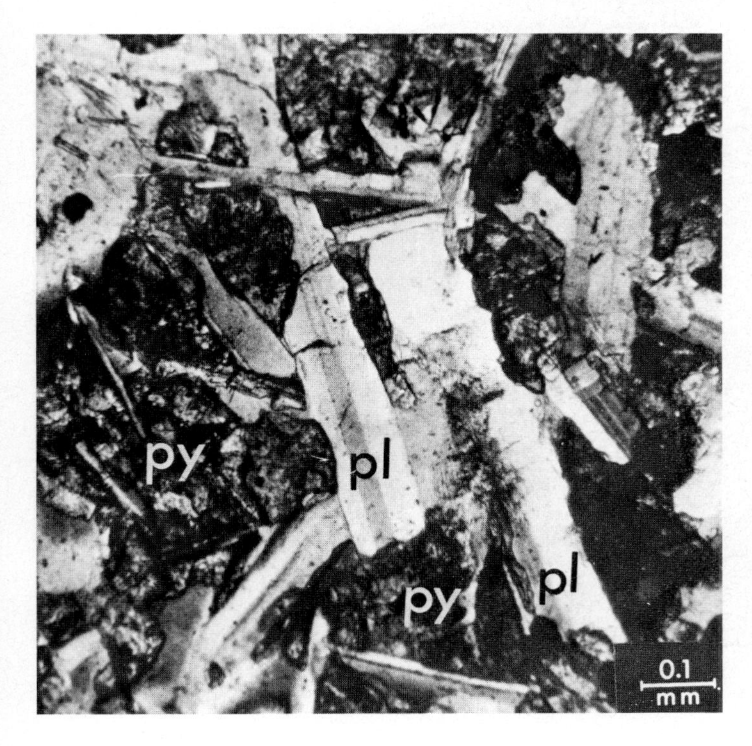

Fig. 361. Pyroxene–plagioclase intersertal textures. Upper part of the Palisade Sill, New Jersey, U.S.A. Pl = plagioclase; py = pyroxene. With crossed nicols.

Fig. 362. Ophitic plagioclase–pyroxene intergrowth texture. A pyroxene phenocryst with cracks and cleavage is invaded by later plagioclase laths, following the cleavage pattern of the host pyroxene.
Deccan plateau basalt, Deccan India. With crossed nicols.

Fig. 363. Ophitic plagioclase–pyroxene intergrowth texture. A pyroxene phenocryst with cleavage and cracks invaded by plagioclase laths which extend from the outside of the pyroxene into the host. Dolerite (coarse-grained basalt). Karroo series, S. Africa. With crossed nicols.

Fig. 364. Ophitic intergrowth of sub-phenocrystalline plagioclase laths invading a pyroxene host from outside inwards to the central part of the pyroxene. Pl = plagioclase; py = pyroxene. Dolerite (coarse-grained basalt). Karroo series, S. Africa. With crossed nicols.

Fig. 365. A pyroxene invaded by plagioclase lamella extending from the outside inwards to the central part of the pyroxene and "sending" a perpendicular branch. Arrow shows the change in width of the plagioclase lamella as it enters the pyroxene. py = pyroxene; pl = plagioclase.
Dolerite (coarse-grained basalt). Karroo series, S. Africa. With crossed nicols.

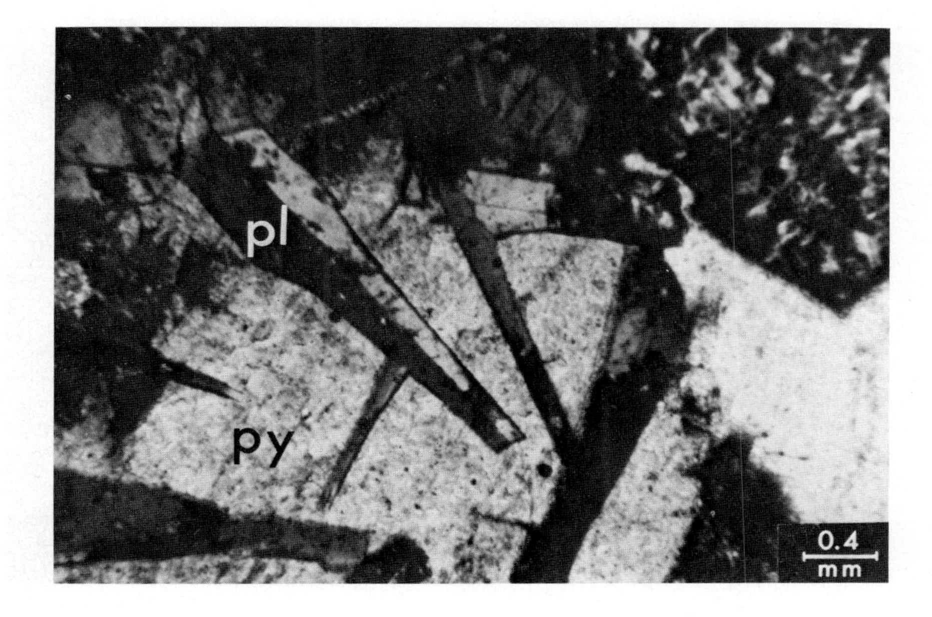

Fig. 366. Ophitic plagioclase–pyroxene intergrowth structure. Plagioclase sub-phenocrystalline laths extending from the margin inwards to the centre of the pyroxene, thinning and branching. py = pyroxene; pl = plagioclase.
Dolerite (coarse-grained basalt). Karroo series, S. Africa. With crossed nicols.

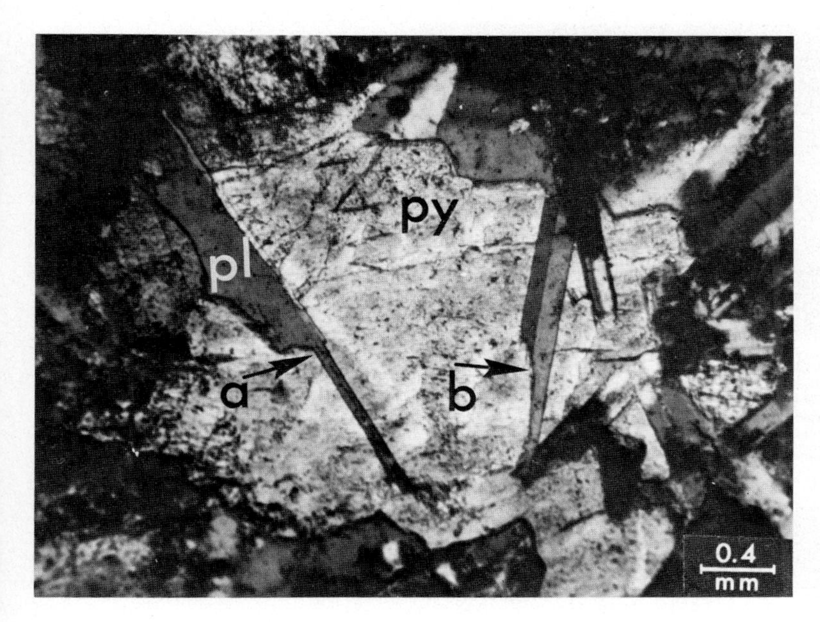

Fig. 367. Ophitic plagioclase–pyroxene intergrowth. Plagioclase sub-phenocrystalline laths extending inwards into the centre of the pyroxene and showing changes in their width. py = pyroxene; pl = plagioclase. Arrow "a" shows changes in the width of the plagioclase lamellae; arrow "b" shows curved outlines of the plagioclase laths.
Dolerite (coarse-grained basalt). Karroo series, S. Africa. With crossed nicols.

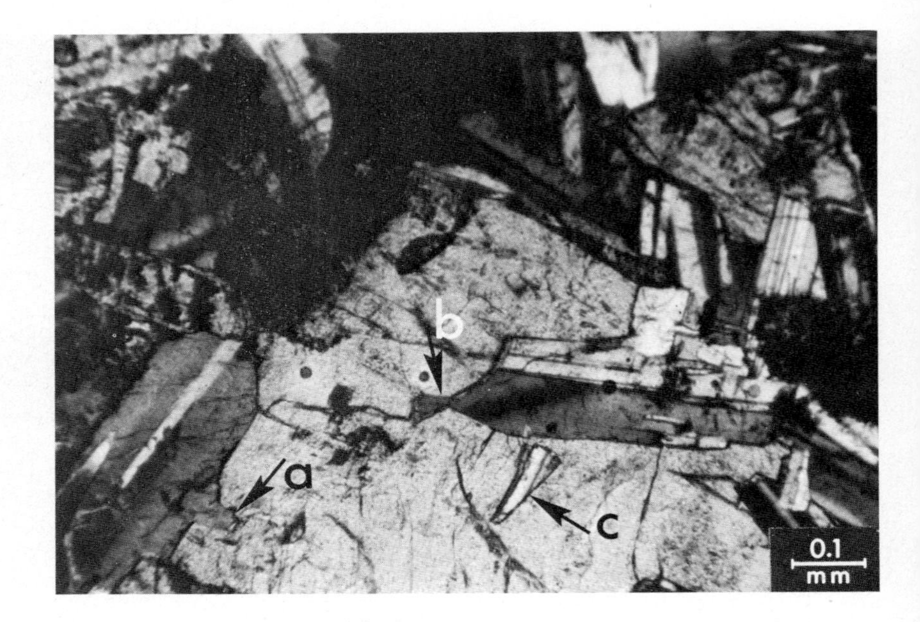

Fig. 368. Sub-phenocrystalline plagioclase laths in ophitic intergrowth with pyroxene. The plagioclase lath (arrow "a") sends a protruberance into the pyroxene. Also, another plagioclase lath shows a curved outline and shows extension into the pyroxene (arrow "b"). A lath of plagioclase seems to be enclosed in the pyroxene – due to the orientation of the section (arrow "c").
Dolerite (coarse-grained basalt). Karroo series, S. Africa. With crossed nicols.

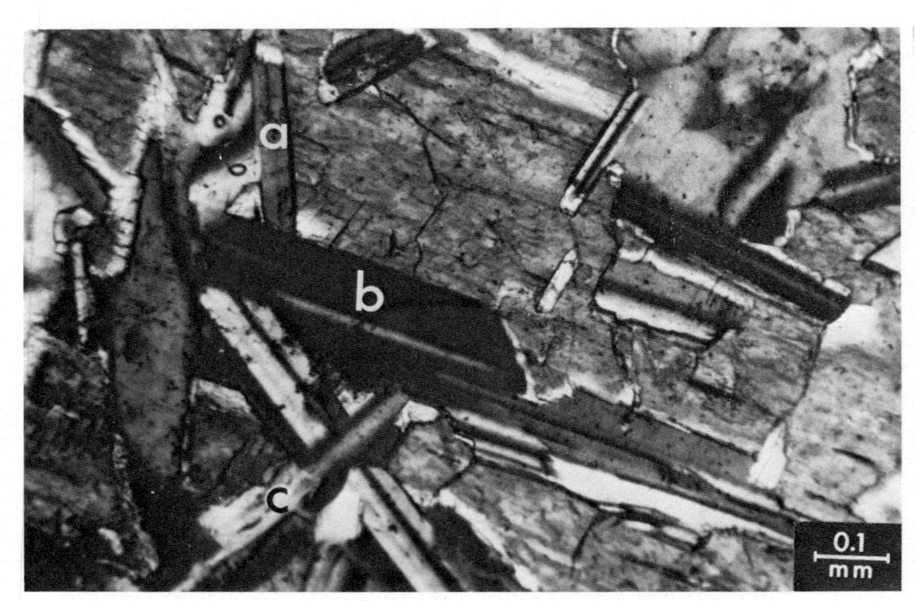

Fig. 369. Plagioclase, sub-phenocrystalline in size, orientated and following intraleptonic (cleavage) or intracrystalline penetrability directions of the host pyroxene. Plagioclase lath "a" is grown into the pyroxene and is at an angle to lath "b" without penetrating into the latter. The plagioclase laths on the whole follow penetrability directions within the pyroxene host. Plagioclase lath "c" penetrates other feldspar laths.
Dolerite (coarse-grained basalt). Karroo series, S. Africa. With crossed nicols.

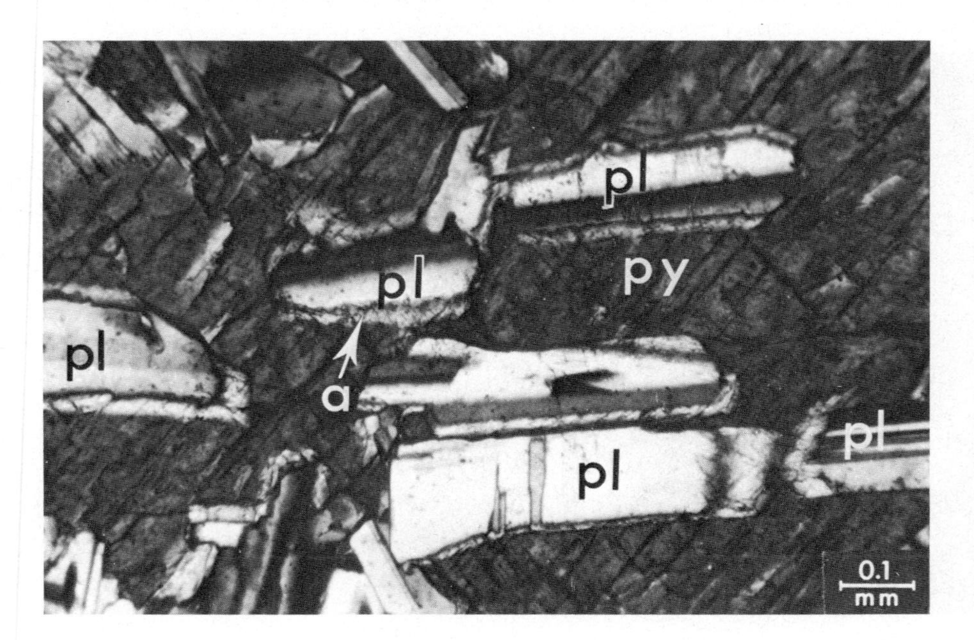

Fig. 370. Orientated sub-phenocrystalline plagioclase laths in ophitic intergrowth with the pyroxene and showing clearly a reaction margin with the pyroxene host. py = pyroxene; pl = orientated plagioclase-laths. Arrow "a" shows the reaction margin of the plagioclase with the pyroxene. The reaction margin seems to be caused by the feldspar forming melts reacting with a pyroxene.
Dolerite (coarse-grained basalt). Karroo series, S. Africa. With crossed nicols.

Fig. 383. Ophitic plagioclase–pyroxene intergrowth. Plagioclase lath "A" and "B" interpenetrate each other as they invade a host pyroxene "C".
Dolerite (coarse-grained basalt). Karroo series, S. Africa.

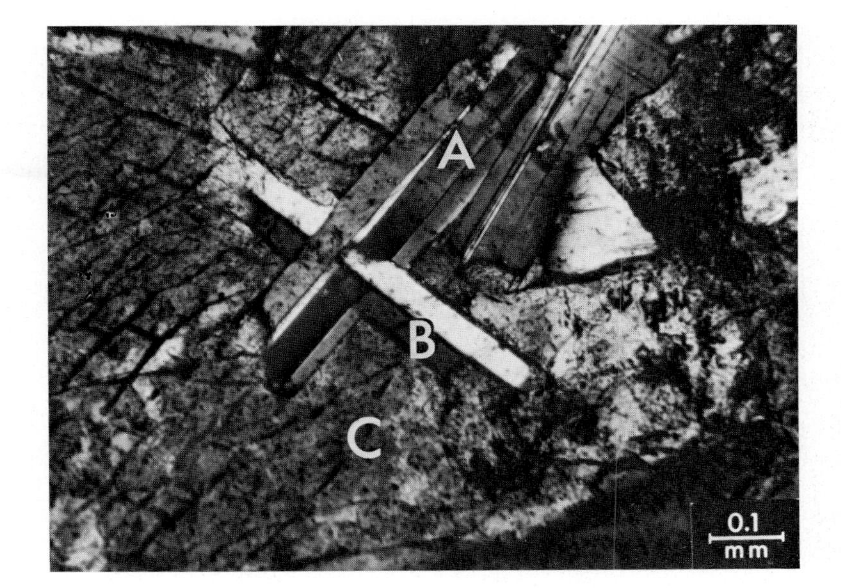

Fig. 384. Plagioclase laths, in ophitic intergrowth with pyroxene, enclose smaller pyroxene crystal grains. Py = pyroxenes; pl = plagioclase; p = pyroxenes (often twinned) enclosed in the plagioclase lath.
Palisade Sill, New Jersey, U.S.A. With crossed nicols.

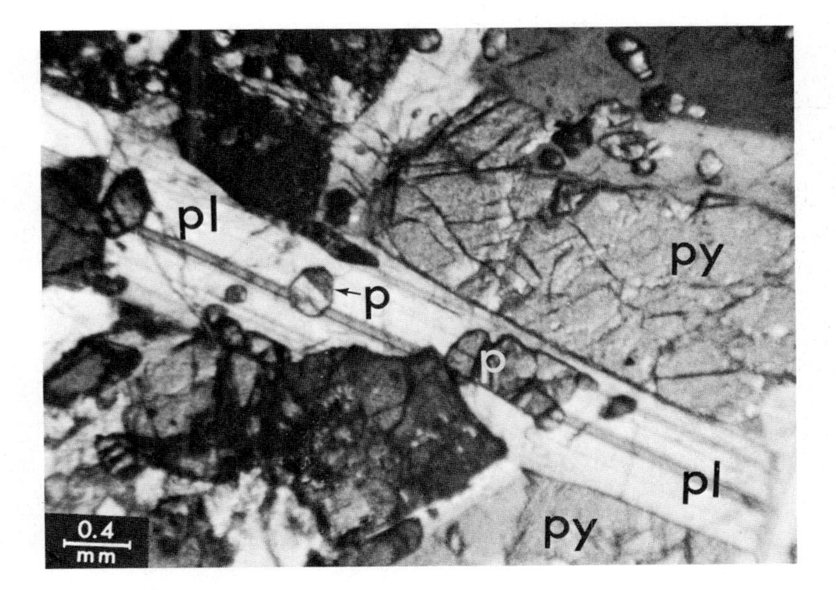

Fig. 385. Both intersertal and ophitic pyroxene–plagioclase intergrowths are shown. Py = pyroxene; pl = plagioclase.
Palisade Sill, New Jersey, U.S.A. With crossed nicols.

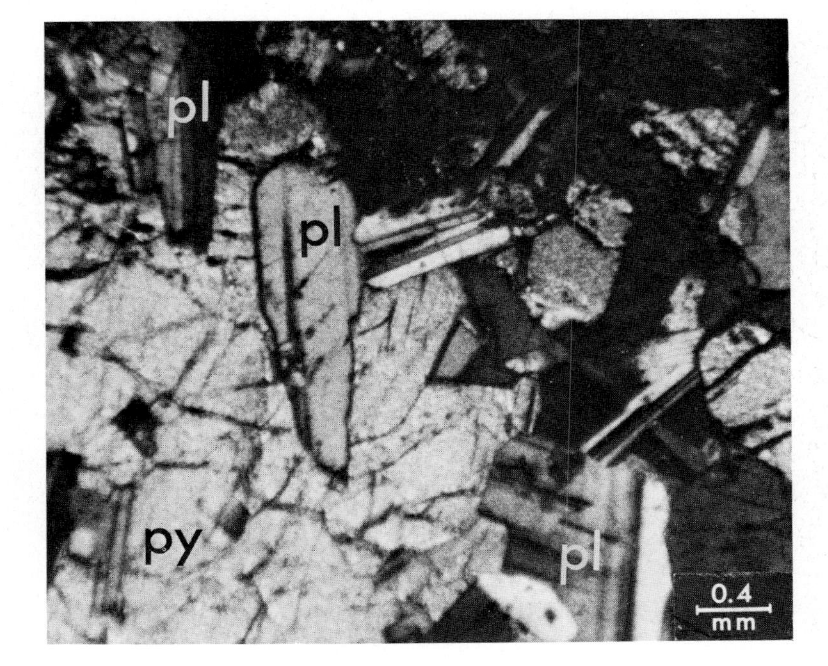

Figs. 386 and 387. Ophitic plagioclase–pyroxene intergrowth, where the feldspar attains a graphic character. Arrow "a" shows the plagioclase invading the pyroxene and arrow "b" an alteration of the feldspar, usually following the contact with the pyroxene. Pl = plagioclase; py = pyroxene.
Dolerite (coarse-grained basalt). Karroo series, S. Africa.

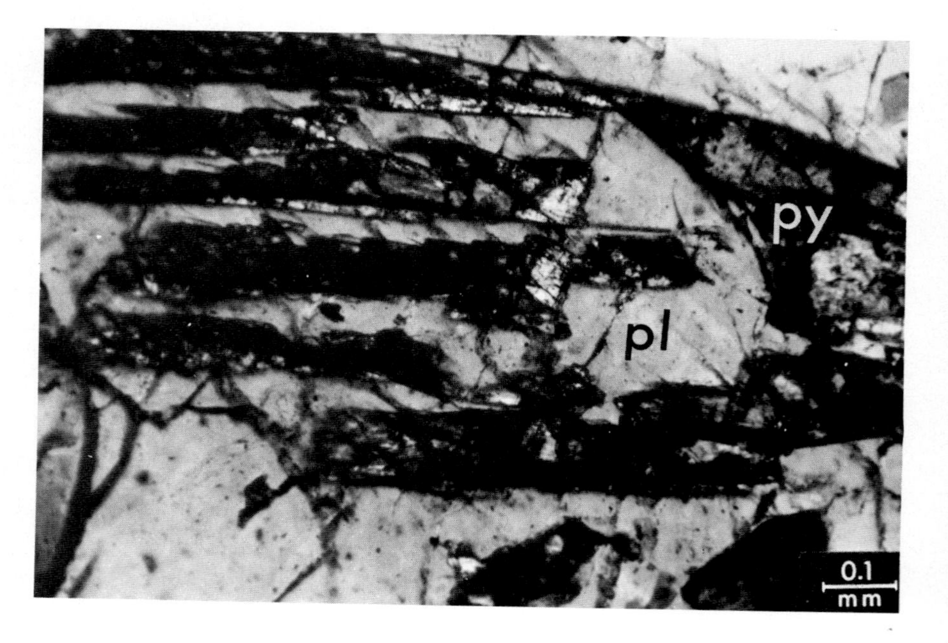

Fig. 388. Pyroxene-feldspar ophitic intergrowth attaining a symplectic graphic-like character. Pl = plagioclase; py = pyroxene.
Dolerite (coarse-grained basalt). Karroo series, S. Africa.

Fig. 389. Ophitic plagioclase–pyroxene intergrowth. The plagioclase is changed into an aggregate of green chlorites. py = pyroxene; ch = initial plagioclase altered into an aggregate of chlorite.
Dolerite (coarse-grained basalt). Karroo series, S. Africa. With crossed nicols.

Fig. 390. This is one of the most well-known textural patterns from the moon (Sea of Tranquility) which shows a gabbroic ophitic intergrowth of pyroxene and plagioclase laths. (Photograph by: NASA, S-69-47907.)

Fig. 391. Plagioclase lath invading the pyroxene along a crack system. This intergrowth pattern is understandable if it is compared with Figs. 230 and 231. Pl = plagioclase; py = pyroxene.
Basalt. Ethiopian Plateau. With crossed nicols.

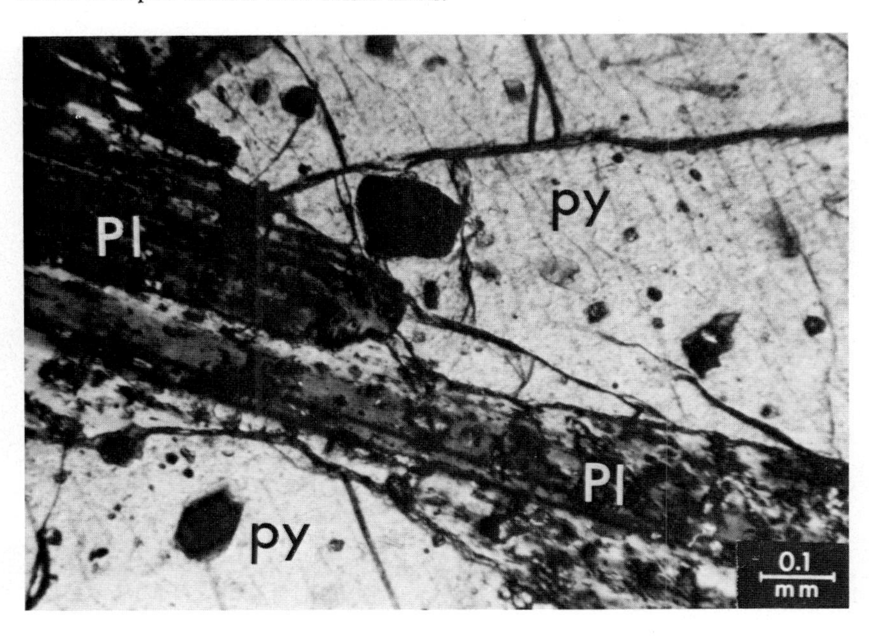

Fig. 392. An agglutination of radiating prismatic pyroxene crystals in a background of basaltic groundmass. G = basaltic groundmass; A = agglutination of radiating prismatic pyroxene microcrystals.
Basalt. Steinl, Kleinbahn, Bohemia, Czechoslovakia. With crossed nicols.

Fig. 393. An agglutination of radiating prismatic pyroxene crystals resembling in their orientation a lining of the walls of a micro-cavity.
Basalt. Steinl, Kleinbahn, Bohemia, Czechoslovakia. With crossed nicols.

Fig. 394. Cummulate of zoned, augitic middle-sized phenocrysts with inter-augitic (intercrystalline) spaces occupied by basaltic groundmass. Au = augites; G = basaltic groundmass; i-g = intercrystalline (inter-augitic) groundmass.
Ethiopian Rift. With crossed nicols.

Fig. 395. Agglutination, crystal cummulates of olivine crystals as an early crystallization phase of a melt crystallization. However, the question arises between early crystallization and xenoliths. No concrete and definite textural criteria could be put forward to differentiate between early crystallization agglutinations and xenolithic bombs. However, the idiomorphic olivine forms exhibited could support an early crystallization hypothesis for the illustrated olivine crystal agglutinations. ic = idiomorphic olivine crystals; G = basaltic groundmass. Basalt. Casseler Grund, Spessart, Germany. With crossed nicols.

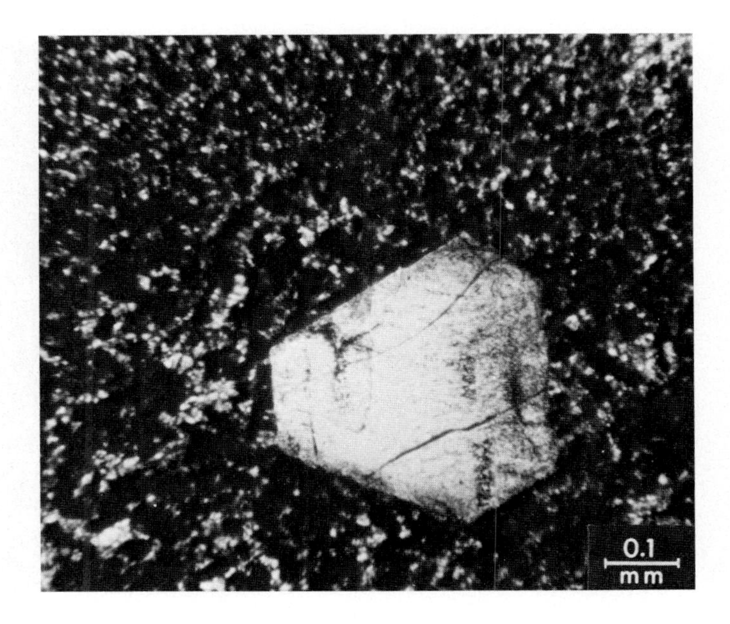

Fig. 396. An idiomorphic "hexagonal" basal section of apatite in a predominantly magnetite-rich basaltic groundmass.
Basalt columns. Minderberg near Linz on Rhine River, Germany. Without crossed nicols.

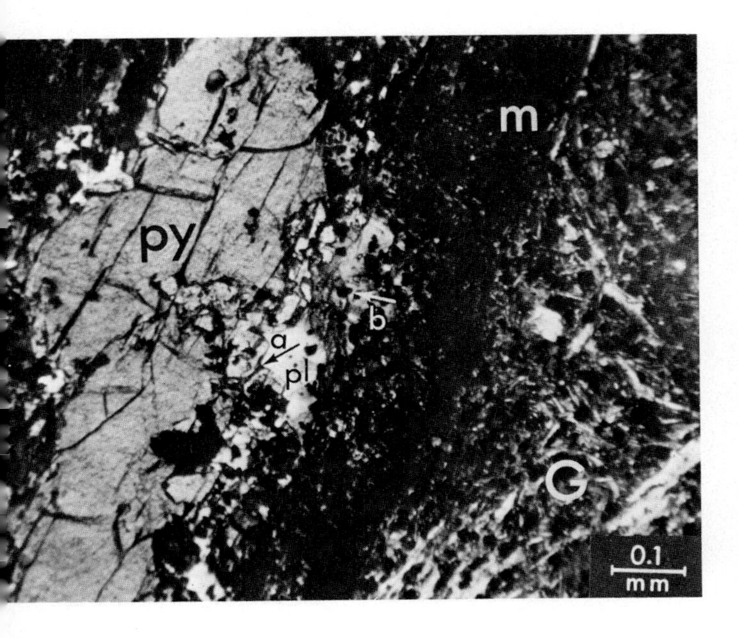

Fig. 397. Pyroxene phenocrysts with a halo of exceptionally magnetite-rich groundmass. Py = pyroxene phenocryst; m = magnetite-rich basaltic groundmass, actually representing magnetite agglutinations; G = basaltic groundmass. Arrow "a" shows groundmass plagioclase corroding the pyroxene phenocryst. Similarly, the same plagioclase, arrow "b" engulfs the components of the magnetite halo that surrounds the pyroxene. pl = groundmass plagioclase.
Basalt columns. Minderberg near Linz on Rhine River, Germany. With crossed nicols.

Fig. 398. Elongated pyroxene with intracrystalline agglutinations of magnetites. The groundmass external to the phenocryst is again rich in magnetites.
Basalt columns. Minderberg near Linz on Rhine River, Germany. With crossed nicols.

Fig. 399. Basalt consisting of fine radiating laths of feldspar with transition to more coarse-grained feldspar laths again showing a radiating arrangement. "a" = fine-grained plagioclase laths; "b" = corase-grained plagioclase laths.
"Spinifex basalt". Mid-Atlantic Ridge. With crossed nicols.

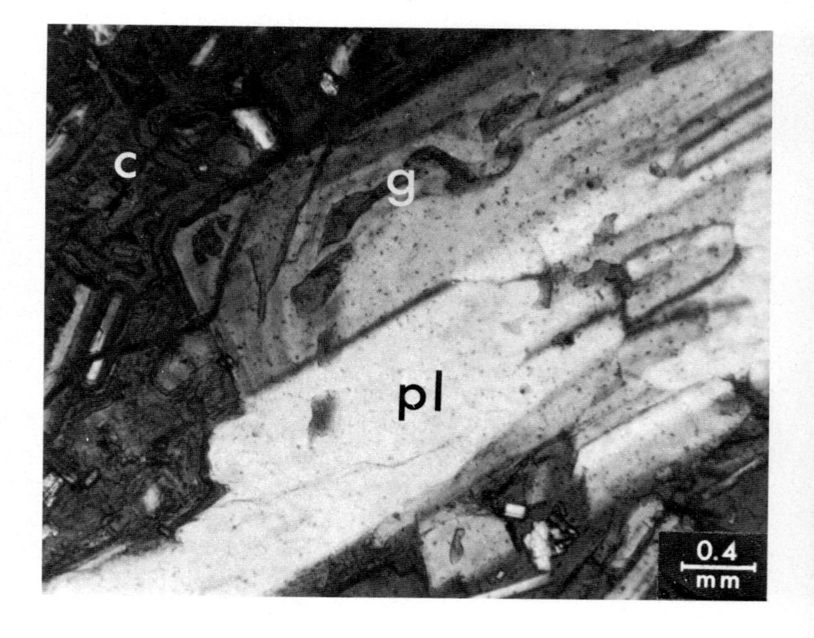

Fig. 400. Plagioclase phenocryst with a glassy phase, zonally within the feldspar, enclosing subphenocrystalline feldspar laths. g = glassy phase zonally within the feldspar phenocryst, pl = plagioclase phenocryst, c = colloform-like glassy mass, gel hyaloids outside the feldspar phase and enclosing plagioclase laths.
Hyalobasalt. Ustica, Sicily, Italy. Half-crossed nicols.

Fig. 401. A glassy groundmass with feldspar crystals and with a spheroidal structure which shows gel hyaloids. f = feldspar crystals; s = spheroid in the glassy groundmass; gh = gel hyaloids in the spheroid. Hyalobasalt. Ustica, Sicily, Italy. Without crossed nicols.

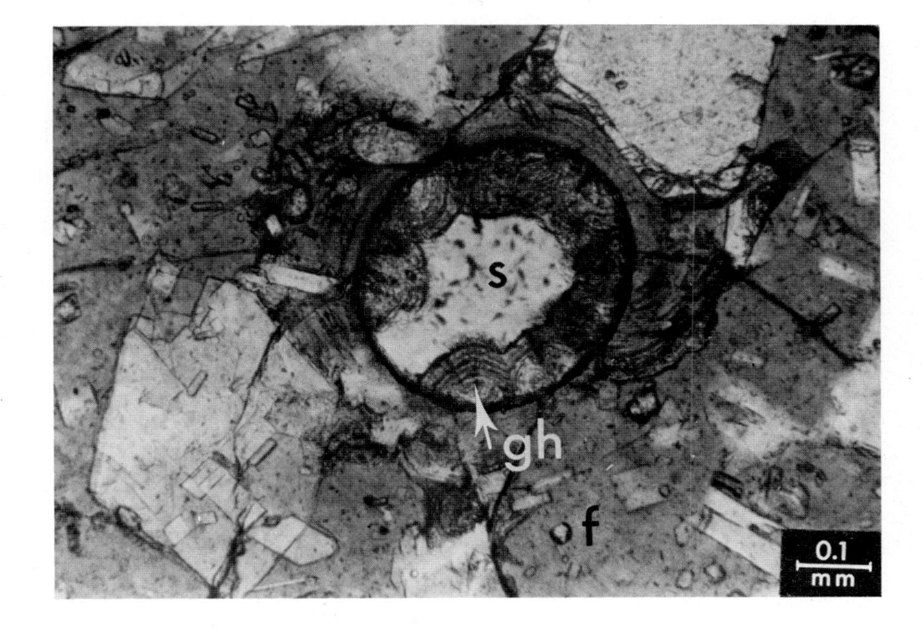

Fig. 402. "Undifferentiated" basaltic groundmass with prismatic pyroxene microcrysts and magnetite (black). py = pyroxene; b-g = undifferentiated basaltic groundmass.
Ethiopian Plateau. With crossed nicols.

Fig. 403. Zoned and twinned augite, invaded by intracrystalline diffusion of groundmass following penetrability directions within the augite phenocrysts. Arrows "a" show intracrystalline groundmass infiltrations following penetrability as defined by the twin planes. Arrow "b" also shows groundmass following crack systems of the augite phenocrysts and transversing the pyroxene.
Augite-rich basalt. Selale Mt. region, Ethiopian Plateau. With crossed nicols.

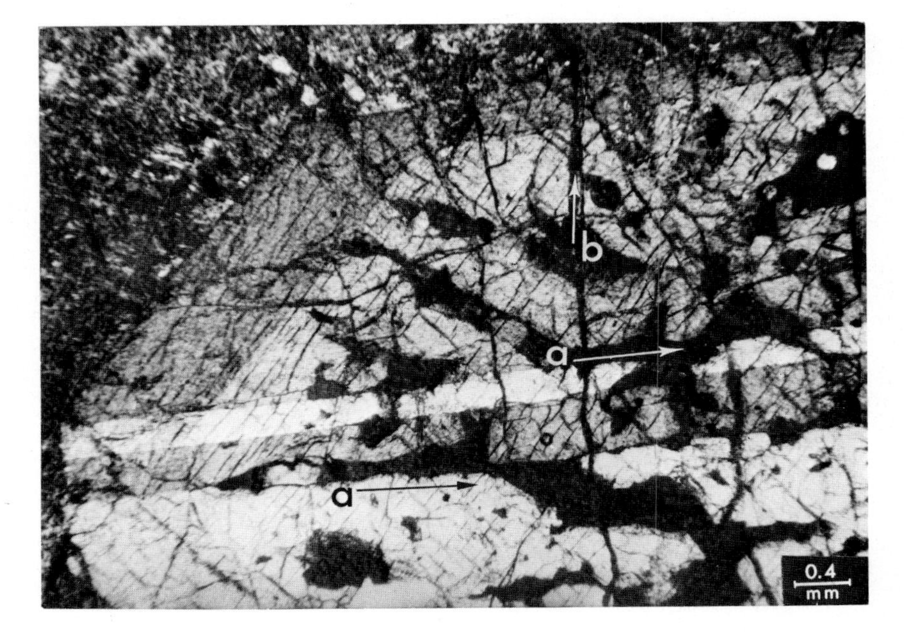

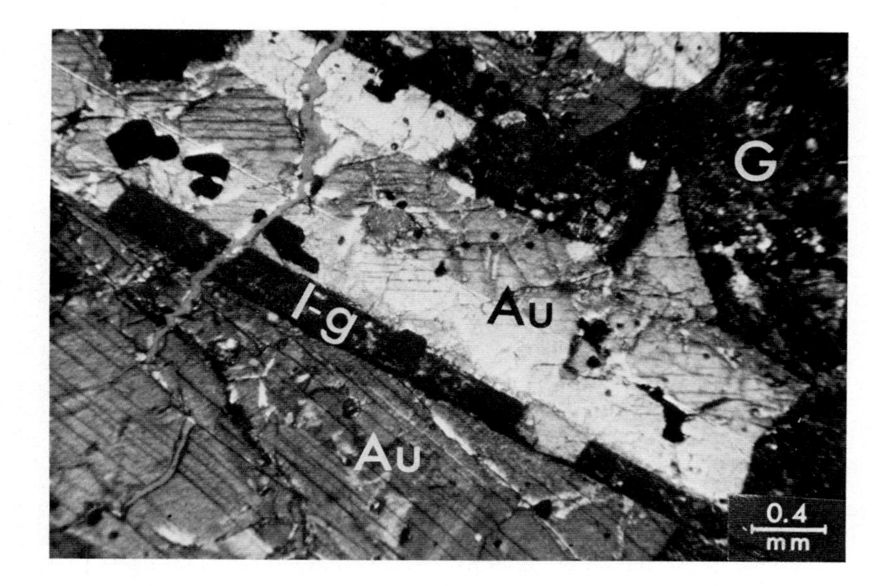

Fig. 404. Twinned augite with basaltic groundmass extending and occupying (inter-twinning, inter-lamellar) space. Au = twinned augite; G = basaltic groundmass; I-g = basaltic groundmass occupying inter-lamellar space.
Augite-rich basalt. Selale Mt. region, Ethiopian Plateau. With crossed nicols.

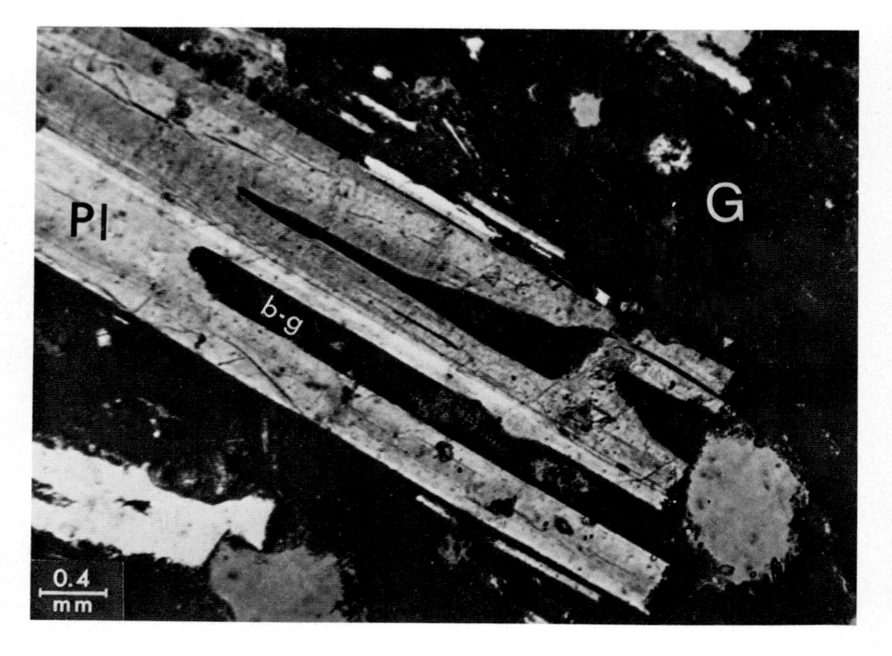

Fig. 405. Plagioclase phenocryst with basaltic groundmass extending and occupying a part that was originally occupied by a plagioclase twin lamella. Pl = plagioclase; G = basaltic groundmass; b-g = basaltic groundmass following and extending into the plagioclase twinning.
Augite-rich basalt. Selale Mt. region, Ethiopian Plateau. With crossed nicols.

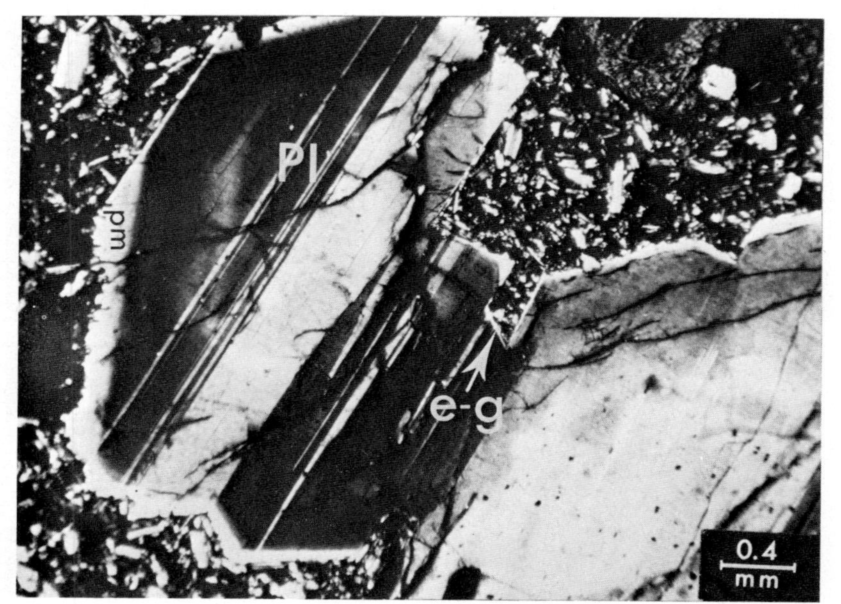

Fig. 406. Plagioclase phenocryst showing a reaction margin with the groundmass and with groundmass as extension protruding along the plagioclase twinning. Pl = plagioclase phenocryst; pm = plagioclase reaction margins; e-g = extensions of groundmass along the plagioclase twinning.
Olivine basalt. Adama (Nazaret), Ethiopia. With crossed nicols.

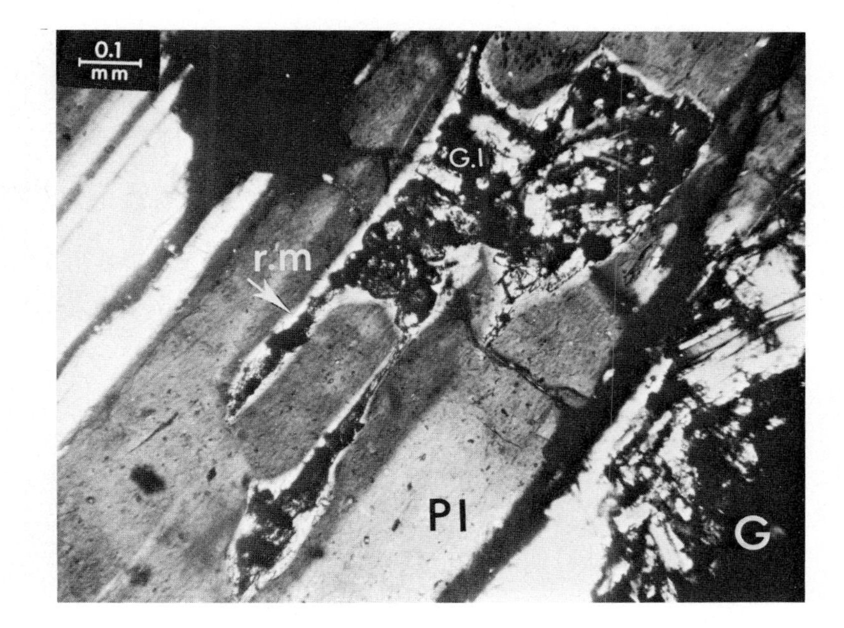

Fig. 407. Plagioclase phenocryst invaded by groundmass; intracrystalline infiltrations of the groundmass often causing a reaction margin on the feldspar. Pl = plagioclase phenocryst; G = basaltic groundmass; G.I = intracrystalline groundmass infiltrations; r.m = reaction margin, caused by the groundmass on the feldspar phenocryst.
Olivine basalt. Adama (Nazaret), Ethiopia. With crossed nicols.

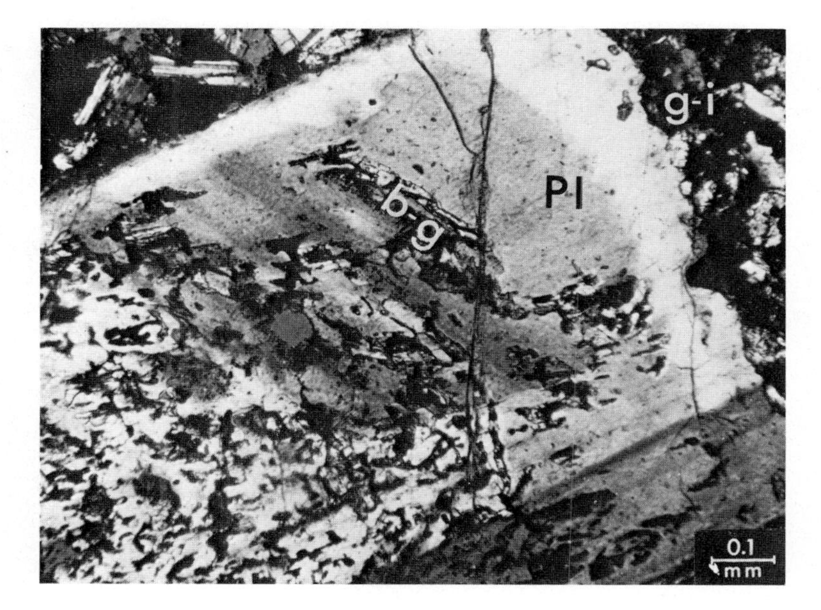

Fig. 408. Zoned plagioclase phenocryst invaded by groundmass, partly following the feldspar zones. Pl = plagioclase; g-i = intracrystalline groundmass infiltration within the feldspar phenocryst; b-g = basaltic groundmass.
Olivine basalt. Adama (Nazaret), Ethiopia. With crossed nicols.

Fig. 409. Intracrystalline groundmass extensions following the cleavage and the crack system within the phenocrystalline augite. Arrows "a" show that the intracrystalline groundmass extends into the augite. Arrows "b" show intracrystalline groundmass extensions following cleavage directions and cracks.
Augite-rich basalt. Selale Mt. region, Ethiopian Plateau. With crossed nicols.

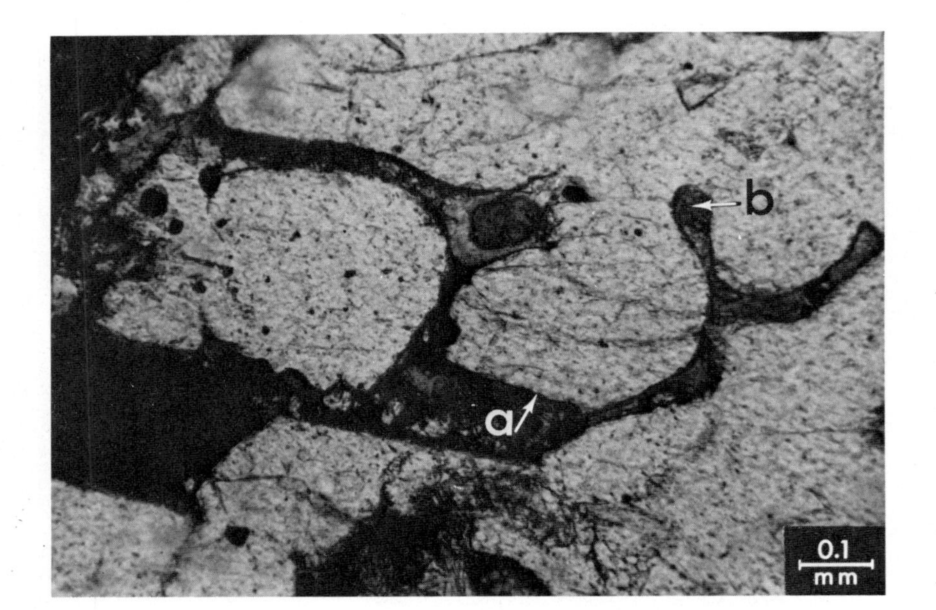

Fig. 410. Basaltic groundmass extending from the exterior of the augite phenocryst into the pyroxene following cleavage penetrability directions and attained myrmekitoid pattern and character. Arrow "a" shows intracrystalline groundmass delimited by cleavage plane. Arrow "b" shows myrmekitic-like pattern of the intracrystalline groundmass.
Basalt. Eisenberg, Bohemia, Czechoslovakia. With crossed nicols.

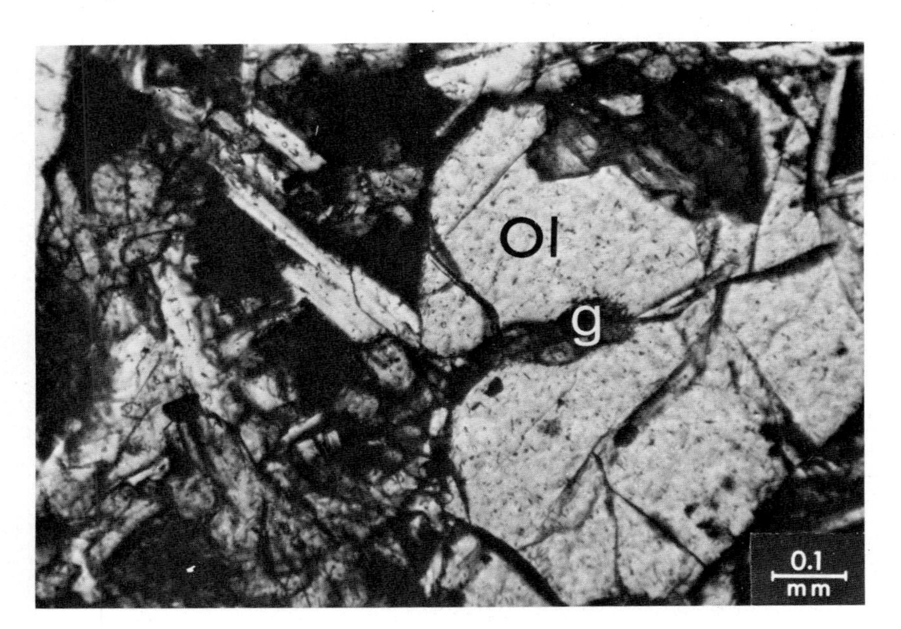

Fig. 411. Olivine phenocryst invaded by groundmass extension. Ol = olivine phenocryst; g = groundmass extension invading the olivine phenocrysts.
Doleritic basalt. Marienberg, Westerwald, Germany. With crossed nicols.

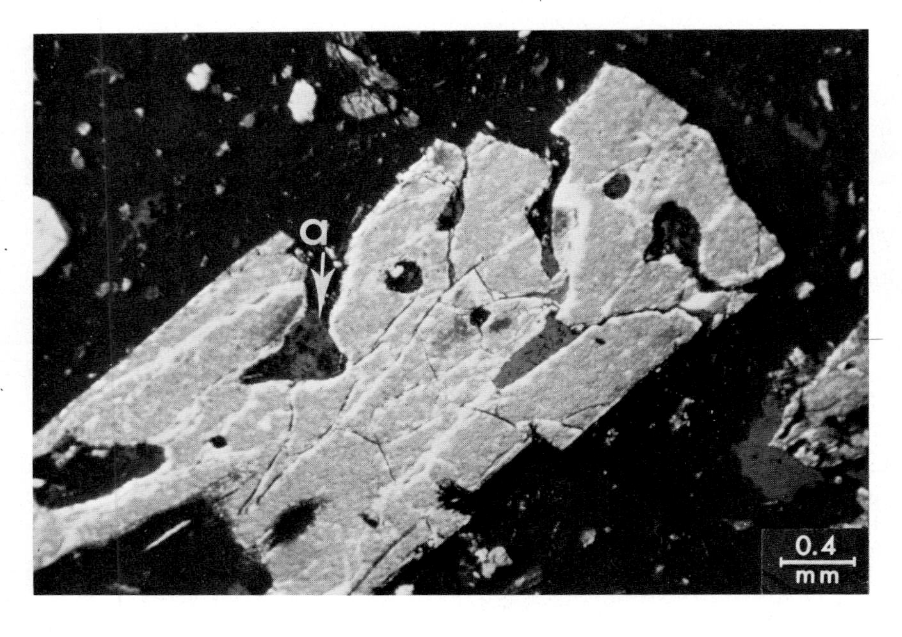

Fig. 412. Olivine phenocryst with the groundmass, corroding and as intracrystalline extensions within the olivine. Arrow "a" shows groundmass extensions corroding and invading the phenocryst.
Basalt. Eisenberg, Bohemia, Czechoslovakia. With crossed nicols.

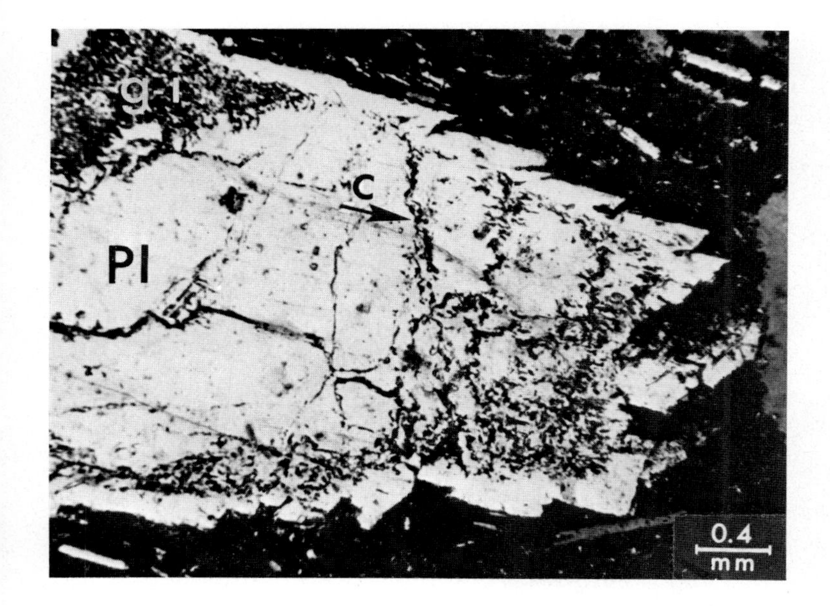

Fig. 413. Plagioclase phenocryst with cracks, some of which are invaded by intracrystalline infiltration of basaltic melts along the phenocryst crack system. Pl = plagioclase phenocryst; c = crack system in the plagioclase; g-i = groundmass infiltrations within the plagioclase phenocryst.
Olivine basalt. Adama (Nazaret) Ethiopia. With crossed nicols.

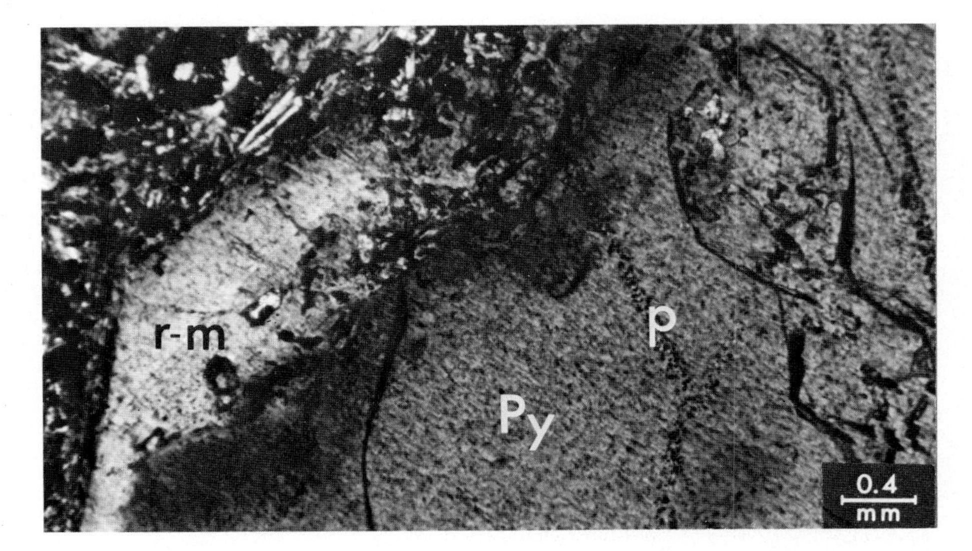

Fig. 414. Pyroxene (component of an olivine–pyroxene bomb in basalt) in contact with basaltic groundmass. Groundmass extensions in the pyroxene and reaction margin of the pyroxene with the groundmass are illustrated. Intracrystalline penetration (infiltration) paths are also present in the pyroxene. Py = pyroxene (component of an olivine pyroxene bomb in basalt); r-m = reaction margin of pyroxene with the groundmass; p = intracrystalline extensions of basaltic melts within the pyroxene phenocryst.
Basalt. Las Planas, Gerona Prov., Spain. With crossed nicols.

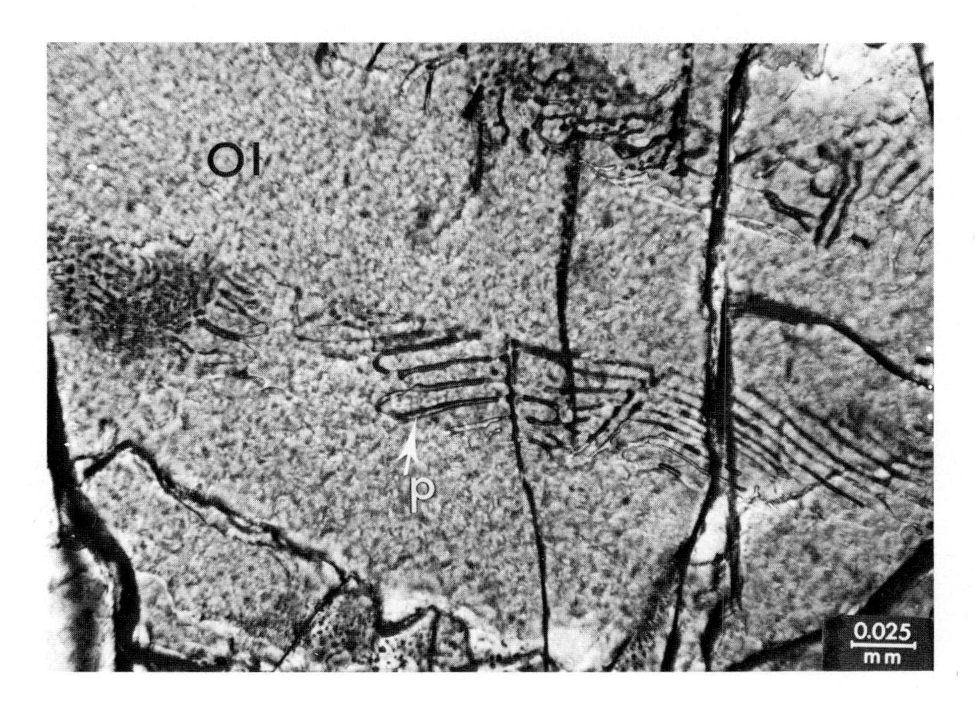

Fig. 415. Intracrystalline infiltration of basaltic melts following paths within the host olivine and exhibiting myrmekitic-like patterns. Ol = olivine host; p = paths of intracrystalline infiltration of melts within the olivine.
"Olivine bomb" in basalt. Jato, Lekempti, W. Ethiopia. With crossed nicols.

Fig. 416. Intracrystalline infiltration of basaltic melts following prevalent intracrystalline directions of penetrability. As arrow "a" and arrow "b" show, the myrmekitoid bodies show a tendency to be arranged perpendicular to each other, or the same "myrmekitoid" individual attains the shape of a right angle (see arrow "c").
Olivine bomb in basalt. Eifel, Germany. With crossed nicols.

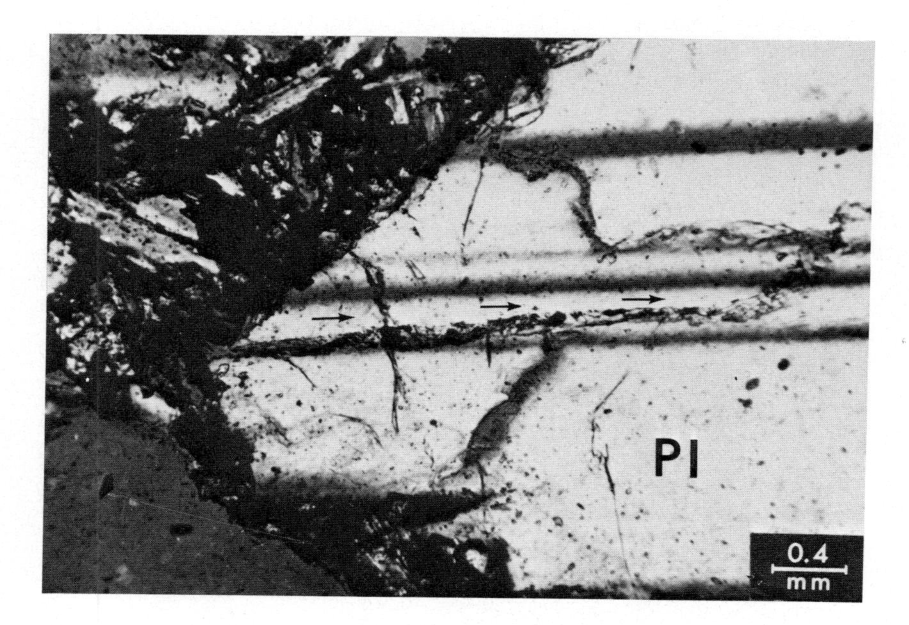

Fig. 417. Intracrystalline infiltration of basaltic melts parallel and crossing the plagioclase host twinning. Pl = plagioclase. Arrows show the path of the basaltic melt within the host plagioclase.
Olivine basalt. Adama (Nazaret) Ethiopia. With crossed nicols.

Fig. 418. Plagioclase phenocryst with melt path of intracrystalline infiltration of basaltic melt. The feldspar and mafic components which crystallize out of the melt-path infiltration show resorption and reaction with the host feldspar.
Olivine basalt. Adama (Nazaret) Ethiopia. With crossed nicols.

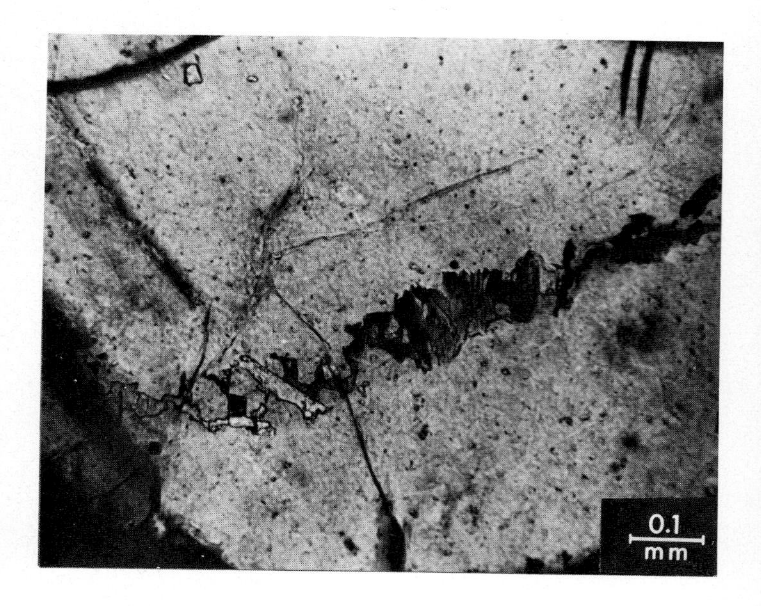

Fig. 431. Plagioclase phenocryst with an external zone of growth which encloses and partly engulfs components of the groundmass. The interrupted line shows the crystal face which internally delimits the external plagioclase zone. pl = plagioclase; e-z = external palgioclase zone. Arrows "a" show components of the groundmass enclosed or partly so, within the external plagioclase zone.
Amygdaloidal basalt. Tramkampstadur, Iceland. With crossed nicols.

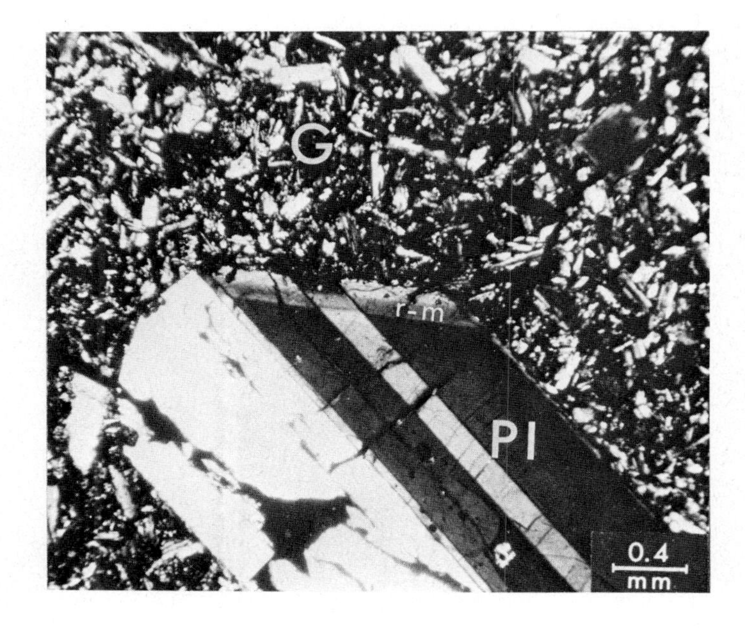

Fig. 432. Plagioclase phenocryst, with reaction margin (attaining a zonal character) in contact with the basaltic groundmass. Pl = plagioclase phenocryst; r-m = reaction margin (in which two zones are noticeable); G = basaltic groundmass.
Amygdaloidal basalt. Tramkampstadur, Iceland. With crossed nicols.

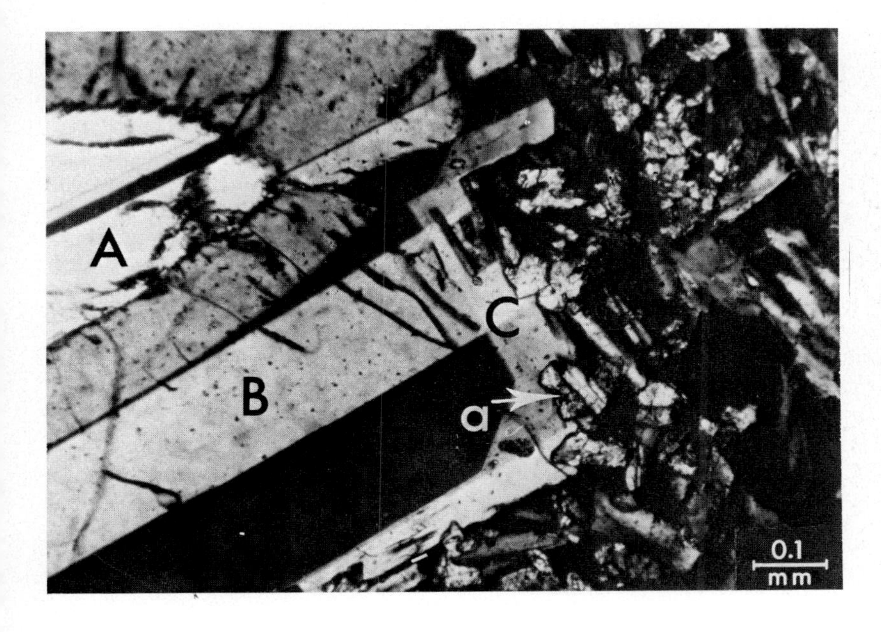

Fig. 433. Plagioclase phenocryst composed of three stages of formation. A = central plagioclase crystal nucleus; B = twinned plagioclase with a definite crystalline outline and C = outer margin of plagioclase which actually corresponds to an external zone in which the groundmass is partly enclosed and partly engulfed. Arrow "a" shows groundmass components enclosed by the external zone of the plagioclase.
Amygdaloidal basalt, Tramkampstadur, Iceland. With crossed nicols.

Fig. 434. Zoned augite phenocryst with ore minerals of the groundmass (magnetites) enclosed in the central part of the pyroxene or interzonally incorporated. Often the enclosed magnetites are accompanied by a reaction rim within the augite. The external augitic zone partly encloses, partly engulfs the external groundmass magnetites. A = zoned augite; e-z = external augitic zone. Magnetites (black). Surrounding magnetites enclosed in the central zone of the augite; there is a reaction rim as indicated by arrow "a".
Augite-rich basalt. Selale Mt. region, Ethiopian Plateau. With crossed nicols.

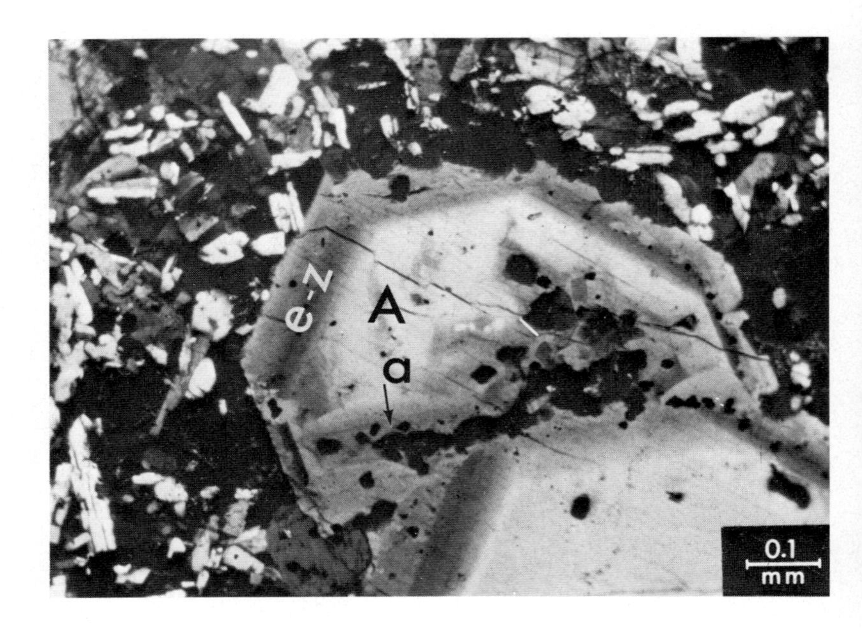

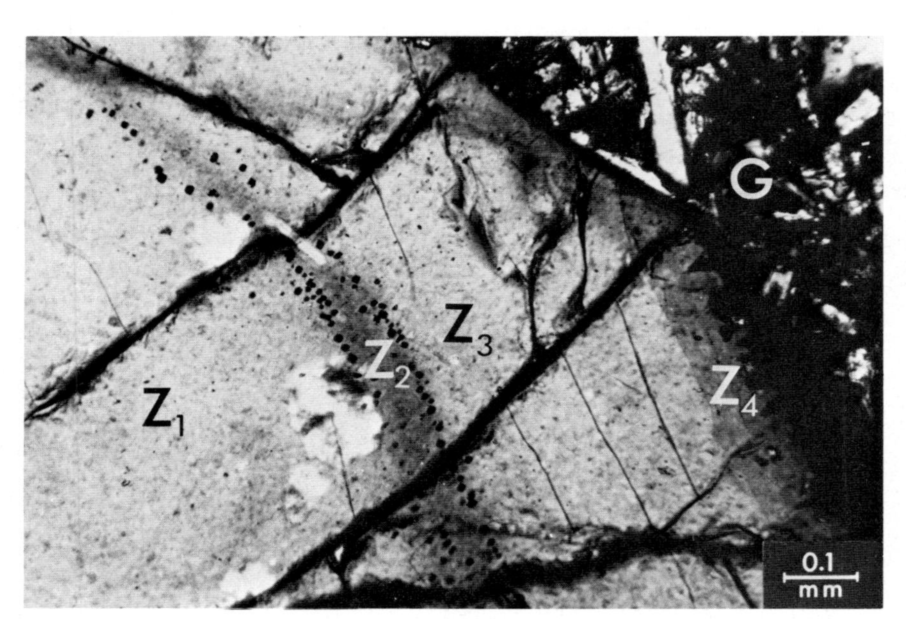

Fig. 435. Zoned augite (Ti-augite) with an internal zone delimited with magnetite granules. Z_1,Z_2,Z_3,Z_4 = zones of the idiomorphic augite. Zone Z_2 is delimited by magnetite granules; G = basaltic groundmass.
Augite-rich basalt. Selale Mt. region, Ethiopian Plateau. With crossed nicols.

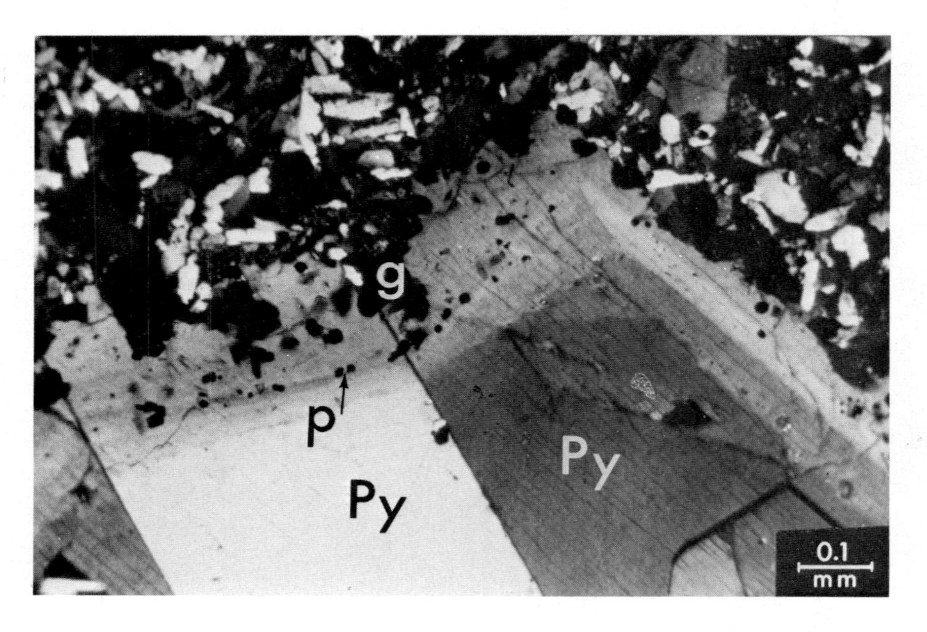

Fig. 436. Augite phenocryst, twinned and zoned with the external zone engulfing and partly enclosing groundmass components. Also ore-pigments interzonally within the pyroxene. Py = pyroxene phenocryst; g = basaltic groundmass enclosed and partly engulfed by external pyroxene zone; p = ore-pigments interzonally arranged within the pyroxene.
Augite-rich basalt. Selale Mt. region, Ethiopian Plateau. With crossed nicols.

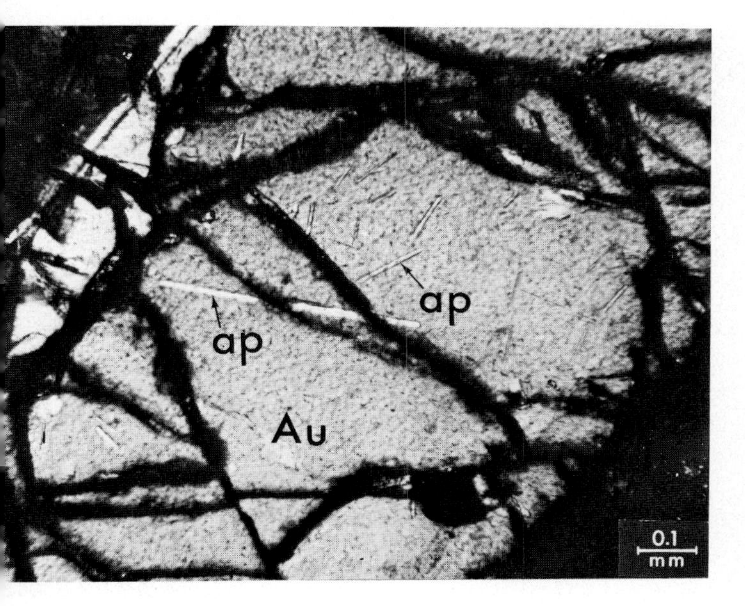

Fig. 437. Fine apatite needles orientated at random within a zoned augite. (The zoning is indicated under Universal stage examination of the augite). Au = augite; ap = apatite needles. Augite-rich basalt, Selale Mt. region, Ethiopian Plateau. With crossed nicols.

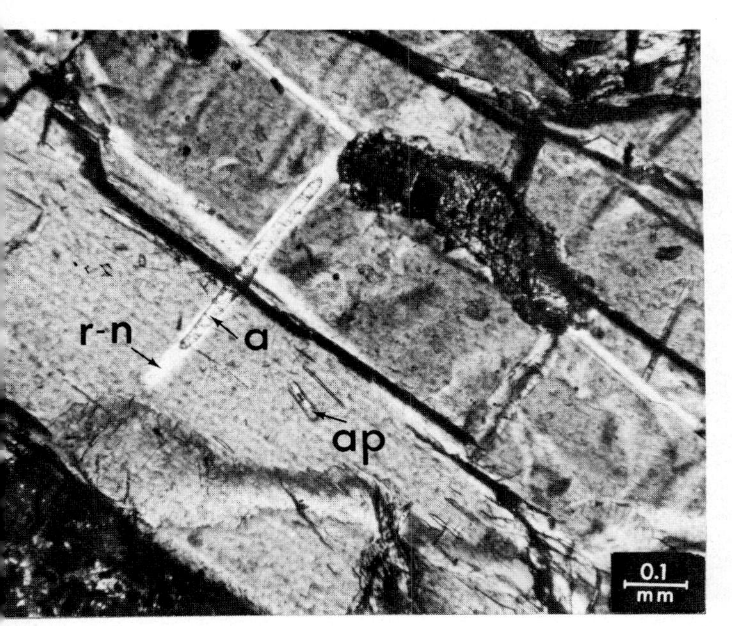

Fig. 438. Fine-zoned augite (Ti-augite) with needles and elongated prismatic forms of rutile. Arrow "a" shows an elongated rutile parallel to the zoning and with a "decoloration" halo of the augite around the rutile; r-n = rutile needles; ap = apatite.
◀ Augite-rich basalt. Selale Mt. region, Ethiopian Plateau. With crossed nicols.

Fig. 439. Groundmass feldspar (g-f) included in an idiomorphic subphenocrystalline plagioclase (Pl). The size and shape of the plagioclase lath enclosed in the plagioclase is comparable to the groundmass plagioclase laths in the groundmass. Basalt. Norrsand, Nordulfön, Sweden. With crossed nicols.

Fig. 440. Idiomorphic plagioclase phenocryst in basaltic groundmass. The internal plagioclase zone is marginally "infiltrated" along cleavage direction by magmatic melts; the resultant crystal granules clearly follow the cleavage pattern (arrow "a"). The outline of the internal zone suggests corrosion. The outer (external) zone of the plagioclase is more free of "infiltration" and results in an idiomorphic crystalline outline of the phenocryst.
Basalt. Schemnitz, Tatra, C.S.S.R. With crossed nicols.

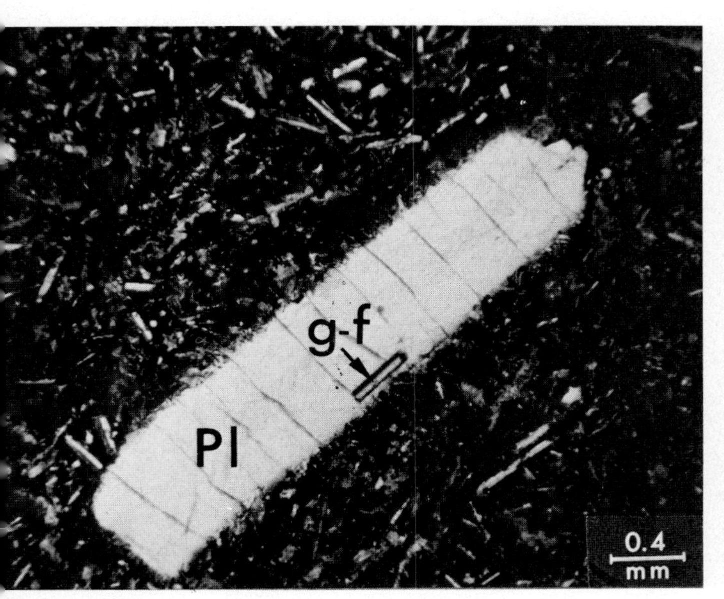

Fig. 439.

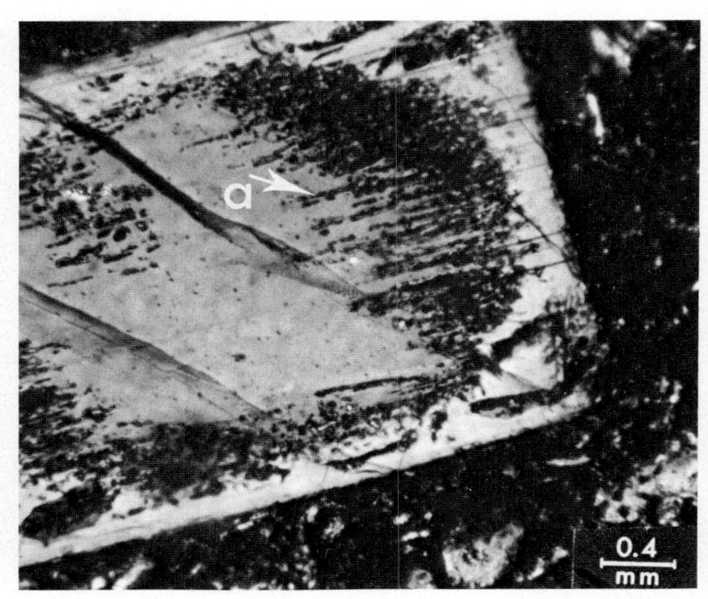

Fig. 440.

Fig. 441. Plagioclase phenocrysts with an internal zone (suggesting a corroded outline) and infiltrated by magmatic melts along possible cleavage directions. Pockets of "infiltrations" are also noticeable in the external plagioclase zone (see arrow "a") which results in an idiomorphic crystal outline.
Basalt. Schemnitz, Tatra, Czechoslovakia. With crossed nicols.

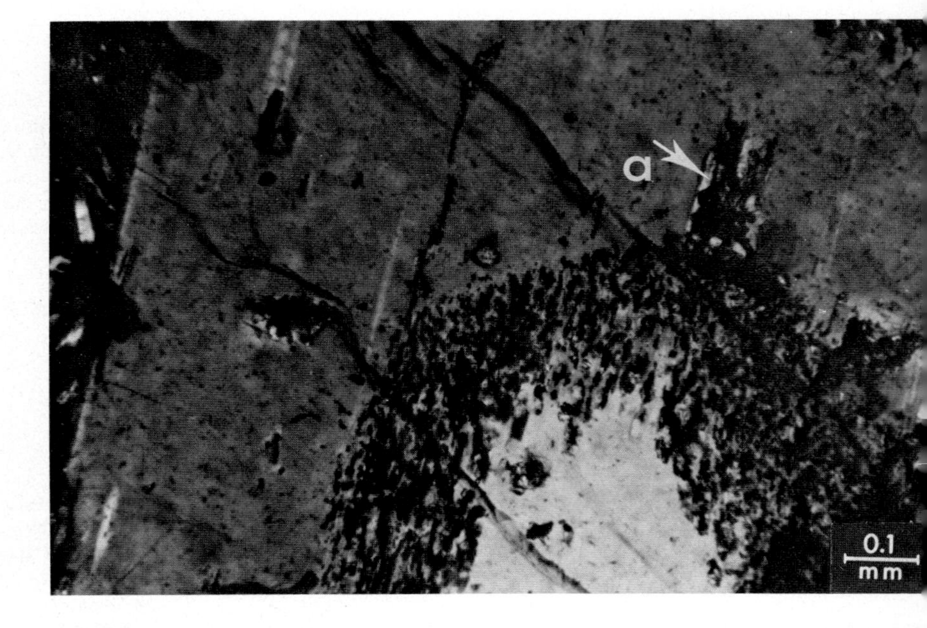

Fig. 442. Zoned plagioclase phenocryst in basaltic groundmass, with groundmass components enclosed within the internal plagioclase zone. Pl-I = plagioclase internal zone with groundmass included, Pl II = plagioclase external zone.
Basalt. Near Addis Ababa, Ethiopia. With crossed nicols.

Fig. 443. Zoned plagioclase with "turbid" internal zone with groundmass inclusions reduced in size. The outer zone is grown on the "corroded" outline of the internal zone and is almost free of groundmass inclusions and externally attains a sub-idiomorphic shape. Pl = internal turbid plagioclase zone with groundmass inclusions reduced in size; E.Pl = external plagioclase zone, free of inclusions.
Basalt. Near Addis-Ababa, Ethiopia. With crossed nicols.

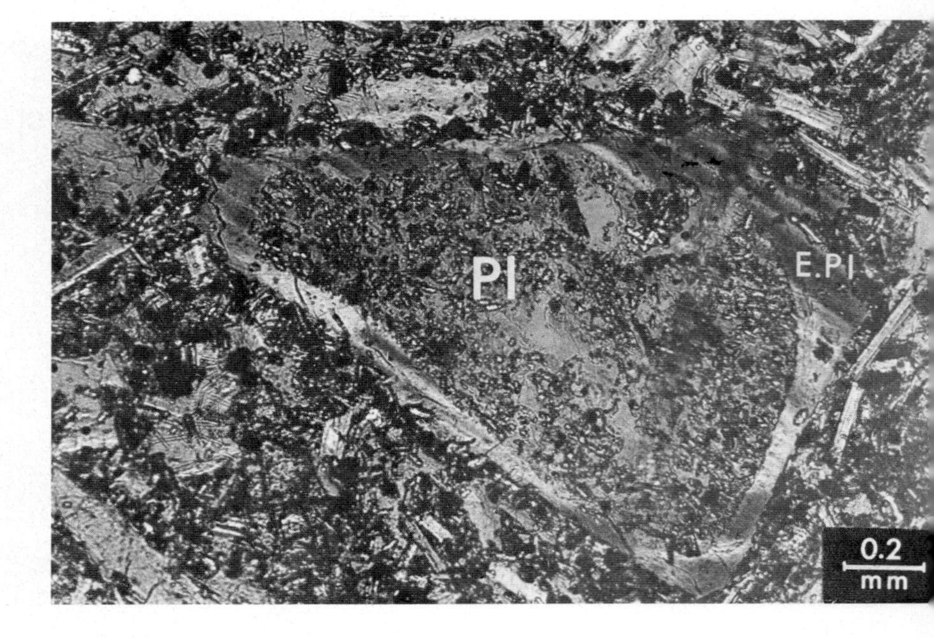

Figs. 444 and 445. Olivine with apatites, without a definite orientation within the olivine host. Ol = olivine; Ap = apatite; b-g = basaltic groundmass.
Olivine-basalt porphyry. Boulder County, Colorado, U.S.A. With crossed nicols.

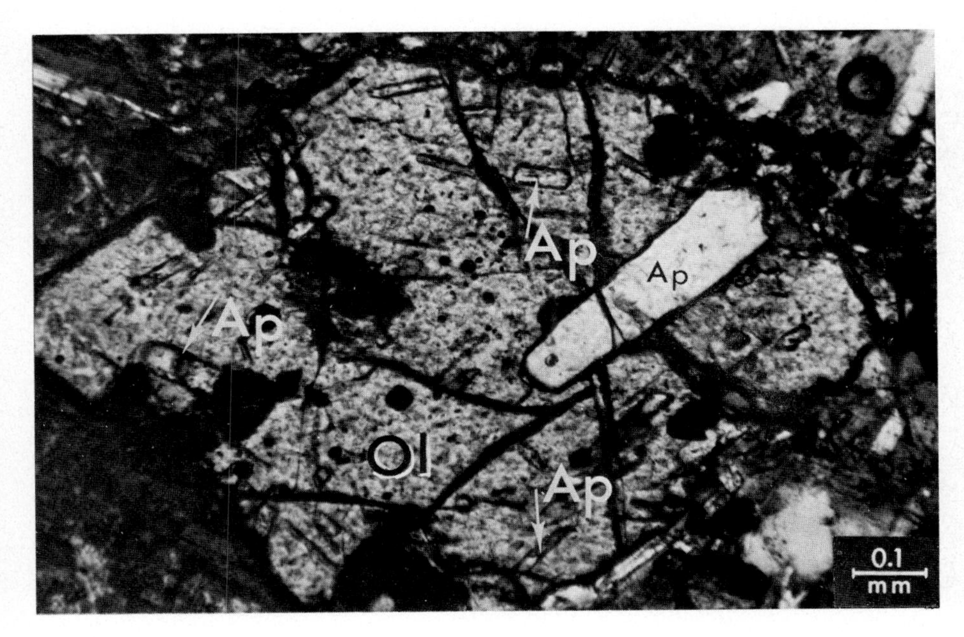

Fig. 446. Olivine crystalloblast (idioblast?) with apatites included. The smaller apatites are zonally arranged within the olivine host. Ol = olivine; Ap = apatite.
Olivine-basalt porphyry. Boulder County, Colorado, U.S.A. With crossed nicols.

Fig. 447. Oscillatory-zoned titano-augite with prismatic apatite parallel to the zoning. A = augite; Ap = prismatic apatite.
Augite-rich basalt. Selale Mt. region, Ethiopian Plateau. With crossed nicols.

Fig. 448. Basaltic hornblende with randomly orientated accessory apatite. H = hornblende; ap = apatite. Olivine-basalt porphyry. Boulder County, Colorado, U.S.A. With crossed nicols.

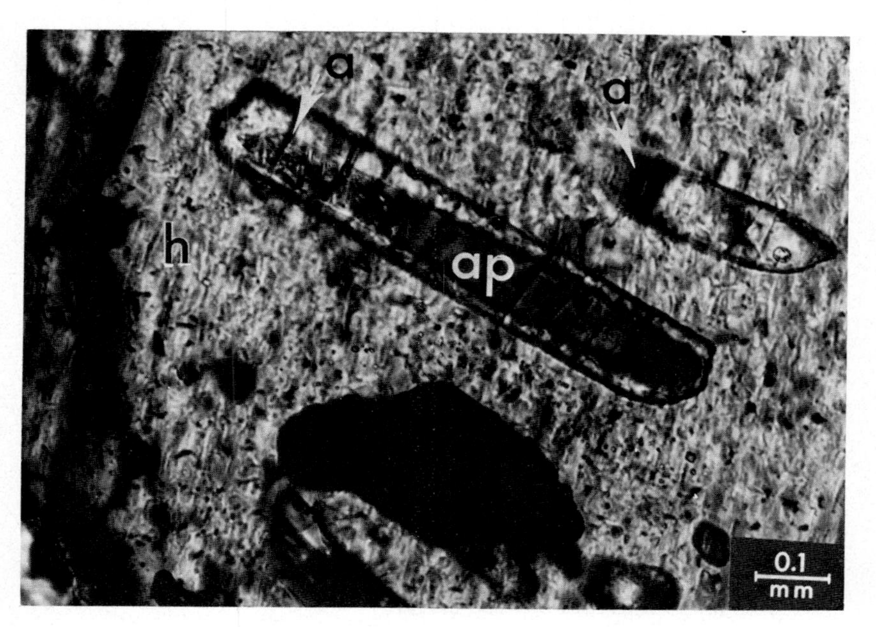

Fig. 449. Hornblende with partly corroded and partly assimilated apatite. h = hornblende; ap = apatite; arrows "a" = apatite corroded by the hornblende crystalloblast.
Olivine-basalt porphyry. Boulder County, Colorado, U.S.A. With crossed nicols.

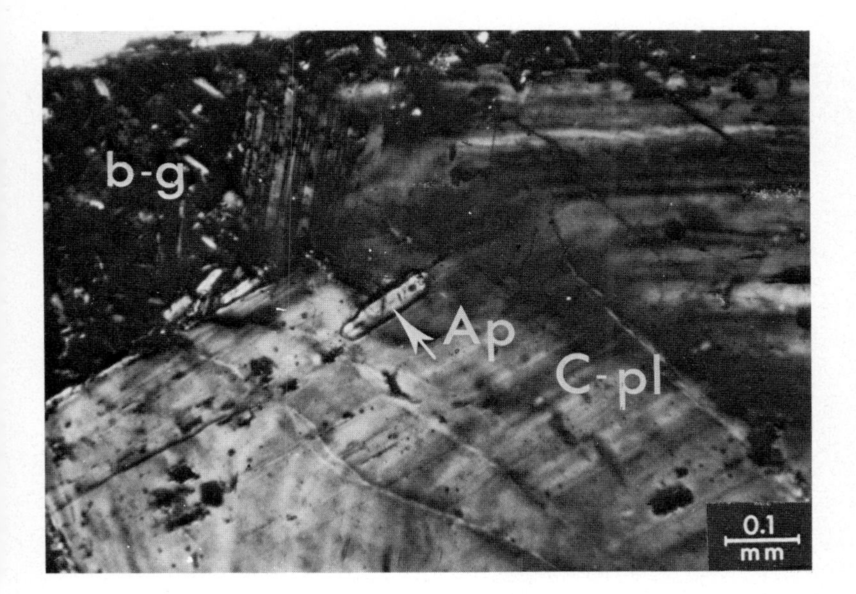

Fig. 450. Plagioclase collocryst with apatite following the collocryst zoning which attains a crystal zoning character. C-pl = plagioclase collocryst; Ap = apatite; b-g = basaltic groundmass. Vesicular basalt. Chaffée County, Colorado, U.S.A. With crossed nicols.

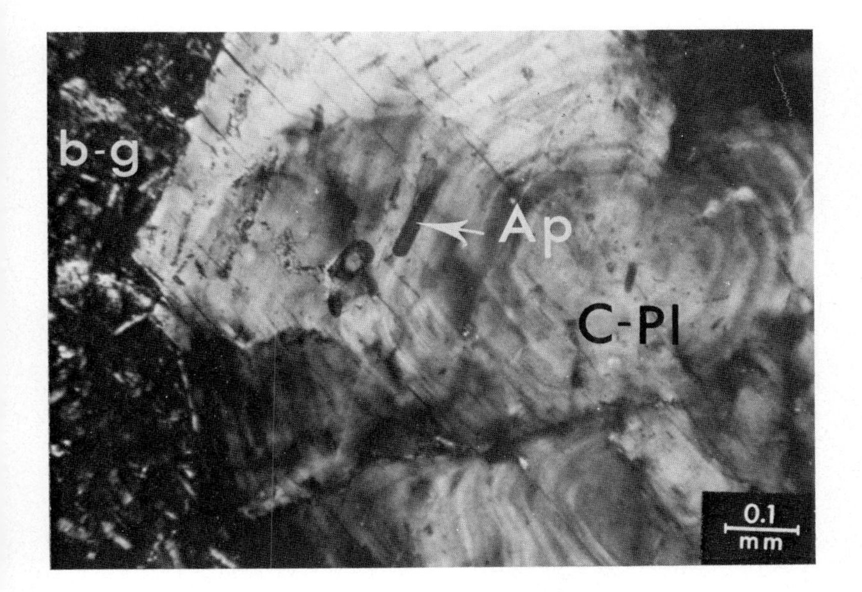

Fig. 451. Plagioclase collocryst with prismatic apatite following the collocryst zonal banding. C-Pl = plagioclase collocryst; Ap = prismatic apatite; b-g = basaltic groundmass. Vesicular basalt. Chaffée County, Colorado, U.S.A. With crossed nicols.

Fig. 452. Idiomorphic to sub-idiomorphic magnetites, prismatic pyroxenes and needles of apatite surrounded by basaltic groundmass feldspars. The groundmass feldspars (plagioclases) do not show the typical prismatic habitus. black = idiomorphic to sub-idiomorphic magnetite; p = groundmass pyroxene; n = needles of apatite; f = groundmass plagioclase.
Olivine-basalt. Adama (Nazaret) Ethiopia. With crossed nicols.

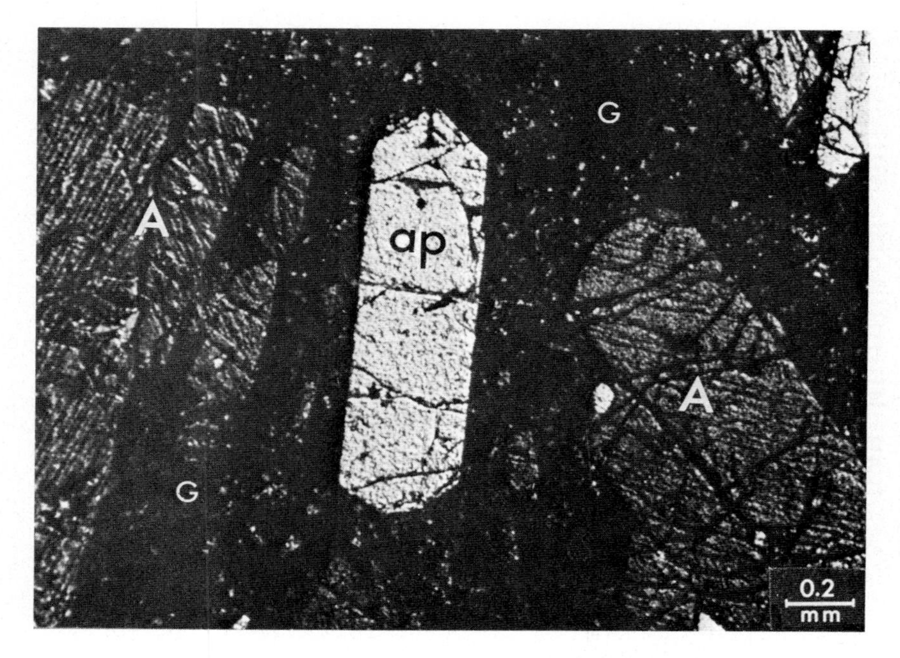

Fig. 453. Basalt with often twinned augite phenocrysts and with idiomorphic apatite sub-phenocrystalline in size. A = augite phenocrysts; ap = apatite, sub-phenocrystalline in size; G = basaltic groundmass.
Basalt. Selale Mt. region, Ethiopian Plateau. With crossed nicols.

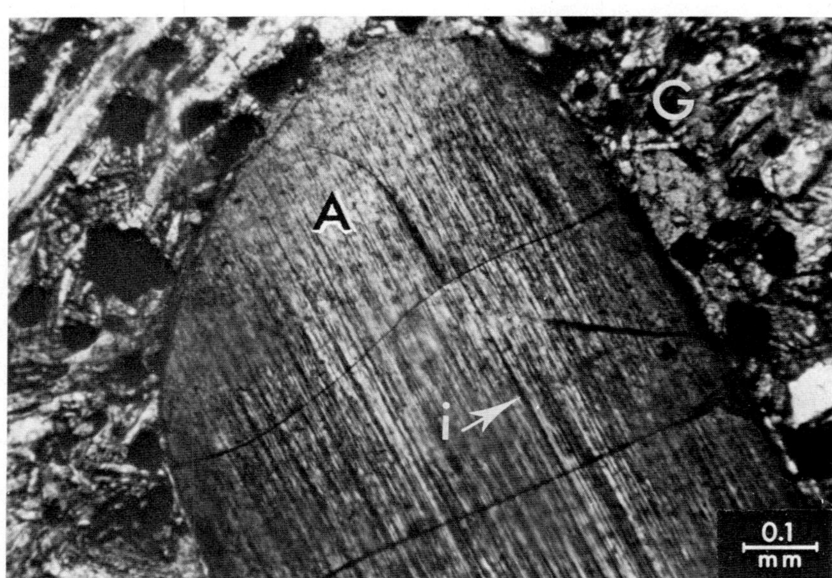

Fig. 454. Sub-phenocrystalline apatite, magmatically corroded (rounded) in basaltic groundmass. The fine lines shown in the photomicrograph of the apatite phenocrysts actually represent orientated inclusions in the apatite, as a basal cross-section shows (see Fig. 455). G = basaltic groundmass; i = orientated inclusions in the apatite, shown as fine lines in the apatite; A = apatite.
Basalt columns. Minderberg near Linz on Rhine River, Germany. With half-crossed nicols.

Fig. 455. Basaltic groundmass exceptionally rich in magnetite (actually representing an agglutination of magnetite crystal grains) with idiomorphic and sub-idiomorphic apatites. s-a = sub-idiomorphic apatite phenocryst with intracrystalline basaltic diffusion; h-a = basal section of apatite, hexagonal in cross-section, with intracrystalline inclusions parallel to the hexagonal prismatic faces of the apatite.
Basalt columns. Minderberg near Linz on Rhine River, Germany. With crossed nicols.

Fig. 456. Sub-phenocrystalline apatite magmatically corroded in basaltic groundmass. A cavity is also shown with pyroxene and magnetites in the sub-phenocrystalline apatite. G = basaltic groundmass; i = orientated inclusions in the apatite; c = cavity in the apatite with pyroxene and magnetites present.
Basalt columns. Minderberg near Linz on Rhine River, Germany. With crossed nicols.

Fig. 457. Basaltic groundmass with a sub-idiomorphic apatite. a = apatite.
Basalt columns. Minderberg near Linz on Rhine River, Germany. With crossed nicols.

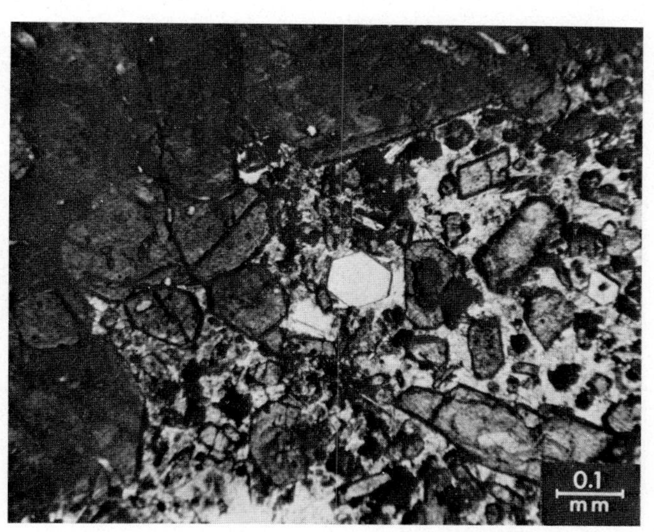

Fig. 458. Hornblende phenocrysts sub-phenocrystalline in size. A basal hexagonal section of apatite is also indicated.
Nephelinite. Pulverbuk, Oberbergen, Baden, Germany. Without crossed nicols.

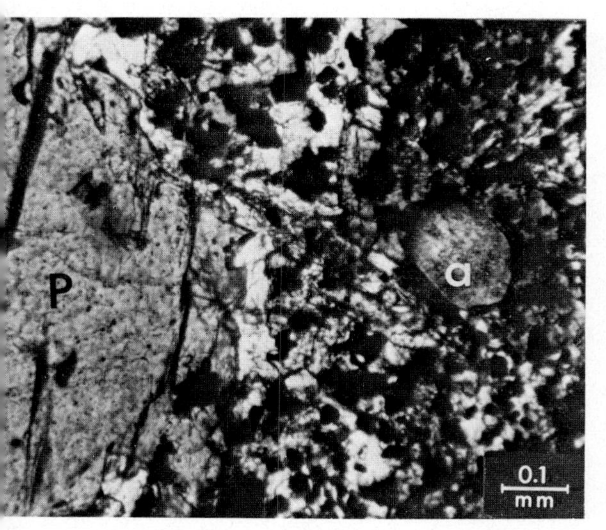

Fig. 459. Pyroxene phenocryst surrounded by groundmass very rich in magnetites. An idiomorphic (almost hexagonal in outline) apatite is also shown associated with the magnetite-rich groundmass. P = pyroxene; a = hexagonal cross-section of apatite.
Basalt columns. Minderberg near Linz on Rhine River, Germany. With crossed nicols.

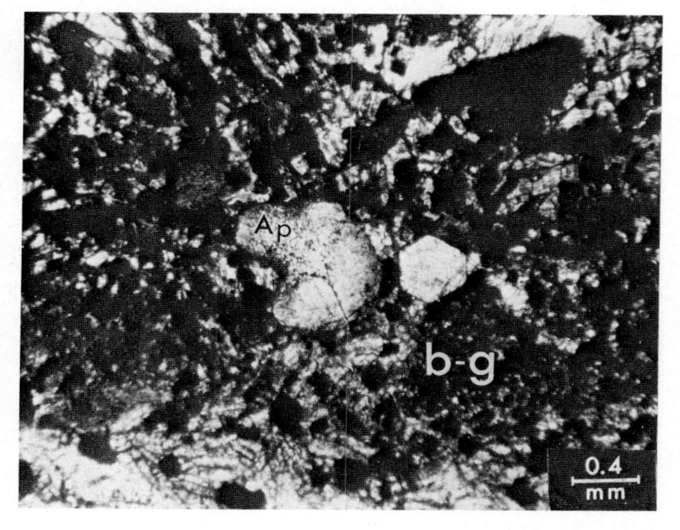

Fig. 460. Idiomorphic apatite and corroded sub-phenocrystalline apatite in basaltic groundmass rich in magnetite crystal grains. Ap = apatite, magnetite (black); b-g = basaltic groundmass.
Basalt columns. Minderberg near Linz on Rhine River, Germany. With crossed nicols.

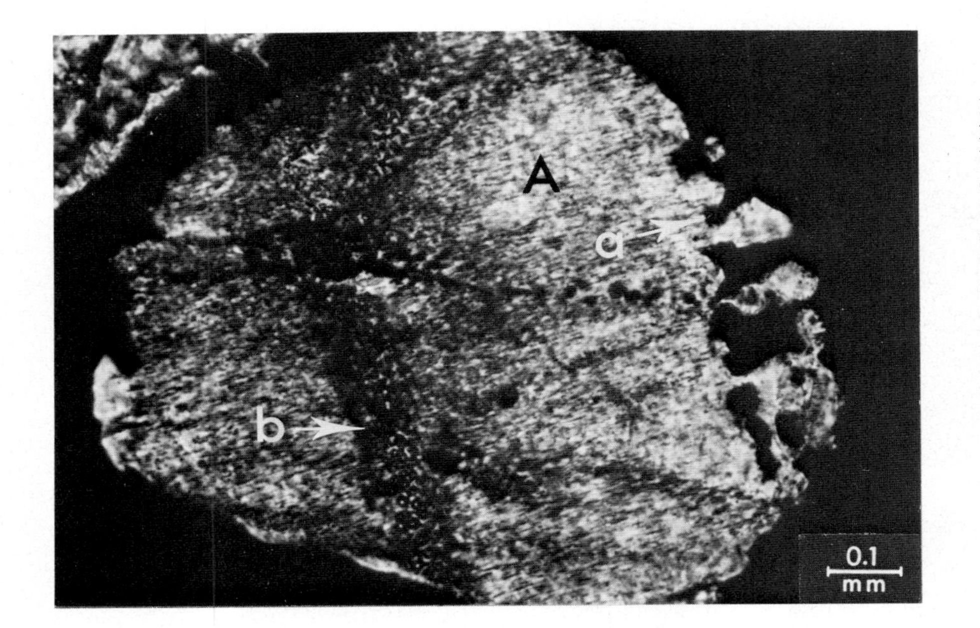

Fig. 461. Apatite, sub-phenocrystalline in size, surrounded and corroded by magnetite. A symplectic pattern of apatite and the intracrystalline diffusion melts is also illustrated. A = apatite; arrow "a" shows magnetite corroding and affecting the apatite; arrow "b" shows symplectic patterns of iron-oxide infiltrations in the apatite. Black = magnetite.
Basalt columns. Minderberg near Linz on Rhine River, Germany. With crossed nicols.

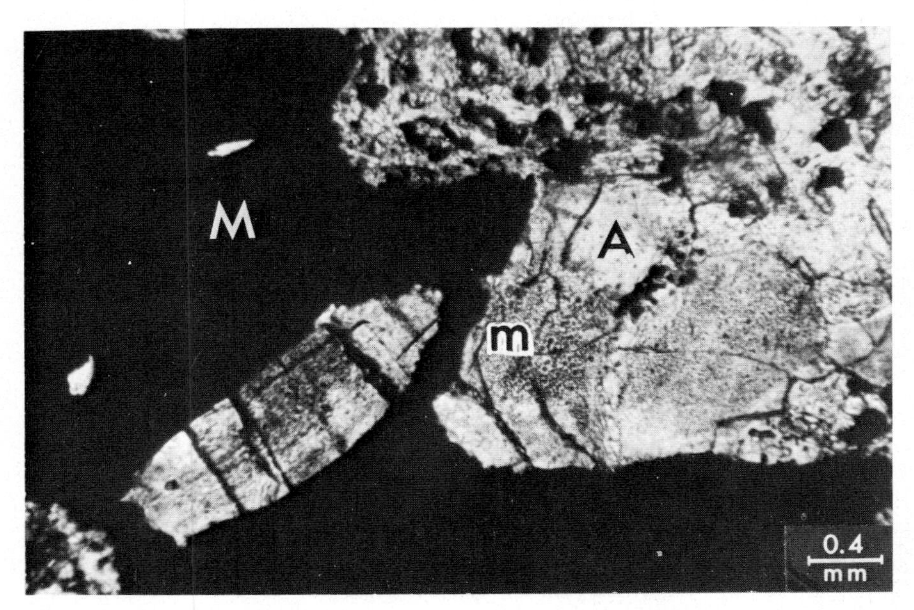

Fig. 462. Association of magnetite with included and partly corroded apatites. A = apatites; M = magnetites; m = magnetite infiltrations in the apatite.
Basalt columns. Minderberg near Linz on Rhine River, Germany. With crossed nicols.

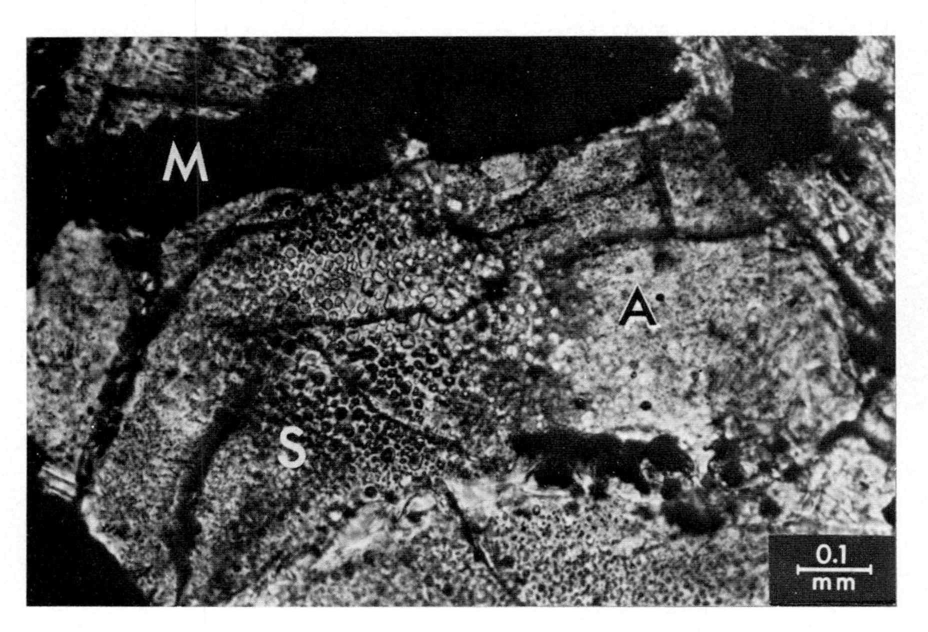

Fig. 463. Detail of Fig. 462. The apatite is invaded by infiltration of basaltic melts (most probably iron oxides) which show a symplectic intergrowth pattern with the host. M = magnetite (black); A = apatite; S = symplectic pattern of intracrystalline diffusions of basaltic melts in the apatite (a droplet pattern of intracrystalline diffusion of basaltic melts in the apatite). Basalt columns. Minderberg near Linz on Rhine River, Germany. With crossed nicols.

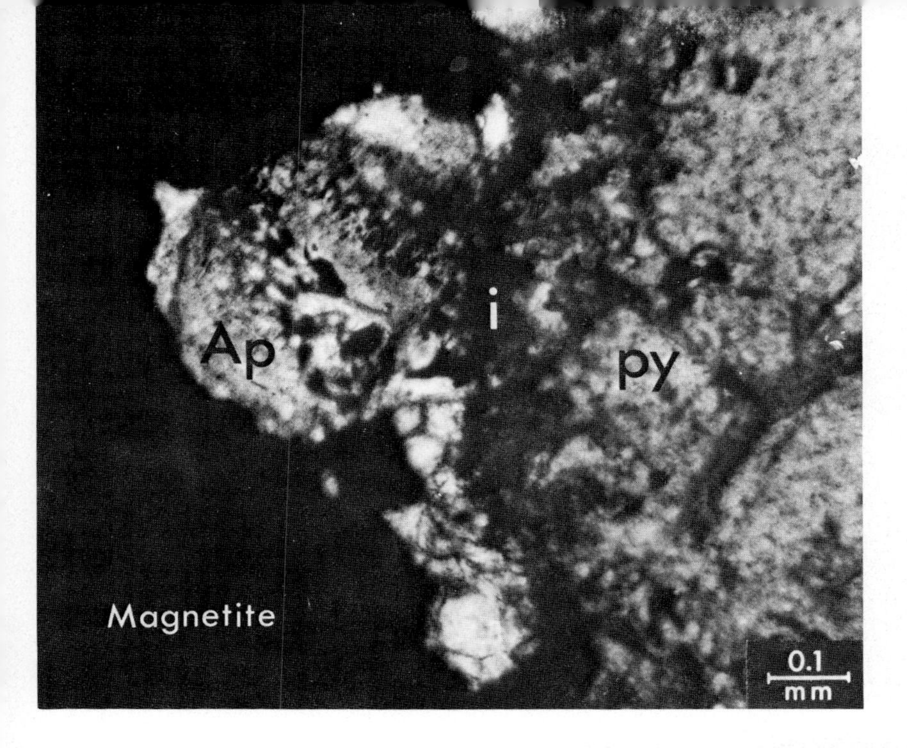

Fig. 464. Apatite invaded by iron-oxides which infiltrate into the host and cause symplectic intergrowths. Ap = apatite; i = iron oxides infiltrated and in symplectic intergrowth with the host apatite; py = pyroxene adjacent to the apatite also invaded by iron oxides.
Basalt columns. Minderberg near Linz on Rhine River, Germany. Without crossed nicols.

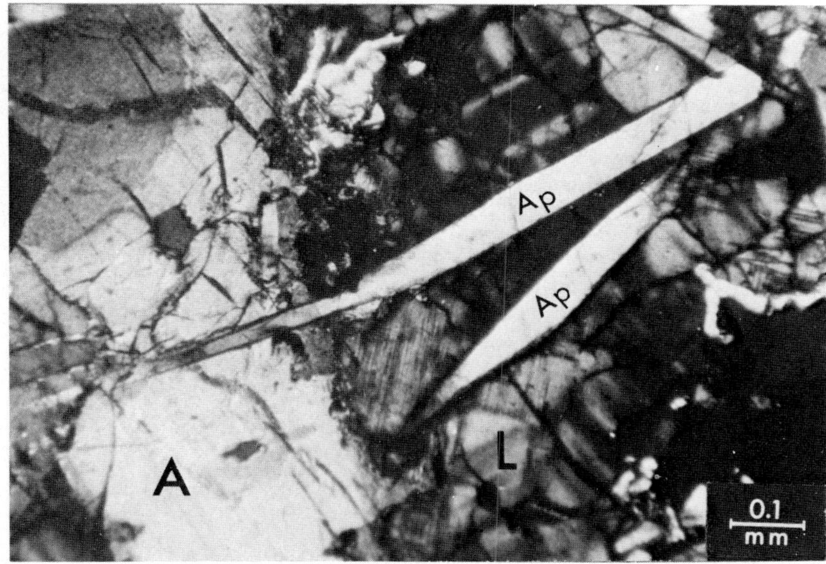

Fig. 465. Elongated prismatic apatite transversing the boundary of phenocrystalline leucite and augite. Other apatites are entirely present in the leucite. A = augite; Ap = apatite; L = leucite.
Basalt. Tottenkappel b. Meiches. Vogelsberg, Germany. With crossed nicols.

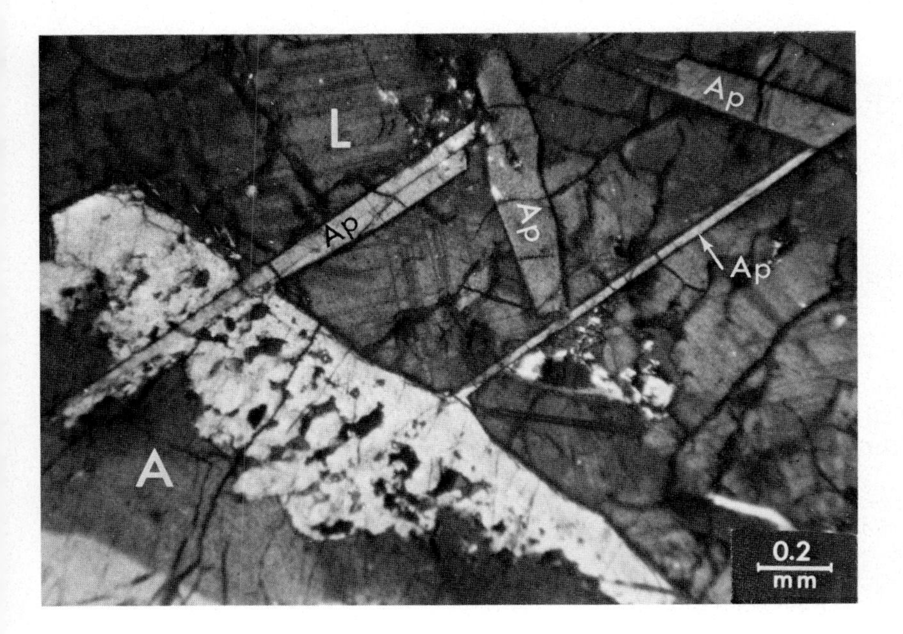

Fig. 466. Prismatic apatites, orientated in leucite basaltic phenocryst, in some cases transverse the boundary of leucite with an adjacent augite and extend as well into the pyroxene. A = augite; Ap = apatite; L = leucite.
Basalt. Tottenkappel b. Meiches. Vogelsberg, Germany. With crossed nicols.

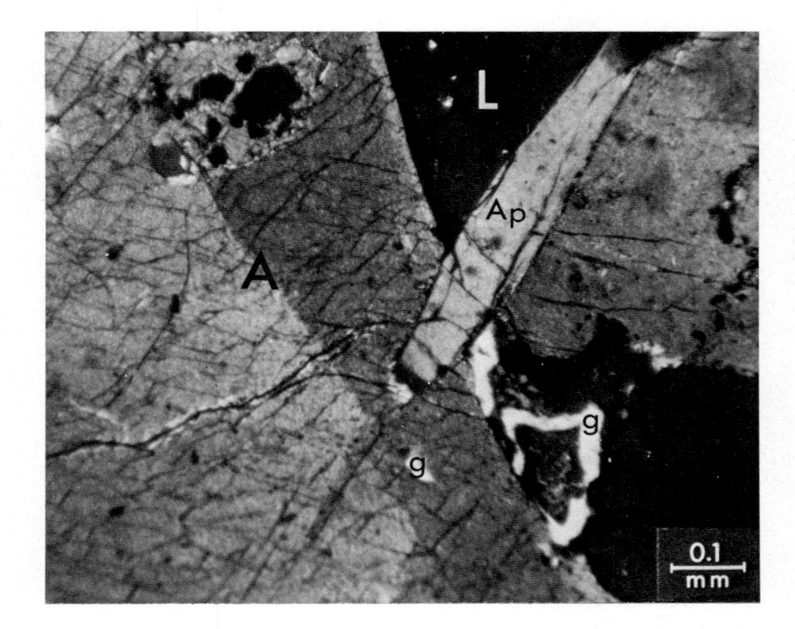

Fig. 467. Prismatic apatite transverses the leucite–augite boundary and also extends into the pyroxene. The apatite cuts across granophyric quartz intergrown with the leucite. A = augite (zoned); ap = apatite, L = leucite; g = granophyric quartz intergrown with the leucite.
Basalt. Tottenkappel b. Meiches. Vogelsberg, Germany. With crossed nicols.

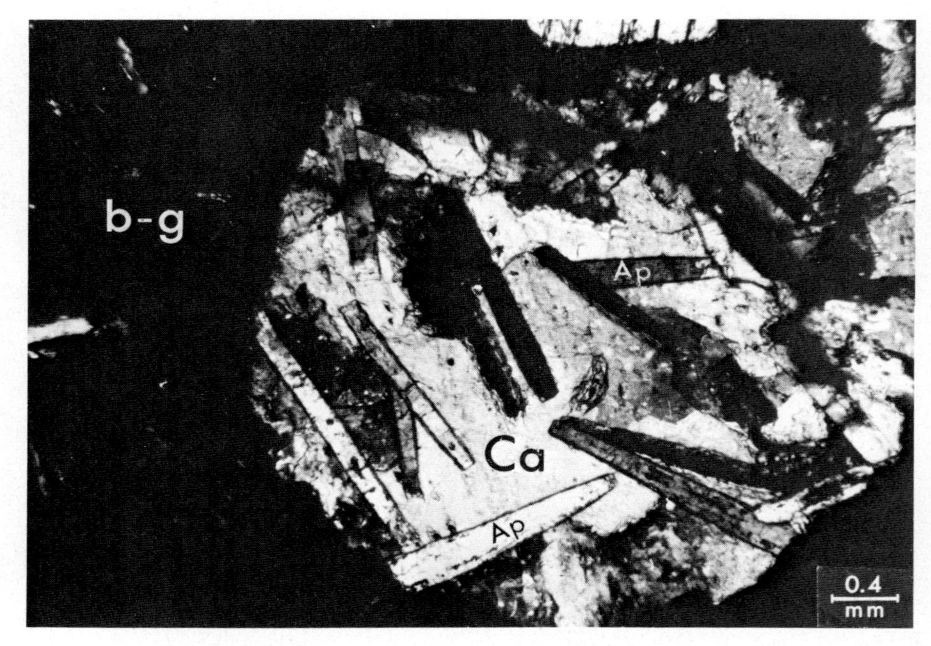

Fig. 468. Amygdaloidal filling in basalt occupied by calcite in which late-phase prismatic apatite phenocryst has grown. Ca = calcite filling the amygdaloidal cavity; b-g = basaltic groundmass; Ap = prismatic apatite crystalloblasts.
Basalt. Haddington, Scotland. With crossed nicols.

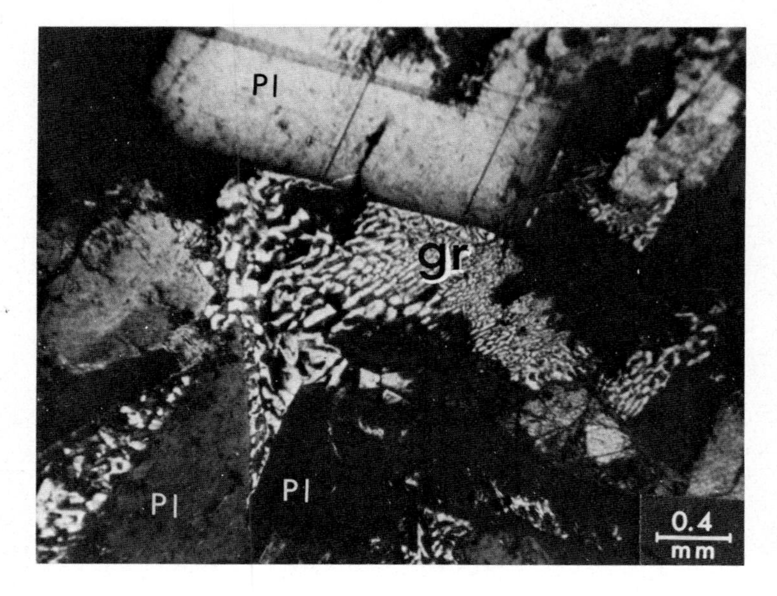

Fig. 469. Micrographic quartz (granophyric) in intergrowth with plagioclase, interstitial between plagioclase phenocrysts. The granophyric quartz–plagioclase intergrowth occupies spaces between the plagioclase crystals. Pl = plagioclase; gr = granophyric quartz (in intergrowth with plagioclase) occupying the spaces between plagioclase phenocryst.
Doleritic basalt. Karroo series, S. Africa. With crossed nicols.

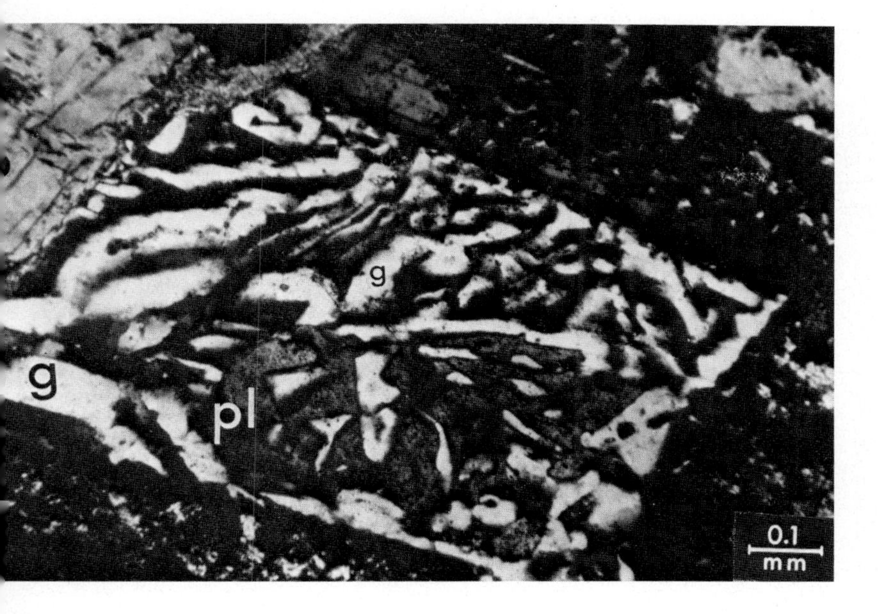

Fig. 470. Granophyric quartz–plagioclase symplectite interstitial between plagioclase phenocrysts. g = granophyric quartz; pl = plagioclase in symplectic intergrowth with quartz; p = adjacent plagioclases.
Doleritic basalt. Karroo series, S. Africa. With crossed nicols.

Fig. 471. Granophyric quartz in intergrowth with plagioclase and enclosing an idiomorphic plagioclase. g = granophyric quartz; p = plagioclase; Pl = plagioclase surrounded by the symplectic quartz–plagioclase intergrowth.
Doleritic basalt. Karroo series, S. Africa. With crossed nicols.

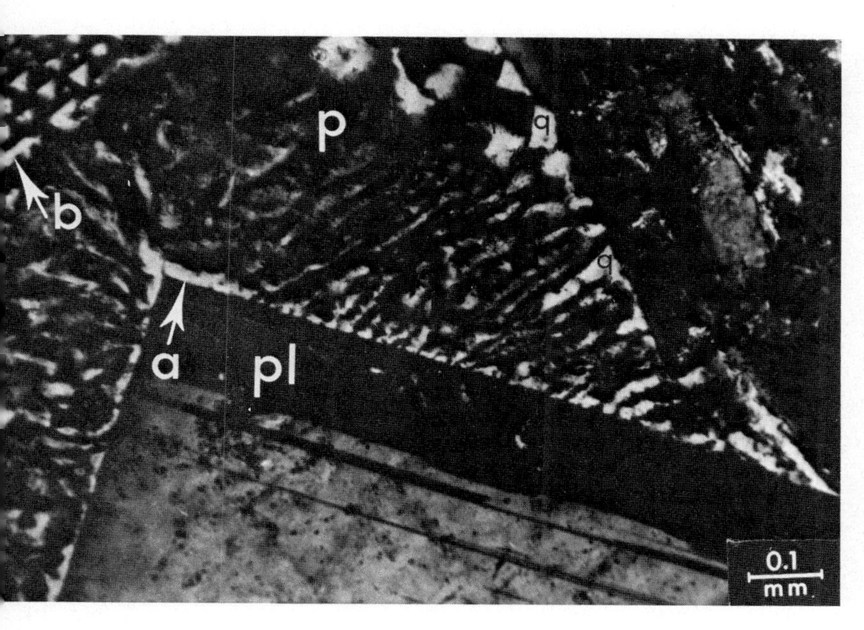

Fig. 472. Granophyric quartz (q) in symplectic intergrowth with plagioclase (p) surrounding a plagioclase phenocryst (pl). Arrow "a" shows granophyric quartz taking advantage of the intergranular (interleptonic) space marginal to the plagioclase phenocryst. Arrow "b" shows cuniform micrographic quartz.
Doleritic basalt. Karroo series, S. Africa. With crossed nicols.

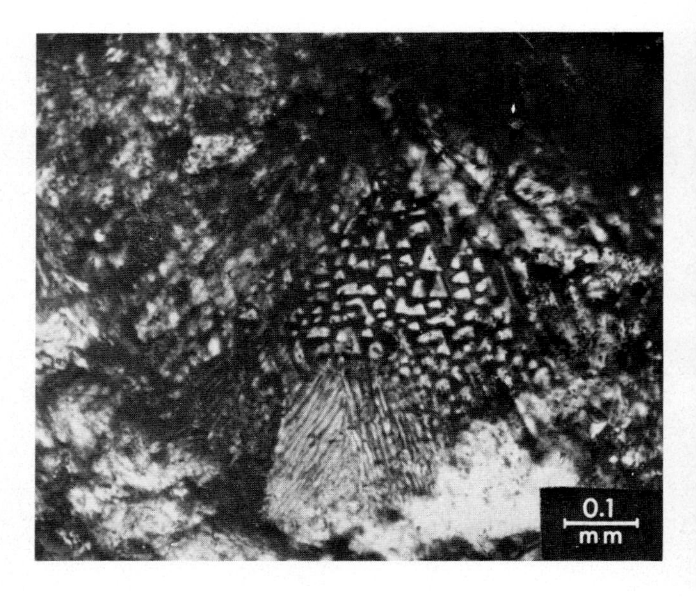

Fig. 473. Topo-developments within the basaltic groundmass of symplectic quartz–feldspar intergrowths, often the quartz which is granophyric in pattern attains micro-hieroglyphic patterns. Triangular (or cuniform) patterns represent quartz. Doleritic basalt. Karroo series, S. Africa. With crossed nicols.

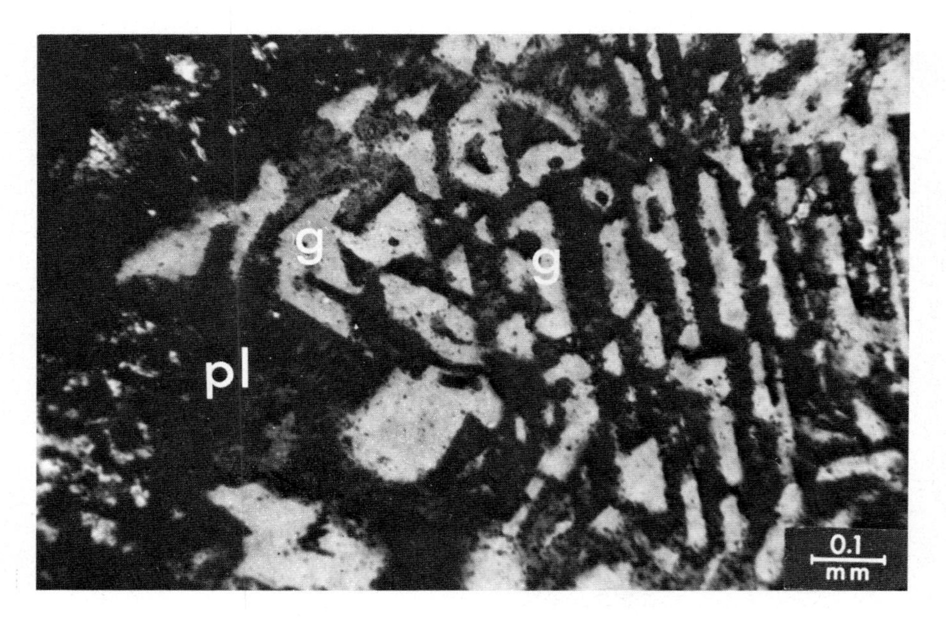

Fig. 474. Cuniform (granophyric) quartz in symplectic intergrowth with plagioclase feldspar. g = granophyric (micrographic) quartz; pl = plagioclase.
Doleritic basalt. Karroo series, S. Africa. With crossed nicols.

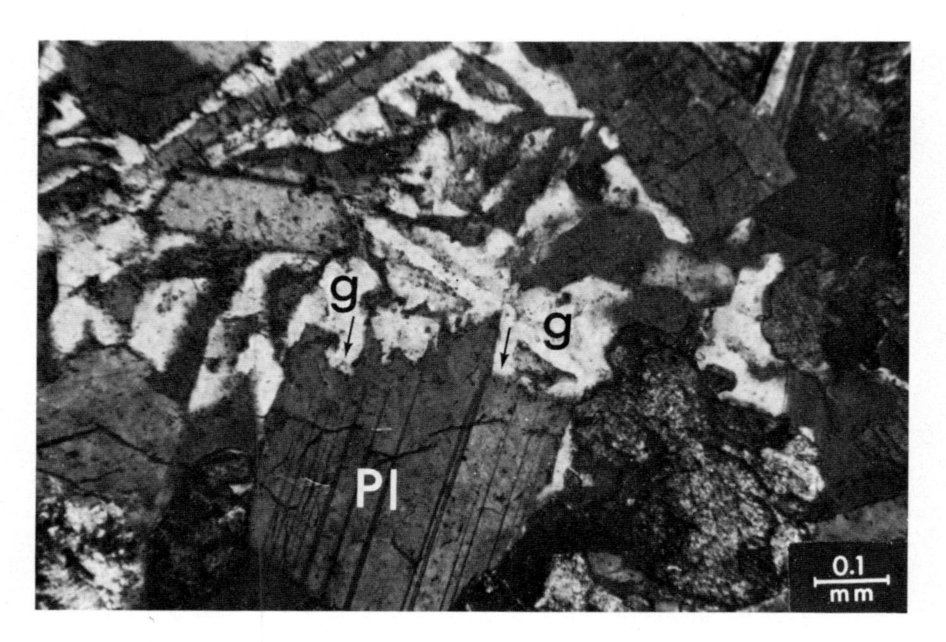

Fig. 475. Granophyric quartz in symplectic intergrowth with plagioclase and invading an adjacent plagioclase along its twinning. g = granophyric quartz in intergrowth with plagioclase; Pl = plagioclase; arrows show granophyric quartz invading the plagioclase along its twinning. Doleritic basalt.
Palisade Sill, New Jersey, U.S.A. With crossed nicols.

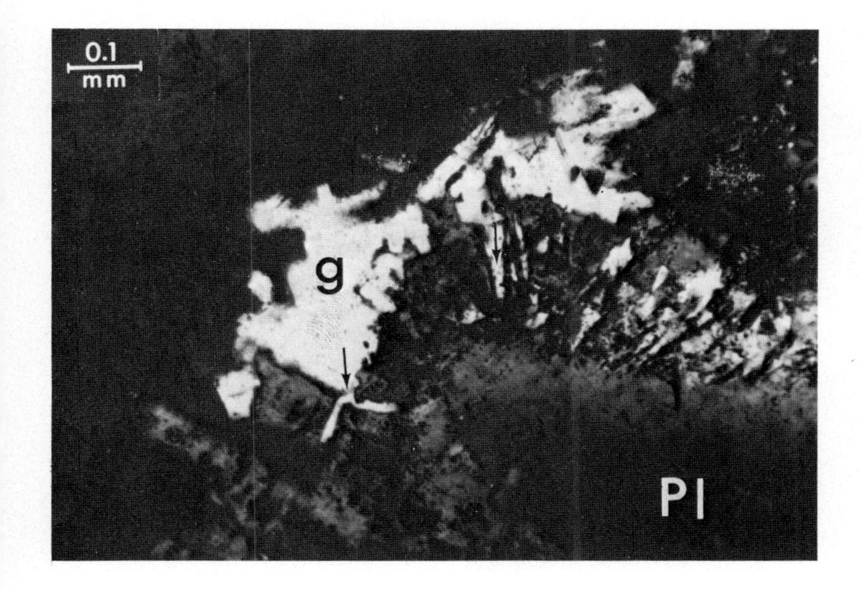

Fig. 476. Intergranular quartz sending extensions attaining a granophyric character into an adjacent plagioclase. Pl = plagioclase; g = intergranular quartz; arrows show extensions of the quartz attaining granophyric character.
Doleritic basalt. Karroo series, S. Africa. With crossed nicols.

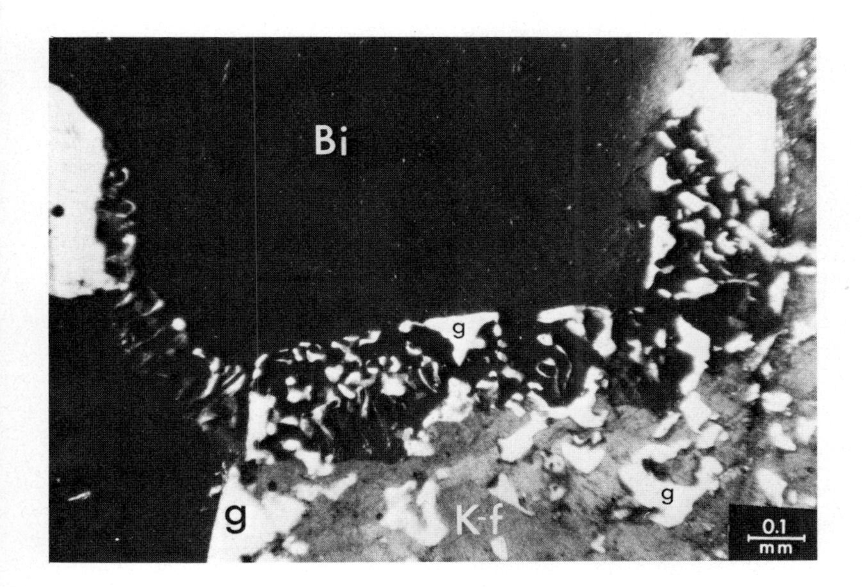

Fig. 477. Biotite and K-feldspar, both with granophyric quartz, often transversing the K-feldspar–biotite boundary and extending into the mica, resulting in a biotite margin with graphic quartz. Bi = biotite; K-f = K-feldspar; g = granophyric quartz.
Pegmatite in the Bushveld complex, South Africa.

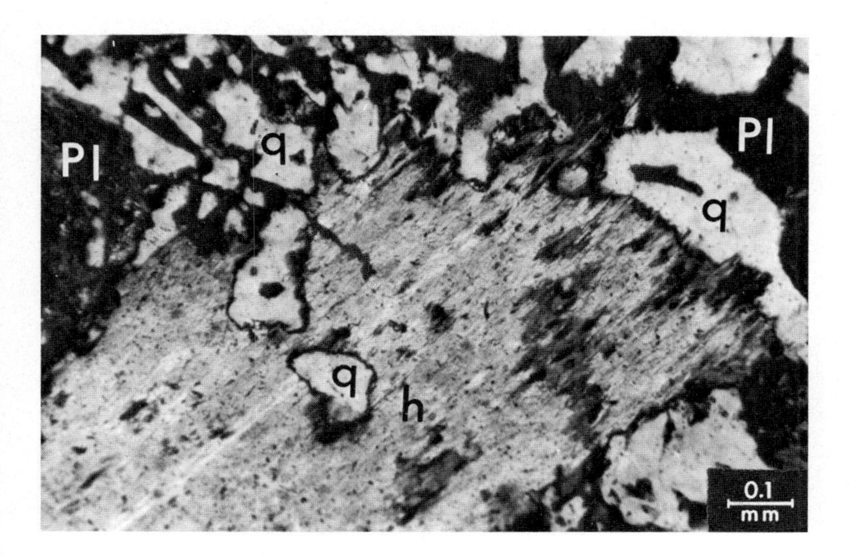

Fig. 478. Granophyric quartz (q) in symplectic intergrowth with plagioclase (Pl) continuous as symplectic intergrowth within an adjacent hornblende (h).
Palisade Sill, New Jersey, U.S.A. With crossed nicols.

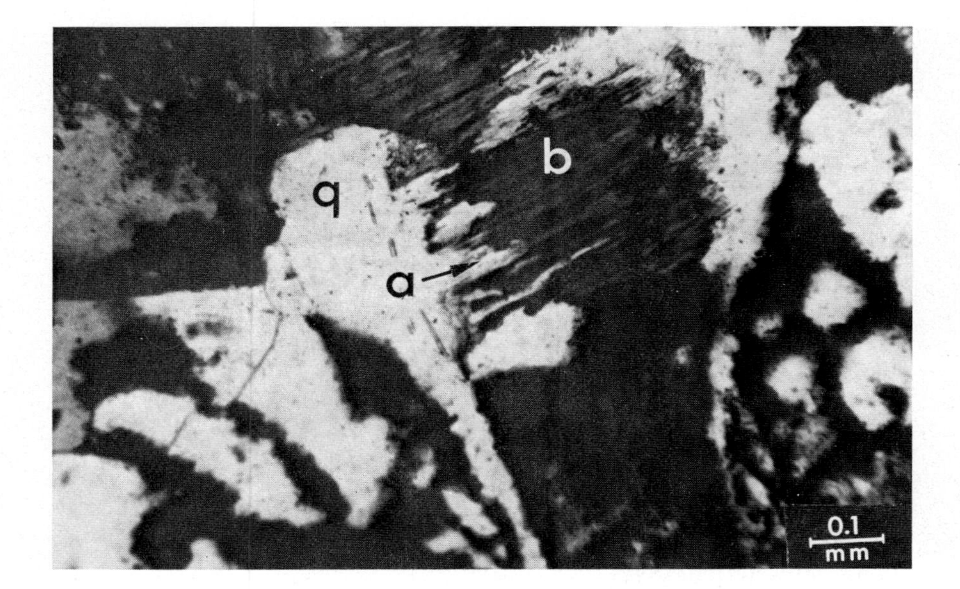

Fig. 479. Granophyric quartz in symplectic intergrowth with plagioclase, sending extension into an adjacent biotite (b). Another quartz grain is in intergrowth with the biotite. b = biotite; q = quartz; arrow "a" shows extension of quartz into the biotite.
Doleritic basalt. Karroo series. S. Africa.

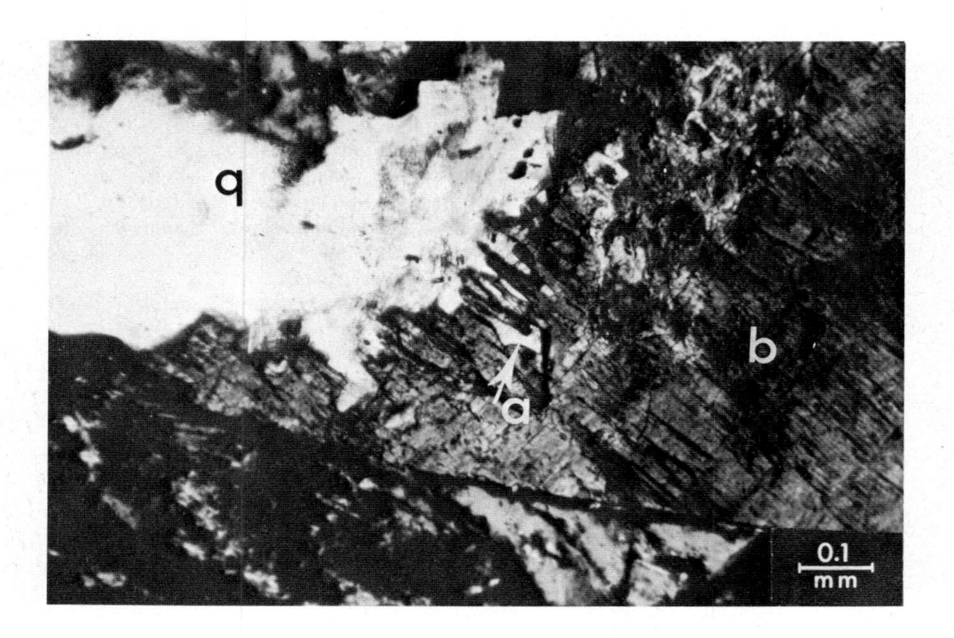

Fig. 480. Intergranular quartz with extension protruding into an adjacent biotite. g = intergranular quartz; b = biotite; arrow "a" shows extensions of quartz into the biotite.
Doleritic basalt. Karroo series, S. Africa.

Fig. 481. Granophyric quartz–plagioclase intergrowth, with the quartz sending extensions into an adjacent pyroxene. Pl = plagioclase; q = quartz; p = plagioclase in symplectic intergrowth with the quartz; py = pyroxene; arrow "a" shows quartz extending into adjacent pyroxene.
Palisade Sill, New Jersey, U.S.A. With crossed nicols.

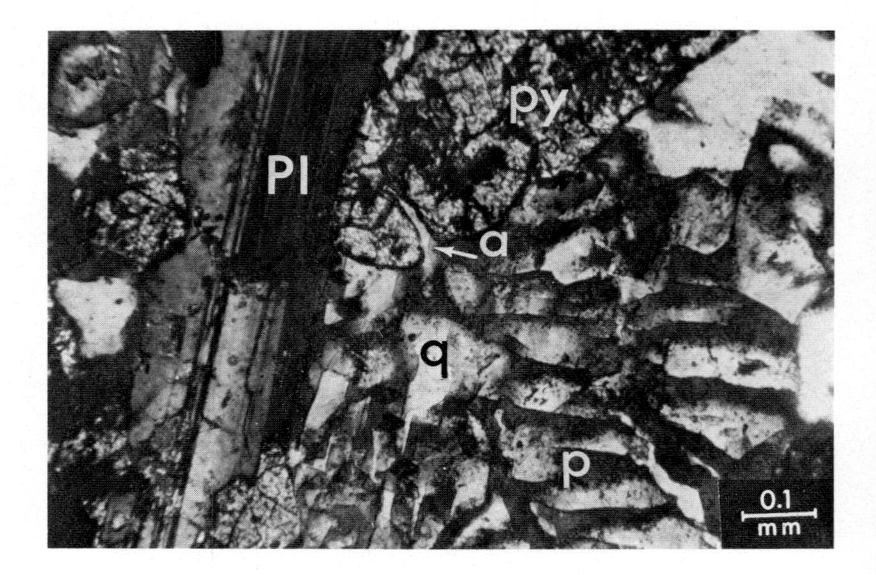

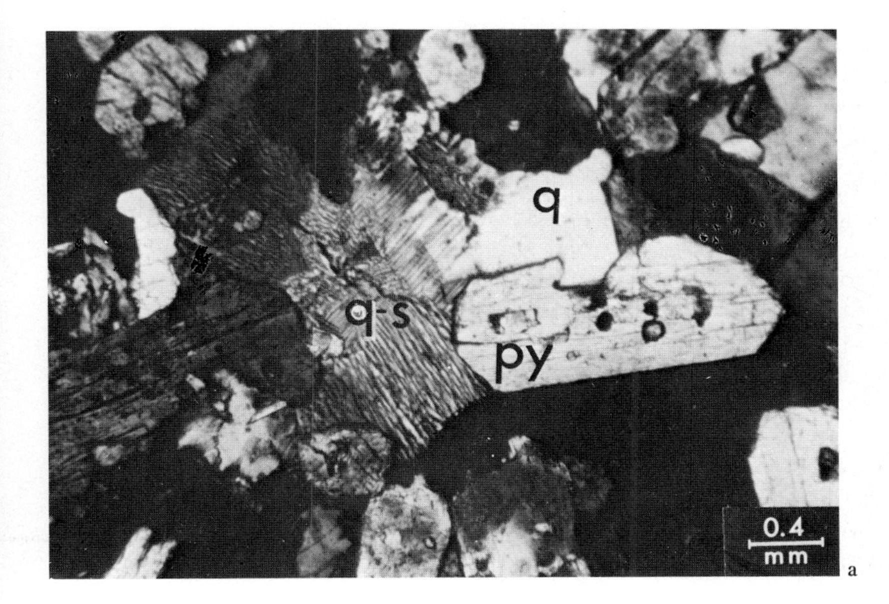

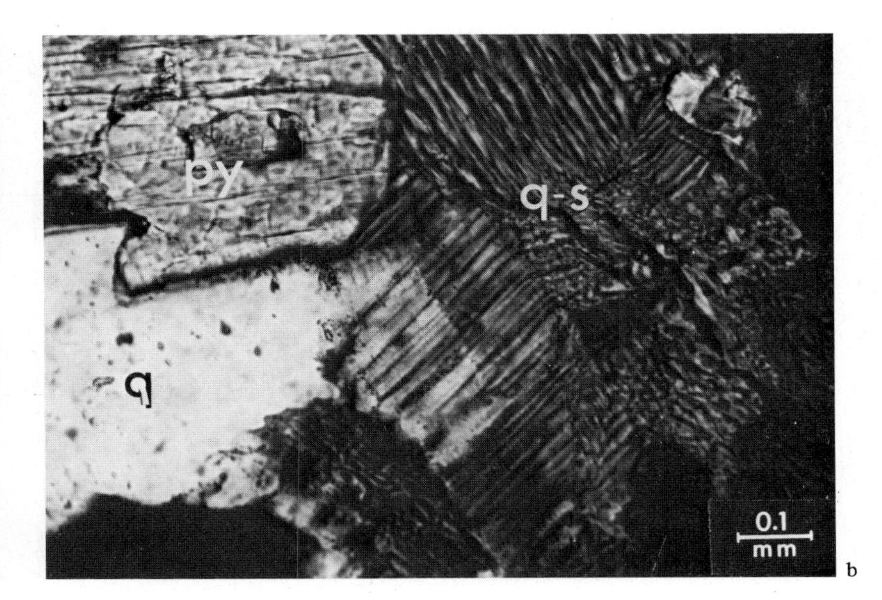

Fig. 482 (a and b). Intergranular quartz (q) with transitions into a granophyric symplectite (q-s) and partly in contact with an adjacent pyroxene (py).
Doleritic basalt. Karroo series, S. Africa.

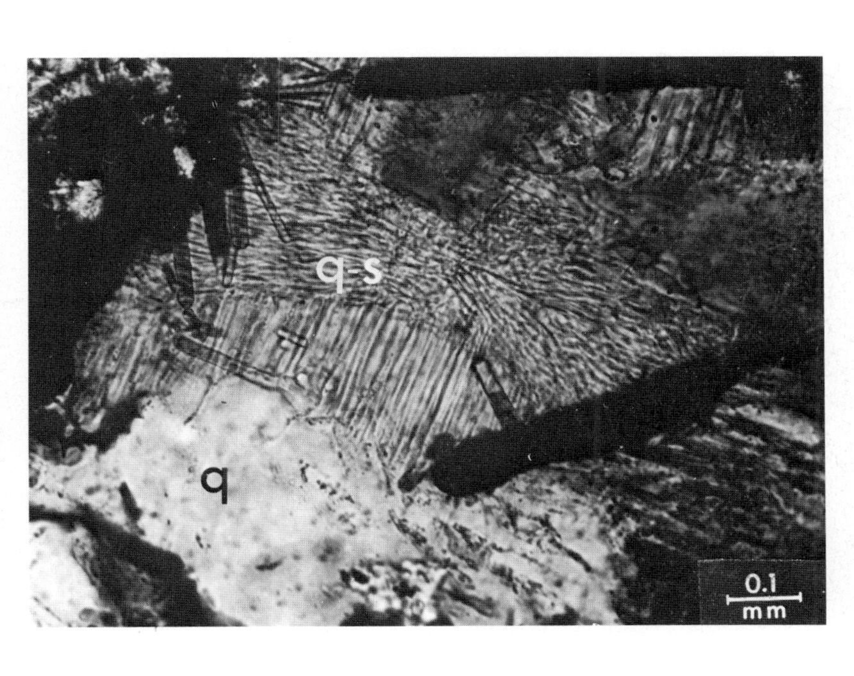

Fig. 483. Quartz (q) with transitions to a granophyric quartz–feldspar symplectite (q-s).
Doleritic basalt. Karroo series, S. Africa.

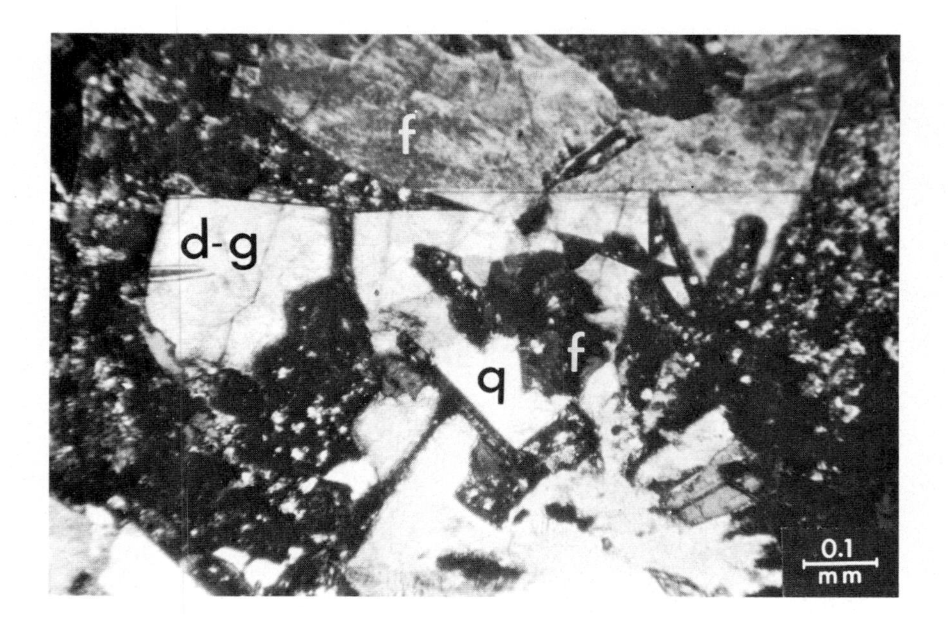

Fig. 484. Interstitial skeleton quartz attaining micrographic character and developed quartz crystal grains. q = micrographic quartz skeletons; d-g = developed quartz crystal grain; f = feldspar.
Doleritic basalt. Karroo series, S. Africa.

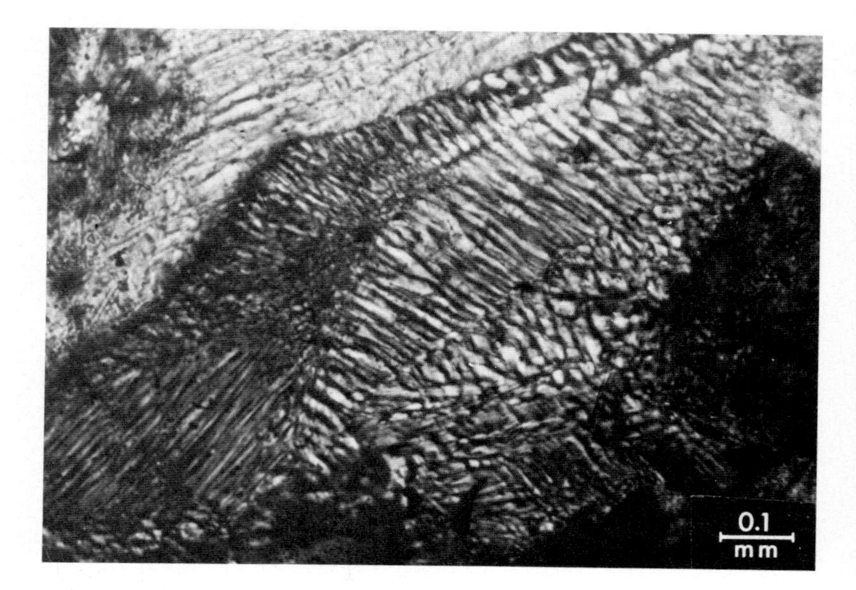

Fig. 485. Symplectic granophyric intergrowth patterns of quartz–feldspar topo-developments within the basaltic groundmass. A fishbone pattern is often exhibited.
Doleritic basalt. Karroo series, S. Africa.

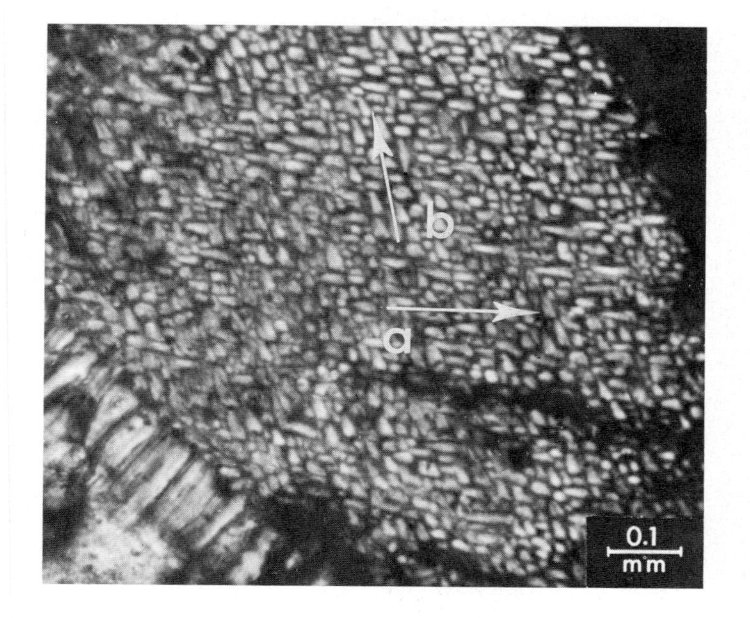

Fig. 486. An intersertal wedge pattern, feldspar–quartz cuniform micrographic symplectic pattern. Two prevailing orientation directions are shown by the wedge-shaped intergrown crystals. Arrows "a" and "b" show the prevailing orientation intergrowth of the wedge-shaped crystal grains.
Doleritic basalt. Karroo series, S. Africa.

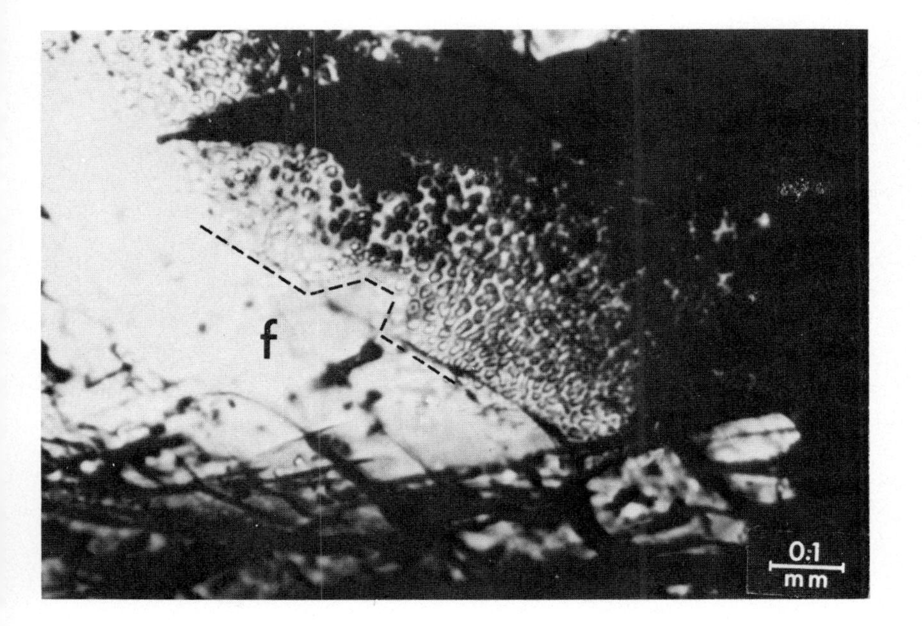

Fig. 487. Feldspar with quartz in granophyric intergrowth. The granophyric intergrowth is restricted to an outer feldspar zone, almost geometrically delimited. f = feldspar (quartz free), granophyric intergrowth.
Sea of Tranquility, Moon.

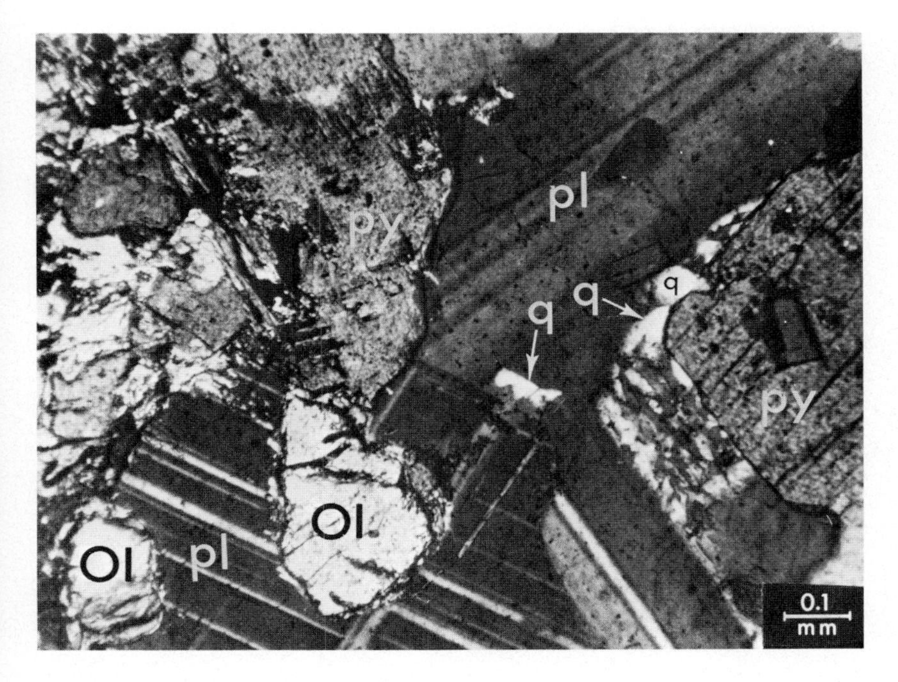

Figs. 488 and 489. Granophyric quartz in intergrowth with plagioclase (Bowen's micropegmatititc quartz intergrowths) coexisting with olivine in the top olivine-rich band of Palisade Sill, New Jersey, U.S.A. Ol = olivine; pl = plagioclase; q = granophyric quartz; py = pyroxene. With crossed nicols.

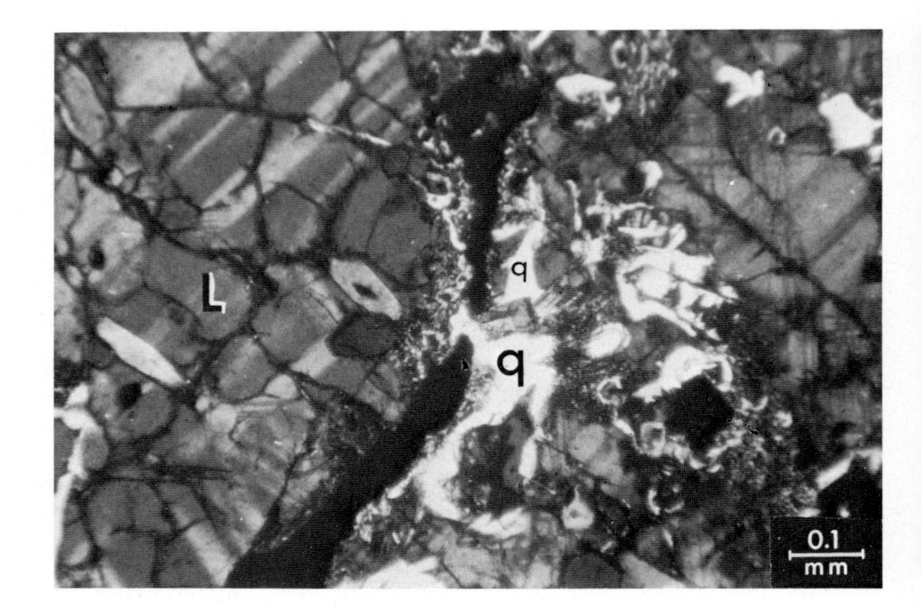

Fig. 490. Granophyric quartz in intergrowth with leucite. q = quartz (granophyric); L = leucite.
Leucite basalt. Totenkappel b. Meiches, Vogelsberg, Germany. With crossed nicols.

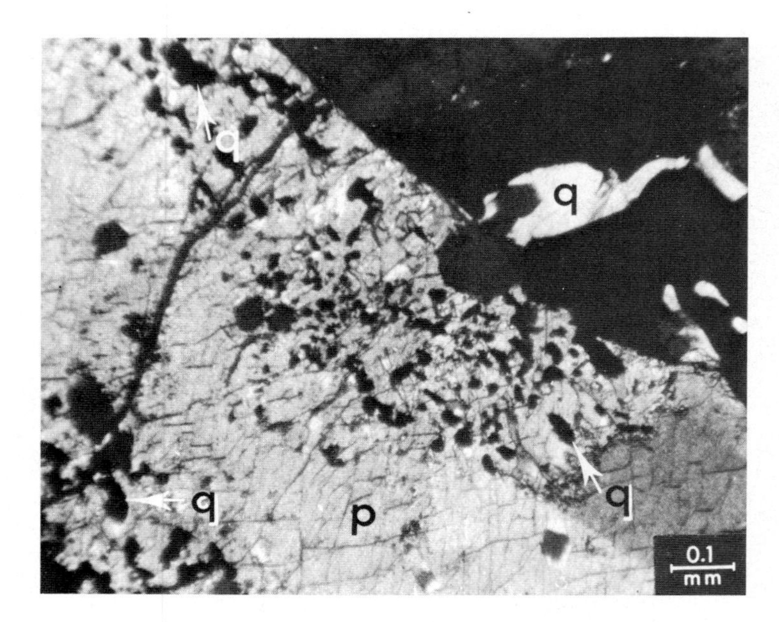

Fig. 491. Granophyric quartz (q) in intergrowth with pyroxene (p).
Leucite basalt. Totenkappel b. Meiches, Vogelsberg, Germany. With crossed nicols.

Fig. 492. Idiomorphic olivine phenocryst (with characteristic olivine outline); serpentinized marginally and along cracks of it, olivine relics are left within the serpentinized mass. Ol = olivine relics of alteration; s = serpentinization of olivine along the margins and cracks of the olivine; G = basaltic groundmass.
Olivine tholeiite. Starz, Dersdorf, Saar region, Germany. With crossed nicols.

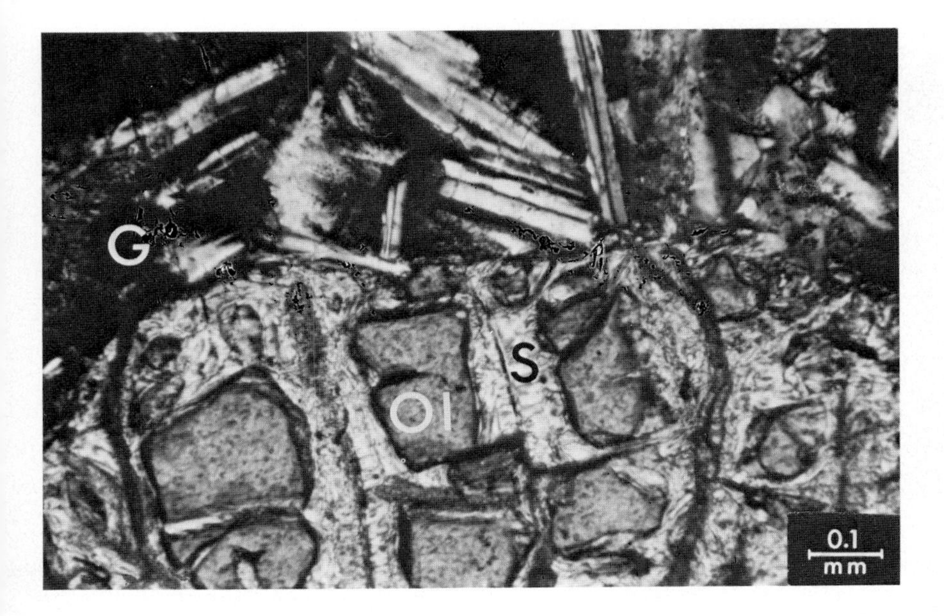

Fig. 493. Basaltic olivine serpentinized (antigorized) along cracks, with relics of olivine left in the serpentine. Ol = olivine relics; s = serpentine (antigorite) along cracks of the olivine; G = basaltic groundmass.
Olivine tholeiite. Starz, Dersdorf, Saar region, Germany. With crossed nicols.

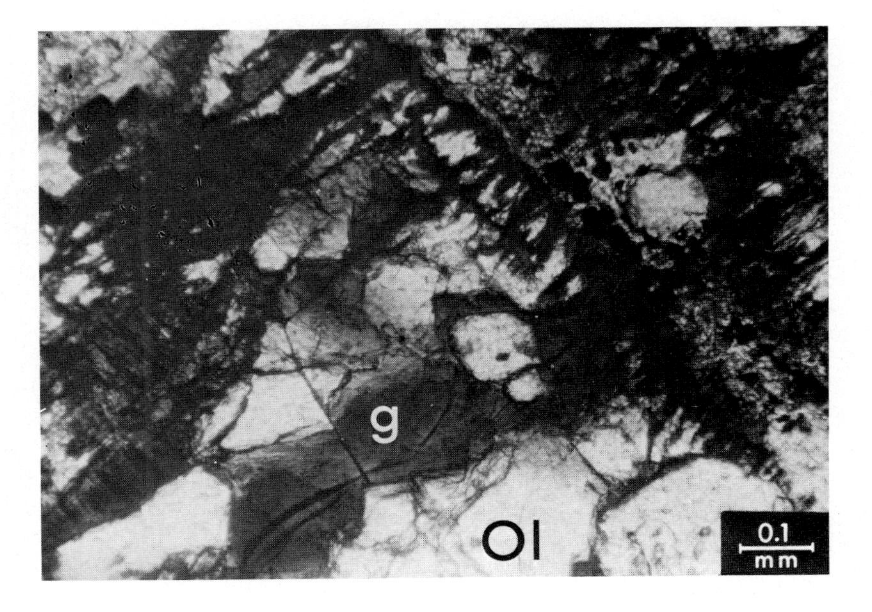

Fig. 494. Olivine alteration into green chlorite following and developing from olivine cracks. Ol = olivine; g = green chlorite alteration product of the olivine, starting and developing from cracks. Olivine basalt. Adama (Nazaret) Ethiopia. With crossed nicols.

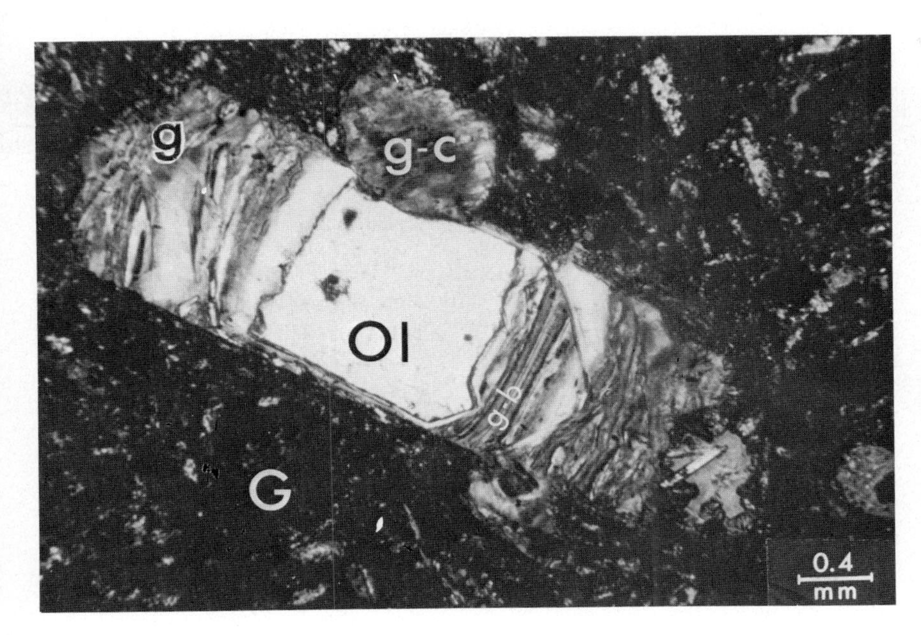

Fig. 495. Olivine with green-chlorite alteration; often the chlorite develops along cracks and attains forms of colloform banding with the fibrous phase developing perpendicular to the crack walls. Ol = olivine; g = green chlorite alteration margins; g-b = green chloritic band (actually a repeated colloform filling with the micro-fibres perpendicular to the walls of the micro-veinlet); G = basaltic groundmass; g-c = a smaller olivine grain completely altered into green chlorite.
Basalt. Hohe Wostrel Aussig, Bohemia, Czechoslovakia. With crossed nicols.

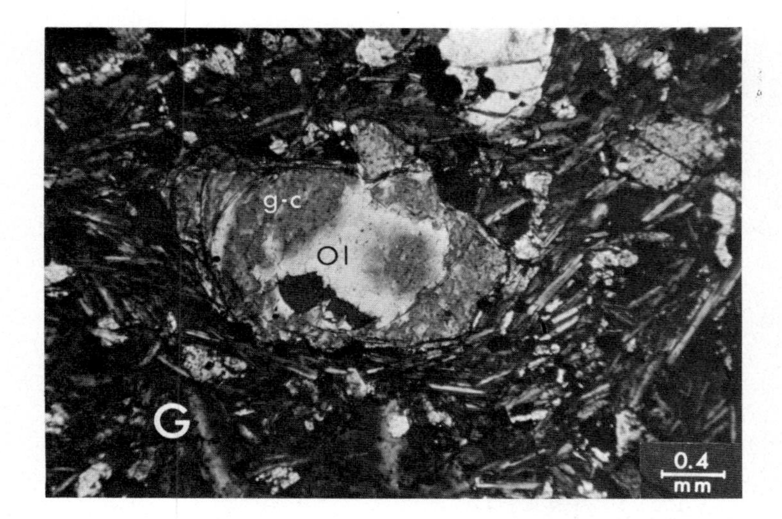

Fig. 496. Olivine grain with a wide margin of green chlorite formation (glauconitization). Ol = olivine; g-c = green chloritic alteration margin of the olivine; G = basaltic groundmass.
Olivine basalt. Adama (Nazaret), Ethiopia. With crossed nicols.

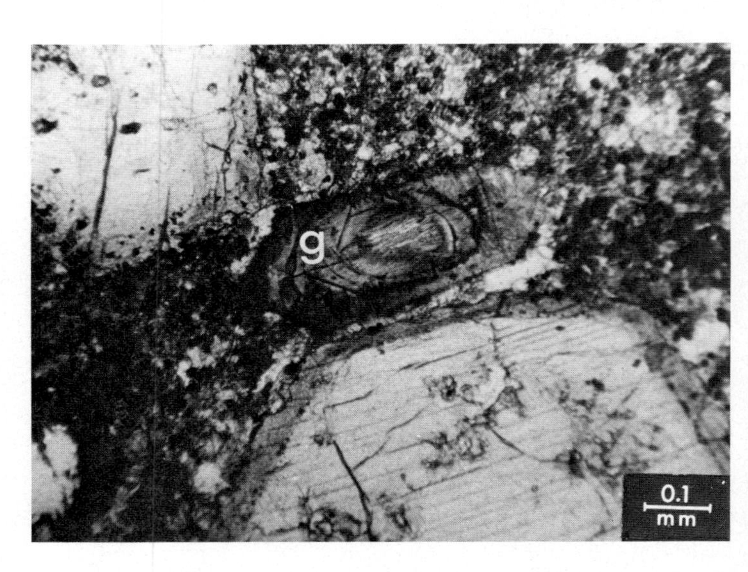

Fig. 497. Olivine grain entirely changed into an aggregate of green chlorites, showing a "zonal pattern" due to the advancing fronts of chloritization of olivines. In contrast to the chloritized olivine, the adjacent augite only shows chloritized green margins. g = green chlorite aggregate of initial olivine grain, showing a zonal pattern due to alteration-front solutions.
Olivine basalt. Adama (Nazaret) Ethiopia. With crossed nicols.

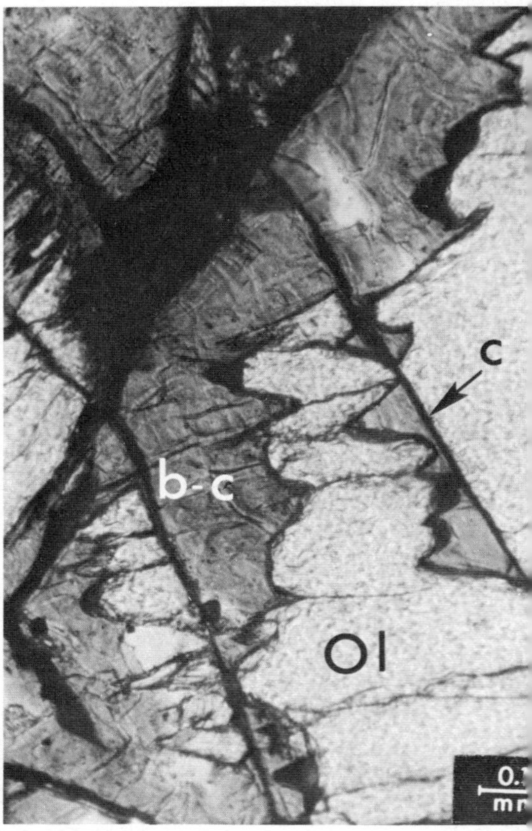

Fig. 499. Detail of Fig. 498. Brown chlorite (strongly pleochroic) replacing the olivine along its cracks. Ol = olivine relics or unaltered; b-c = brown chlorite (pleochroic), c = cracks along the olivine.
Olivine basalt. Ethiopian part of Great Rift Valley. With crossed nicols.

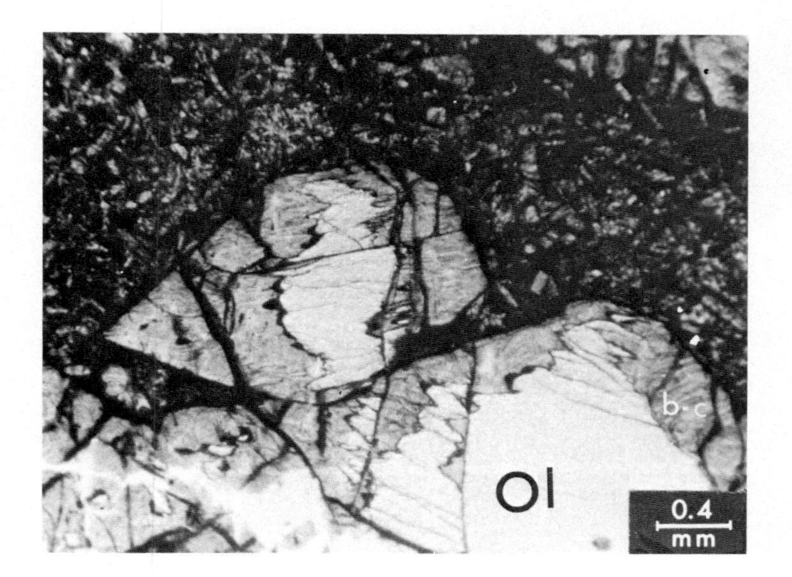

◀ Fig. 498. Brown chlorite (with strong brown pleochroism) replacing the olivine marginally and along cracks. Ol = olivine relics; b-c = brown chlorite replacing the olivine.
Olivine basalt. Ethiopian part of Great Rift Valley. With crossed nicols.

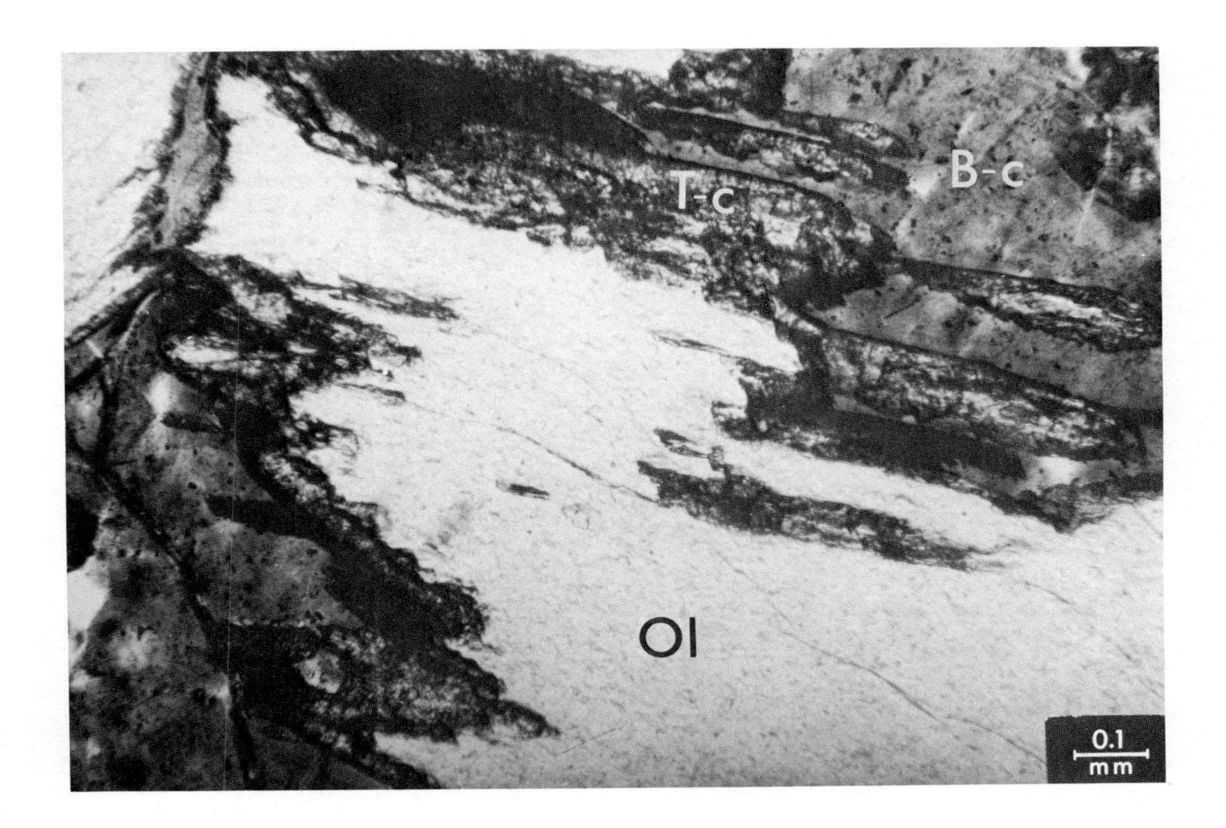

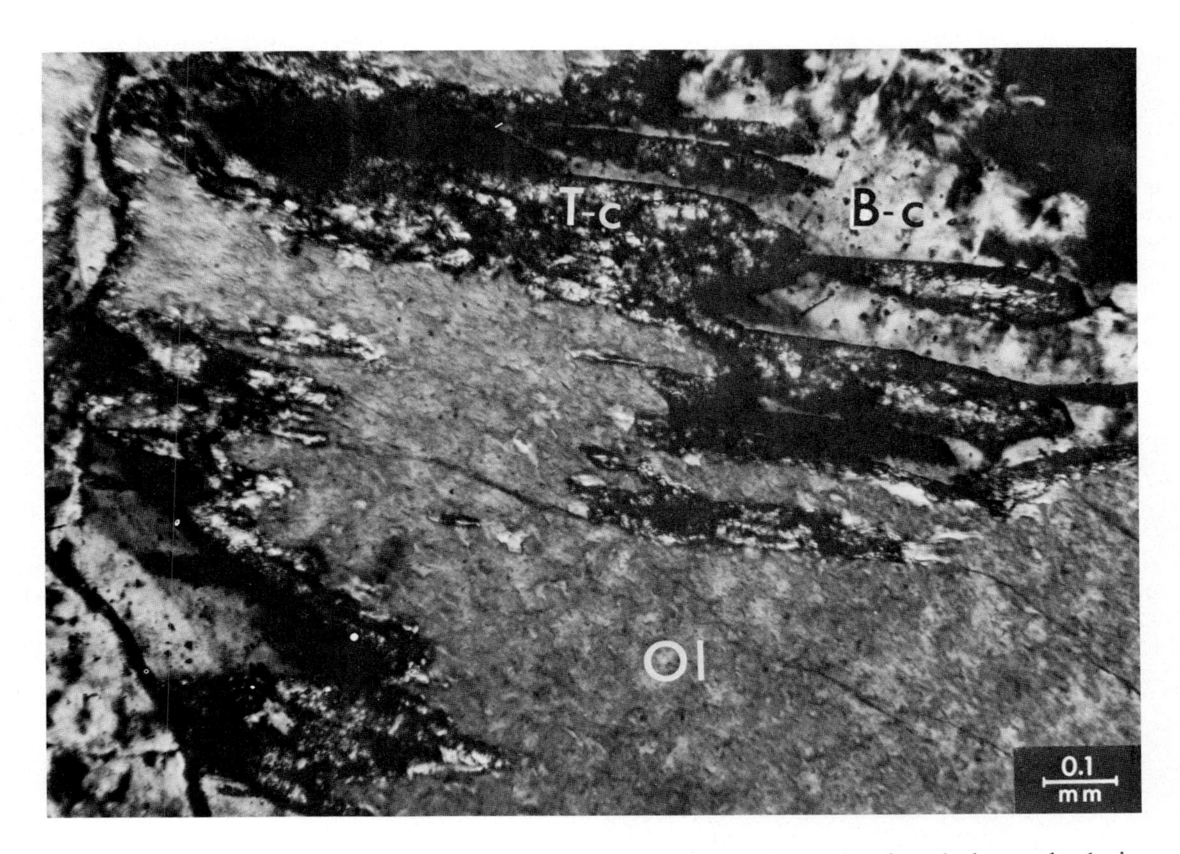

Figs. 500 and 501. Transition phases between olivine (Ol) and its alteration product into the brown-pleochroic chlorite (B-c). The transition phase (T-c) is an intermediate phase in the chloritization of the olivine.
Olivine basalt. Ethiopian part of Great Rift Valley. With and without crossed nicols.

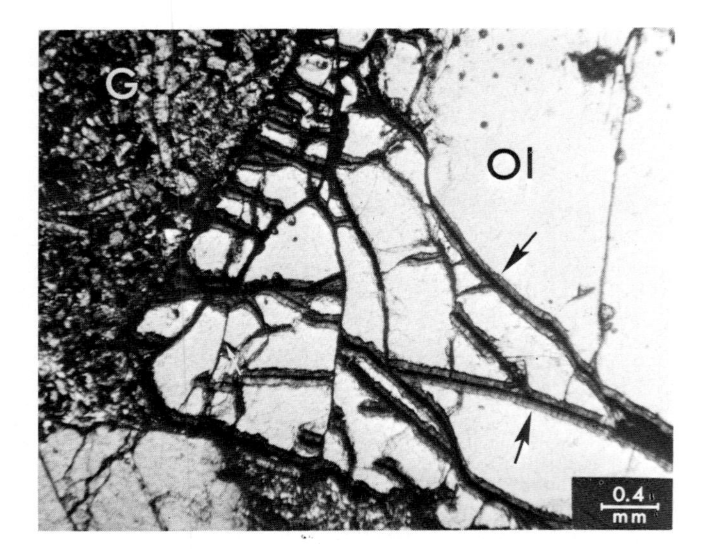

Fig. 502. Alteration of the olivine along a crack system of the olivine. Ol = olivine; G = basaltic groundmass. Arrows show the alteration of olivine following a crack pattern of the host. Olivine basalt. Addis Ababa, Rift Valley, Ethiopia.

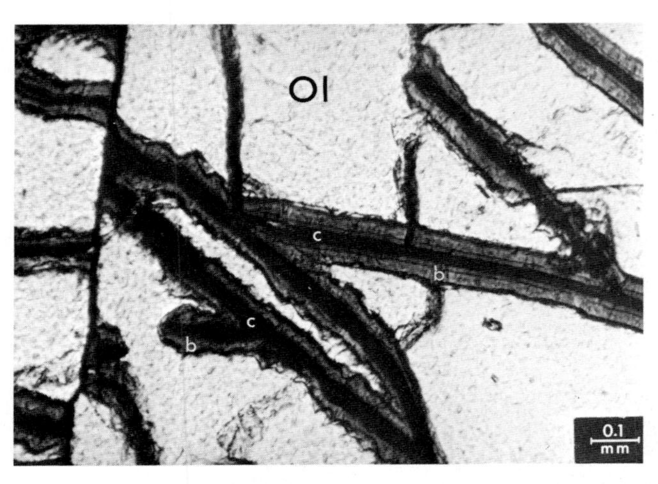

Fig. 503. Detail of Fig. 502, showing the alteration cracks of the olivine in greater detail. The olivine cracks are occupied centrally by opaque (sub-metallic in character) iron-oxides with a fibrous pleochroic brown chloritic material between the central iron-oxide mass and the olivine. Ol = olivine; c = central, iron-oxide mass occupying the centre of the olivine alteration cracks; b = brown chloritic mass consisting of minute, elongated fibres perpendicular to the walls of the cracks.
Olivine basalt. Addis Ababa, Rift Valley, Ethiopia.

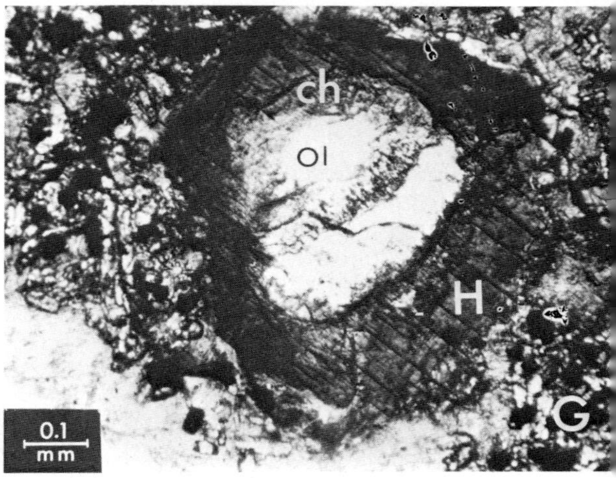

Fig. 505. Olivine with a green-chlorite alteration and with brown-hornblende alteration. It should be noticed that a distinct cleavage develops in the olivine alteration product. Ol = olivine; ch = green-chlorite alteration product of the olivine; H = hornblende alteration product of the olivine (with distinct cleavage); G = basaltic groundmass.
Olivine basalt. Addis Ababa, Rift Valley, Ethiopia. With crossed nicols.

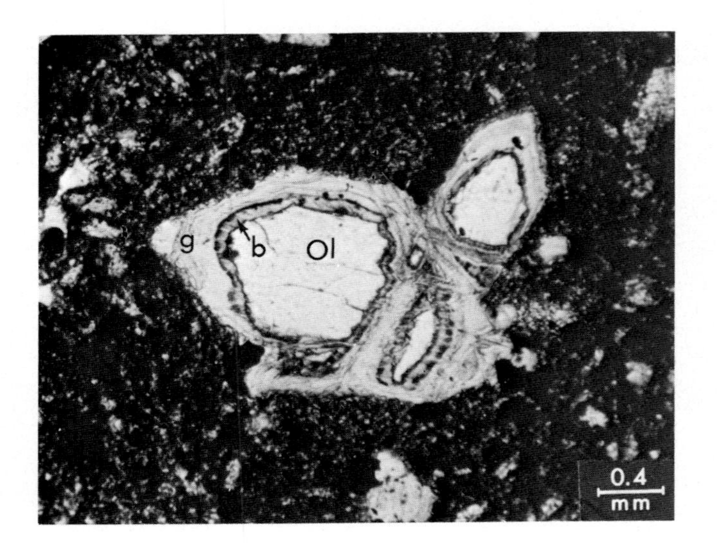

◀ Fig. 504. Olivine with a brown fringe of alteration and with an outer margin of green-chlorite formation. Ol = olivine; b = brown-chloritic fringe immediately surrounding the olivine; g = green-chlorite outer alteration product of the olivine.
Basalt. Hohe Wostrel Aussig, Bohemia, Czechoslovakia. With crossed nicols.

Fig. 506. Picritic basalt with idiomorphic phenocrystalline olivine transversed by veinlets of antigorite and with a veinlet of "antigorite", circular in outline, transversing the olivine (the circular structure is its circular section). Ol = olivine (idiomorphic phenocryst of olivine in picritic basalt); An = antigorite veinlets transgressing the olivine phenocryst; arrow shows circilur "antigoritic" veinlet transgressing the olivine.
Picritic basalt. Agia Marina Syliatou, Cyprus. With crossed nicols.

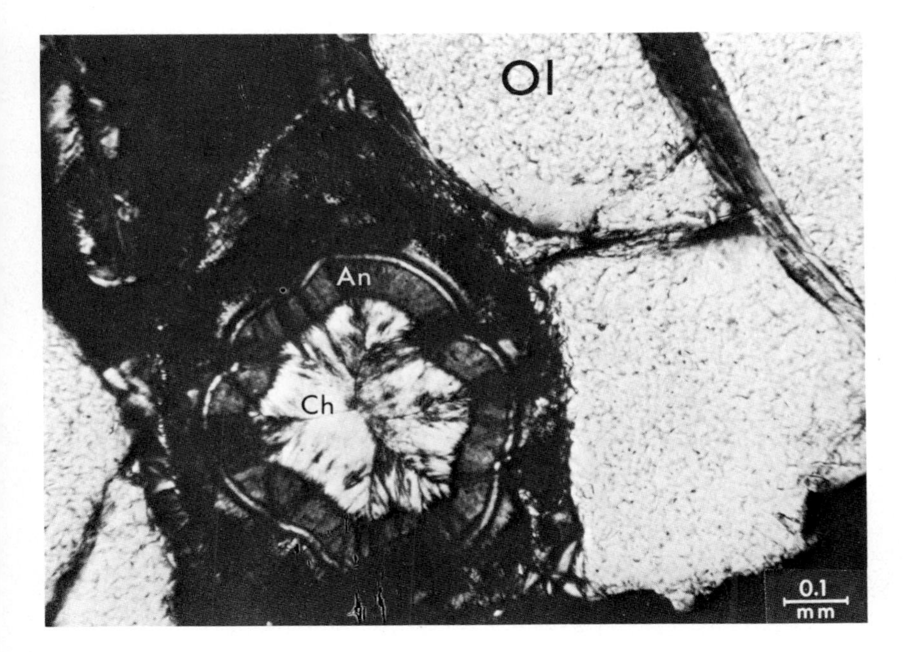

Fig. 507. A circular section perpendicular to the length of a veinlet invading a basaltic picrite (see Fig. 506). The circular section of the veinlet shows antigorite (An) in the periphery and the radiating chlorite (Ch) in the central part of the vein. Ol = olivine phenocryst.
Picritic basalt. Agia Marina Syliatou, Cyprus. With crossed nicols.

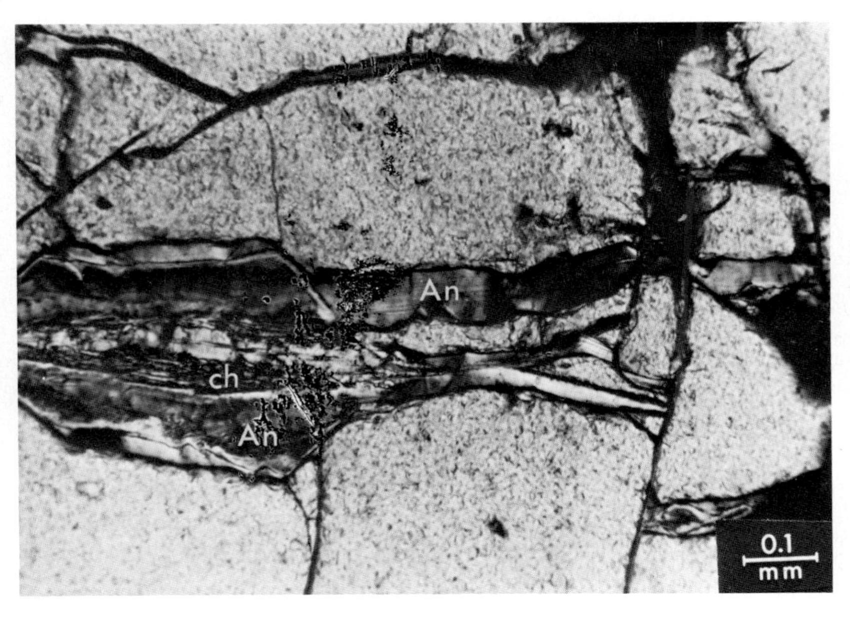

Fig. 508. A longitudinal section of a composite veinlet invading an olivine phenocryst of a picritic basalt. The margins of the veinlet (in longitudinal section) consist of banded antigorite (the banding parallel to the longitudinal section). The central part of the composite veinlet consists of fine crystalline chlorites. An = Antigorite, forming the walls of the veinlet; ch = chlorite, forming the centre of the veinlet.
Picritic basalt. Agia Marina Syliatou, Cyprus. With crossed nicols.

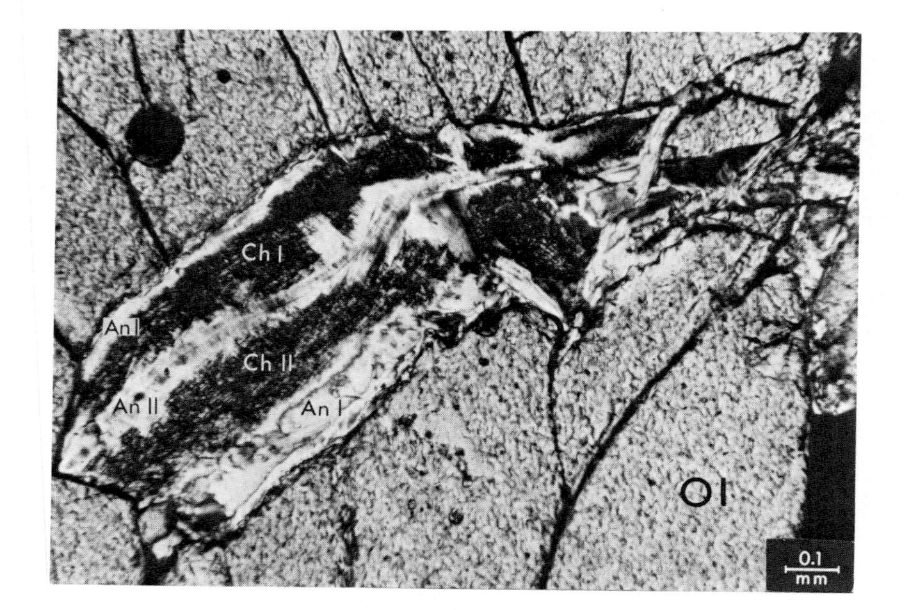

Fig. 509. An alteration body veinform in character, extending from cracks of an olivine into the host crystal, consisting of alterations of antigorite (An I), fine chlorite (Ch I), banded antigorite (An II), fine crystalline chlorite (Ch II) and again antigorite (corresponding to antigorite An I). Ol = olivine phenocryst of the picritic basalt.
Picritic basalt. Agia Marina Syliatou, Cyprus. With crossed nicols.

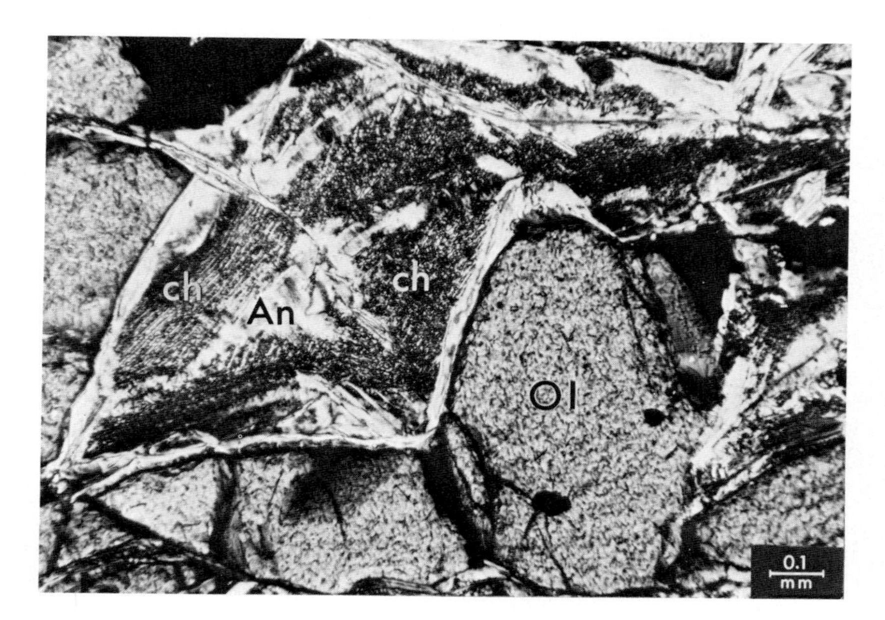

Fig. 510. Is comparable to Fig. 509 and shows alternating repetitions of antigorite and fine chlorite. Longitudinal section of veinform body in picritic olivine. An = antigorite; ch = fine chlorite; Ol = olivine.
Picritic basalt. Agia Marina Syliatou, Cyprus. With crossed nicols.

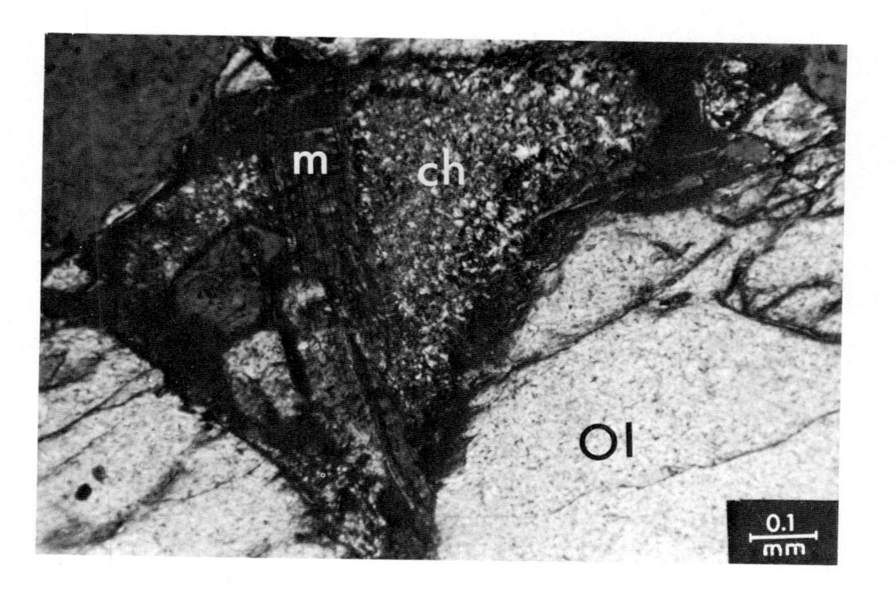

Fig. 511. Olivine picritic phenocryst, with a fine chloritic mass invading the olivine in veinform and with a micaceous neocrystallization in the chloritic mass. Ol = olivine phenocryst; ch = fine chlorite extension in the olivine; m = micaceous (biotite) neocrystallization in the chloritic mass.
Agia Marina Syliatou, Cyprus. With crossed nicols.

Fig. 512. Idiomorphic olivine phenocryst with an iddingsite margin. Ol = olivine idiomorphic grain; id = iddingsite alteration margin of olivine.
Olivine basalt. Nidda, Hessen, Germany. With crossed nicols.

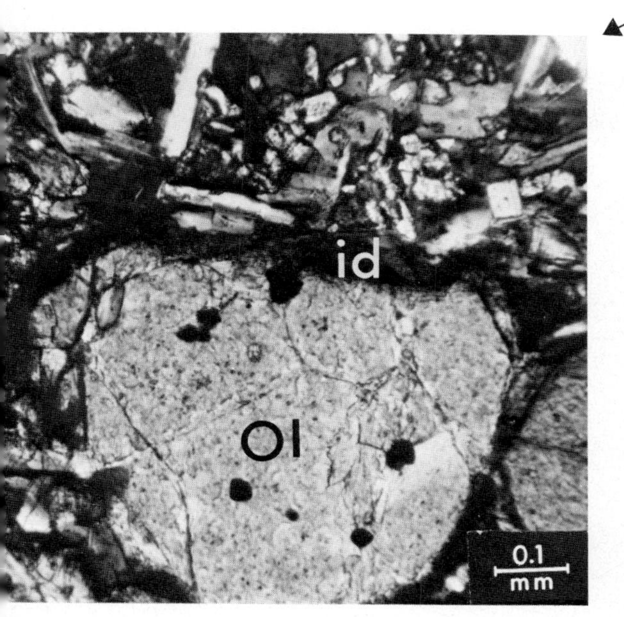

Fig. 513. Olivine grains with iddingsite alteration, marginally and along cracks of the olivine. Ol = olivine grain, id = iddingsite margins and iddingsite following cracks of the olivine.
Olivine basalt. Nidda Hessen, Germany. With crossed nicols.

Fig. 515. Olivine almost entirely iddingsitized. The iddingsitization of olivine proceeded along cracks of the olivine. Ol = olivine relics within the extensive substitution of olivine by iddingsite; id = iddingsite replacing the olivine. Arrows show iddingsitization of olivine along cracks.
Olivine basalt. Debra-Sina, Ethiopia. Without crossed nicols.

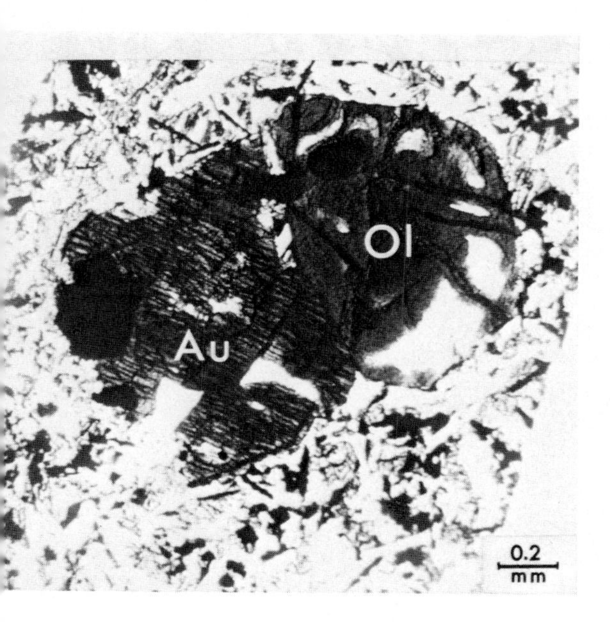

Fig. 514. Both olivine and augite are completely transformed into iddingsite. Ol = iddingsite pseudomorph after olivine; Au = iddingsite pseudomorph after augite.
Olivine basalt. Debra-Sina, Ethiopia. Without crossed nicols.

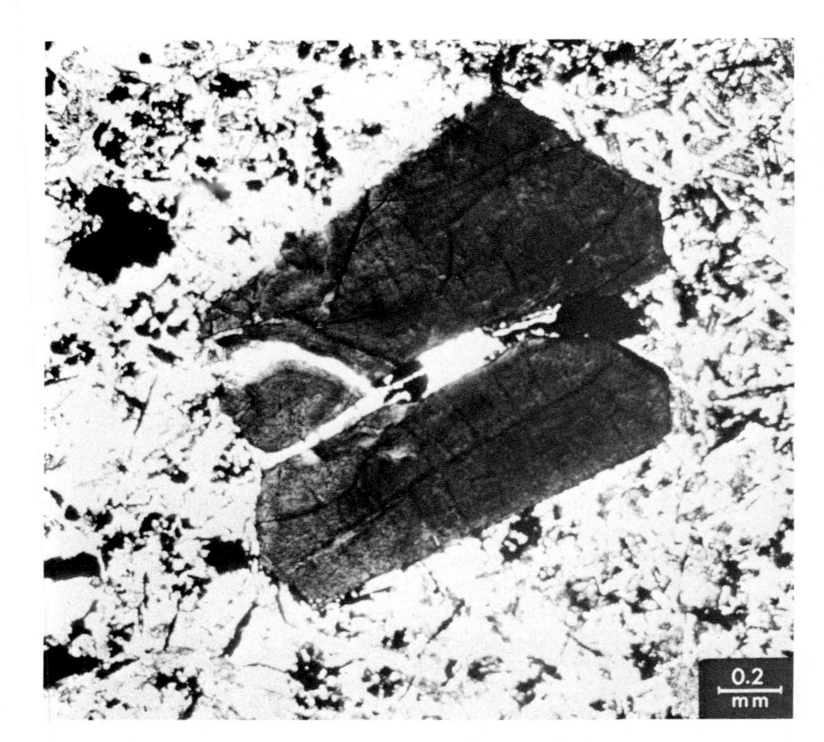

Fig. 516. Complete iddingsitization of olivine grains, i[n] basaltic groundmass.
Olivine basalt. Debra-Sina, Ethiopia. Without crosse[d] nicols.

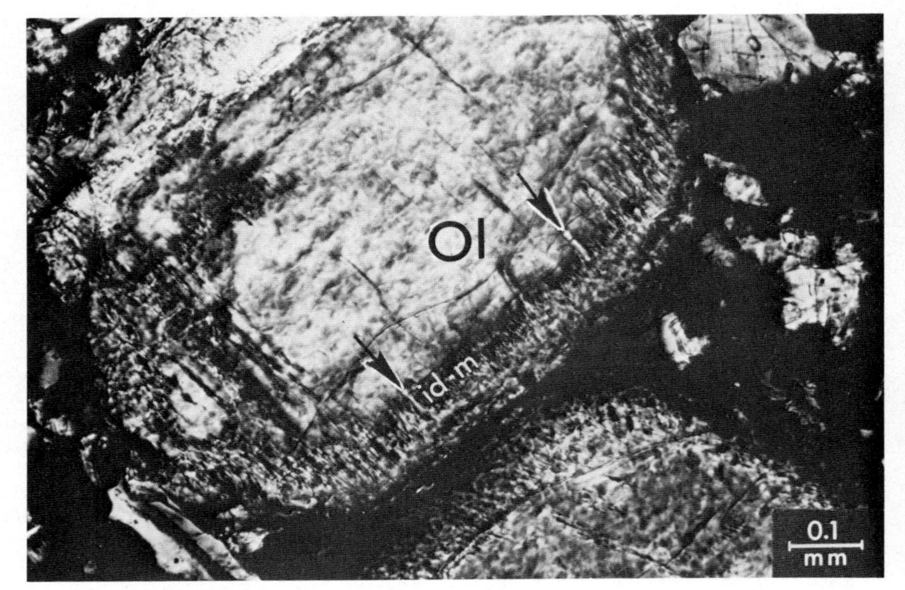

Fig. 517. Olivine with margins of iddingsite formation; along intracrystalline directions of penetrability, extensions of the marginal iddingsite protrude into the unaltered olivine. Ol = olivine; id-m = iddingsite alteration margin. Arrows show marginal iddingsite extensions into the olivine.
Olivine basalt. Addis Ababa, Rift Valley, Ethiopia. With crossed nicols.

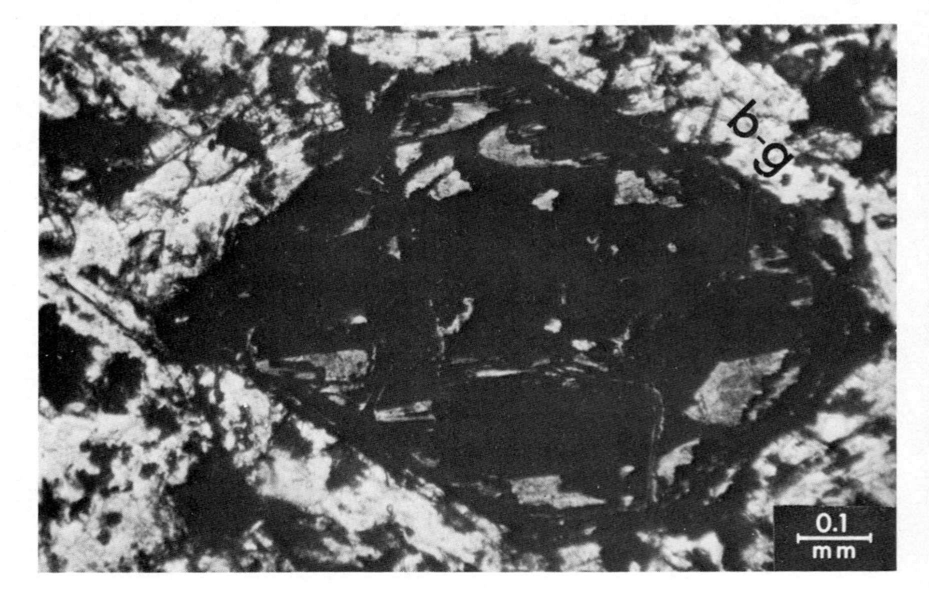

Fig. 518. Olivine idiomorph almost completely replaced by iron oxides, some relic olivine parts are present. Black = initially olivine, idiomorphic phenocryst replaced by iron oxide; b-g = basaltic groundmass.
Olivine basalt. Ethiopian Rift. With crossed nicols.

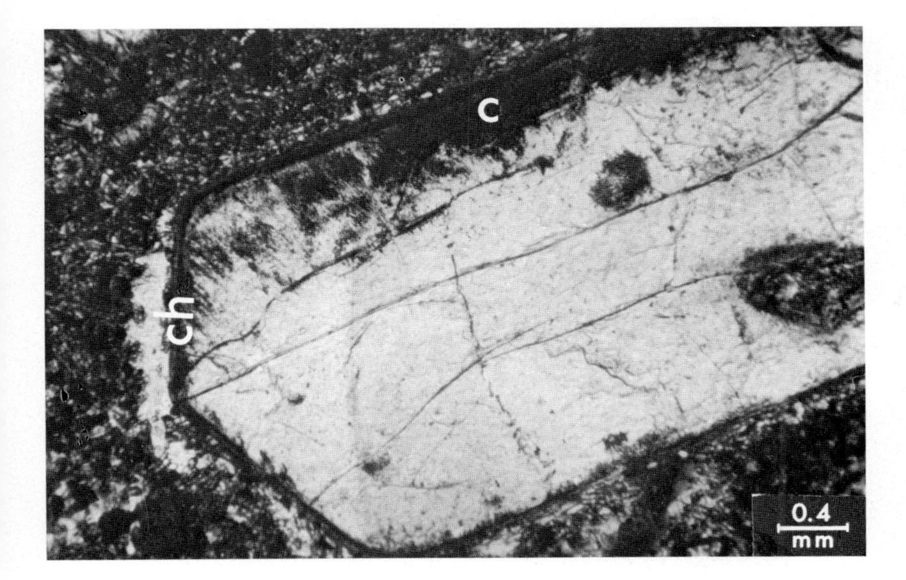

Fig. 519. Marginal chloritization of olivine with intracrystalline clouding of the olivine. ch = chloritic margin of olivine; c = clouding of the olivine.
Olivine basalt. Nidda Hessen, Germany. With crossed nicols.

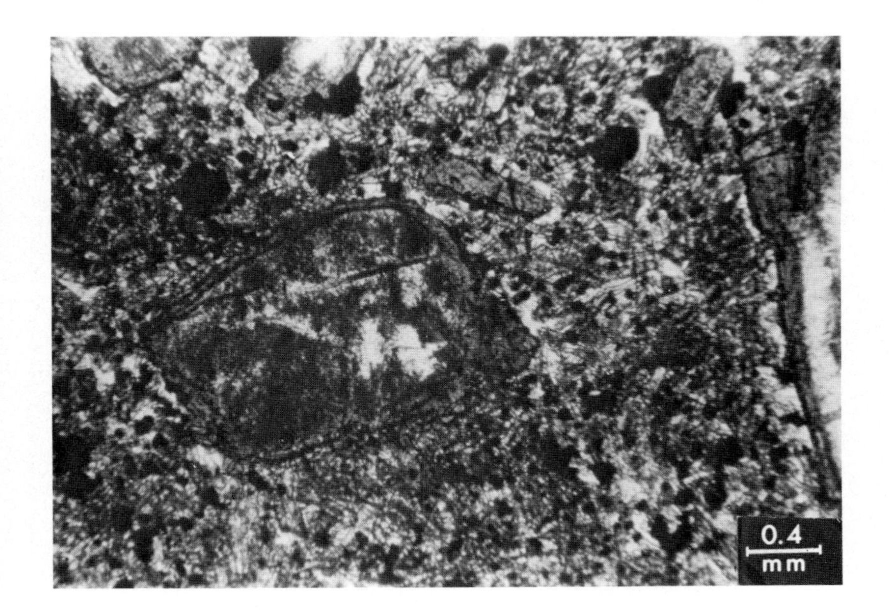

Fig. 520. Marginal chloritization of olivine with extensive intracrystalline clouding of the olivine.
Olivine basalt. Nidda, Hessen, Germany. With crossed nicols.

Fig. 521. Olivine almost in the extinction position with a margin consisting of serpentine (antigorite). Magnesite is present as alteration product of the olivine between the olivine and the serpentine. Ol = olivine; S = serpentine; Ma = magnesite.
Dunite, Yubdo, Wollaga, Ethiopia. With crossed nicols.

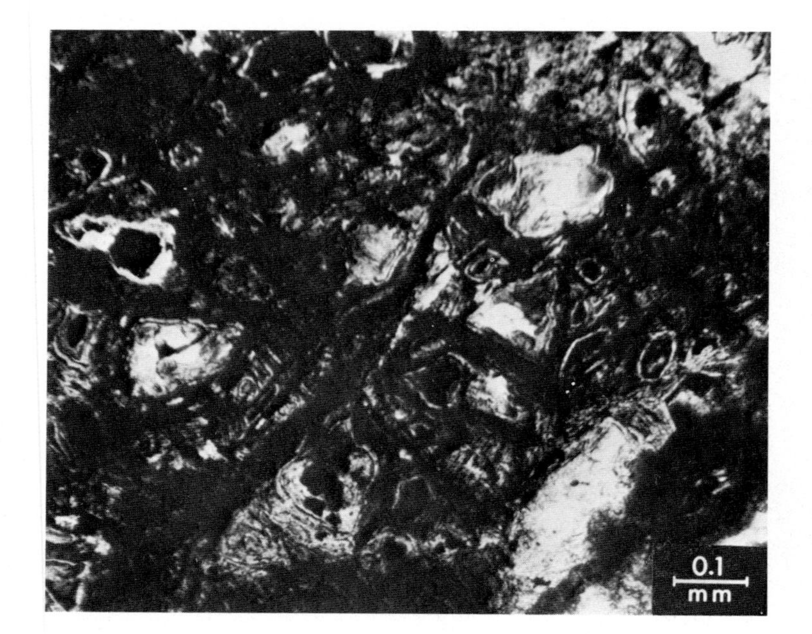

Fig. 522. Initial dunite with granular olivine altered due to leaching out of Mg with the result of formation of a granular structure consisting of recrystallized silica and limonite.
Mandoudi, Euboea, Greece. With crossed nicols.

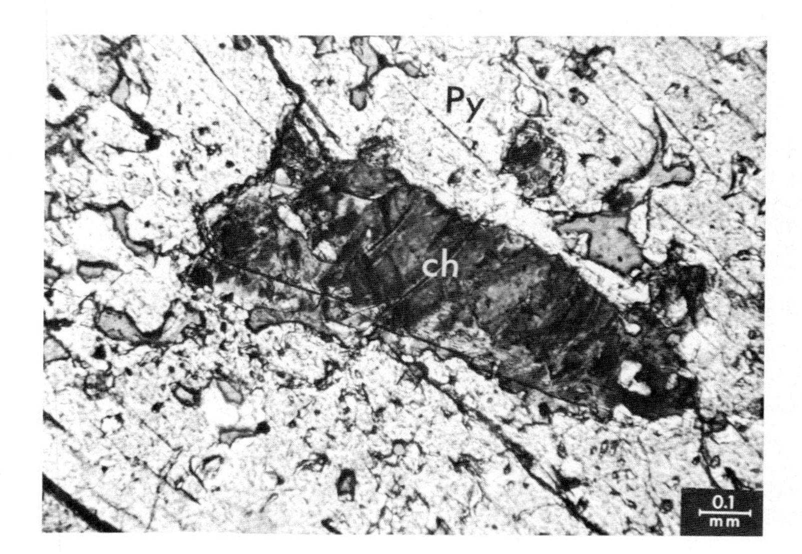

Fig. 523. Alteration of pyroxene, resulting in the formation of green chlorites. The chloritization can be seen as a metasomatism. Py = pyroxene; ch = chlorite formation.
Basalt. Mugher-Selale, Ethiopia. With crossed nicols.

Fig. 524. Augitic microcavity with infilling of fine crystalline chlorite. In the central part of the microcavity the chlorite becomes more coarse crystalline. Py = pyroxene; f-ch = fine crystalline chlorite at walls of microcavity; c-ch = coarser-grained chlorite at central part of microcavity.
Mugher-Selale, Ethiopia. With crossed nicols.

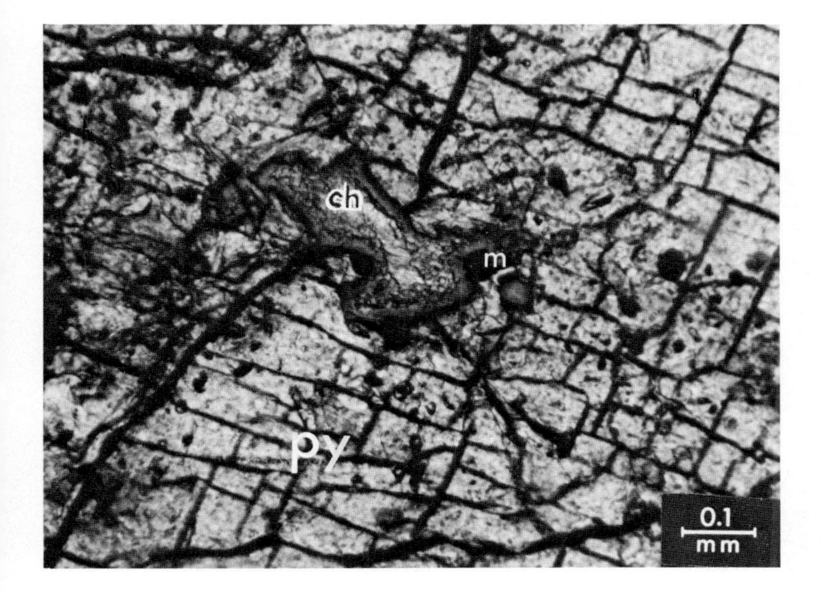

Fig. 525. Chloritization of pyroxene. The process should be seen more or less as a metasomatic replacement of the pyroxene, as is indicated by the presence of magnetite both in the chlorite and crossing the chlorite–magnetite boundary. py = pyroxene; ch = chlorite; m = magnetite.
Augite basalt. Mugher-Selale, Ethiopia. With crossed nicols.

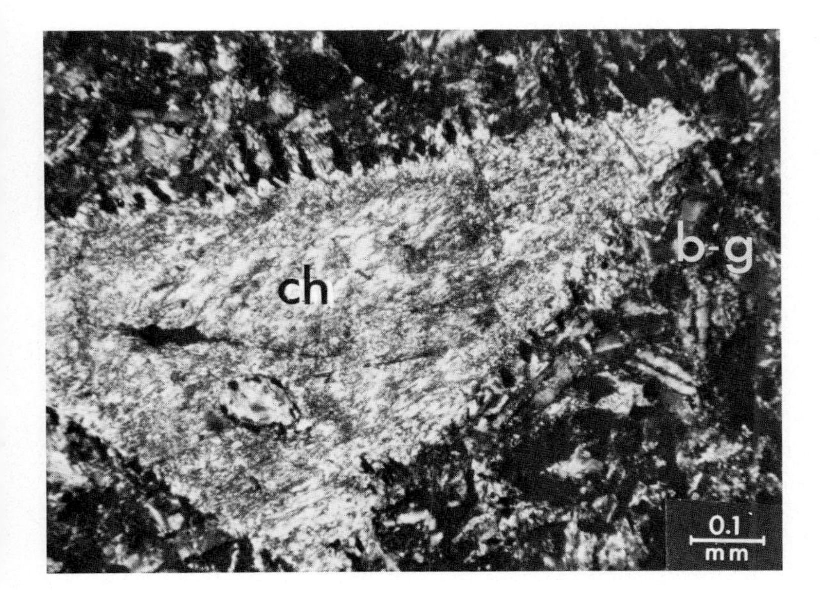

Fig. 526. Initial pyroxene phenocryst completely replaced by chlorite. ch = chlorite which has completely replaced the initial pyroxene phenocryst. b-g = basaltic groundmass.
Basalt. Albany, Maine, U.S.A. With crossed nicols.

Fig. 527. Zoned plagioclase with the inner zone indicating more intense chloritization. pl-i = plagioclase internal zone; pl-e = plagioclase external zone; ch = chloritization of internal plagioclase zone.
Olivine basalt. Khidane Meheret (between Entoto and British Embassy), Addis Ababa, Ethiopia. With crossed nicols.

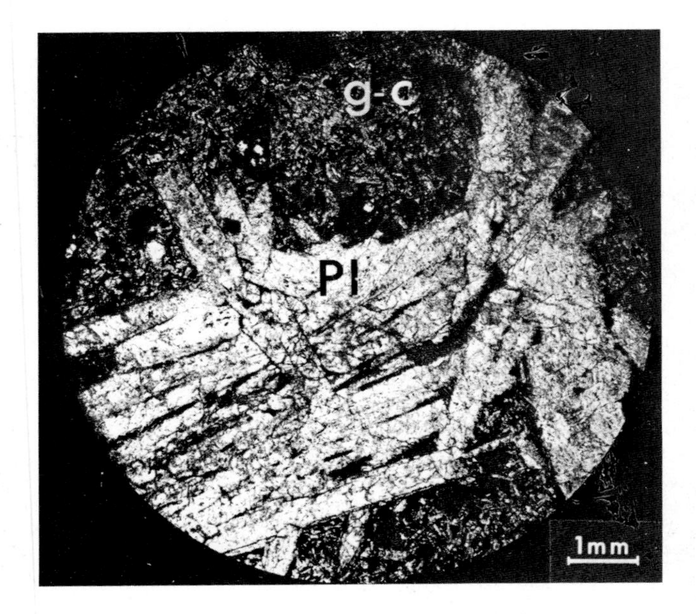

Fig. 528. Plagioclase phenocryst altered (chloritized) in a groundmass. Pl = plagioclase chloritized; g-c = chloritized basaltic groundmass.
Olivine basalt. Debra-Sina. Ethiopian part of Great Rift Valley. With crossed nicols.

Fig. 529. Plagioclase phenocrysts with crystal cavities filled with chlorite. Pl = plagiolcase phenocrysts; ch = chlorite filling crystal cavity. Khidane Meheret (between Entoto and British Embassy), Addis Ababa, Ethiopia. With crossed nicols.

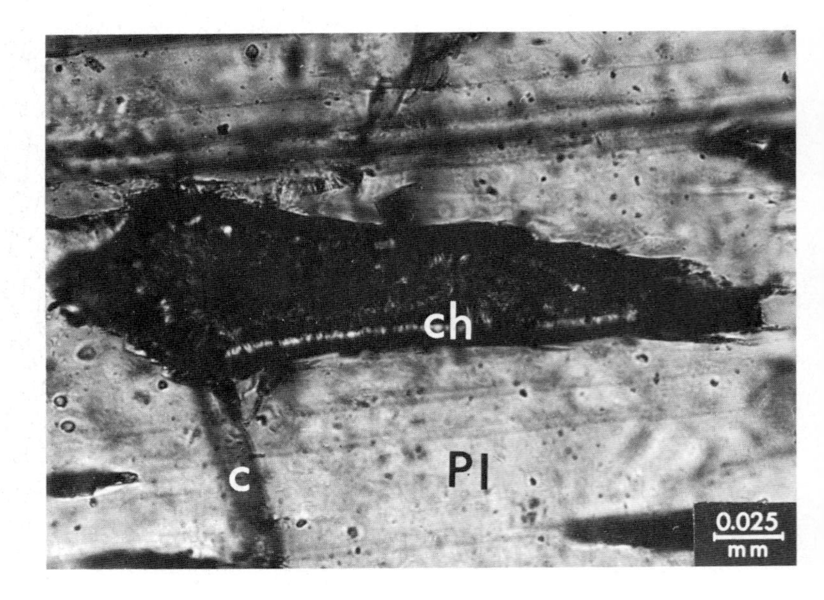

Fig. 530. Detail of Fig. 529. Crystal microcavity with chlorite lining on the walls and with a feeding microchannel shown. Pl = plagioclase; ch = chloritic lining; c = feeding channel of the chloritic microcavity.
With crossed nicols.

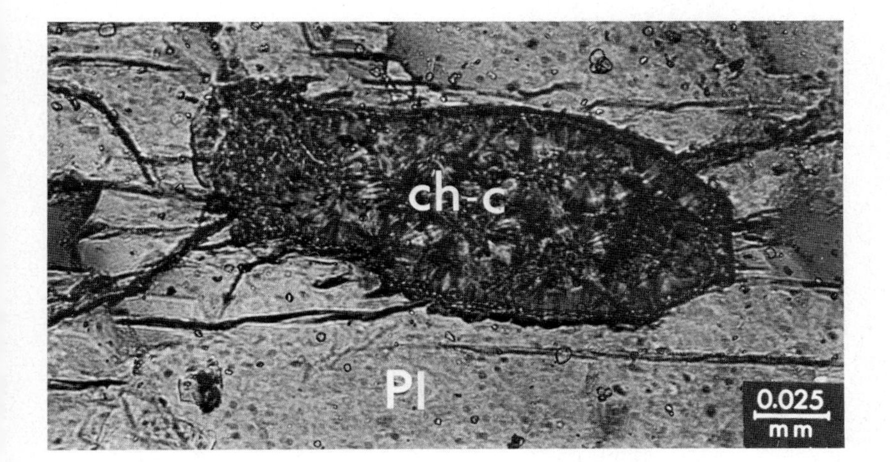

Fig. 531. Plagioclase with crystal microcavity filled with chlorite. Pl = plagioclase; ch-c = chloritic microcavity. Khidane Meheret (between Entoto and British Embassy), Addis Ababa, Ethiopia. Without crossed nicols.

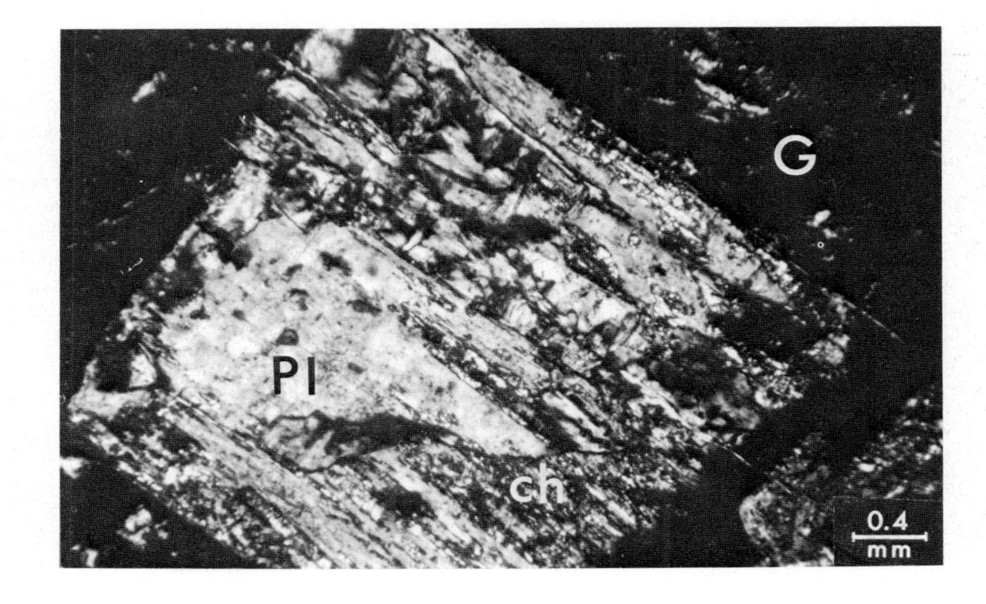

Fig. 532. Plagioclase phenocrystal replacement by chloritic aggregates. Pl = plagioclase; ch = chloritic aggregates replacing the feldspar; G = basaltic groundmass. Amygdaloidal melaphyre. Oberstein, Nahe River, Germany. With crossed nicols.

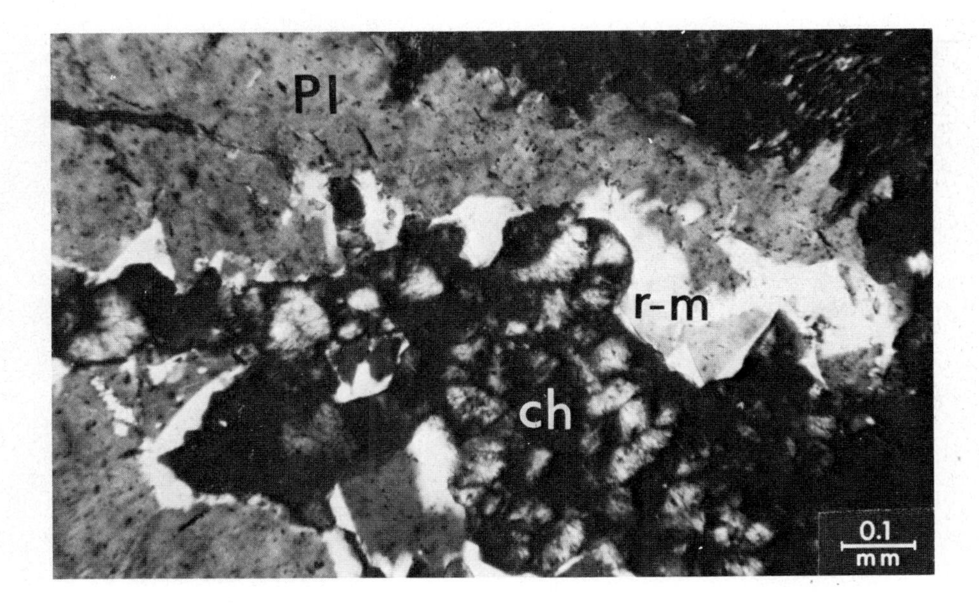

Fig. 533. Glauconitization of basalt with green-chlorite formation replacing the groundmass and the plagioclase basaltic phenocrysts causing a reaction margin with the feldspar. Pl = plagioclase; ch = chlorite (glauconitization); r-m = reaction margin of the plagioclase with the chlorite. Oceanic floor, Atlantic Ocean. With crossed nicols.

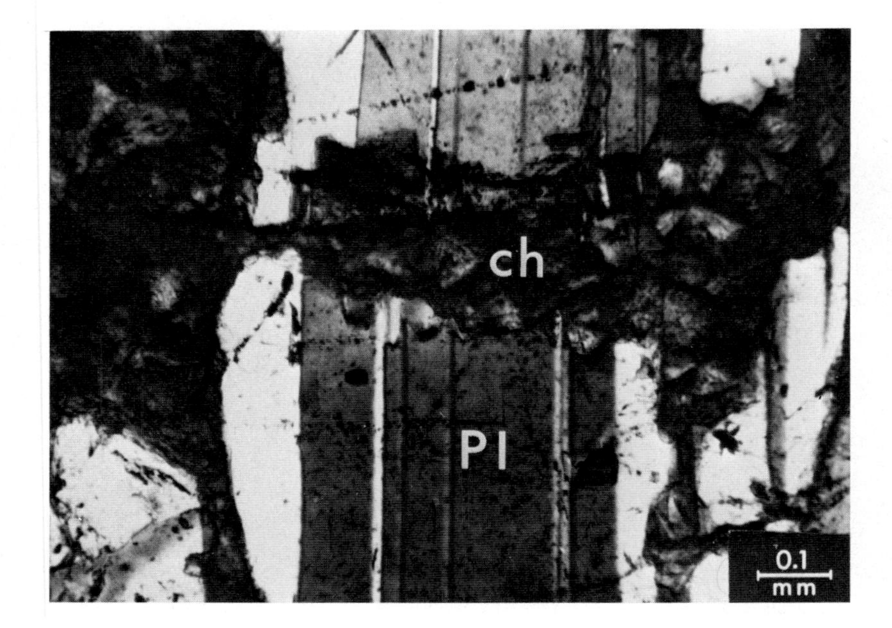

Fig. 534. Glauconitization of basalt with green-chlorite formation replacing the groundmass and invading and replacing the plagioclase phenocryst. Ch = green chlorite replacing the basaltic groundmass; Pl =plagioclase.
Oceanic floor, Atlantic Ocean. With crossed nicols.

Fig. 535. Plagioclase phenocryst with the basaltic groundmass partly replaced by chlorite which also partly invades the plagioclase phenocryst. Pl = plagioclase; ch = chlorite, extensively present in the groundmass and filling the spaces between the plagioclase groundmass laths; Ch-p = chlorite infiltrating the plagioclase phenocryst along cracks; black = magnetite. Amygdaloidal basalt.
Tramkampstadur, Iceland. With crossed nicols.

Fig. 536. The groundmass of an olivine picritic basalt with chloritic orientated cavities (the long axis is perpendicular to the plane of the photograph and occupies the interspaces between the plagioclase laths). The cavities have an antigorite margin and chloritic infillings. Chloritic cavity occupying the entire space between laths of feldspar. Ch = chloritic cavity, occupying the interspaces between the groundmass plagioclase laths; An = antigorite margin of chloritic cavity; pl = plagioclase feldspar laths.
Picritic basalt. Agia Marina, Syliatou, Cyprus. With crossed nicols.

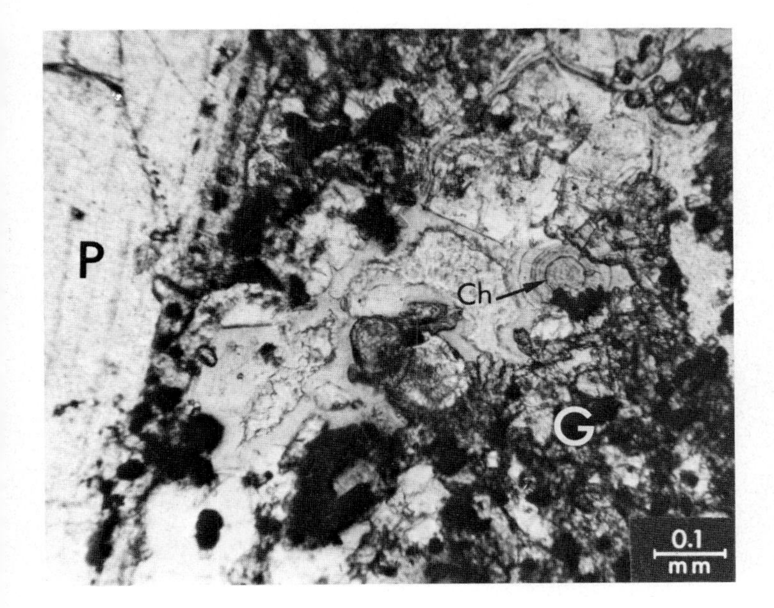

Fig. 537. Colloform chlorite as groundmass, intergranular space-filling and substitution. G = basaltic groundmass; Ch = colloform chlorite; P = pyroxene. Basalt. Mugher-Selale, Ethiopia. With crossed nicols.

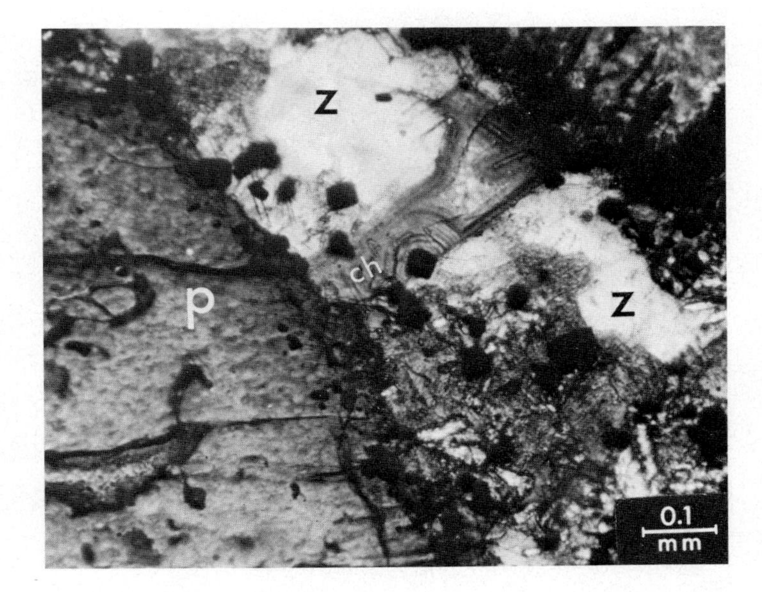

Fig. 538. Groundmass magnetite as nuclei for the formation of colloform chlorite. Also zeolites are present as groundmass space filling. Black = magnetite; ch = chlorite colloform; z = zeolites as groundmass space filling; p = pyroxene.
Basalt. Mugher-Selale, Ethiopia. With crossed nicols.

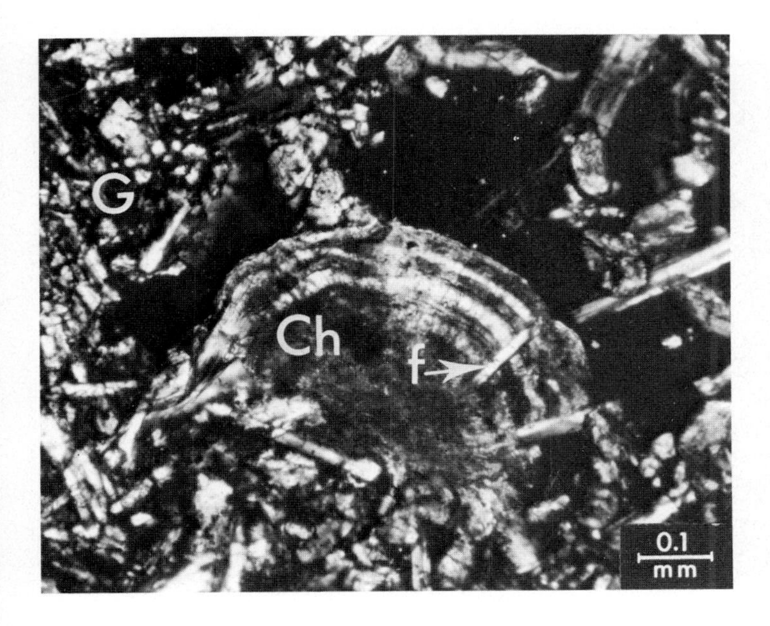

Fig. 539. Banded colloform chloritic structure as groundmass-space filling and with feldspar laths perpendicular to the banded chlorites. Ch = chlorites (banded); f = feldspar lath; G = basaltic groundmass. Amygdaloidal basalt. Tramkampstadur, Iceland. With crossed nicols.

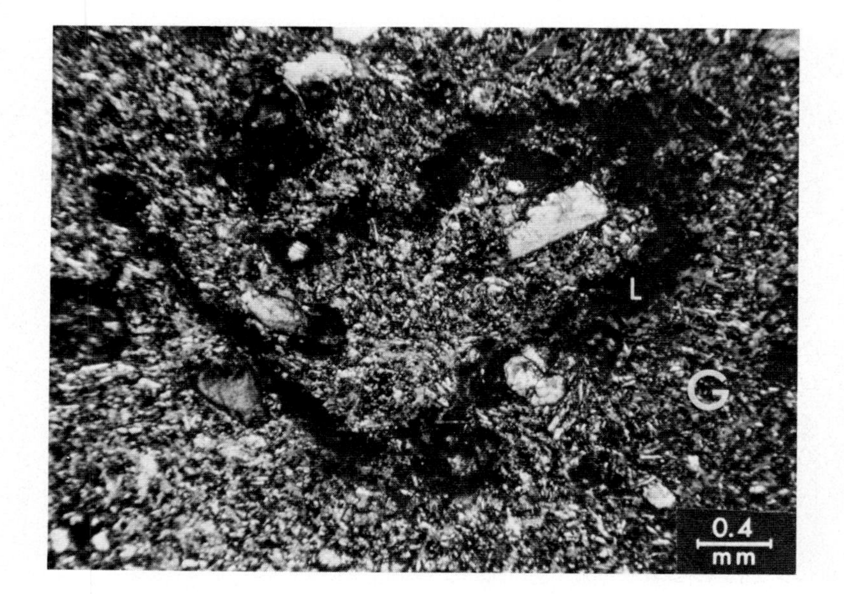

Fig. 540. A limonitic diffusion ring-structure in basalt due to chemical weathering which involves iron mobilization and deposition. G = basaltic groundmass; L = diffused limonite deposition due to chemical weathering of basalt.
Basalt spheres. Oberkassel, Siebengebirge, Germany. Without crossed nicols.

Fig. 541. Spheroidal weathering of basalt with exfoliation and lateritic soil formation. B-n = unweathered basaltic relic, e-x = exfoliation accompanying spheroidal weathering; L = lateritic soil formation often with spheroidal relic patterns.
Olivine basalt. Debra-Sina, Ethiopia. Spheroidal structures about 20 cm in diameter.

Fig. 542. Spheroidal basalt weathering with a pilling off of layers.
Basalt. Nicosia–Troodos, Cyprus.

Fig. 543. Spheroidal basalt weathering, starting and developing along basalt joints. Unweathered basaltic relics, spheroidal in shape, in laterite of basaltic derivation. U-b = unweathered basalt; L = lateritic basaltic soil; arrows show basaltic joints. Weathered basalt. About 15 km west of Ambo, Ethiopia.

Fig. 544. Lateritic covers due to basalt weathering. The lateritization of basalt is in a more advanced stage than that shown in Fig. 543. A spheroidal basaltic relic and relic spheroidal pattern, completely lateritized, are shown. A = spheroid basaltic relic in completely lateritized basalt; B = a spheroidal pattern in laterite representing initially spheroidal basalt. Spheroidal structure about 30 cm in diameter. Laterized basalt. About 15 km west of Ambo, Ethiopia.

Fig. 545. Completely lateritized basalt forming covers about 20 m thick.
Lateritic cover of basalts. Siere-Wollaga, Ethiopia.

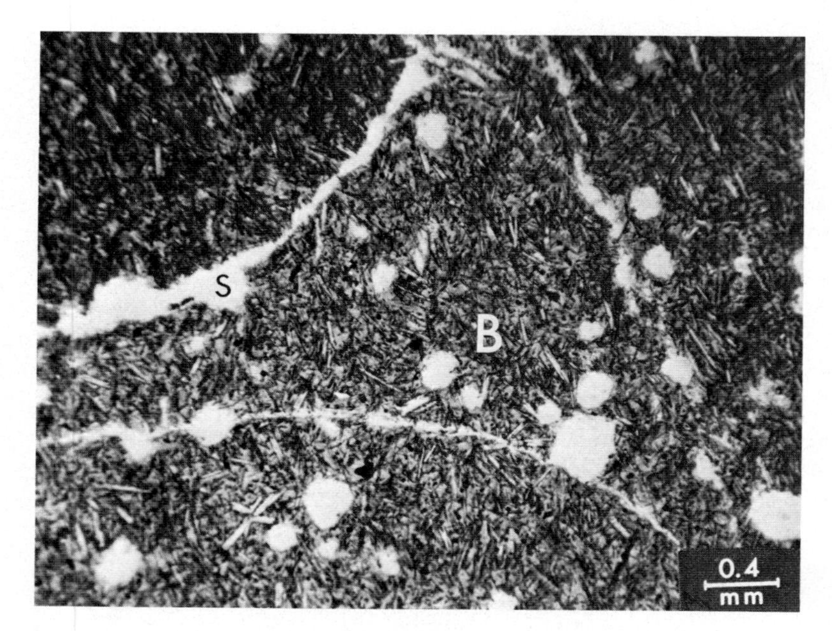

Fig. 546. Weathered fine-grained basalt showing "Sonnenbrenner" structure. B = basalt (fine-grained); S = Sonnenbrenner cracks as curved fissures in the basalt.
Basalt. Troodos, Cyprus. Without crossed nicols.

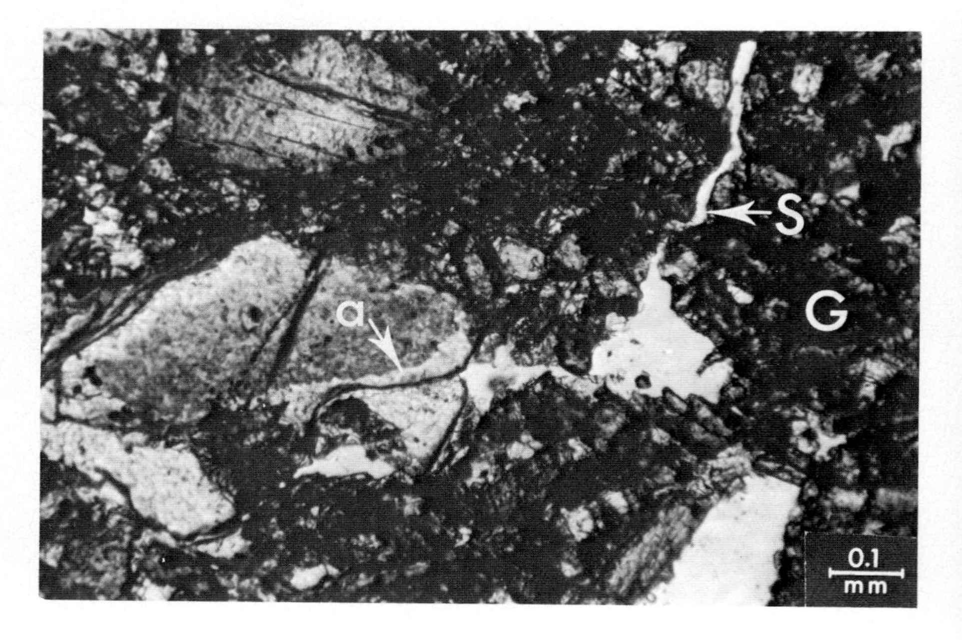

Fig. 547. "Sonnenbrenner" (comparable to stylolitic effects) structure transversing the basaltic groundmass and through an olivine phenocryst "producing noticeable" chemical effects on the olivine along the "Sonnenbrenner" crack. G = basaltic groundmass; S = "Sonnenbrenner" cracks transgressing the olivine phenocryst and the groundmass; arrow "a" shows the chemical effects on the olivine phenocryst as it is transgressed by the "Sonnenbrenner" cracks.
Basalt, "Sonnenbrenner". Westerburg, Westerwald, Germany. Without crossed nicols.

Fig. 548. A fine crack transversing a feldspar phenocryst and showing chemical effects along the feldspar crack. G = basaltic groundmass; f = feldspar phenocryst; arrow "a" shows chemical effects accompanying micro-crack transversing the feldspar phenocryst.
Basalt, "Sonnenbrenner". Rothenkopf, Möchröden, Oberfranken, Germany. With crossed nicols.

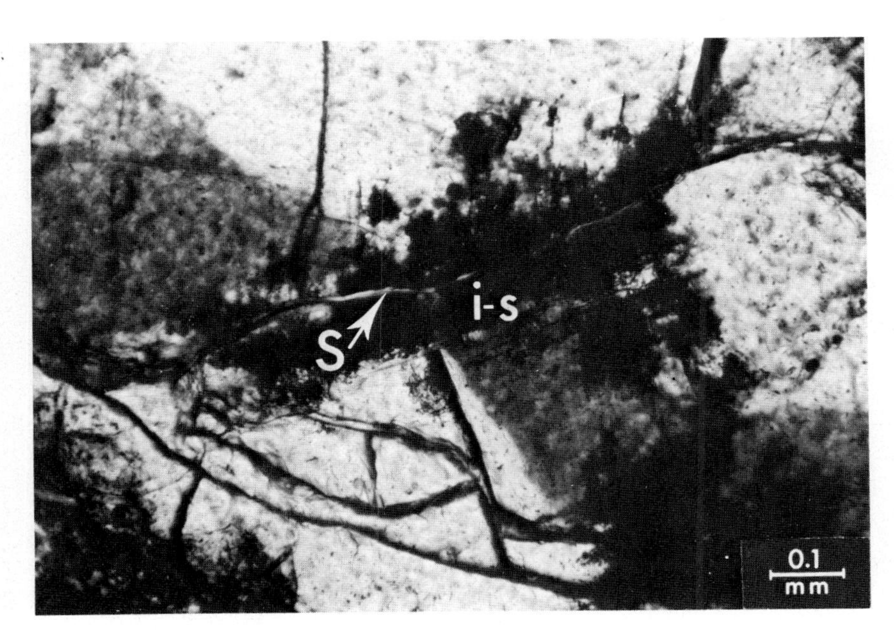

Fig. 549. "Sonnenbrenner" crack transversing basaltic phenocrysts and producing a limonitization effect along the "Sonnenbrenner" crack. S = "Sonnenbrenner" crack system transversing a plagioclase phenocryst; i-s = iron staining associated with the "Sonnenbrenner" cracks.
Basalt, "Sonnenbrenner". Rothenkopf, Möchröden, Oberfranken, Germany. Without crossed nicols.

Fig. 550. Olivine phenocryst and basaltic groundmass transversed by "Sonnenbrenner" crack which is partly filled with limonite. Ol = olivine phenocrysts; G = basaltic groundmass; Li = limonitic material, partly occupying the "Sonnenbrenner" structure; S = "Sonnenbrenner" cracks.
Basalt, "Sonnenbrenner". Westerburg, Westerwald, Germany. Without crossed nicols.

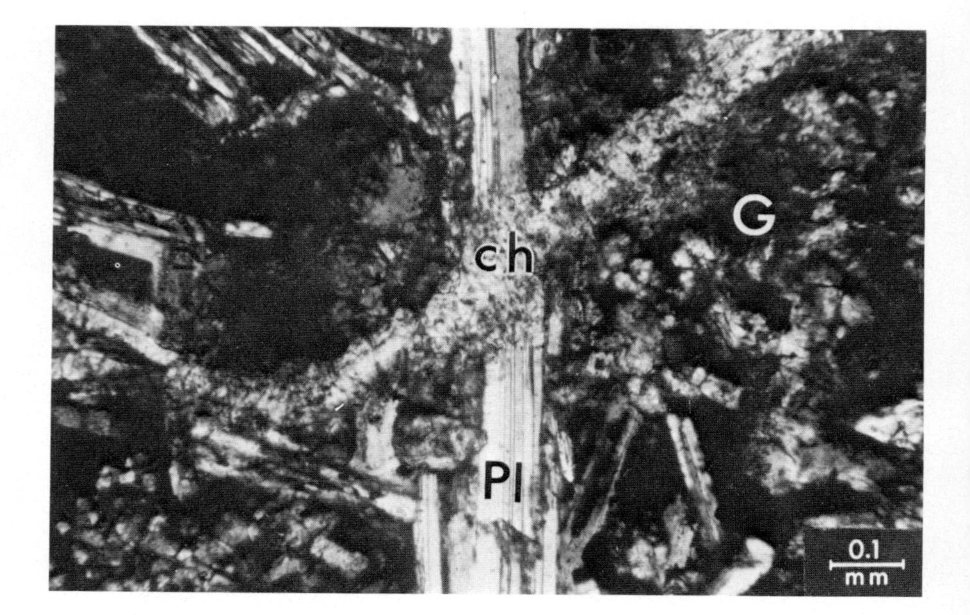

Fig. 551. Basaltic groundmass and elongated feldspar phenocrysts transversed by a veinlet of chloritic material. G = basaltic groundmass; Pl = plagioclase phenocryst; ch = diffused veinform chloritic mass transversing both the groundmass and the feldspar phenocryst.
Doleritic basalt. Marienberg, Westerwald, Germany. With crossed nicols.

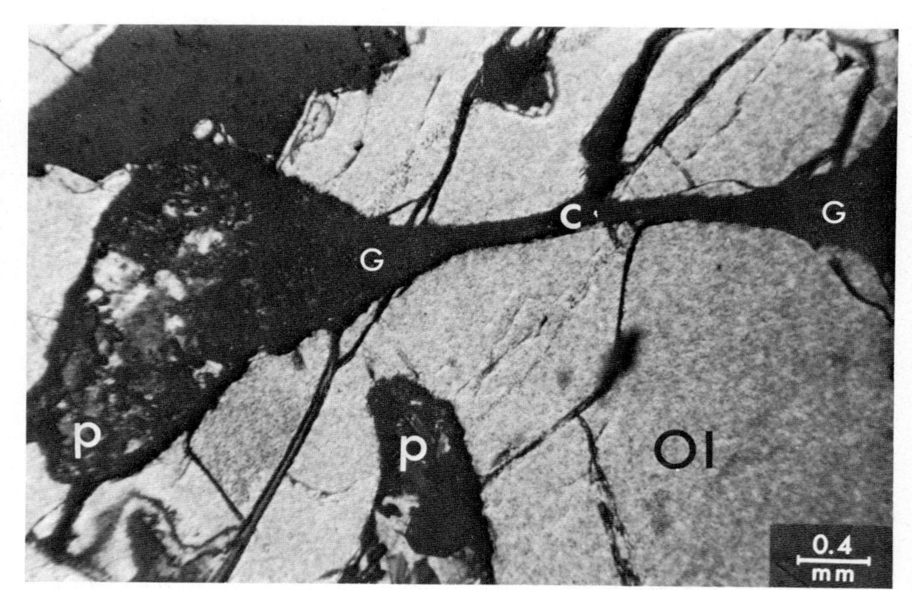

Fig. 552. Olivine phenocryst with basaltic groundmass, extensions of which form groundmass enclaves "pockets" along cracks within the olivine phenocryst. Ol = olivine phenocryst; G = basaltic groundmass, extensions of which follow cracks of the olivine and form pockets in the olivine; c = cracks of the olivine invaded by the groundmass; p = groundmass pocket as "enclave" in the olivine.
Alkali olivine basalt. Bona–Oberkassel, Siebengebirge, Rhineland, Germany. With crossed nicols.

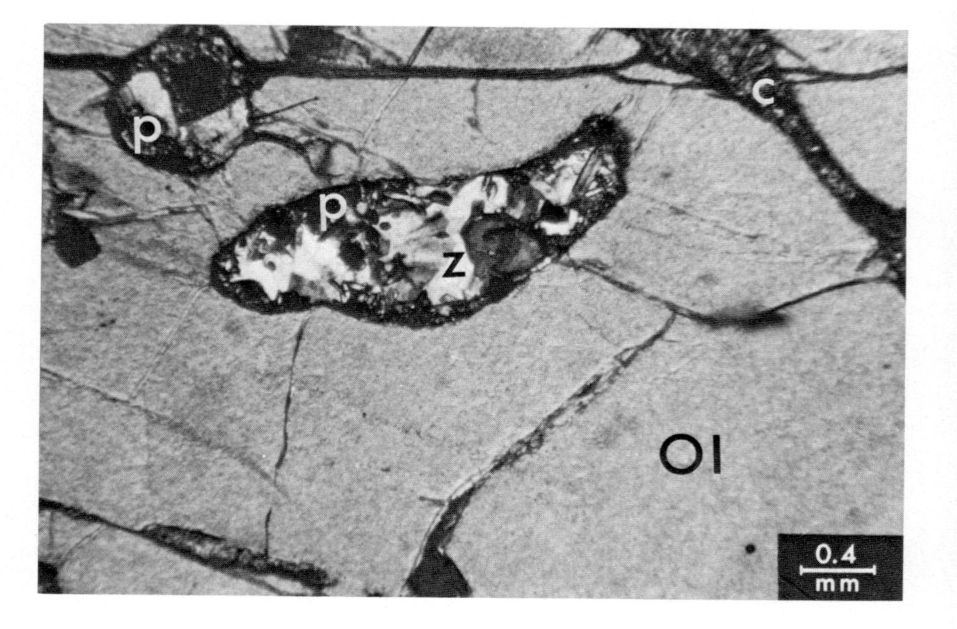

Fig. 553. Olivine phenocryst with groundmass pockets (enclaves) in the phenocryst. Also groundmass following cracks of the olivine phenocryst. Ol = olivine; p = pockets of groundmass in the olivine phenocrysts; c = cracks of the olivine occupied by groundmass; z = zeolites as microcavity fillings within the components of groundmass enclaves in the phenocryst.
Olivine basalt. Bonn–Oberkassel, Siebengebirge, Rhineland, Germany. With crossed nicols.

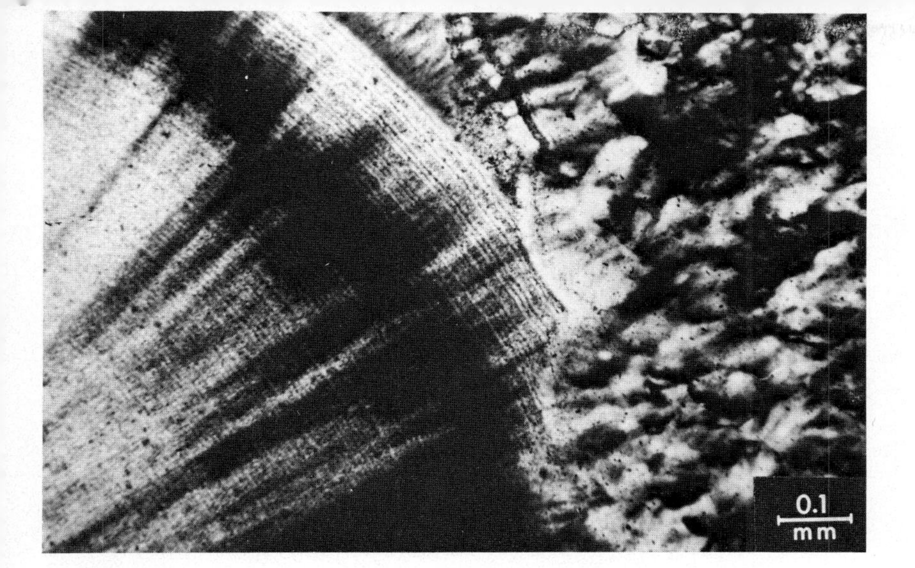

Fig. 571. Detail of Fig. 570. Fine-banded colloform structure shown in detail. With crossed nicols.

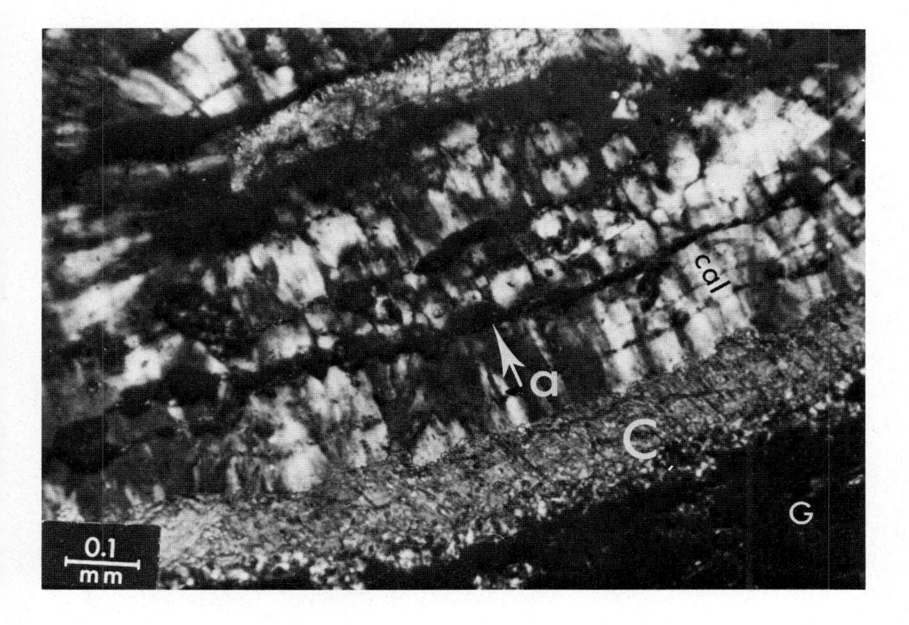

Fig. 572. Basaltic groundmass with a gas cavity filled from the margin to the centre as follows: calcite; banded colloform radiating chalcedony, with interzonal pigments – indicating the relic colloform pattern of the initial gel-silica; colloform chalcedony with bands; and centrally in the gas-cavity again calcite. C = calcite; cal = banded colloform chalcedony interzonal with the colloform band; arrow "a" = pigment rests showing relic gel patterns; G = basaltic groundmass.
Amygdaloidal basalt, Teigarhorni, Iceland. With crossed nicols.

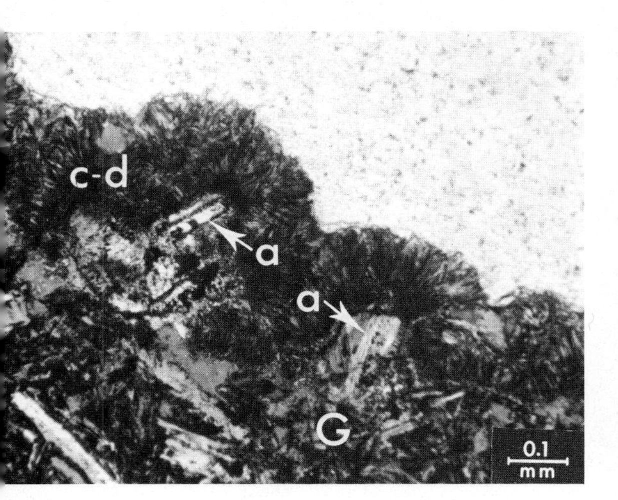

Fig. 573. Basaltic groundmass with gas cavity occupied with marginal deposition of colloform chlorite (fine radiating chloritic needles) and by calcite occupying the central part of the gas cavity. Arrows "a" show needle-formed groundmass components (plagioclase laths) at the walls of the gas cavity). G = basaltic groundmass; c-d = colloform chlorite at the margins of the basaltic gas-cavity.
Amygdaloidal melaphyre. Oberstein, Nahe River, Germany. Without crossed nicols.

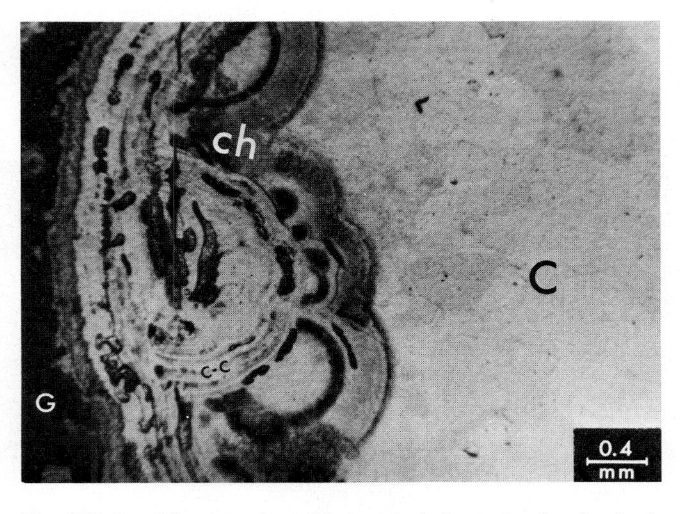

Fig. 574. Basaltic gas cavity with marginal alternating bands of colloform chlorite and calcite. The central part of the cavity is occupied by calcite. G = basaltic groundmass; ch = colloform chloritic bands, and c-c = colloform calcite; C = calcite, crystalline, occyping the central part of the gas cavity.
Amygdaloidal melaphyric. Oberstein, Nahe River, Germany. Without crossed nicols.

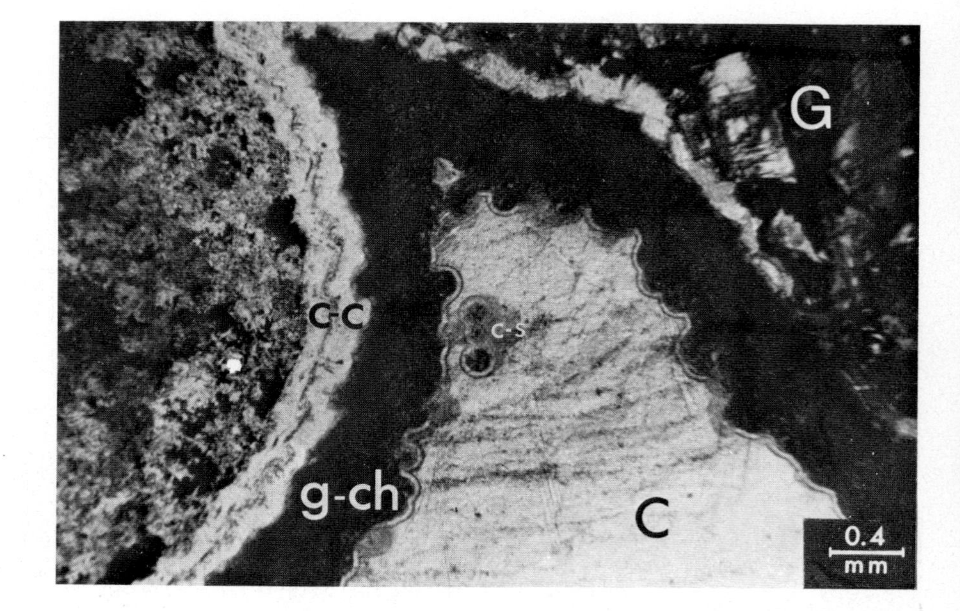

Figs. 575 and 576. Basaltic groundmass with a gas cavity filled with the following sequence of "deposited" material, from the margin of the gas cavity to centre: calcite-colloform, green-chloritic band, central calcite (crystal aggregates with colloform sphericules of chlorite). G = basaltic groundmass; c-c = colloform calcite band; g-ch = green-chlorite colloform; C = crystalline calcite; c-s = colloform spheroids of chlorite in the calcitic mass.
Augite porphyry. Val Bufaure, South Tirol. With crossed nicols.

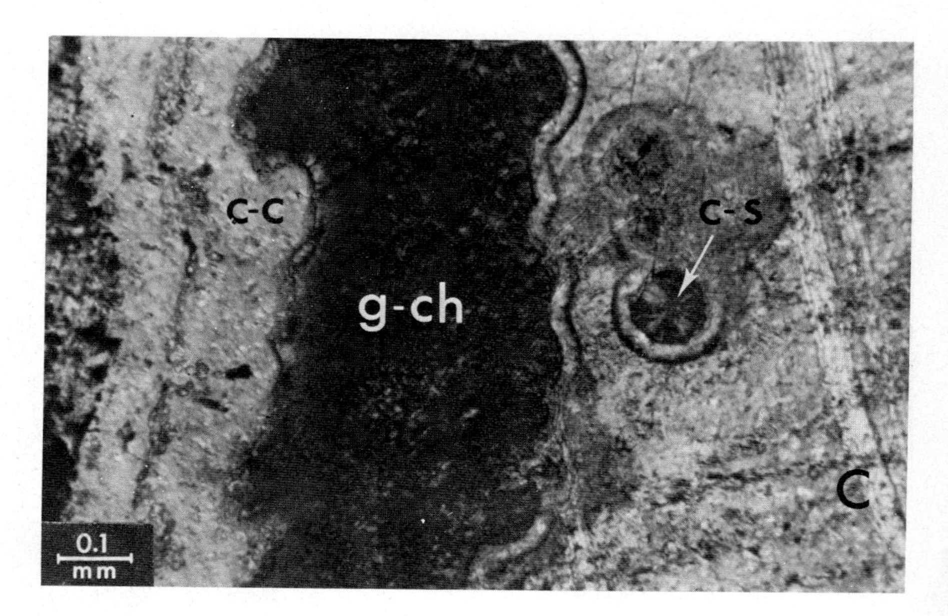

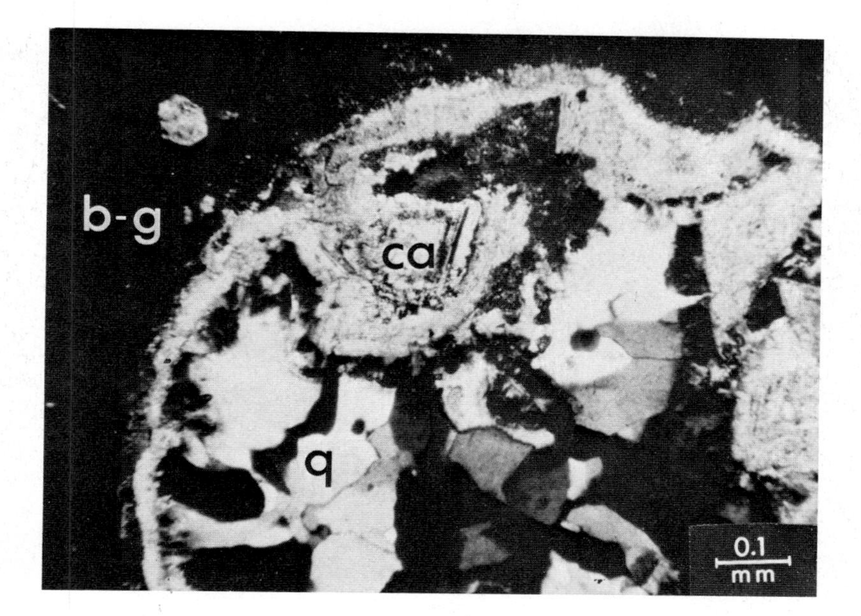

Fig. 577. Amygdaloidal cavity occupied centrally by quartz and marginally by calcite. b-g = basaltic groundmass; q = quartz; ca = calcite.
Amygdaloidal melaphyre. Oberstein, Nahe River, Germany. With crossed nicols.

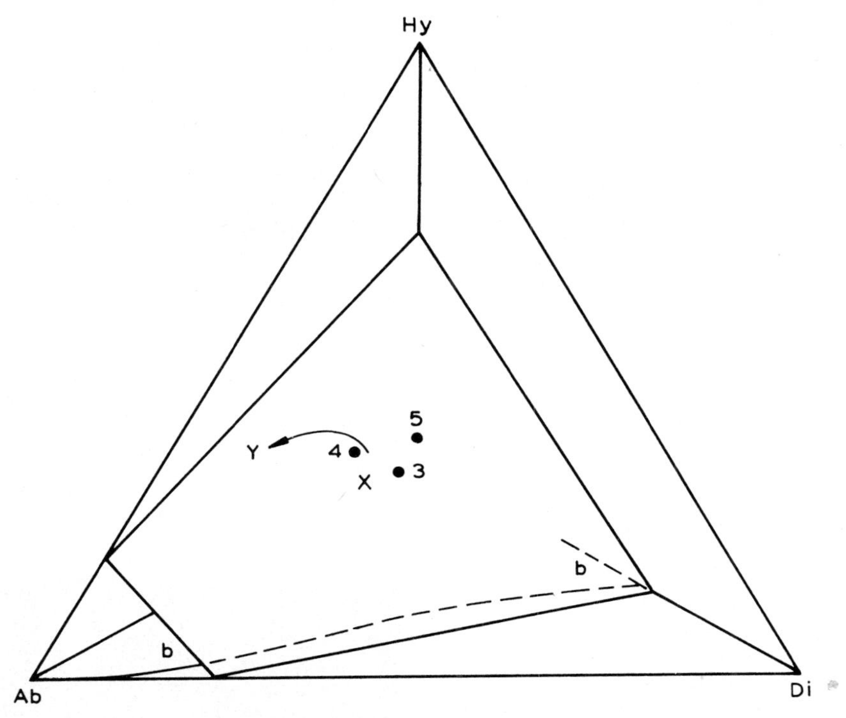

Fig. 598. Tetrahedron illustrating the approximate composition of basaltic lavas. The plane represents the boundary surface which separates those lavas that precipitated plagioclase first from those that precipitated pyroxene first. X–Y indicates the change in the residual liquid with advancing crystallization. Points 3,4 and 5 indicate the composition of the Deccan trap rock, the Oregon traps and the Karroo dolerites, respectively. (After T. Barth, *Am. J. Sci.*, 1936.) Hy = hypersthene; Ab = albite; Di = diopside.

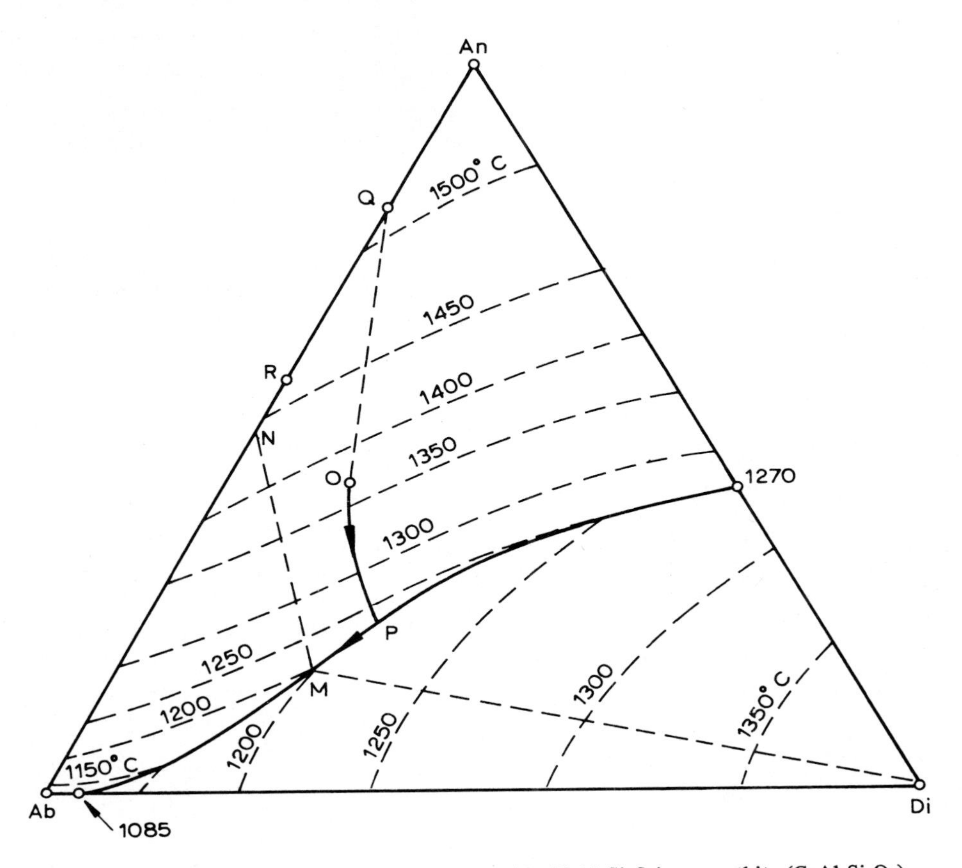

Fig. 599. Equilibrium diagram of the system diopside ($CaMgSi_2O_6$) – anorthite ($CaAl_2Si_2O_8$) – albite ($NaAlSi_3O_8$). (After Bowen, *Am. J. Sci.*, 1915.) Ab = albite; An = anorthite; Di = diopside.

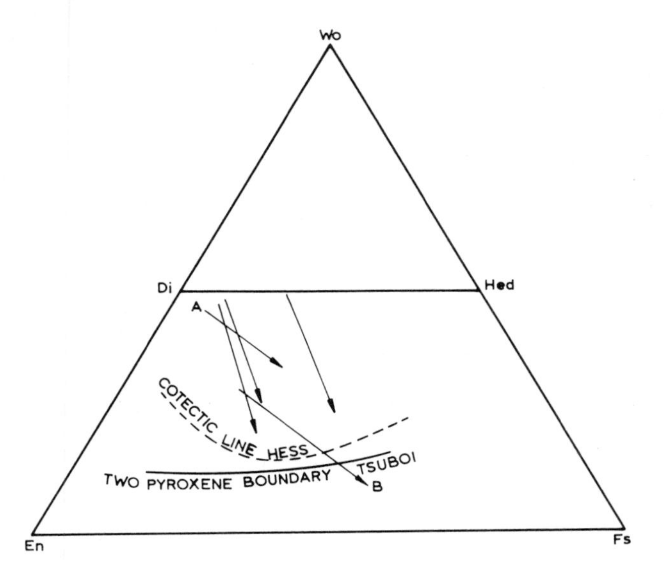

Fig. 600. Paths of crystallization of pyroxenes from basalts collected in several localities. After T. Barth (Wahlstrom has added a portion of the cotectic line for the pyroxenes as determined by Hess. Molecular percentages are shown). (*Am. J. Sci.*, 5th series, 31, 1936). Di = diopside; Hed = hedenbergite; Wo = wellastonite; Fs = ferrosilite.

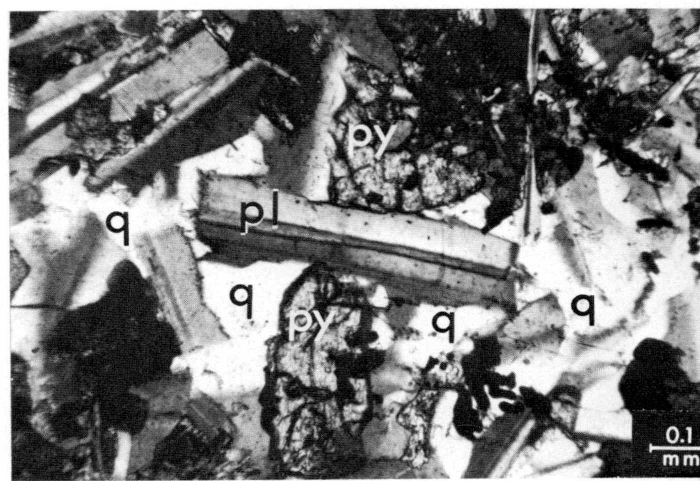

Fig. 601. Plagioclase–pyroxene intersertal growths with interstitial later quartz between the plagioclases and the pyroxenes. Pl = plagioclase; py = pyroxene; q = quartz. Top Palisade Sill, New Jersey, U.S.A. With crossed nicols.

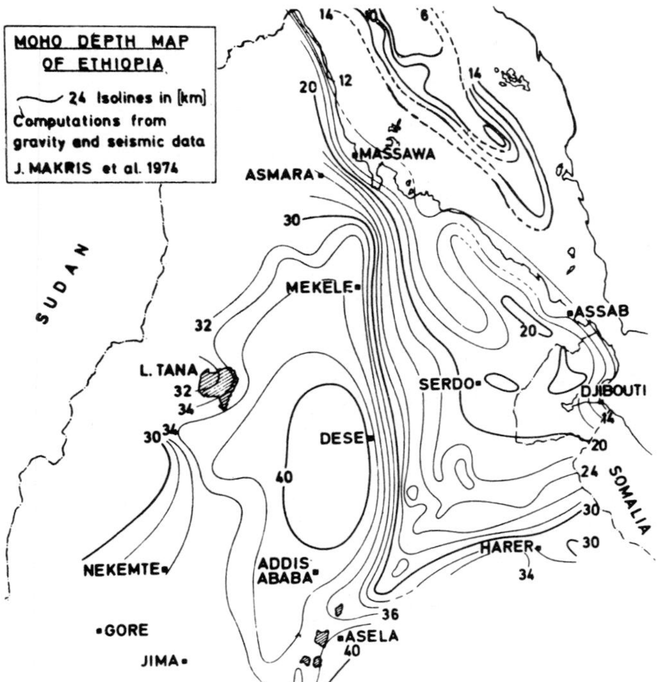

Fig. 602. Moho depth map of Ethiopia. Isolines in km. Computations from gravity and seismic data by Makris et al. (1975).

Fig. 603. Melted salt poured out of a volcanic crater. Dankalia (Dallol area). Eritrea, Ethiopia. Specimen: courtesy Eng. O. Wollak.

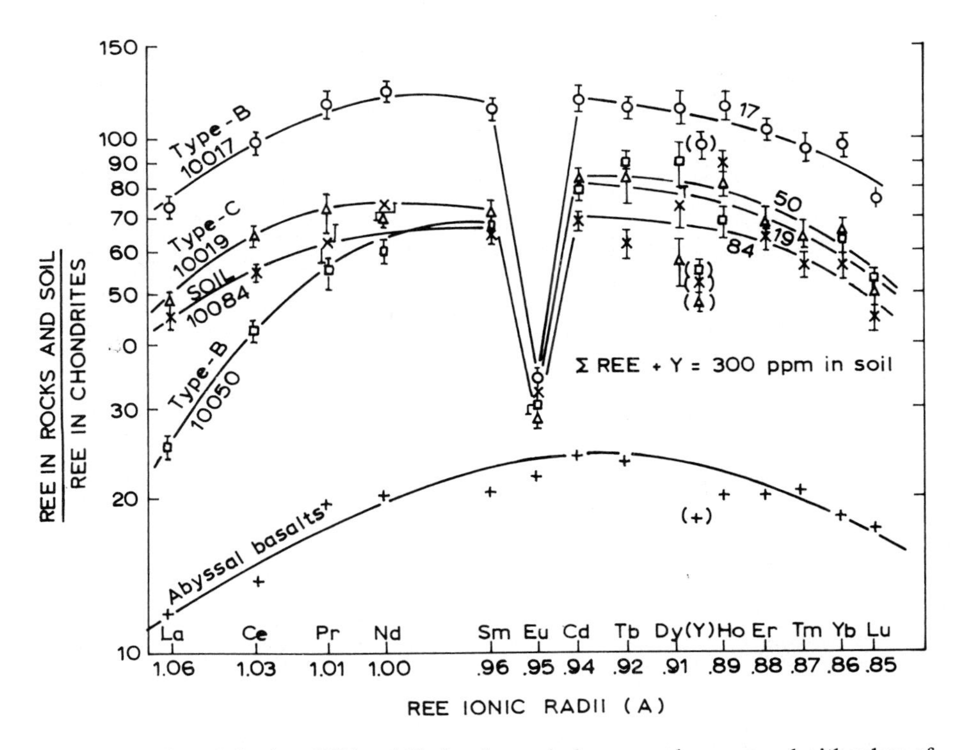

Fig. 604. Ratios of absolute REE and Y abundances in lunar samples compared with values of abyssal terrestrial basalts. (Only the crosses represent the terrestrial samples, the other symbols refer to lunar samples). It is to be noticed that Eu is severely depleted in lunar rocks and soils by about 60% relative to the adjacent elements Sm and Gd. (Simplified after Schmitt et al. – *Science,* 167 (3918), 1970).

References

Ade-Hall, J.M. and Lawley, E.A., 1970. An unsolved problem – opaque petrological differences between Tertiary basaltic dykes and lavas. In: *Mechanism of Igneous Intrusion. Geol. J.*, Spec. Issue 2: 217–230.

Agiorgitis, G., Schroll, E. and Stepan, E., 1970. *K/Rb, Ca/Sr und K/Ti-Verhältnisse in basaltoiden Gesteinen der Ostalpen und benachbarter Gebiete.* Springer, Vienna – New York.

Agricola, 1530. *De Re Metallica.*

Alsac, C., 1959. Étude pétrographique des pillow-lavas de la pointe Guilben, près de Paimpol (Côtes du Nord). *Bull. Soc. Fr. Mineral. Cristallogr.*, 82: 363–366.

Andersen Jr., A.T., Crewe, A.V., Goldsmith, J.R., Moore, P.B., Newton, J.C., Olsen, E.J., Smith, J.V. and Wyllie, P.J., 1970. Petrologic history of Moon suggested by petrography mineralogy and crystallography. *Science*, 167: 587–589.

Andersen, O., 1915. The system anorthite–forsterite–silica. *Am. J. Sci.*, Ser. 4, 39: 407–454.

Anderson, D.L., 1970. Petrology of the mantle. *Min. Soc. Am. Spec. Pap.*, 3: 85–93.

Aoki, K., 1962. The clinopyroxenes of the Otaki dolerite sill. *J. Jap. Assoc. Mineral., Petrol. Econ. Geol.*, 47: 41–45.

Aoki, K., 1970. Petrology of kaersutite-bearing ultramafic and mafic inclusions in Iki Island, Japan. *Contrib. Mineral. Petrol.*, 25: 270–283.

Artemjev, M.E. and Artyushkov, E.V., 1971. Structure and isostasy of the Baikal Rift and the mechanism of rifting. *J. Geophys. Res.*, 76: 1197–1212.

Augustithis, S.S., 1959/1960. Dr. Thesis, University of Hamburg, 1956. Über Blastese in Gesteinen unterschiedlicher Genese. *Hamb. Beitr. Angew. Mineral., Kristallphys. Petrog.*, 2 (1959/1960): 40–68.

Augustithis, S.S., 1962. Researches of blastic processes in granitic rocks and later graphic quartz in pegmatites (pegmatoids) from Ethiopia. *Nova Acta Leopoldina*, N.F., 25 (156: 1–17.

Augustithis, S.S., 1963. Oscillatory zoning of plagioclase phenocrysts of the olivine basalt of Debra Sina, Ethiopia. *Chem. Erde*, 23: 71–81.

Augustithis, S.S., 1964. On the phenomenology of phenocrysts (tecoblasts, zoned phenocrysts, holo-phenocrystalline rock types). In: S.S. Augustithis (Editor), *Special Bulletin of Petrogenesis*. The Institute of Petrogenesis and Geochemistry, Athens.

Augustithis, S.S., 1965. On the origin of olivine and pyroxene nodules in basalts. *Chem. Erde*, 24 (2): 197–210.

Augustithis, S.S., 1967. On the phenomenology and geochemistry of differential leaching and element agglutination processes. *Chem. Geol.*, 2: 311–329.

Augustithis, S.S., 1972. Mantle fragments in basalt (On the origin and significance of olivine bombs in basalts). *Geol. Bull. Greece*, 3: 93–101.

Augustithis, S.S., 1973. *Atlas of the Textural Patterns of Granites, Gneisses and Associated Rock Types*. Elsevier, Amsterdam, 373 pp.

Augustithis, S.S., 1974. Oscillatory zoning of augitic pyroxenes in plateau basalt. *Collected Abstr., Int. Mineral. Assoc. Meet. Berlin*, 1974, p. 110.

Augustithis, S.S. and Ottemann, J., 1966. On diffusion rings and spheroidal weathering. *Chem. Geol.*, 1: 201–209.

Augustithis, S.S., Mposkos, E. and Vgenopoulos, A., 1974. Geochemical and mineralogical studies of euxinite and its alteration products in graphic pegmatites from Harrar, Ethiopia. *I.A.E.A. Symp. Form. Uranium Ore Deposits, May 6–10, 1974, Athens*. International Atomic Energy Agency, Vienna, pp. 61–71.

Bailey, J.C., Chapness, P.E., Dunham, A.C., Esson, J., Fyfe, W.S., McKenzie, W.S., Stumpfl, E.F. and Zussmann, J., 1970. Mineralogical and petrological investigations of lunar samples. *Science*, 167 (3918): 592–594.

Baker, M.J., 1966. Blocks of plutonic aspect in a basaltic lava from Faial, Azores. *Geol. Mag.*, 103: 51–56.

Bankwitz, P., 1964. Bemerkungen zum Verhalten magmatischer Massen in der Erdkruste und im Erdmantel. *Monatsber. Dtsch. Akad. Wiss. Berlin*, 6: 538–543.

Barker, P.F., 1970. Plate tectonics of the Scotia Sea region. *Nature*, 228: 1293–1296.

Barth, T.F.W., 1936. The crystallization of basalt. *Am. J. Sci.*, 5th Ser., 31: 321–351.

Barth, T.F.W., 1952. *Theoretical Petrology*. Wiley, New York, N.Y. and Chapman and Hall, London, 387 pp.

Barth, T.F.W., 1962. Die Menge der Kontinentalsedimente und ihre Beziehung zu den Eruptivgesteinen. *Neues Jahrb. Mineral., Monatsh.*, 1962: 59–67.

Beeson, M.H. and Jackson, E.D., 1970. Origin of the garnet pyroxenite xenoliths at Salt Lake Crater, Oahu. *Min. Soc. Am., Spec. Pap.*, 3: 95–112.

Beloussov, V.V., 1967. Against continental drift. *J. Sci.*, 67 (168): 7.

Beloussov, V.V., 1969. Continental drifts. In: P.J. Hart (Editor). *The Earth's Crust and Upper Mantle. Geophys. Monogr.*, 13. Am. Geophys. Union, Washington, D.C., pp. 539–544.

Benioff, H., 1954. Orogenesis and deep crustal structure. *Bull. Geol. Soc. Am.*, 65: 385–400.

Best, M.G., 1970. Kaersutite–peridotite inclusions and kindred megacrysts in basanitic lavas, Grand Canyon, Arizona. *Contrib. Mineral. Petrol.*, 27: 25–44.

Binns, R.A., 1969. High pressure megacrysts in basaltic lavas near Armidale, New South Wales. *Am. J. Sci.*, Z67-A (Schairer vol.): 33–49.

Bird, J.M. and Dewey, J.F., 1970. Lithosphere plate. Continental margin tectonics and the evolution of the Appalachian orogen. *Geol. Soc. Am. Bull.*, 81: 1031–1060.

Bjornson, S., 1967. *Iceland and the Mid-Oceanic Ridges*. Soc. Sci. Icelandica.

Black, P.M. and Brothers, R.N., 1965. Olivine nodules in olivine nephelinite from Tokatoka, Northland. *N. Z. J. Geol. Geophys.*, 8: 62–80.

Blackett, P.M.S., Bullard, E.C. and Runcorn, S.K. (Editors), 1965. Symposium on Continental Drift. *Phil. Trans. R. Soc. London, Ser. A*, 258.

Bloxam, T.W., 1960. Pillow structure in spilitic lavas at Downan Point, Ballantrae. *Trans. Geol. Soc. Glasgow*, 24: 29–52.

Bocalletti, M., Elter, P. and Guazzone, G., 1971. Plate tectonic models for the development of the western Alps and northern Apennines. *Nature (London), Phys. Sci.*, 234: 108–111.

Borcher, H., 1967. Vulkanismus und oberer Erdmantel in ihrer Beziehung zum äusseren Erdkern und zur Geotektonik. *Boll. Geofis. Teor. Appl.*, 9: 194–213.

Borodin, L.S. and Gladkikh, V.S., 1967. Geochemistry of zirconium in differentiated alkali basalt series. *Geochem. Int.*, 4: 925–935. (Transl. from *Geokhimiya*, 10: 1023–1034.)

Bose, M.K., 1967. Differentiation of alkali basaltic magma. *Nature*, 207: 1187–1188.

Bott, M.H.P., 1971a. *The Interior of the Earth*. Arnold, London, 316 pp.

Bott, M.H.P., 1971b. Evolution of young continental margins and formation of shelf basins. *Tectonophysics*, 11: 319–327.

Bott, M.H.P., 1971c. The mantle transition zone as possible source of global gravity anomalies. *Earth Planet. Sci. Lett.*, 11: 28–34.

Bowen, N.L., 1914. The ternary system: diopside–forsterite–silica. *Am. J. Sci.*, Ser. 4, 38: 207–264.

Bowen, N.L., 1915. Crystallization of haplobasaltic, haplodioritic and related magmas. *Am. J. Sci.*, Ser. 4, 40: 161–185.

Bowen, N.L., 1922. The reaction principle in petrogenesis. *J. Geol.*, 30: 177–198.

Bowen, N.L., 1928. *The Evolution of the Igneous Rocks*. Princeton University Press, Princeton, N.J., 334 pp.

Bowen, N.L., 1937. Recent high temperature research on silicates and its significance in igneous geology. *Am. J. Sci.*, Ser. 5, 33: 1–21.

Bowen, N.L., 1945. Phase equilibria bearing on the origin and differentiation of alkaline rocks. *Am. J. Sci.*, 243A: 75–89.

Bowen, N.L. and Schairer, J.F., 1935. The system MgO–FeO–SiO_2. *Am. J. Sci.*, Ser. 5, 29: 197.

Bowen, N.L. and Tuttle, O.F., 1949. The system MgO–SiO_2–H_2O. *Bull. J. Soc. Am.*, 60: 439.

Breithaupt, A., 1849. Über regelmässige Verwachsungen von Kristallen zweier und dreier Mineralspezies. *Neues Jahrb. Mineral., Geol. Palaeontol., Abh.*, 89.

Brothers, R.M., 1960. Olivine nodules from New Zealand. *Rep. Int. Geol. Congr. 21st*, Sess. Norden, 13: 68–81.

Brousse, R. and Rudel, A., 1964. Bombes de peridotites, de norites, de charnockites et de granulites dans les scories du Puy Beaunit. *C.R. Acad. Sci., Paris*, 259: 185–188.

Brown, G. and Stephen, I., 1959. A structural study of iddingsite from New South Wales, Australia. *Am. Mineral.*, 44: 251.

Brown, M.G., Emeleus, H.C., Holland, G.J. and Phillips, R., 1970. Petrographic, mineralogic and X-ray fluorescence analysis of lunar igneous-type rocks and spherules. *Science*, 167 (3918): 599–601.

Bullard, Sir Edward, 1969. The origin of the oceans. *Sci. Am.*, Sept.: 66–75.

Carey, S.W., 1958. A tectonic approach to continental drift. In: S.W. Carey (Editor), *Continental Drift. Symp. Continental Drift, Univ. Tasmania, Hobart*, pp. 177–355.

Carr, J.M., 1954. Zoned plagioclases in layered gabbroes of the Skaergraad intrusion East Greenland. *Mineral. Mag.* and *J. Mineral. Soc., London*, 30 (225): 367–375.

Carswell, D.A. and Dawson, J.B., 1970. Garnet peridotite xenoliths in South African kimberlite pipes and their petrogenesis. *Contrib. Mineral. Petrol.*, 25: 163–184.

Challis, C.A., 1963. Layered xenoliths in a dyke, Awatere Valley, New Zealand. *Geol. Mag.*, 100: 11–16.

Chayes, F., 1964. A petrographic distinction between Cenozoic volcanics in and around the open ocean. *J. Geophys. Res.*, 69 (8).

Chayes, F., 1965. Titania and alumina content of oceanic and circumoceanic basalts. *Mineral. Mag.*, 34 (268), London.

Chayes, F. and Zies, E.G., 1961. Staining of alkali feldspars from volcanic rocks. *Carnegie Inst. Washington, Yearb.*, 60: 172–173.

Clark, S.P., 1961a. A redetermination of equilibrium relations between kyanite and sillimanite. *Am. J. Sci.*, 259: 641–650.

Clark, S.P., 1961b. Temperatures in the continental crust. *Carnegie Inst. Washington, Yearb.*, 60: 187–190.

Coleman, R.G., 1971. Plate tectonic emplacement of upper mantle peridotites along continental edges. *J. Geophys. Res.*, 76: 1212–1222.

Collee, A.L.G., 1963. A fabric study of lherzolites with special reference to ultrabasic nodular inclusions in the lavas of Auvergne (France). *Leidse Geol. Meded.*, 28: 3–102.

Coombs, D.C., 1963. Trends and affinities of basaltic magmas and pyroxenes as illustrated on the diopside–olivine–silica diagram. *Mineral. Soc. Am., Spec. Pap.*, 1: 227–250.

Craig, H., Boats, G. and White, O.E., 1956. The isotopic geochemistry of thermal waters. In: *Nuclear Processes in Geologic Settings. Nat. Res. Council, Nucl. Sci. Ser., Rep.*, 19: 29–44.

Cundari, A. and Le Maitre, R.W., 1970. On the petrology of the leucite-bearing rocks of the Roman and Birunga volcanic regions. *J. Petrol.*, 11: 33–47.

Dahm, K.P., 1967. Pillow Lavas in der Deutschen Demokratischen Republik. *Ber. Dtsch. Ges. Geol. Wiss., Reihe B*, 12-3: 257–265.

Daly, R.A., 1924. *Igneous Rocks and their Origin*. McGraw-Hill, New York, N.Y., 458 pp.

Daly, R.A., 1925. *Proc. Am. Phil. Soc.*, 64: 283.

Daly, R.A., 1928. *Our Mobile Earth*. New York, N.Y., 134 pp.

Deer, W.A., Howie, R.A. and Zussmann, J., 1962. *Rock Forming Minerals*. Longmans, London, 1788 pp.

Den Tex, E., 1963. Gefügekundliche und geothermometrische Hinweise auf die tiefe, exogene Herkunft iherzolithischer Knollen aus Basaltlaven. *Neues Jahrb. Mineral., Monatsh.*, 1963: 225–236.

Deutsche Forschungsgemeinschaft, 1972. *Das Unternehmen Erdmantel*. Franz Steiner Verlag, Wiesbaden.

De Vries, R.C., Roy, R. and Osborn, E.G., 1954. The system TiO_2–SiO_2. *Trans. Br. Ceram. Soc.*, 53: 525–540.

Dewey, J.F. and Bird, J.M., 1970. Plate tectonics and geosynclines. *Tectonophysics*, 10 (5/6): 625–638.

Dewey, J.F. and Horsfield, B., 1970. Plate tectonics, orogeny and continental growth. *Nature*, 225: 521–525.

Dietz, R.S., 1961. Continent and ocean evolution by spreading of the sea floor. *Nature*, 190: 854–857.

Dietz, R.S., 1963. Collapsing continental rises: an actualistic concept of geosynclines and mountain building. *J. Geol.*, 71: 314–333.

Dittmann, J., 1960. Syntheseversuche zur Metasomatose von Magnesiumsilikaten. *Hamb. Beitr. Angew. Mineral., Kristallphys. Petrog.*

Dixey, F., 1956. The East African Rift system. *Bull. Colonial Geol.* and *Min. Res., London*, Supp. No. 1.

Downes, M.J., 1974. Sector and oscillatory zoning in calcic augites from Mt. Etna, Sicily, Italy. *Contrib. Mineral. Petrol.*, 47: 187–196.

Drescher-Kaden, F.K., 1948. *Die Feldspar–Quartz-Reaktionsgefüge der Granite und Gneise*. Springer, Berlin–Göttingen–Heidelberg, 259 pp.

Drescher-Kaden, F.K., 1969. *Granitprobleme*. Akademie-Verlag, Berlin, 586 pp.

Duffield, W.A., 1969. Concentric structure in elongate pillows. Amador County, California. *U.S. Geol. Surv., Prof. Pap.*, 650-D: 19–25.

Dunham, K.C., 1950. Petrography of the nickeliferous norite of St. Stephen, New Brunswick. *Am. Mineral.*, 35: 711.

Eaton, J.P. and Murata, K.J., 1960. How volcanoes grow. *Science*, 132 (3432): 925–932.

Eckermann, H. von, 1944. Some notes on the reaction series. *Geol. Fören. Förhandl.*, 66 (2): 283–287.

Edwards, A.B., 1938. The formation of iddingsite. *Am. Mineral.*, 23: 277.
Ellis, J., 1946. A theory for the origin of coronas around olivine. *Trans. Geol. Soc. S. Afr.*, 48: 103.
Elliston, J.N., 1963. Sediments of the Warramunga geosyncline. *Syntaphral tectonics and diagenesis*. A Symposium. University of Tasmania, Hobart.
Engel, A.E.J. and Engel, C.G., 1970. Lunar rock compositions and some interpretations. *Science*, 167 (3918): 527–528.
Ernst, Th., 1961. Probleme des Sonnenbrandes basaltischer Gesteine. *Z. Dtsch. Geol. Ges.*, 112: 178–182.
Ernst, Th., 1962. Folgerungen für die Entstehung der Basalte aus dem speziellen Auftreten der Pyroxene in diesen Gesteinen. *Geol. Rundsch.*, 51: 364–374 (English, French and Russian summaries).
Ernst, Th., 1963. Basaltmagmen-Entstehung und Peridotit. *Neues Jahrb. Mineral., Monatsh.*, 203–205.
Ernst, T., 1964. Do peridotitic inclusions in basalts represent mantle material? *I.U.G.S. Upper Mantle Symp., New Delhi*, pp. 180–185.
Ernst, T., 1975. Erdmantel-Bericht und Theorie. *Fortschr. Mineral.*, 52: 100–140.
Ernst, T. and Drescher-Kaden, F.K., 1941. Über den Sonnenbrand der Basalte, 2. *Z. Angew. Mineral.*, 3: 73–141.
Ferguson, J.B. and Merwin, H.E., 1919. The tertiary system $CaO-MgO-SiO_2$. *Am. J. Sci., Ser. 4*, 48: 81–123.
Francis, T.J.G., 1969. Upper mantle structure along the axis of the Mid-Atlantic Ridge near Iceland. *Geophys. J.*, 17: 507–520.
Freund, R., 1970. Plate tectonics of the Red Sea and East Africa. *Nature*, 228: 453.
Friedman, G.M., 1955. Petrology of the Memesagamesing Lake norite mass, Ontario, Canada. *Am. J. Sci.*, 253: 590.
Frisch, T., 1970. The detailed mineralogy and significance of an olivine-two pyroxene gabbro nodule from Lanzarote, Canary Islands. *Contrib. Mineral. Petrol.*, 28: 31–41.
Galli, M., 1963. Studi petrographici sulla formazione ofiolitica dell'Appennino ligure. *Period. Mineral.*, 32: 2–3.
Garswell, D.A. and Dawson, J.B., 1970. Garnet-peridotite xenoliths in South African Kimberlite pipes and petrogenesis. *Contrib. Mineral. Petrol.*, 25: 163–184.
Gass, I.G., 1958. Ultrabasic pillow lavas from Cyprus. *Geol. Mag.*, 95: 241–251.
Gass, I.G., 1968. Is the Troodos Massif of Cyprus a fragment of Mesozoic ocean floor? *Nature*, 220: 39–42.
Gass, I.G. and Masson-Smith, D., 1963. The geology and gravity anomalies of the Troodos massif, Cyprus. *Philos. Trans. R. Soc. London, Ser. A*, 255: 417–467.
Gast, P.W., 1968. Trace element fractionation and the origin of the tholeiitic and alkaline magma types. *Geochim. Cosmochim. Acta*, 32: 1057–1086.
Gay, P. and Lemaitre, R.W., 1961. Some observations on iddingsite. *Am. Mineral.*, 46: 92–111.
Gibb, F.G.F., 1969. Cognate xenoliths in the Tertiary ultrabasic dykes of south-west Skye. *Mineral. Mag.*, 37: 504–514.
Girdler, R.W., 1971. Rifting in Africa. *Comments Earth Sci. Geophys.*, 2(2): 44–48.
Gjelsvik, T., 1952. Metamorphosed dolerites in the gneiss area of Sunnmore on west coast of Southern Norway. *Norsk. Geol. Tidsskr.*, 30: 32–134.
Goldschmidt, V.M., 1954. *Geochemistry*. Oxford University Press, London, 730 pp.
Gregory, J.W., 1896. The Great Rift Valley (1894 – Contributions to the physical geography of British East Africa. *Geogr. J.*, 4: 290).
Gregory, J.W., 1920. The African Rift Valleys. *Geogr. J.*, 56: 31.
Gregory, J.W., 1921. Rift Valleys and Geology of East Africa.
Haggerty, S.E., Boyd, F.R., Bell, P.M., Finger, L.W. and Bryan, W.B., 1970. Iron-titanium oxides and olivine from 10020 and 10071. *Science*, 167(3918): 613–615.
Hamad, S. el D., 1963. The chemistry and mineralogy of the olivine nodules of Calton Hill, Derbyshire. *Mineral Mag.*, 33: 483–497.
Hamilton, D.L. and Anderson, G.M., 1967. Effects of water and oxygen pressure on the crystallization of basaltic magmas. In: H. Hess and A. Poldervaart (Editors), *Basalts*. Wiley, New York, N.Y., 1: 445–482.
Hargraves, R.B., Hollister, L.S. and Otalova, G., 1970. Compositional zoning and its significance in pyroxenes from three coarse grained lunar samples. *Science*, 167(3918): 631–633.
Harker, A., 1950. *Metamorphism*. Methuen, London, 362 pp.
Harumoto, A., 1953. Melilite-nepheline basalt, its olivine nodules and other inclusions from Nagahama, Japan. *Mem. Coll. Sci., Univ. Kyoto, Ser. B*, 20: 69–88.
Hashimoto, H., Mannami, M. and Naiki, T., 1961. Dynamical theory of electron diffraction for the electron microscopic images of crystal lattices. *Phil. Trans. R. Soc., London*, 253: 459–516.
Hashimoto, H. et al., 1974. High resolution electron microscopy of Labradorite feldspar. Reprint from *Electron Microscopy in Mineralogy*, pp. 333–344.
Hashimoto, H., Kumao, A., Endoh, H., Nissen, H.U., Ono, A. and Watanabe, E., 1975. Lattice image contrast by many beam dynamical theory and structure determination of labradorite feldspar. *Proc. EMAG 75 Conf., Bristol.*
Heezen, B.C., Tharp, M. and Ewing, M., 1959. The floor of the oceans, 1. The North Atlantic. *Geol. Soc. Am., Spec. Pap.*, 65.
Henckel, 1725. *Pyritologia.*
Heritsch, H. and Riechert, L., 1960. Strukturuntersuchung an einer basaltischen Hornblende von Cernosin, CSR. *Mineral. Petrogr. Mitt.*, 3: 235–245.
Herron, E.M., 1972. Sea floor spreading and the Cenozoic history of the east central Pacific. *Geol. Soc. Am. Bull.*, 83: 1671–1691.
Herz, N., 1951. Petrology of the Baltimore gabbro, Maryland. *Bull. Geol. Soc. Am.*, 62: 979–1016.
Hess, H.H., 1941. Pyroxenes of common mafic magmas. *Am. Mineral.*, 26: 515–533, 573–594.
Hess, H.H., 1962. History of ocean basins. In: A.E.J. Engel et al. (Editors), *Petrologic Studies – Buddington Memorial Volume*. Geol. Soc. Am., New York, N.Y., pp. 599–620.
Hess, H.H. and Poldervaart, A., 1964. *Basalts – The Poldervaart Treatise on Rocks of Basaltic Composition*, Vol. 2. Interscience, New York.
Holmes, A., 1965. *Principles of Physical Geology*. Nelson, Edinburgh, 2nd ed., 1288 pp.
Hopgood, A.M., 1962. Radical distribution of soda in a pillow of spilitic lava from the Franciscan, California. *Am. J. Sci.*, 260: 383–396.
Huang, W.T. and Merrit, C.A., 1954. Petrography of the troctolite of the Wichita Mountains, Oklahoma. *Am. Mineral.*, 39: 549–565.
Huckenholz, H.G., 1964/1965. Der petrogenetische Werdegang der Klinopyroxene in den tertiären Vulkaniten der Hocheifel. I. Die Klinopyroxene der Alkaliolivin–Basalt–Trachyt–Assoziation. *Beitr. Mineral. Petrogr.*, 11: 138–195.
Hytönen, K. and Schairer, J.F., 1961. The plane enstatite–anorthite–diopside and its relation to basalts. *Carnegie. Inst., Wash. Yearb.*, 60: 125–141.
Ilić, M., 1967. Position of ophiolites in the geotectonic evolution of the Dinaric folded area. *Acta. Geol. Acad. Sci. Hung.*, 11(1–3): 77–93.
Illies, J.H., 1970. Graben tectonics as related to crust–mantle interaction. In: J.H. Illies and S. Mueller (Editors), *Graben Problems*. Schweizerbart, Stuttgart.

Isacks, B.L., Oliver, J. and Sykes, L.R., 1968. Seismology and the new global tectonics. *J. Geophys. Res.*, 73: 5855–5899.

Ishikawa, T., 1953. Xenoliths included in the lavas from Volcano Tarumai, Hokkaido, Japan. *J. Fac. Sci. Hokkaido Univ., Ser. IV*, 8: 225–244.

Jackson, E.D., 1968. The character of the lower crust and upper mantle beneath the Hawaiian Islands. *Rep. Int. Geol. Congr., 23rd, Prague*, 1: 135–150.

Jackson, E.D. and Wright, T.L., 1970. Xenoliths in the Honolulu Volcanic Series, Hawaii. *J. Petrol.*, 11: 405–430.

Johnston, R. and Gibb, F.G.F., 1973. Multiple-twinned and reversezoned pigeonite in Apollo 14 basalt 14310. *Mineral. Mag.*, 39: 248–251.

Joplin, G.A., 1960. On the tectonic environment of basic magma. *Geol. Mag.*, 97: 363–568.

Judd, 1886. *Q. J. Geol. Soc.*, 42: 54.

Keller, J., 1969. Origin of rhyolites by anatectic melting of granitic crustal rocks. The example of rhyolitic pumice from the island of Kos (Aegean Sea). *Bull. Volcanol.*, 33: 942–959.

Kennedy, W.Q., 1933. Trends of differentiation of basaltic magmas. *Am. J. Sci., 5th Ser.*, 25: 239–256.

Kennedy, W.Q. and Anderson, E.M., 1938. Crustal layers and the origin of magmas. *Bull. Volcanol., Ser. 2*, 3: 24.

Krishnan, M.S., 1961. *The Surface and Interior of the Earth.* Mahadevan Volume. Osmania University Press, Hyderabad.

Kuno, H., 1954. Geology and petrology of Omuro-yama volcano group, North Izu. *J. Fac. Sci., Imp. Univ. Tokyo*, 9: 241–265.

Kuno, H., 1959a. Discussion of paper by J.F. Lovering, "The nature of the Mohorovicic Discontinuity". *J. Geophys. Res.*, 64: 1071–1072.

Kuno, H., 1959b. Origin of Cenozoic petrographic provinces of Japan and surrounding areas. *Bull. Volcanol., Ser. 2*, 20: 37–76.

Kuno, H., 1967a. Mafic and ultramafic nodules from Itinomegata, Japan. In: P.J. Wyllie (Editor), *Ultramafic and Related Rocks* Wiley, New York, N.Y., 337.

Kuno, H., 1967b. Silicate systems related to basaltic rocks. In: H.H. Hess and A. Poldervaart, *Basalts*. Wiley, New York, N.Y., 1: 360–397.

Kuno, H., 1968. Differentiation of basaltic magmas in basalts. In: H.H. Hess and A. Poldervaart (Editors), *Basalts*. New York, N.Y., 2: 623–688.

Kuryleva, N.A., 1958. On the petrography of Siberian kimberlites. *Mem. All-Union Mineral. Soc., Ser. 2*, 87: 233–237.

Kushiro, I., 1962. Clinopyroxene solid solutions. I. The $CaAl_2/SiO_6$ component. *Jap. J. Geol. Geogr.*, 33: 213–220.

Kushiro, I. and Aoki, K.I., 1968. Origin of some eclogite inclusions in kimberlite. *Am. Mineral.*, 53: 1347–1367.

Kushiro, I. and Schairer, J.F., 1963. New data on the system diopside–forsterite–silica. *Carnegie Inst., Washington, Yearb.*, 62: 95–103.

Kushiro, I., Yasuo Nakamuta and Hiroshi Haramura, 1970. Crystallization of some lunar mafic magmas and generation of rhyolitic liquid. *Science*, 167(3918): 610–612.

Lacroix, A., 1912. *Compt. Rend.*, 154: 252.

Lacroix, A., 1923. *Minéralogie de Madagascar*, III: 46.

Larson, R.L. and Chase, C.G., 1970. Relative velocities of the Pacific, North America and Cocos Plates in the Middle America region. *Earth Planet. Sci. Lett.*, 7: 425–428.

Laves, F., Nissen, H.U. and Bollmann, W., 1965. On Schiller and submicroscopical lamellae of labradorite. $(Na,Ca)(Si,Al)_3O_8$. *Naturwissenschaften*, 52.

Le Bas, M.J., 1955. Magmatic and amygdaloidal plagioclases. *Geol. Mag.*, 92: 291–296.

Le Bas, M.J., 1962. The role of aluminum in igneous clinopyroxenes with relation to their parentage. *Am. J. Sci.*, 260: 267–288.

Lebedev, L.M., 1967. *Metacolloids in Endogenic Deposits.* Plenum Press, New York, N.Y., 298 pp.

Lemaitre, Mme. O., Brousse, R., Goni, J. and Remond, G., 1966. Sur l'importance de l'apport de fèr dans la transformation de l'olivine et iddingsite. *Bull. Soc. Franç. Mineral. Cristallogr.*, 89: 477–483.

Lemaitre, R.W., 1965. The significance of the gabbroic xenoliths from Gough Island, South Atlantic. *Mineral. Mag.*, 34: 303–317.

Le Pichon, X., 1968. Sea-floor spreading and continental drift. *J. Geophys. Res.*, 73: 3661–3697.

Le Pichon, X., 1969. Models and structure of the oceanic crust. *Tectonophysics*, 7: 385–401.

Le Pichon, X., 1970. Correction to paper by Xavier Le Pichon "Sea-floor spreading and continental drift". *J. Geophys. Res.*, 75: 2793.

Le Pichon, X., Francheteau, J. and Bonnin, J., 1973. *Plate Tectonics. Develop. Geotectonics*, 6. Elsevier, Amsterdam, 300 pp.

Levicki, O.D., 1955. On the question of the importance of colloidal solutions during the deposition of ores. Basic problems in Science. On magmatogenic ore-deposits. *Izv. Akad. Nauk S.S.S.R.*, 1955, 312–334.

Lindgreen, W., 1933. *Mineral Deposits*. McGraw-Hill, New York, N.Y., 989 pp.

Makris, J., 1974. Afar and Iceland – a geophysical comparison. 1: 379–390.

Makris, J., Menzel, H., Zimmermann, J. and Gouin, P., 1975. Gravity field and crustal structure of north Ethiopia. In: A. Pilger and A. Rösler (Editors), *Afar Depression of Ethiopia. Inter-Union Comm. Geodynamics, Sci. Rep.*, 14. Schweizerbart, Stuttgart, pp. 135–144.

Machatscki, F., 1953. *Spezielle Mineralogie auf geochemischer Grundlage*. Springer, Vienna, 378 pp.

Mason, B., 1952. *Principles of Geochemistry*. Wiley, New York, N.Y., 310 pp.

Mason, B. and Melson, W.G., 1970. Comparison of lunar rocks with basalts and stony meteorites. *Proc. Apollo 11 Lunar Sci. Conf.*, 1: 661–751.

Masuda, A., 1964. Lead isotope composition in volcanic rocks of Japan. *Geochim. Cosmochim. Acta*, 28: 291–305.

Masuda, A., 1966. Lanthanides in basalts of Japan with three distinct types. *Geochem. J.*, 1: 11–26.

McConnell, R.B., 1969. Evolution of Rift Systems in Africa. *Nature*, 227: 99–700.

McKenzie, D.P., 1970. Plate tectonics and continental drift. *Endeavour* (I.C.I. Ltd.), 29: 39–44.

McLaren, A.C., 1974. Transmission electron microscopy of the feldspars. In: W.S. McKenzie and J. Zussmann (Editors), *The Feldspars. Proc. NATO Av. Study Inst.*, Manchester University Press, pp. 378–423.

McLaren, A.C. and Marshall, D.B., 1974. Transmission electron microscope study of the domain structure associated with b-, c-, d-, e- and fe reflections in plagioclase feldspar. *Contrib. Mineral. Petrol.*, 44: 237–249.

McNiel, R.D., 1963. Slip complexes in the Warramunga Geosyncline. In: *Syntaphral Tectonics and Diagenesis*. A Symposium. Geol. Dep., Univ. Tasmania, Hobart.

Mehnert, K.R., 1968. *Migmatites*. Elsevier, Amsterdam, 393 pp.

Menard, H.W., 1958. Development of median elevations in ocean basins. *Bull. Geol. Soc. Am.*, 69: 1179–1186.

Menard, H.W., 1960. The East Paficic Rise. *Science*, 132 (3441): 1737–1746.

Menard, H.W., 1969. Growth of drifting volcanoes. *J. Geophys. Res.*, 74: 4827–4857.

Middlemost, E.A.K., 1972. A simple classification of volcanic rocks. *Bull. Volcanol. Ser. 2*, 26: 382–397.

Mihirk, B., 1967. The upper mantle and alkalic magmas. *Norsk. Geol. Tidsskr.*, 47: 121–129.

Milashev, V.A., 1960. Cognate inclusions in the kimberlite pipe "Obnazhennaya" (Olenck basin). *Mem. All-Union Mineral. Soc.*, 89: 284–299.

Mitchell, A.H.G. and Bell, J.D., 1970. Volcanic episodes in island arcs. *Proc. Geol. Soc.*, 1662: 9–12.

Miyashiro, A., 1975. Volcanic Rock Series and Tectonic Setting. *Ann. Rev. Earth Planet. Sci.*, 3: 251–269.

Montgomery, H.B., 1962. Description and origin of pillow lavas, Maybrun Mines Property, Kenora District, Ontario, Canada. *J. Geol.*, 70: 619–620.

Moore, J.G., 1966. Rate of palagonitization of submarine basalt adjacent to Hawaii. *U.S. Geol. Surv., Prof. Pap.*, 550-D: 163–171.

Morgan, W.J., 1971. Plate motions and deep mantle convection. In: R. Shagam (Editor), *Hess Volume. Geol. Soc. Am., Mem.*, 132.

Muan, A. and Osborn, E.F., 1956. Phase equilibria at liquidus temperatures in the system $MgO-FeO-Fe_2O_3-SiO_2$. *J. Am. Ceram. Soc.*, 39: 121–140.

Muir, I.D. and Tilley, C.D., 1964. Basalts from the northern part of the rift zone of the Mid-Atlantic Ridge. *J. Petrol.*, 5: 409–434.

Murata, K.J. and Richter, D.H., 1966. The settling of olivine in Kilauean magma as shown by the lavas of the 1959 eruption. *Am. J. Sci.*, 264: 194–203.

Murthy, M.V.N., 1958. Coronites from India and their bearing on the origin of coronas. *Bull. Geol. Soc. Am.*, 68: 23.

Nashar, B., 1963. The origin of the megacrysts in some low grade schist from Tennant Creek Northern Territory. *Syntaphral Tectonics and Diagenesis*. A Symposium. Geol. Dept., Univ. Tasmania, Hobart.

Nissen, H.U., 1974. Exsolution lamellae in plagioclase feldspars: electron microscopy and X-ray microanalysis. *Proc. Int. Congr. Electron Microscopy, 9th, Canberra*, 1: 468–469.

Noe-Nygaard, A. and Rasmussen, J., 1956. The making of the basalt plateau of the Faroes. *Proc. Int. Geol. Congr., 20th Sess., Mexico*, pp. 399–407.

Oelsen, W. and Maetz, H., 1940/1941. *Mitt. Kaiser-Wilhelm Inst. Eisenforsch., Düsseldorf*, 23.

O'Hara, M.J., 1968. Are ocean floor basalts primary magma? *Nature*, 220: 683–686.

O'Hara, M.J., 1969. The origin of ariegite nodules in basalt. *Geol. Mag.*, 105: 322–330.

Oji, Y., 1961. Olivine in the alkali basalts of western San-in and northern Kyushu. *J. Jap. Assoc. Mineral. Petrogr., and Econ. Geol.*, 45: 133–136.

Olsen, A., 1974. Schiller effects and exsolution in sodium-rich plagioclases. *Contrib. Mineral. Petrol.*, 47: 141–152.

Osborn, E.F., 1942. *The System $NaAlSiO_4-Ca_2MgSi_2O_7$ and $NaAlSiO_4-Ca_2MgSi_2O_7-CaMgSi_2O_5$ and their Petrologic Application*. Thesis, Hokkaido Univ., Sapporo, 63 pp. (manuscript).

Osborn, E.F., 1949. Coronite, labradorite, anorthosite and dykes of andesine anorthosite, New Glasgow, P.Q. *Trans. R. Soc. Can.*, 43: 85.

Osborn, E.F. and Tait, D.B., 1952. The system diopside–forsterite–anorthite. *Am. J. Sci.*, Bowen volume: 413–433.

Oxburgh, E.R., 1964. Petrological evidence for the presence of amphiboles in the upper mantle and its petrogenetic and geophysical implications. *Geol. Mag.*, 101: 1–19.

Pieruccini, R., 1961. Si di un particolare processo pseudomorfico, la transformazione dell' augite in termini plagioclasici. Ed. la Sicilia–Messina.

Pilger, A. and Rösler, A. (Editors), 1975. *Afar Depression of Ethiopia. Inter-Union Comm. Geodynamics, Sci. Rep.*, 14. Schweizerbart, Stuttgart.

Pliny, 1634. *Historia Naturalis*. Philemon Holland, London.

Poldervaart, A. and Hess, H.H., 1951. Pyroxenes in the crystallization of basaltic magma. *J. Geol.*, 59: 472–489.

Radcliffe, S.V., Heuer, A.J., Fisher, R., Christie, J.M. and Griggs, D.T., 1970. High voltage transmission electron microscopy study of lunar surface material. *Science*, 167 (3918): 638–640.

Ramdohr, P., 1960. *Die Erzmineralien und ihre Verwachsungen*. Akademie-Verlag, Berlin.

Rance, H., 1967. Major lineaments and torsional deformation of the earth. *J. Geophys. Res.*, 72: 2213–2217.

Rankin, G.A. and Wright, F.E., 1915. The ternary system $CaO-Al_2O_3-MgO$. *J. Am. Chem. Soc.*, 38: 568–588.

Read, H.H., 1957. *The Granite Controversy*. Murby, London, 430 pp.

Richter, D.H. and Murata, K.J., 1961. Xenolithic nodules in the 1800–1801 Kaupulchu flow of Hualalai Volcano. *U.S. Geol. Surv., Prof. Pap.*, 424-B: 215–217.

Ricker, R.W. and Osborn, E.F., 1954. Additional phase equilibrium data for the system $CaO-MgO-SiO_2$. *J. Am. Ceram. Soc.*, 37: 137–139.

Rittmann, A., 1958. A physico-chemical interpretation of the terms magma, migma crust and sub-stratum. *Bull. Volcanol., Ser. 2*, 19: 85–102.

Rittmann, A., 1967. Die Bimodalität des Vulkanismus und die Herkunft der Magmen. *Geol. Rundsch.*, 57: 277–295.

Rittmann, A., 1969. Magma und Kruste. *Ber. Dtsch. Ges. Geol. Wiss., Reihe A*, 14 (3): 321–332.

Robinson, P.R., 1969. High-titania alkali-olivine basalt of north central Oregon, USA. *Contrib. Mineral. Petrol.*, 22: 349–360.

Roedder, E., 1951a. The role of liquid immiscibility in igneous petrogenesis: a discussion. *J. Geol.*, 64: 84–88.

Roedder, E., 1951b. Low temperature liquid immiscibility in the system $K_2O-FeOAl_2O_3-SiO_2$. *Am. Mineral.*, 36: 282–286.

Roman, C. et al., 1970. Abundances of 30 elements in Lunar rocks, soil and core samples. *Science*, 167 (3918): 512–515.

Rosenbusch, H., 1898/1923. *Elemente der Gesteinslehre*. Schweizerbart, Stuttgart, 779 pp.

Ross, C.S., Foster, M.D. and Myers, A.T., 1954. Origin of dunites and of olivine rich inclusions in basaltic rocks. *Am. Mineral.*, 39: 693.

Ross, M., 1970. Lunar clinopyroxenes: chemical composition, structural state and texture. *Science*, 167 (3918): 628–630.

Rossy, M., 1969. Sur la nature de quelques pillow-lavas du Crétacé supérieur du pays Basque Espagnol. *C.R. Acad. Sci., Paris, Sér. D*, 269: 542–543.

Rüger, L., 1932. Hundert Jahre geologischer Forschung am Rheintalgraben. *Badische Geol. Abh*. Verlag C.F. Müller, Karlsruhe.

Runcorn, S.K. (Editor), 1962. *Continental Drift*. Academic Press, New York, N.Y., 352 pp.

Ryall, A. and Bennett, D.L., 1968. Crustal structure of southern Hawaii related to volcanic processes in the upper mantle. *J. Geophys. Res.*, 73: 4561–4582.

Sabatier, G., 1950. La cristallisation, par chauffage, des gels mixtés de silice et de magnésie. *Compt. Rend. Acad. Sci., Paris*, 230: 1962.

Sacks, B.B., Oliver, J. and Sykes, L., 1968. Seismology and the new global tectonics. *J. Geophys. Res.*, 73: 5855–5899.

San Miguel, A., 1972. Some remarks on lunar rock textures and its geological implications. *Publ. Univ. Barcelona*, 27: 83–111.

Sarbadhikari, T.R., 1958. An orthopyroxene-bearing rock from near Tinpahar in Rajmahal Hills area. *Q. J. Geol., Min. Metall. Soc. India*, 30: 221–227.

Sarbadhikari, T.R., 1965. On the difference in twinning between phenocrysts and groundmass plagioclase of basalts. *Am. Mineral.*, 50: 1466–1469.

Savelli, C. and Wedepohl, K.H., 1967. Somma-Vessuv-Magmen durch Dolomit-Assimilation. *Naturwissenschaften*, 54: 644.

Schairer, J.F., 1957. Melting relations of the common rock-forming oxides. *Naturwissenschaften*, 40: 215–235.

Schairer, J.F. and Bailey, D.K., 1962. The system $Na_2O-Al_2O_3-Fe_2O_3-SiO_2$ and its bearing on the alkaline rocks. *Carnegie Inst. Washington, Yearb.*, 61: 91–96.

Schairer, J.F. and Bowen, N.L., 1935. Preliminary report on equilibrium relations between feldspathords, alkali-feldspars and silica. *Trans. Am. Geophys. Union*, 16: 325.

Schairer, J.F. and Bowen, N.L., 1938a. The system leucite diopside silica. *Am. J. Sci., 5th Ser.*, 35A: 289–309.

Schairer, J.F. and Bowen, N.L., 1938b. Crystallization equilibrium in nepheline albite silica mixtures with fayalite. *J. Geol.*, 46: 397–411.

Schairer, J.F. and Thwaite, R.D., 1950. $Na_2O-4SiO_2-Al_2O_3-Fe_2O_3$. *Carnegie Inst. Washington, Yearb.*, 49: 46–47.

Schairer, J.F. and Thwaite, R.D., 1952. The quaternary system $K_2O-MgO-Al_2O_3-SiO_2$. *Carnegie Inst. Washington, Yearb.*, 51: 51–53.

Schairer, J.F., Yagi, K. and Yoder, H.S. Jr., 1962. The system nepheline–diopside. *Carnegie Inst. Washington, Yearb.*, 61: 96–98.

Scheidegger, A.E., 1953. On some physical aspects of the theory of the origin of mountain belts and island arcs. *Can. J. Phys.*, 31: 1148.

Schmitt, R.A., Hiroshi Wakita and Plinio Rey, 1970. Abundance of 30 elements in Lunar rocks, soil and core samples. *Science*, 167: 512–515.

Schneider, A.W., 1964. Contribuiçao à Petrologia dos Derrames Basalticos da Bacia do Parane. *Publ. Univ. Rio Grande do Sul, Publ. Avulso*, 1: 76.

Schorer, G., 1970. Sanduhrban und Optik von Titanaugiter alkalibasaltischer Gesteine des Vogelberges. *Neues Jahrb. Mineral. Monatsh.*, 1970 (7).

Schreyer, W., 1969. Über den Aufbau der Erde aus der heutigen Sicht der Petrologie. *Fortschr. Mineral.*, 46: 29–41.

Sederholm, J.J., 1910. Die regionale Umschmelzung (Anatexis) erläutert an typischen Beispielen. *Compt. Rend. Int. Geol. Congr., 11me, Stockholm*, pp. 573–586.

Shand, S.J., 1945. Coronas and coronites. *Bull. Geol. Soc. Am.*, 56: 247–266.

Shaw, D.M., 1968. A review of K/Rb fractionation trends by co-variance analysis. *Geochim. Cosmochim. Acta*, 32: 573–602.

Sheppard, R.A., 1962. Iddingsitization and recurrent crystallization of olivine in basalts from the Simcoe Mountains, Washington. *Am. J. Sci.*, 260: 67–74.

Sigurdsson, H., 1968. Petrology of acid xenoliths from Surtsey. *Geol. Mag.*, 105: 440–453.

Singer, A. and Navrot, J., 1970. Diffusion rings in altered basalt. *Chem. Geol.*, 6: 31–41.

Sirin, A.N., 1962. A variety of columnar jointing in a lava flow and the conditions of formation. *Tr. Labor. Vulkanol., Akad. Nauk S.S.S.R.*, 21: 50–56.

Smith, J.V., 1959. The crystal structure of protenstatite, $MgSiO_3$. *Acta Crystallogr.*, 12: 515–519.

Smith, W.W., 1961. Structural relationships within pseudomorphs after olivine. *Mineral. Mag.*, 32: 823–825.

Sobolev, N.V., 1964. Xenoliths of eclogite with ruby. *Proc. Acad. Sci. U.S.S.R., Earth-Sci. Sect.*, 157: 1382–1384.

Sobolev, N.V., 1968. The xenoliths of eclogites from the kimberlite pipes of Yakutia as fragments of the upper mantle substance. *Rep. Int. Geol. Congr., 23rd, Prague*, pp. 155–163.

Solomon, M., 1966. Origin of pillow structure in lavas. *Nature*, 211: 399.

Spry, A., 1962. The origin of columnar jointing particularly in basalt flows. *J. Geol. Soc., Australia*, 8: 191–216.

St. Amand, P., 1957. Circum-Pacific orogeny. *Publ. Dominion Astrophys. Obs., Victoria (B.C.)*, 20 (2): 403–411.

Staub, R., 1924. Der Bau der Alpen. *Beitr. Geol. Karte Schweiz, N.F.*, 52: 272.

Stanik, E., 1970. The relationship between the chemism of olivine and the chemism of parent volcanic rocks. *Acta Univ. Carol., Geol.*, 1970 (3).

Steiner, A., 1958. Petrogenetic implications of the 1954 Ngauruhoe (New Zealand) lava and its xenoliths. *N.Z. J. Geol. Geophys.*, 1: 325–363.

Stewart, D.B., Walker, G.W., Wright, T.L. and Fahey, J.J., 1966. Physical properties of calcic labradorite from Lake County, Oregon. *Am. Mineral.*, 51: 177–197.

Sun, M.S., 1957. The nature of iddingsite in some basaltic rocks of New Mexico. *Am. Mineral.*, 42: 525.

Suzuki, T., 1962. On the internal structure of pyrite. *Tohoku Univ. Sci. Rep., 3rd Ser.*, 8: 69–136.

Sviatlovsky, A.E., 1975. *Regional Volcanology*. Nedra, Moscow, 223 pp.

Symons, D.T.A., 1967. The magnetic and petrologic properties of basalt column. *Geophys. J. R. Astron. Soc.*, 12: 473–490.

Talbot, J.L., Hobbs, B.E., Wilshire, H.G. and Sweatman, T.R., 1963. Xenoliths and xenocrysts from lavas of the Kerguelen Archipelago. *Am. Mineral.*, 48: 159–179.

Taylor, S.R., Capp, A.C., Graham, A.L. and Blake, D.H., 1969. Trace element abundance in andesites. II. Saipan, Bougainville and Fiji. *Contrib. Mineral. Petrol.*, 23: 1–26.

Tazieff, H., 1968. Relations tectoniques entre l'Afar et la Mer Rouge. *Bull. Soc. Geol. Fr.*, 7 (10): 468–477.

Tazieff, H., 1970. Tectonics of the northern Afar (or Danakil Rift). In: J.H. Illies and S. Mueller (Editors), *Graben Problems*. Schweizerbart, Stuttgart, pp. 280–283.

Tilley, C.E., 1947. The dunite-mulonites of St. Paul's Rocks (Atlantic). *Am. J. Sci.*, 245: 483.

Tilley, C.E., 1950. Some aspects of magmatic evolution. *Q. J. Geol. Soc., London*, 106: 37–61.

Tilley, C.E., 1961. The occurrence of hypersthene in Hawaiian basalts. *Geol. Mag.*, 98: 257–260.

Trask, N.J., 1969. Ultramafic xenoliths in basalt. Nye County, Nevada. *U.S. Geol. Surv., Prof. Pap.*, 650: 43–48.

Tryggvason, T., 1957. The gabbro bombs at Lake Graenavatn. *Bull. Geol. Inst. Uppsala*, 38: 1–5.

Turnock, A.C., 1962. Preliminary results on melting relations on synthetic pyroxenes on diopside–hedenbergite join. *Carnegie Inst. Washington, Yearb.*, 61: 81–82.

Tyrrell, G.W., 1926. Petrography of Jan Mayen. *Trans. R. Soc. Edinburgh*, 54: 762.

Uffen, R.J., 1959. On the origin of rock magma. *J. Geophys. Res.*, 64: 117–122.

U.M.C./UNESCO, 1965a. Report of the U.M.C./UNESCO Seminar on the East African Rift System. *Upper Mantle Rep.*, 6 (I): 145 pp.

U.M.C./UNESCO, 1965b. Report on the geology and geophysics of the East African Rift System. *Upper Mantle Rep.*, 6 (II): 116 pp.

Urry, W.D., 1949. Significance of radio-activity in geophysics – thermal history of the earth. *Trans. Am. Geophys. Union*, 30: 171–180.

Vallance, T.G., 1965. On the chemistry of pillow lavas and the origin of silicates. *Mineral. Mag.*, 34: 471–481.

Van Andel, T.J., 1968. The structure and development of rifted mid-oceanic rises. *J. Mar. Res.*, 26 (2): 144–161.

Van Bemmelen, R.W., 1966. On mega-undations: a new model for the earth's evolution. *Tectonophysics*, 3 (2): 83–127.

Van Bemmelen, R.W., 1972. *Geodynamic Models: An Evolution and a Synthesis. Develop. Geotectonics*, 2. Elsevier, Amsterdam, 267 pp.

Vance, J.A., 1962. Zoning in igneous plagioclase: normal and oscillatory zoning. *Am. J. Sci.*, 260: 746–760.

Vance, J.A., 1969. On Synneusis. *Contrib. Mineral. Petrol.*, 24: 7–29.

Vening Meinesz, F.A., 1964. *The Earth's Crust and Mantle. Develop. Solid Earth Geophys.*, 1. Elsevier, Amsterdam, 124 pp.

Verhoogen, J., 1960. Temperatures within the earth. *Am. Sci.*, 48: 134–159.

Verhoogen, J., 1962. Distribution of titanium between silicates and oxides in igneous rocks. *Am. J. Sci.*, 260: 211–220.

Vine, F.J., 1966. Ocean floor spreading: new evidence. *Science*, 154: 1405.

Wahlstrom, E.E., 1950. *Introduction to Theoretical Igneous Petrology*. Wiley, New York, N.Y., 365 pp.

Walker, G.P.L., 1960. Zeolite zoning and dike distribution in relation to the structure of the basalts of eastern Iceland. *J. Geol.*, 68: 515–528.

Washington, H.S., 1906. The Roman comagmatic region. *Carnegie Inst. Washington Publ.*, 57.

Washington, H.S., 1914a. *Am. J. Sci.*, 38: 79.

Washington, H.S., 1914b. *J. Geol.*, 22: 752.

Washington, H.S., 1922. Ocean trapps and other plateau basalts. *Bull. Geol. Soc. Am.*, 33.

Washington, H.S., 1923a, Kilauaea and general petrology of Hawaii. *Am. J. Sci.*, 6: 338–365.

Washington, H.S., 1923b. Hualalai and Mauna Loa. *Am. J. Sci.*, 6: 100–126.

Washington, H.S., 1923c. Petrology of the Hawaiian Islands. I. Kohala and Mauna Kea, Hawaii. *Am. J. Sci.*, 5: 465–502.

Washington, H.S., 1923. The formation of Aa and Pahoehoe. *Am. J. Sci.*, 6: 403–423.

Wedepohl, K.H., 1963. Einige Überlegungen zur Geschichte des Meerwassers. *Fortschr. Geol. Rheinl. Westfalen,* 10: 129.

Wedepohl, K.H., 1967. Geochemie. Walter de Gruyter, Berlin, 221 pp.

Wegener, A., 1924. *Origin of Continents and Oceans*. Dutton, New York.

Welch, 1764. *Systematic Description of the Mineral Kingdom.*

White, D.E., 1957. Magmatic connate and metamorphic waters. *Bull. Geol. Soc. Am.*, 68: 1659–1682.

Wilkinson, J.F.G., 1956a. Clinopyroxene of alkali olivine-basalt magma. *Am. Mineral.*, 41: 724–743.

Wilkinson, J.F.G., 1956b. The olivines of a differentiated teschenite sill near Gunnedah, New South Wales. *Geol. Mag.*, 93: 441–455.

Wilkinson, J.F.G., 1967. The petrography of basaltic rocks. In: H.H. Hess and A. Poldervaart (Editors), *Basalts*. Wiley, New York, N.Y., 1: 163–226.

Willis, B., 1936. *East African Plateaux and Rift Valleys*. Carnegie Institute of Washington, Washington.

Wilshire, H.G., 1958. Alteration of olivines and orthopyroxenes in basic lavas and shallow intrusions. *Am. Mineral.*, 43: 120.

Wilson, J. Tuzo, 1966. Did the Atlantic Ocean close and then reopen? *Nature*, 221.

Wilson, J. Tuzo, 1967. Rift Valleys and Continental Drift. *Trans. Leicester Lit. Philos. Soc.*, 61.

Wilson, M.E., 1960. Origin of pillow structure in early Pre-Cambrian lavas of western Quebec. *J. Geol.*, 68: 97–102.

Wiseman, J.D.H., 1966. St. Paul Rocks and the problem of the upper mantle. *Geophys. J. R. Astron. Soc.*, 11: 519–525.

Wright, J.B., 1964. Olivine nodules in a phonolite of the East Otago alkaline province, New Zealand. *Nature*, 210: 519.

Wright, J.B., 1969. Olivine nodules in trachyte from the Jos Plateau, Nigeria. *Nature*, 223: 285–286.

Wyllie, P.J., Cox, K.G. and Biggar, G.M., 1962. The habit of apatite in synthetic systems and igneous rocks. *J. Petrol.*, 3: 238–243.

Yagi, K., 1953. A reconnaissance of systems acmite–diopside and acmite–nepheline. *Carnegie Inst. Washington, Yearb.*, 61: 133–134.

Yagi, K. and Onuma, K., 1967. The join $CaMgSi_2O_6$–$CaTiAl_2O_6$ and its bearing on the titanaugites. *J. Fac. Sci. Hokkaido Univ., Ser. IV*, 13: 463–483.

Yashima, R., 1961a. Phenocryst pigeonite from Sengamori Fukushima Prefecture. *J. Jap. Assoc. Mineral., Petrol. Econ. Geol.*, 45: 9–13.

Yashima, R., 1961b. *The Geology and Petrology of the Sano Gabbro–Diorite Complex and its Surrounding Area.* Thesis, Univ. Tokyo (unpublished).

Yoder, H.S., 1950a. The jadeite problem. *Am. J. Sci.*, 248: 225–248, 312–334.

Yoder, H.S., 1950b. High-low quartz inversion up to 10,000 bars. *Trans. Am. Geophys. Union*, 31: 827–835.

Yoder, H.S., 1952. The MgO–Al_2O_3–SiO_2–H_2O system and the related metamorphic facies. *Am. J. Sci.*, Bowen Vol.: 569.

Yoder, H.S. Jr. and Tilley, C.E., 1962. Origin of basalt magmas, an experimental study of natural and synthetic rock systems. *J. Petrol.*, 3: 342–532.

Yoder, H.S. Jr., 1976. *Generation of Basaltic Magma*. National Academy of Science, Washington, D.C., 265 pp.

Yoder, H.S. Jr., Tilley, C.E. and Schairer, J.F., 1963. Pyroxene quadrilateral. *Carnegie Inst. Washington, Yearb.*, 62: 84–92.

Yoder, H.S. Jr., Tilley, C.E. and Schairer, J.F., 1964. Isothermal sections of pyroxene quadrilateral. *Carnegie Inst. Washington, Yearb.*, 63: 121–129.

Zies, E.G., 1962. A titaniferous basalt from the island of Pantelleria. *J. Petrol.*, 3: 177–180.

Author index

Subject index to the text part

Subject index to the illustrations